THIRD EDITION

GAS *Turbine* Combustion

Alternative Fuels and Emissions

THIRD EDITION

GAS *Turbine* Combustion

Alternative Fuels and Emissions

Arthur H. Lefebvre
and
Dilip R. Ballal

CRC Press
Taylor & Francis Group
Boca Raton London New York

CRC Press is an imprint of the
Taylor & Francis Group, an **informa** business

Reprint 2018

CRC Press
Taylor & Francis Group
6000 Broken Sound Parkway NW, Suite 300
Boca Raton, FL 33487-2742

Printed in India by Saurabh Printers Pvt. Ltd.

ISBN 978-1-1385-6958-4

Library of Congress Cataloging-in-Publication Data

Lefebvre, Arthur H. (Arthur Henry), 1923-2003.
Gas turbine combustion : alternative fuels and emissions / authors, Arthur H. Lefebvre, Dilip R. Ballal. -- 3rd ed.
p. cm.
"A CRC title."
Includes bibliographical references and index.
ISBN 978-1-4200-8604-1 (hardcover. : alk. paper)
1. Gas-turbines--Combustion. 2. Gas-turbines--Fuel. I. Ballal, Dilip R. II. Title.

TJ778.L417 2010
621.43'3--dc22 2010005450

Visit the Taylor & Francis Web site at
http://www.taylorandfrancis.com

and the CRC Press Web site at
http://www.crcpress.com

Contents

Preface xvii
Authors xix

1. Basic Considerations 1
1.1 Introduction 1
1.2 Early Combustor Developments 2
1.2.1 Britain 3
1.2.2 Germany 5
1.2.2.1 Jumo 004 5
1.2.2.2 BMW 003 6
1.2.3 The United States 6
1.3 Basic Design Features 8
1.4 Combustor Requirements 9
1.5 Combustor Types 10
1.5.1 Tubular 11
1.5.2 Tuboannular 11
1.5.3 Annular 12
1.6 Diffuser 14
1.7 Primary Zone 15
1.8 Intermediate Zone 16
1.9 Dilution Zone 17
1.10 Fuel Preparation 18
1.10.1 Pressure-Swirl Atomizers 18
1.10.2 Airblast Atomizer 19
1.10.3 Gas Injection 20
1.11 Wall Cooling 20
1.11.1 Wall-Cooling Techniques 22
1.12 Combustors for Low Emissions 23
1.13 Combustors for Small Engines 26
1.14 Industrial Chambers 28
1.14.1 Aeroderivative Engines 31
References 33
Bibliography 33

2. Combustion Fundamentals 35
2.1 Introduction 35
2.1.1 Deflagration 35
2.1.2 Detonation 35
2.2 Classification of Flames 36
2.3 Physics or Chemistry? 37

2.4 Flammability Limits ... 37
2.5 Global Reaction-Rate Theory ... 38
2.5.1 Weak Mixtures ... 39
2.5.2 Rich Mixtures ... 39
2.6 Laminar Premixed Flames ... 41
2.6.1 Factors Influencing Laminar Flame Speed ... 41
2.6.1.1 Equivalence Ratio ... 41
2.6.1.2 Initial Temperature ... 42
2.6.1.3 Pressure ... 42
2.7 Laminar Diffusion Flames ... 42
2.8 Turbulent Premixed Flames ... 43
2.9 Flame Propagation in Heterogeneous Mixtures of Fuel Drops, Fuel Vapor, and Air ... 45
2.10 Droplet and Spray Evaporation ... 49
2.10.1 Heat-Up Period ... 50
2.10.2 Evaporation Constant ... 51
2.10.3 Convective Effects ... 52
2.10.4 Effective Evaporation Constant ... 52
2.10.5 Spray Evaporation ... 54
2.10.6 Some Recent Developments ... 54
2.11 Ignition Theory ... 55
2.11.1 Gaseous Mixtures ... 55
2.11.2 Heterogeneous Mixtures ... 56
2.12 Spontaneous Ignition ... 64
2.13 Flashback ... 69
2.14 Stoichiometry ... 70
2.15 Adiabatic Flame Temperature ... 71
2.15.1 Factors Influencing the Adiabatic Flame Temperature ... 71
2.15.1.1 Fuel/Air Ratio ... 71
2.15.1.2 Initial Air Temperature ... 71
2.15.1.3 Pressure ... 72
2.15.1.4 Inlet-Air Vitiation ... 72
Nomenclature ... 73
References ... 74
Bibliography ... 77

3. Diffusers ... 79
3.1 Introduction ... 79
3.2 Diffuser Geometry ... 81
3.3 Flow Regimes ... 82
3.4 Performance Criteria ... 83
3.4.1 Pressure-Recovery Coefficient ... 84
3.4.2 Ideal Pressure-Recovery Coefficient ... 84

3.4.3 Overall Effectiveness 85
3.4.4 Loss Coefficient 85
3.4.5 Kinetic-Energy Coefficient 86
3.5 Performance 86
3.5.1 Conical Diffusers 87
3.5.2 Two-Dimensional Diffusers 88
3.5.3 Annular Diffusers 89
3.6 Effect of Inlet Flow Conditions 90
3.6.1 Reynolds Number 91
3.6.2 Mach Number 91
3.6.3 Turbulence 92
3.6.4 Swirl 93
3.7 Design Considerations 93
3.7.1 Faired Diffusers 93
3.7.2 Dump Diffusers 97
3.7.2.1 Influence of Liner Depth Ratio 98
3.7.3 Splitter Vanes 99
3.7.4 Vortex-Controlled Diffuser 100
3.7.5 Hybrid Diffuser 101
3.7.6 Diffusers for Tubular and Tuboannular Combustors 104
3.7.7 Testing of Diffusers 105
3.8 Numerical Simulations 106
Nomenclature 108
References 109

4. Aerodynamics 113
4.1 Introduction 113
4.2 Reference Quantities 114
4.3 Pressure-Loss Parameters 114
4.4 Relationship between Size and Pressure Loss 117
4.5 Flow in the Annulus 118
4.6 Flow through Liner Holes 120
4.6.1 Discharge Coefficient 120
4.6.2 Initial Jet Angle 123
4.7 Jet Trajectories 124
4.7.1 Experiments on Single Jets 124
4.7.2 Penetration of Multiple Jets 126
4.8 Jet Mixing 129
4.8.1 Cylindrical Ducts 129
4.8.2 Rectangular Ducts 131
4.8.2.1 Influence of Density Ratio 132
4.8.3 Annular Ducts 133

4.9 Temperature Traverse Quality ... 133
4.10 Dilution Zone Design ... 135
4.10.1 Cranfield Design Method ... 136
4.10.2 NASA Design Method ... 137
4.10.3 Comparison of Cranfield and NASA Design Methods ... 137
4.11 Correlation of Pattern Factor Data ... 138
4.12 Rig Testing for Pattern Factor ... 140
4.13 Swirler Aerodynamics ... 140
4.14 Axial Swirlers ... 142
4.14.1 Swirl Number ... 143
4.14.2 Size of Recirculation Zone ... 144
4.14.3 Flow Reversal ... 145
4.14.4 Influence of Swirler Exit Geometry ... 146
4.15 Radial Swirlers ... 146
4.16 Flat Vanes Versus Curved Vanes ... 147
Nomenclature ... 147
References ... 149

5. Combustion Performance ... 153
5.1 Introduction ... 153
5.2 Combustion Efficiency ... 153
5.2.1 The Combustion Process ... 154
5.3 Reaction-Controlled Systems ... 154
5.3.1 Burning Velocity Model ... 155
5.3.2 Stirred Reactor Model ... 159
5.4 Mixing-Controlled Systems ... 160
5.5 Evaporation-Controlled Systems ... 161
5.6 Reaction- and Evaporation-Controlled Systems ... 165
5.7 Flame Stabilization ... 167
5.7.1 Definition of Stability Performance ... 167
5.7.2 Measurement of Stability Performance ... 168
5.7.3 Water Injection Technique ... 170
5.8 Bluff-Body Flameholders ... 173
5.8.1 Experimental Findings on Bluff-Body Flame Stabilization ... 173
5.8.1.1 Homogeneous Mixtures ... 173
5.8.1.2 Heterogeneous Mixtures ... 177
5.8.2 Summary of Experimental Findings ... 179
5.9 Mechanisms of Flame Stabilization ... 179
5.9.1 Homogeneous Mixtures ... 181
5.9.2 Heterogeneous Mixtures ... 182
5.10 Flame Stabilization in Combustion Chambers ... 183

5.10.1 Influence of Mode of Fuel Injection 184
5.10.2 Correlation of Experimental Data 185
5.11 Ignition 188
5.12 Assessment of Ignition Performance 189
5.13 Spark Ignition 190
5.13.1 The High-Energy Ignition Unit 190
5.13.2 The Surface Discharge Igniter 191
5.13.2.1 Igniter Performance 192
5.13.2.2 Igniter Design 193
5.13.2.3 Igniter Life 194
5.14 Other Forms of Ignition 195
5.14.1 Torch Igniter 195
5.14.2 Glow Plug 196
5.14.3 Hot-Surface Ignition 196
5.14.4 Plasma Jet 197
5.14.5 Laser Ignition 197
5.14.6 Chemical Ignition 198
5.14.7 Gas Addition 199
5.14.8 Oxygen Injection 199
5.15 Factors Influencing Ignition Performance 199
5.15.1 Ignition System 200
5.15.1.1 Spark Energy 200
5.15.1.2 Spark Duration 200
5.15.1.3 Sparking Rate 202
5.15.1.4 Igniter Location 202
5.15.2 Flow Variables 203
5.15.2.1 Air Pressure 203
5.15.2.2 Air Temperature 204
5.15.2.3 Air Velocity 205
5.15.2.4 Turbulence 206
5.15.3 Fuel Parameters 207
5.15.3.1 Fuel Type 207
5.15.3.2 Fuel/Air Ratio 207
5.15.3.3 Spray Characteristics 208
5.15.3.4 Fuel Temperature 209
5.16 The Ignition Process 209
5.16.1 Factors Influencing Phase 1 210
5.16.2 Factors Influencing Phase 2 210
5.16.3 Factors Influencing Phase 3 211
5.17 Methods of Improving Ignition Performance 211
5.17.1 Correlation of Experimental Data 212
Nomenclature 214
References 216

6. Fuel Injection 221
6.1 Basic Processes in Atomization 221
6.1.1 Introduction 221
6.1.2 Breakup of Drops 222
6.1.2.1 Drop Breakup in Turbulent Flow Fields 223
6.2 Classical Mechanism of Jet and Sheet Breakup 223
6.2.1 Breakup of Fuel Jets 224
6.2.2 Breakup of Fuel Sheets 226
6.3 Prompt Atomization 227
6.4 Classical or Prompt? 228
6.5 Drop-Size Distributions 228
6.5.1 Graphical Representation of Drop-Size Distributions 228
6.5.2 Mathematical Distribution Functions 230
6.5.3 Rosin–Rammler 231
6.5.4 Modified Rosin–Rammler 233
6.5.5 Mean Diameters 234
6.5.6 Representative Diameters 234
6.5.7 Prediction of Drop-Size Distributions 236
6.6 Atomizer Requirements 237
6.7 Pressure Atomizers 237
6.7.1 Plain Orifice 238
6.7.2 Simplex 238
6.7.3 Dual Orifice 239
6.7.4 Spill Return 240
6.8 Rotary Atomizers 241
6.9 Air-Assist Atomizers 242
6.10 Airblast Atomizers 243
6.10.1 Plain-Jet Airblast 243
6.10.2 Prefilming Airblast 244
6.10.3 Piloted Airblast 245
6.10.4 Airblast Simplex 246
6.11 Effervescent Atomizers 249
6.12 Vaporizers 251
6.13 Fuel Nozzle Coking 254
6.14 Gas Injection 256
6.15 Equations for Mean Drop Size 256
6.16 SMD Equations for Pressure Atomizers 258
6.16.1 Plain Orifice 258
6.16.2 Pressure Swirl 258
6.17 SMD Equations for Twin-Fluid Atomizers 261
6.18 SMD Equations for Prompt Atomization 264
6.18.1 Comments on SMD Equations 265
6.19 Internal Flow Characteristics 266

6.20 Flow Number 266
6.21 Discharge Coefficient 268
6.21.1 Plain-Orifice Atomizers 268
6.21.2 Pressure-Swirl Atomizers 270
6.21.3 Film Thickness 270
6.22 Spray Cone Angle 273
6.22.1 Plain-Orifice Atomizers 273
6.22.2 Pressure-Swirl Atomizers 274
6.22.2.1 Theoretical Aspects 274
6.23 Radial Fuel Distribution 276
6.24 Circumferential Fuel Distribution 280
6.24.1 Pressure-Swirl Atomizers 280
6.24.2 Airblast Atomizers 282
Nomenclature 284
References 285

7. Combustion Noise 293
7.1 Introduction 293
7.2 Direct Combustion Noise 294
7.2.1 Theory 294
7.2.2 Core Noise Prediction Methods 296
7.3 Combustion Instabilities 297
7.3.1 Descriptions of Acoustic Oscillations 297
7.3.1.1 Growl 297
7.3.1.2 Howl 298
7.3.2 Characteristic Times 298
7.3.3 Influence of Fuel Type 299
7.3.4 Influence of Combustor Operating Conditions 299
7.3.5 Influence of Ambient Conditions 300
7.3.6 Aerodynamic Instabilities 300
7.3.7 Fuel-Injector Instabilities 302
7.3.8 Compressor-Induced Oscillations 303
7.3.9 LPM Combustor Noise 304
7.3.10 Test Rig Simulations 304
7.4 Control of Combustion Instabilities 305
7.4.1 Passive Control 305
7.4.2 Active Control 306
7.4.2.1 Open-Loop Systems 307
7.4.2.2 Closed-Loop Systems 307
7.4.3 Examples of Active Control 307
7.4.4 Influence of Control Signal Frequency 309
7.5 Modeling of Combustion Instabilities 309
References 310
Bibliography 313

8. Heat Transfer ... 315
8.1 Introduction ... 315
8.2 Heat-Transfer Processes ... 316
8.3 Internal Radiation ... 318
8.3.1 Radiation from Nonluminous Gases ... 318
8.3.2 Radiation from Luminous Gases ... 320
8.4 External Radiation ... 321
8.5 Internal Convection ... 322
8.6 External Convection ... 323
8.7 Calculation of Uncooled Liner Temperature ... 324
8.7.1 Method of Calculation ... 325
8.7.2 Significance of Calculated Uncooled Liner Temperatures ... 328
8.8 Film Cooling ... 329
8.8.1 Wigglestrips ... 329
8.8.2 Stacked Ring ... 330
8.8.3 Splash-Cooling Ring ... 331
8.8.4 Machined Ring ... 331
8.8.5 Rolled Ring ... 331
8.8.6 Z Ring ... 332
8.9 Correlation of Film-Cooling Data ... 333
8.9.1 Theories Based on Turbulent Boundary-Layer Model ... 334
8.9.2 Theories Based on Wall-Jet Model ... 335
8.9.3 Calculation of Film-Cooled Wall Temperature ... 337
8.9.4 Film Cooling with Augmented Convection ... 341
8.9.5 Impingement Cooling ... 342
8.9.6 Transpiration Cooling ... 343
8.10 Practical Applications of Transpiration Cooling ... 343
8.10.1 Transply ... 344
8.10.2 Lamilloy ... 345
8.10.3 Effusion Cooling ... 346
8.11 Advanced Wall-Cooling Methods ... 346
8.11.1 Angled Effusion Cooling ... 346
8.11.2 Tiles ... 347
8.12 Augmented Cold-Side Convection ... 349
8.13 Thermal Barrier Coatings ... 349
8.14 Materials ... 351
8.14.1 Metal Alloys ... 352
8.14.2 Ceramics ... 353
8.14.3 Mechanical Integrity ... 354
8.15 Liner Failure Modes ... 354
Nomenclature ... 355
References ... 356

9. Emissions 359
9.1 Introduction 359
9.2 Concerns 360
9.3 Regulations 362
9.3.1 Aircraft Engines 362
9.3.2 Stationary Gas Turbines 364
9.4 Mechanisms of Pollutant Formation 366
9.4.1 Carbon Monoxide 366
9.4.1.1 Influence of Equivalence Ratio 368
9.4.1.2 Influence of Pressure 369
9.4.1.3 Influence of Ambient Air Temperature 369
9.4.1.4 Influence of Wall-Cooling Air 369
9.4.1.5 Influence of Fuel Atomization 370
9.4.2 Unburned Hydrocarbons 370
9.4.3 Smoke 370
9.4.3.1 Influence of Pressure 371
9.4.3.2 Influence of Fuel Type 371
9.4.3.3 Influence of Fuel Atomization 372
9.4.4 Oxides of Nitrogen 374
9.4.4.1 Thermal Nitric Oxide 374
9.4.4.2 Nitrous Oxide Mechanism 378
9.4.4.3 Prompt Nitric Oxide 378
9.4.4.4 Fuel Nitric Oxide 378
9.4.5 Influence of Pressure on Oxides of Nitrogen Formation 379
9.4.6 Influence of Fuel Atomization on Oxides of Nitrogen Formation 381
9.5 Pollutants Reduction in Conventional Combustors 382
9.5.1 Carbon Monoxide and Unburned Hydrocarbons 383
9.5.2 Smoke 384
9.5.3 Oxides of Nitrogen 387
9.5.3.1 Water Injection 387
9.5.3.2 Selective Catalytic Reduction 390
9.5.3.3 Exhaust Gas Recirculation 390
9.6 Pollutants Reduction by Control of Flame Temperature 391
9.6.1 Variable Geometry 391
9.6.2 Staged Combustion 393
9.7 Dry Low-Oxides of Nitrogen Combustors 398
9.7.1 Solar Dry Low-Emissions Concepts 398
9.7.2 Siemens Hybrid Burner 400
9.7.3 General Electric DLN Combustor 401
9.7.3.1 Primary 402
9.7.3.2 Lean-Lean 403

9.7.3.3 Secondary 403
9.7.3.4 Premix 403
9.7.4 ABB EV Burner 403
9.7.5 Rolls Royce RB211 Industrial Burner 406
9.7.6 EGT DLN Combustor 407
9.7.7 General Electric LM6000 Combustor 409
9.7.8 Allison AGT100 Combustor 412
9.7.9 Developments in Japan 413
9.8 Lean Premix Prevaporize Combustion 415
9.8.1 Fuel–Air Premixing 416
9.9 Rich-Burn, Quick-Quench, Lean-Burn Combustor 418
9.10 Catalytic Combustion 421
9.10.1 Design Approaches 422
9.10.2 Design Constraints 423
9.10.3 Fuel Preparation 423
9.10.4 Catalyst Bed Construction 424
9.10.5 Postcatalyst Combustion 425
9.10.6 Design and Performance 425
9.10.7 Use of Variable Geometry 427
9.10.8 Future 428
9.11 Correlation and Modeling of Oxides of Nitrogen and Carbon Monoxide Emissions 428
9.11.1 Oxides of Nitrogen Correlations 429
9.11.1.1 Odgers and Kretschmer 429
9.11.1.2 Lewis 431
9.11.1.3 Rokke et al. 431
9.11.1.4 Rizk and Mongia 431
9.11.2 Carbon Monoxide Correlations 432
9.12 Concluding Remarks 434
Nomenclature 435
References 435

10. Alternative Fuels 443
10.1 Introduction 443
10.2 Types of Hydrocarbons 444
10.2.1 Paraffins 444
10.2.2 Olefins 445
10.2.3 Naphthenes 446
10.2.4 Aromatics 446
10.3 Production of Liquid Fuels 448
10.3.1 Removal of Sulfur Compounds 448
10.3.2 Contaminants 449
10.3.2.1 Asphaltenes 449
10.3.2.2 Gum 449
10.3.2.3 Sediment 450

10.3.2.4 Ash....450
10.3.2.5 Water....451
10.3.2.6 Sodium....452
10.3.2.7 Vanadium....452
10.3.3 Additives....453
10.3.3.1 Gum Prevention....454
10.3.3.2 Corrosion Inhibition/Lubricity Improvers....454
10.3.3.3 Anti-Icing....454
10.3.3.4 Antistatic–Static Dissipators....454
10.3.3.5 Metal Deactivators....455
10.3.3.6 Antismoke....455
10.4 Fuel Properties....456
10.4.1 Relative Density....456
10.4.1.1 API Gravity....457
10.4.1.2 Molecular Mass....458
10.4.2 Distillation Range....458
10.4.3 Vapor Pressure....458
10.4.4 Flash Point....459
10.4.5 Volatility Point....460
10.4.6 Viscosity....460
10.4.7 Surface Tension....462
10.4.8 Freezing Point....462
10.4.9 Specific Heat....463
10.4.10 Latent Heat....465
10.4.11 Thermal Conductivity....465
10.5 Combustion Properties of Fuels....466
10.5.1 Calorific Value....466
10.5.2 Enthalpy....468
10.5.3 Spontaneous-Ignition Temperature....469
10.5.4 Limits of Flammability....470
10.5.5 Smoke Point....471
10.5.5.1 Luminometer Number....472
10.5.5.2 Smoke Volatility Index....472
10.5.6 Pressure and Temperature Effects....472
10.5.6.1 Subatmospheric Pressure....473
10.5.6.2 Low Temperature....473
10.5.6.3 High Temperature....473
10.6 Classification of Liquid Fuels....474
10.6.1 Aircraft Gas Turbine Fuels....475
10.6.1.1 Airframe....475
10.6.1.2 Engine Fuel System....476
10.6.1.3 Combustion Chamber....476
10.6.2 Aircraft Fuel Specifications....476
10.6.3 Industrial Gas Turbine Fuels....476

10.7 Classification of Gaseous Fuels 478
10.7.1 Gaseous Fuel Impurities 480
10.8 Alternative Fuels 481
10.8.1 Pure Compounds 481
10.8.1.1 Hydrogen 481
10.8.1.2 Methane 483
10.8.1.3 Propane 483
10.8.1.4 Ammonia 484
10.8.1.5 Alcohols 484
10.8.2 Supplemental Fuels 485
10.8.3 Slurry Fuels 485
10.9 Synthetic Fuels 486
10.9.1 Fuels Produced by Fischer–Tropsch Synthesis of Coal/Biomass 486
10.9.2 Biofuels 488
10.9.3 Alternative Fuel Properties 489
10.9.4 Combustion and Emissions Performance 490
10.9.4.1 Fischer–Tropsch Fuels 490
10.9.4.2 Biodiesel Fuels 498
10.9.4.3 Highly Aromatic (Broad Specification) Alternative Fuels 504
References 508

Author Index 513
Subject Index 519

Preface

Developments in gas turbine technology continue to meet the propulsion, power, fuel efficiency, and low pollutant emissions needs of the twenty-first century. Ten years have passed since the publication of the second edition, which continues to be widely used in many parts of the world. Professor Arthur Lefebvre passed away in 2003. Last year, when the publisher approached me with a proposal for preparing the third edition, I could not refuse. After all, Professor Lefebvre was my teacher, friend, and a co-researcher for 35 years; I was involved in numerous discussions during the writing of the first and second editions; and finally, I learned a great deal from all the material presented in the book.

The book has a clear purpose; it is directed primarily toward those who design, manufacture, and operate gas turbines in applications ranging from aeronautical to power generation. It serves as a graduate-level textbook, design manual, and research reference in the field of gas turbine combustion. The text is essentially self-contained and assumes only a modest prior knowledge of physics and chemistry. In preparation for the twenty-first century, the second edition was thoroughly revised and updated with numerous changes.

As I examined each chapter of the second edition, I found the text as up-to-date and refreshing as ever, proving that improvements in gas turbine combustion have been gradual and evolutionary. So minimum revisions were required in the areas of multifuel capabilities, flame flashback, high off-design combustion efficiency, and liner failure studies with reduced film cooling. In the quest to achieve higher fuel efficiency and decrease carbon dioxide emissions, compressor pressure ratios and turbine inlet temperatures gradually increased in the last decade. Yet gaseous and particulate emissions decreased by one third or more and are well below the emissions regulations in effect as of July 2006. Thus, Chapter 9 on emissions was updated.

The most significant change has been the addition of a new Chapter 10, "Alternative Fuels" and the book's subtitle *Alternative Fuels and Emissions.* Today, the ever-rising cost of petroleum fuel is prompting research into developing alternative liquid fuels based on coal, biomass, and other feedstock. Depleting global resources of petroleum fuel combined with increasing terrorist activities are leading various industrialized and developing countries to develop domestic sources of fuel for assured supply and energy security. These domestically produced alternative fuels have to be capable of using the available infrastructure of fuel refining, transportation, distribution, and consumption. The future of the alternative fuel industry in the forthcoming decade depends upon the right fuel properties and handling characteristics for the engines and infrastructure already in place; environmental impact, which includes competition with food, water, and land; CO_2 life cycle analysis and carbon footprint issues; and economics of return on investment, production, and sustainability.

Accordingly, Chapter 10 presents the physical and chemical properties of conventional (petroleum-based) liquid and gaseous fuels for gas turbines. Next, properties of alternative (synthetic) fuels and conventional–alternative fuel blends are reviewed. The influence of these different fuels and their blends on combustor performance, design, and emissions is described. Reference is made to the special requirements of aircraft fuels and the problems encountered with fuels for industrial gas turbines.

By not deviating from the highly successful formula of the first and second editions, I have been able to bring all the relevant material up to date through 2009. I hope that this third edition will find favorable acceptance among the gas turbine combustion, fuels, and emissions community.

Finally, I would like to thank my wife Shubhangi for her help and encouragement during the preparation of this book.

Dilip R. Ballal
Dayton, Ohio

Authors

Arthur H. Lefebvre (1923–2003) was Emeritus professor at Purdue University. With industrial and academic experience spanning more than four decades, he wrote over 150 technical papers on both fundamental and practical aspects of atomization and combustion. The honors he received include the ASME Gas Turbine and ASME R. Tom Sawyer Awards, ASME George Westinghouse Gold Medal, and the IGTI Scholar Award. He was also the first recipient of the AIAA Propellants and Combustion Award.

Dilip R. Ballal is the Hans von Ohain Distinguished Chair Professor at the University of Dayton. He has four decades of experience in the field of gas turbine fuels and combustion, engine aerodynamics, heat transfer, and fluid mechanics. During this period, he worked on numerous fundamental and practical problems related to gas turbine fuels and combustion. He is a Life-Fellow of both ASME and AIAA, ASME Senior Vice President and editor of the *ASME Journal of Engineering for Gas Turbines and Power.* He has served on the National Academies Study Committees on "Gas Turbine Propulsion" and "Aeroengine Fuel Efficiency." His honors include the ASME-IGTI Aeroengine Technology Award, the AIAA Airbreathing Propulsion Award, the AIAA Propellants and Combustion Award, and the AIAA Energy Systems Award.

1

Basic Considerations

1.1 Introduction

The primary purpose of this introductory chapter is to discuss the main requirements of gas turbine combustors and to describe, in general terms, the various types and configurations of combustors employed in aircraft and industrial engines. The principal geometric and aerodynamic features that are common to most types of gas turbine combustors are briefly reviewed, with special attention being given to fuel preparation and liner-wall cooling to reflect the important role these topics continue to play in combustor development. Reference is made to most of the key issues involved in combustor design and development, but the descriptive material is necessarily brief because these and other important aspects of combustor performance are described more fully in subsequent chapters.

Bearing in mind the pressures and exigencies of wartime Britain and Germany, and the lack of knowledge and experience available to the designer, it is perhaps hardly surprising that the first generation of gas turbine combustors were characterized by wide variations in size, geometry, and the mode of fuel injection. With the passage of time and the post-war lifting of information exchange, some commonalities in design philosophy began to emerge. By around 1950, most of the basic features of conventional gas turbine combustors, as we know them today, were firmly established.

Since that time, combustor technology has developed gradually and continuously, rather than through dramatic change, which is why most of the aero-engine combustors now in service tend to resemble each other in size, shape, and general appearance. This close family resemblance stems from the fact that the basic geometry of a combustor is dictated largely by the need for its length and frontal area to remain within the limits set by other engine components, by the necessity for a diffuser to minimize pressure loss, and by the requirement of a liner (flame tube) to provide stable operation over a wide range of air/fuel ratios. During the past half century, combustion pressures have risen from 5 to 50 atmospheres, inlet air temperatures from 450 to 900 K, and outlet temperatures from 1100 to 1850 K. Despite the continually increasing severity of operating conditions, which are greatly exacerbated by the concomitant increases in compressor outlet velocity, today's combustors exhibit close to 100% combustion efficiency over their normal operating range, including idling, and demonstrate substantial reductions in pollutant emissions. Furthermore, the life expectancy of aero-engine liners has risen from just a few hundred hours to many tens of thousands of hours.

Although many formidable problems have been overcome, the challenge of ingenuity in design still remains. New concepts and technology are needed to further reduce pollutant emissions and to respond to the growing requirement of many industrial engines for multifuel capability. Gas turbines are "omnivorous" machines, capable of operating efficiently on a wide variety of cheap fuels, solid, liquid, and gaseous, with the exception of aircraft engines. Today, the ever-rising cost of petroleum fuel is prompting research into developing alternative liquid fuels and this is posing new combustor design challenges. Another problem of increasing importance is that of acoustic resonance, which occurs when combustion instabilities become coupled with the acoustics of the combustor. This problem could be crucial to the future development of lean premixed combustors.

It is clearly important that combustor developments should keep pace with improvements in other key engine components. Thus, reduction of combustor size and weight will remain an important requirement for aero engines, whereas the continuing trend toward higher turbine inlet temperatures will call for a closer adherence to the design temperature profile at the turbine inlet. Simultaneously, the demand for greater reliability, increased durability, and lower manufacturing, development, and maintenance costs seems likely to assume added importance in the future. To meet these challenges, the search goes on for new materials and new methods of fabrication to simplify basic combustor design and reduce cost. The search has already led to the development of advanced wall-cooling techniques and the widespread use of thermal barrier refractory coatings within the combustion liner.

1.2 Early Combustor Developments

The material contained in this book is largely a chronicle of developments in gas turbine combustion during the last half century. For both British and German engineers, the development of a workable combustor was an obstacle that had to be overcome in their independent and concurrent efforts to achieve a practical turbojet engine. It proved to be a formidable task for both groups and, in Whittle's case, combustion problems dominated the first three years of engine development. The following abridged account of the early history of gas turbine combustion in Britain, Germany, and the United States is intended to cover the period from the start of World War II until around 1950, by which time it was generally accepted that the piston engine had reached its limit as a propulsion system for high-speed flight and the gas turbine was firmly established as the powerplant of choice for aircraft applications.

1.2.1 Britain

One method of preparing a liquid fuel for combustion is to heat it above the boiling point of its heaviest hydrocarbon ingredient, so that it is entirely converted into vapor before combustion. This was the method adopted by Whittle for his first turbojet engine. This engine employed 10 separate tubular combustors in a reverse-flow arrangement to permit a short engine shaft. Whittle tried several vaporizer tube configurations, more than 30 in all, one of which is illustrated in Figure 1.1. This figure shows that fuel was heated in tubes located in the flame zone. The fuel was maintained at high pressure so that vaporization could not occur until it had been injected through a nozzle and its pressure reduced to that of the combustion zone. Whittle experienced considerable difficulties with this system, due mainly to problems of thermal cracking and coking up of the vaporizer tubes, as well as difficulties in controlling the fuel flow rate.

After many trials and setbacks, Whittle adopted a combustor whose main attraction was the replacement of vaporizer tubes by a pressure-swirl atomizer having a wide spray cone angle. Another interesting feature of the new combustor was that most of the primary-zone airflow entered the combustion zone through a large air swirler located at the upstream end of the liner around the fuel nozzle, as shown in Figure 1.2. This swirler served to create a toroidal flow reversal that entrained and recirculated a portion of the hot combustion products to mix with the incoming air and fuel. This arrangement not only anchored the flame, but also provided the rapid mixing of fuel vapor, air, and combustion products needed to achieve high heat-release rates. The additional air required to complete combustion and reduce the gas temperature to a value acceptable to the turbine was supplied through stub

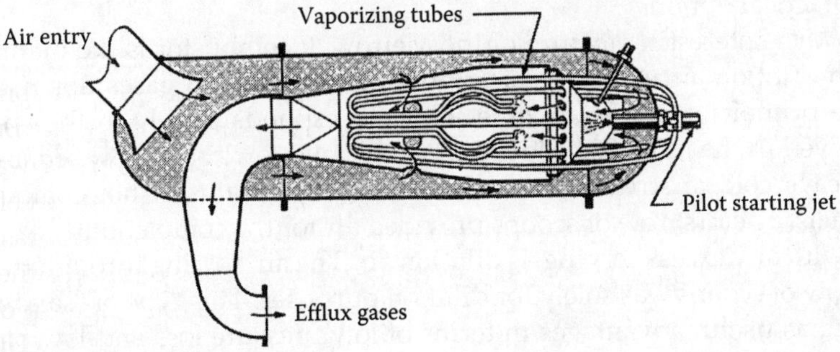

FIGURE 1.1
Early Whittle vaporizer combustor.

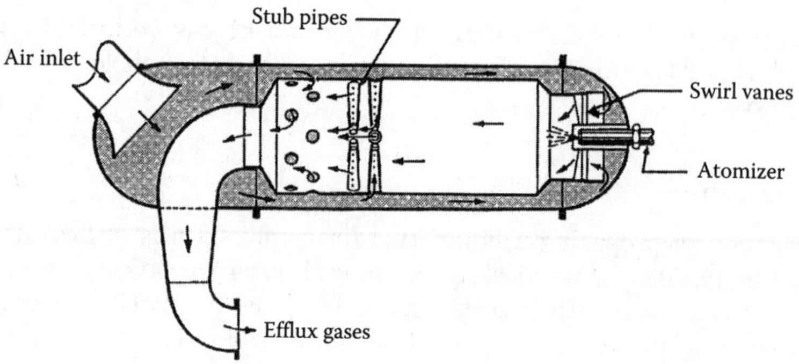

FIGURE 1.2
Early Whittle atomizer combustor.

pipes that projected radially inward and through holes pierced in the liner walls. After suitable development, this combustor was adopted for the Power Jets W1 engine, which made the first British turbojet-powered flight on the evening of May 15, 1941.

Another early British engine was the De Havilland Goblin, which was the first engine to power the Lockheed P-40. (It was later replaced by General Electric's I-40 engine, which provided 33% more thrust.) The Goblin is of historical interest because it was the first British engine to use "straight-through" combustors, as opposed to the "reverse-flow" type employed on all previous engines. The first British annular combustor appeared on the Metropolitan Vickers Beryl engine. A noteworthy feature of this combustor was the use of upstream fuel injection. This system was also used in other engines, the earliest example being the German Jumo 004. The main advantage claimed for upstream fuel injection was a longer residence time of the fuel droplets in the combustion zone, which provided more time for fuel evaporation. Its main drawback stemmed from the immersion of key components in the flame. Cooling arrangements for the atomizer feed arm could be provided, but it was difficult to eliminate entirely the problem of carbon deposition on the atomizer face. For this reason, upstream fuel injection is no longer regarded as a practical option.

Another interesting feature of the Metrovick combustor is the manner in which dilution air was introduced into the combustion gases downstream of the primary combustion zone. This corresponds closely to the method employed in the Jumo 004. Figure 1.3 shows two rows of narrow scoops that interleave cold airstreams between co-flowing streams of hot combustion products. The first row of scoops provided air for the completion of combustion, with any excess serving as dilution air. The air flowing through the second row of scoops was solely for dilution purposes. This type of "sandwich" mixer has useful advantages in terms of low pressure loss and low pattern factor. However, it carries a high weight penalty, which is clearly a serious drawback for aircraft engines, and the scoops are also prone to burnout

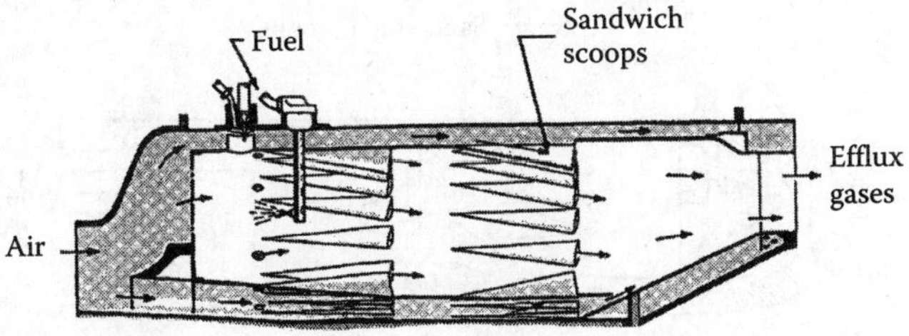

FIGURE 1.3
Metrovick annular combustor.

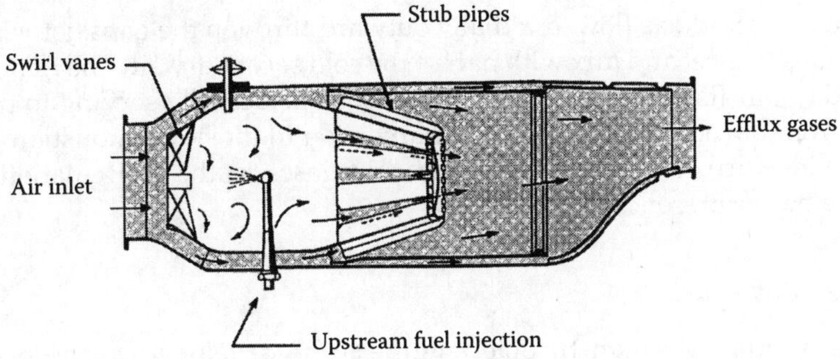

FIGURE 1.4
Jumo 004 tubular combustor.

because of their exposure to the high-velocity combustion gases. Sandwich mixers are no longer used, except in a highly abbreviated form where their main function is to strengthen the liner and raise the flow discharge coefficient (typically to around 0.8, as opposed to around 0.6 for a plain hole).

1.2.2 Germany

1.2.2.1 Jumo 004

This engine is of great historical interest because it was the world's first mass-produced turbojet and one that saw extensive service in World War II. It was among the first engines to employ axial flow turbomachinery and straight-through combustors. Each of the six tubular combustors was supplied with fuel at pressures up to 5.2 MPa (750 psi) from a pressure-swirl atomizer, which sprayed the fuel upstream into the primary combustion zone. Figure 1.4 shows schematically the basic combustor design. The primary air flowed into the liner through six swirl vanes, the amount of air being sufficient to achieve near-stoichiometric combustion at the engine design point. Mixing between combustion products and dilution air was achieved using an assembly of stub pipes that were welded to a ring at their upstream end and to the outer perimeter of a 10-cm diameter dished baffle at their downstream end. The hot

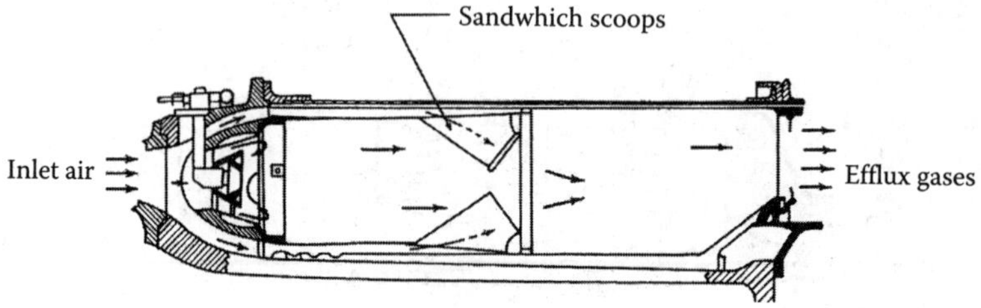

FIGURE 1.5
BMW 003 annular combustor.

combustion products flowed radially outward through the gaps between the stub pipes to meet and mix with part of the cold secondary air. The remaining secondary air flowed through the stub pipes, incidentally serving to protect them from burnout because of their immersion in the hot combustion gases, to provide further mixing of hot and cold gases in the recirculation zone created by the presence of the baffle.

1.2.2.2 BMW 003

The only other German turbojet engine to be developed to the production stage during World War II was the BMW 003. This engine employed an annular combustor fitted with 16 equispaced, downstream-spraying, pressure atomizers. Each fuel nozzle was surrounded by a baffle and the primary combustion air flowed both through and around it. The method used to inject the dilution air was a sandwich mixer arrangement, which interleaved streams of cold secondary air between parallel streams of hot combustion products. Dilution air flowed through 40 scoops attached to the outer liner, alternating in circumferential locations with 40 similar scoops attached to the inner liner. The end result, as shown in Figure 1.5, was a combustor having a relatively low pressure loss, but also a fairly high length/height ratio.

1.2.3 The United States

During the development of the W1 engine, the decision was made to build a larger Whittle engine of 1600 lb thrust to be designated as the W2B. In 1941, a W2B engine, complete with drawings, was delivered to the General Electric Company (GE); within six months this company had built two more engines with the same design. In 1947, Pratt and Whitney (P&W), having been fully preoccupied with piston engine production throughout the war, made its first entry into the turbojet arena by licensing the Nene engine from Rolls Royce. Having established a foothold in the turbojet engine business, GE and P&W lost no time in producing their own independent combustor designs. For example, GE's Whittle-derived J31 engine employed a reverse-flow

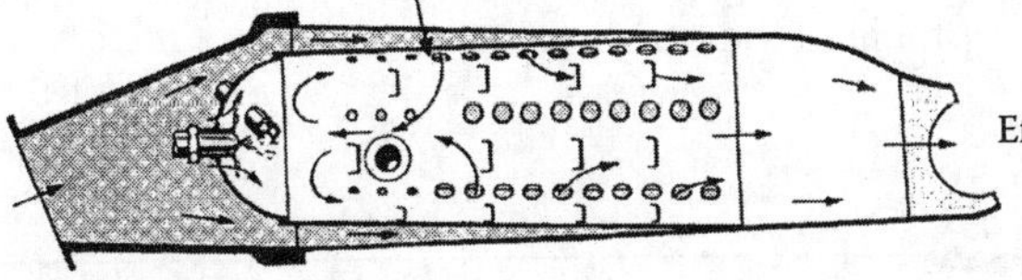

FIGURE 1.6
General Electric J33 tubular combustor.

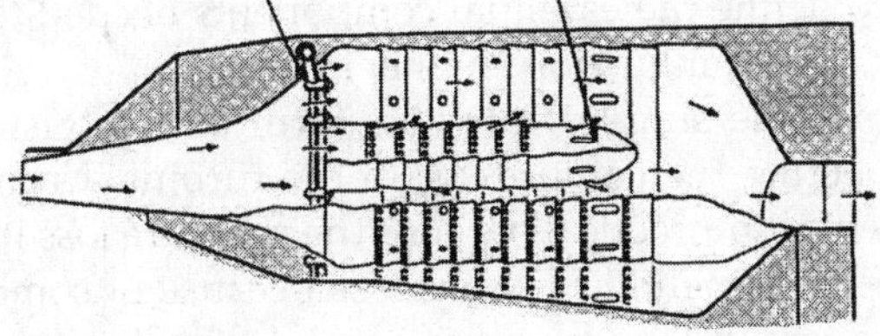

FIGURE 1.7
Pratt & Whitney J57 tuboannular combustor.

combustor, but a straight-through version (see Figure 1.6) was adopted for the J33 and for subsequent engines such as the J35 and J47. For its J57 engine, shown in Figure 1.7, P&W employed eight tubular liners located within an annular casing. Each liner had a perforated tube along its central axis that extended about halfway down the liner. In effect, this central tube converted the tubular liner into a small annular combustor, supplied with fuel from six equispaced, pressure-swirl nozzles.

By 1943, Westinghouse had developed successful axial-flow turbojet engines without any European input. An annular combustor was selected for its J30 engine, whereas a dual-annular configuration was adopted for the J34. This dual-annular concept was ahead of its time, and interest in it lapsed until it was resurrected by GE in the 1970s to serve as a low-emissions combustor for their CFM56-B engine.

By the end of the 1940s, the development work carried out in the UK, Germany, and the United States had established the basic design features of aero-engine combustors that have remained largely unchanged. The main components are a diffuser for reducing the compressor outlet air velocity to avoid high-pressure losses in combustion, a liner (or flame tube) that is arranged to be concentric within the outer combustor casing, means for supplying the combustion zone with atomized or vaporized fuel and, with tubular liners, interconnectors (or cross-fire tubes) through which hot gases can flow from a lighted liner to an adjacent unlighted liner.

Within the liner itself, the distribution of air is arranged to ensure that the primary combustion zone operates at a much higher fuel/air ratio than the overall combustor fuel/air ratio. More air is admitted downstream of the primary zone to complete the combustion process and to dilute the combustion products to a temperature acceptable to the turbine.

1.3 Basic Design Features

It is of interest to examine briefly the considerations that dictate the basic geometry of the "conventional" gas turbine combustor. It is also instructive because it helps to define the essential components needed to carry out the primary functions of a combustion chamber.

Figure 1.8a shows the simplest possible form of combustor—a straight-walled duct connecting the compressor to the turbine. Unfortunately, this simple arrangement is impractical because the pressure loss incurred would be excessive. The fundamental pressure loss because of combustion is proportional to the square of the air velocity and, for compressor outlet velocities of the order of 170 m/s, this loss could amount to almost one-third of the pressure rise achieved in the compressor. To reduce this pressure loss to an acceptable level, a diffuser is used to lower the air velocity by a factor of about 5, as shown in Figure 1.8b. Having fitted a diffuser, a flow reversal must then be created to provide a low-velocity region in which to anchor the flame. Figure 1.8c shows how this may be accomplished with a plain baffle. The only remaining defect in this arrangement is that to produce the desired temperature rise, the overall chamber air/fuel ratio must normally be around 30–40, which is well outside the limits of flammability for hydrocarbon–air mixtures. Ideally, the air/fuel ratio in the primary combustion zone should

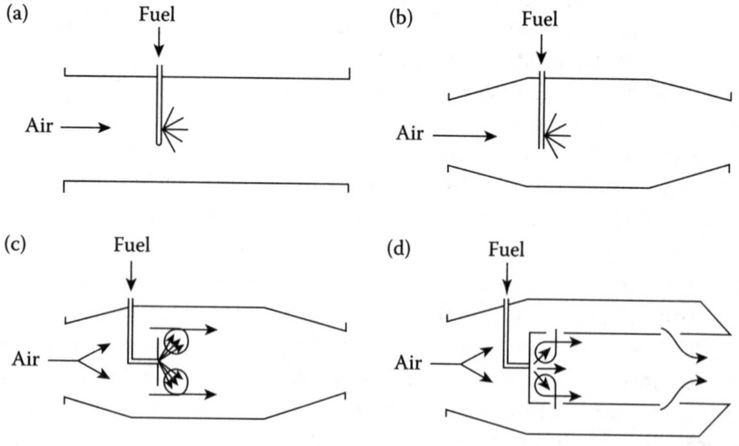

FIGURE 1.8
Derivation of conventional combustor configuration.

be around 18, although higher values (around 24) are sometimes preferred if low emissions of nitric oxides (NO_X) is a prime consideration. To deal with this problem, combustion is sustained by a recirculatory flow of burned products that provide a continuous source of ignition for the incoming fuel–air mixture. The air not required for combustion is admitted downstream of the combustion zone to mix with the hot burned products, thereby reducing their temperature to a value that is acceptable to the turbine.

Figure 1.8 thus illustrates the logical development of the conventional gas turbine combustion chamber in its most widely used form. As would be expected, there are many variations on the basic pattern, shown in Figure 1.8d, but, in general, all chambers incorporate an air casing, diffuser, liner, and fuel injector as key components.

The choice of a particular type and layout of combustion chamber is determined largely by engine specifications, but it is also strongly influenced by the desirability of using the available space as effectively as possible. On large aircraft engines, the chamber is almost invariably of the straight-through type, in which the air flows in a direction essentially parallel to the axis of the chamber. For smaller engines, the reverse-flow annular combustor provides a more compact unit and allows close coupling between the compressor and turbine. In most combustors, the fuel is injected into the burning zone in the form of a well-atomized spray, obtained either by forcing it through a fine orifice under pressure, or by utilizing the pressure differential across the liner wall to create a stream of high-velocity air that shatters the fuel into fine droplets before transporting it into the primary combustion zone.

1.4 Combustor Requirements

A gas turbine combustor must satisfy a wide range of requirements whose relative importance varies among engine types. However, the basic requirements of all combustors may be listed as follows:

1. High-combustion efficiency (i.e., the fuel should be completely burned so that all its chemical energy is liberated as heat)
2. Reliable and smooth ignition, both on the ground (especially at very low ambient temperatures) and, in the case of aircraft engines, after a flameout at high altitude
3. Wide stability limits (i.e., the flame should stay alight over wide ranges of pressure and air/fuel ratio)
4. Low pressure loss
5. An outlet temperature distribution (pattern factor) that is tailored to maximize the lives of the turbine blades and nozzle guide vanes

6. Low emissions of smoke and gaseous pollutant species
7. Freedom from pressure pulsations and other manifestations of combustion-induced instability
8. Size and shape compatible with engine envelope
9. Design for minimum cost and ease of manufacturing
10. Maintainability
11. Durability
12. Petroleum, synthetic, and biomass-based multifuel capability.

For aircraft engines, size and weight are important considerations, whereas for industrial engines more emphasis is placed on other items, such as long operating life and multifuel capability. For all types of engines, the requirements of low fuel consumption and low pollutant emissions are paramount.

1.5 Combustor Types

The choice of a particular combustor type and layout is determined largely by the overall engine design and by the need to use the available space as effectively as possible. There are two basic types of combustor, tubular and annular. A compromise between these two extremes is the "tuboannular" or "can-annular" combustor, in which a number of equispaced tubular liners are placed within an annular air casing. The three different combustor types are illustrated in Figure 1.9.

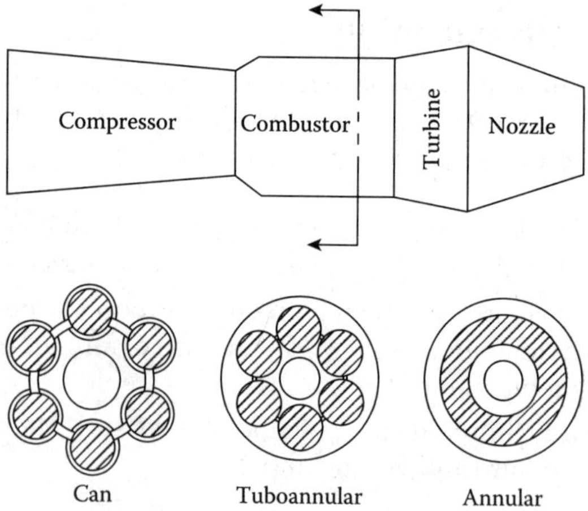

FIGURE 1.9
Illustration of three main combustor types.

1.5.1 Tubular

A tubular (or "can") combustor is comprised of a cylindrical liner mounted concentrically inside a cylindrical casing. Most of the early jet engines, such as the Whittle W2B, Jumo 004, and the RR Nene, Dart, and Derwent, featured tubular combustors, usually in numbers varying from 6 to 16 per engine.

The main advantage of tubular systems is that relatively little time and money is incurred in their development. However, their excessive length and weight prohibit their use in aircraft engines, and their main application is to industrial units where accessibility and ease of maintenance are prime considerations. A multi-can combustor layout is shown in Figure 1.10.

1.5.2 Tuboannular

As engine pressure ratios started to climb in the late 1940s, the tuboannular or can-annular combustor began to find increasing favor on both sides of the Atlantic. With this design, a group of tubular liners, usually from 6 to 10, is arranged inside a single annular casing, as illustrated in Figure 1.11. This concept attempts to combine the compactness of the annular chamber with the mechanical strength of the tubular chamber. A drawback to the tuboannular combustor, which it shares with tubular configurations, is the need for interconnectors (cross-fire tubes). Engines fitted with tuboannular combustors include the Allison 501-K, the GE J73 and J79, the P&W J57 and J75, and the RR Avon, Conway, Olympus, Tyne, and Spey.

Compared with the annular design, the tuboannular chamber has an important advantage in that much useful chamber development can be carried out with very modest air supplies, using just a small segment of the

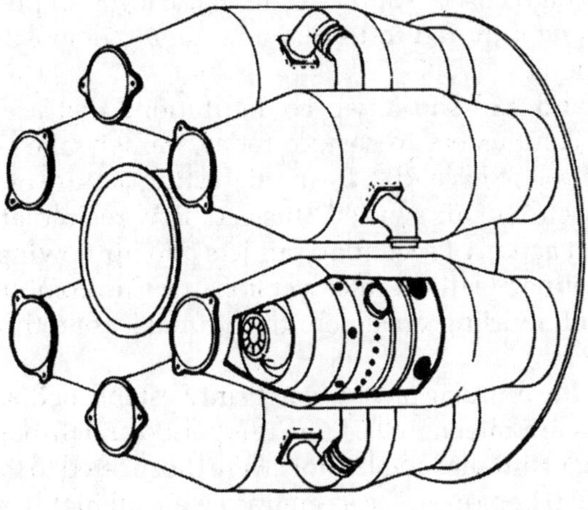

FIGURE 1.10
Multi-can combustor arrangement. (From Odgers, J. and Kretschmer, D., *Northern Research and Engineering Corporation Report* No. 1344-1, 1980.)

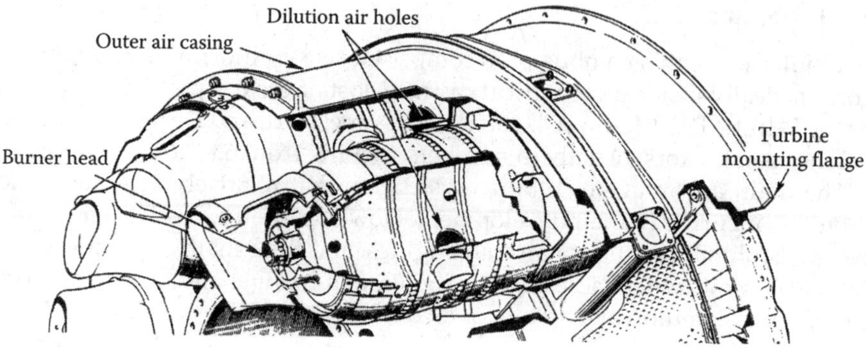

FIGURE 1.11
Tuboannular combustor arrangement. (Courtesy of Rolls Royce plc.)

total chamber containing one or more liners. Its drawbacks emerge when trying to achieve a satisfactory and consistent airflow pattern; in particular, the design of the diffuser can present serious difficulties.

1.5.3 Annular

In this type, an annular liner is mounted concentrically inside an annular casing. In many ways it is an ideal form of chamber, because its clean aerodynamic layout results in a compact unit of lower pressure loss than other combustor types. Its main drawback stems from the heavy buckling load on the outer liner. Thus, in the early days of turbojet development, the use of annular liners was confined to engines of low pressure ratio, such as the BMW 003, the Metrovick Beryl, and the Westinghouse J30. Another drawback is the very high cost of supplying air at the levels of pressure, temperature, and flow rate required to test large annular combustion chambers at full-load conditions.

Figures 1.12 and 1.13 show two configurations that are representative of the annular combustors in service today, namely, the General Electric CF6-50 and the Rolls Royce RB211. An interesting feature of the RB211 combustor is the absence of air swirlers. Instead, flow recirculation is achieved by the combined action of secondary air jets and air flowing over the backplate along the liner wall. In later versions of this combustor, an appreciable amount of swirling air enters the primary zone through modified airblast atomizers.

By the 1960s, the annular layout was firmly established as the automatic choice for all new aircraft engines. From this period and throughout the 1980s, the most important annular combustors were those fitted to the GE CF6, P&W JT9D, and RR RB211 engines. These engines were all highly successful, both technically and commercially. Improvements in wide-body aircraft, along with continuing market pressures to reduce cost, called for engines in the

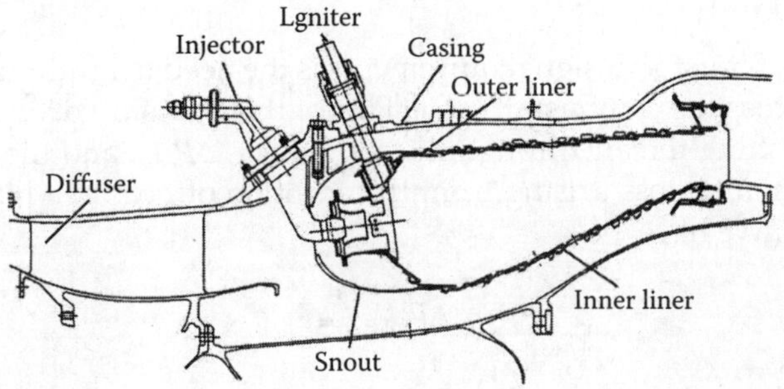

FIGURE 1.12
CF6-50 annular combustor.

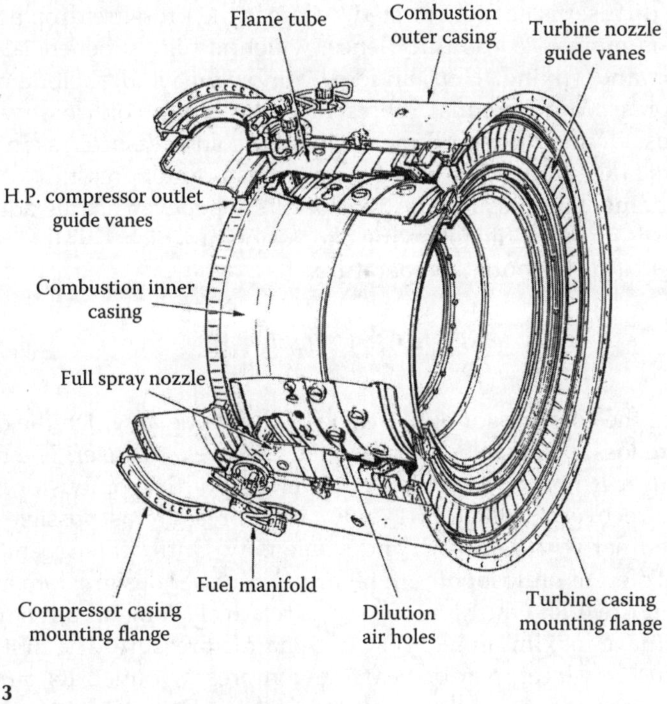

FIGURE 1.13
RB211 annular combustor. (Courtesy of Rolls Royce plc.)

80,000–100,000 lb thrust class with growth potential up to around 115,000 1b. To meet this demand, Rolls Royce developed the Trent engine, which is a direct descendant of the RB211 series. The GE90 and the P&W4084 both achieve similar performance and thrust levels. All three engines are fitted with annular combustors, which embody the latest advances in fuel injection and wall-cooling techniques.

1.6 Diffuser

Among the combustor design requirements is the need to minimize the pressure drop across the combustor, ΔP_{3-4}. Part of this pressure drop is incurred in simply pushing the air through the combustor, ΔP_{cold}, and the remainder is the fundamental loss arising from the addition of heat to a high-velocity stream, ΔP_{hot}. We have

$$\Delta P_{3-4} = \Delta P_{cold} + \Delta P_{hot}. \quad (1.1)$$

The cold loss represents the sum of the losses arising in the diffuser and the liner. From the viewpoint of overall engine performance, the distinction between diffuser pressure loss and liner pressure loss is immaterial. However, from a combustion standpoint, it is important because pressure loss in the diffuser is entirely wasted, whereas the pressure drop across the liner wall is manifested as turbulence, which is highly beneficial to both combustion and mixing. Thus, an ideal combustor would be one in which the liner pressure differential represented the entire cold loss, with zero pressure loss in the diffuser. Typical values of cold pressure loss in modern combustors range from 2.5 to 5% of the combustor inlet pressure.

The fundamental pressure loss that occurs whenever heat is added to a flowing gas is given by the following expression in which T_3 is the inlet temperature and T_4 is the outlet temperature.

$$\Delta P_{hot} = 0.5\rho U^2 [T_4/T_3 - 1]. \quad (1.2)$$

To reduce the compressor outlet velocity to a value at which the combustor pressure loss is tolerable, it is customary to use a diffuser. The function of the diffuser is not only to reduce the velocity of the combustor inlet air, but also to recover as much of the dynamic pressure as possible, and to present the liner with a smooth and stable flow. Until quite recently, there were two different philosophies in regard to diffuser design; both are illustrated in Figure 1.14. One is to employ a relatively long aerodynamic diffuser to achieve maximum recovery of dynamic pressure. The first section of the diffuser is located at or near the compressor outlet. Its purpose is to achieve some reduction in velocity, typically about 35%, before the air reaches the snout, at which point it divides and flows into three separate diffusing passages. Two of these passages convey air to the inner and outer liner annuli in roughly equal proportions. The central diffuser passage discharges the remaining air into the dome region, which provides air for atomization and dome cooling.

The other main diffuser type is the so-called "dump" or "step" diffuser. It consists of a short conventional diffuser in which the air velocity is reduced

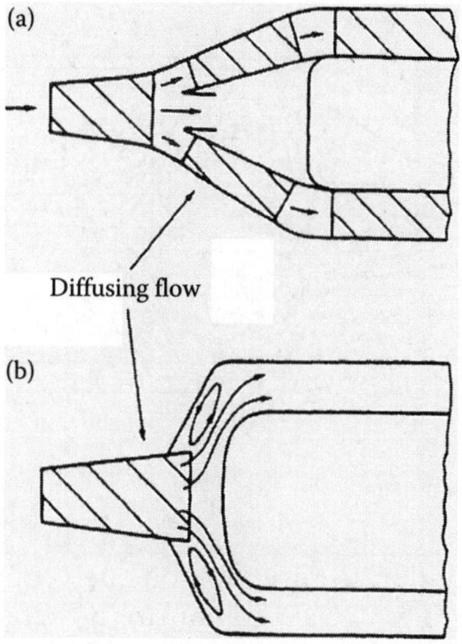

FIGURE 1.14
Two basic types of annular diffusers: (a) aerodynamic, (b) dump.

to almost half its inlet value. At exit, the air is then "dumped" and left to divide itself between air for the inner and outer annuli and dome air.

Both faired and dump diffusers have been widely used in aero-engine combustors. Dump diffusers are now generally preferred [2] because of their higher tolerance to variations in inlet velocity profile and hardware dimensions. Thus, whereas most Rolls Royce annular combustors have faired diffusers, the latest annular design for the Trent engine features a dump diffuser.

1.7 Primary Zone

The main function of the primary zone is to anchor the flame and provide sufficient time, temperature, and turbulence to achieve essentially complete combustion of the incoming fuel–air mixture. The importance of the primary-zone airflow pattern to the attainment of these goals cannot be overstated. Many different types of flow patterns are employed, but one feature that is common to all is the creation of a toroidal flow reversal that entrains and recirculates a portion of the hot combustion gases to provide continuous ignition to the incoming air and fuel. Some early combustors used air swirlers to create the toroidal flow pattern, whereas others had no swirler and relied solely on air injected through holes drilled in the liner wall at the upstream end of the liner (see, for example, Figures 1.2 and 1.6). Both methods are capable of generating flow recirculation in the primary zone.

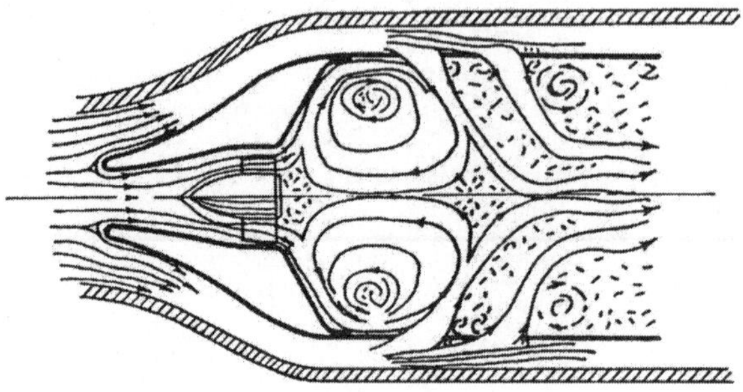

FIGURE 1.15
Lucas primary-zone airflow pattern.

An important contribution to primary-zone aerodynamics was made by the Lucas combustion group in their combustor designs for the Whittle W2B and Welland engines. The basic airflow patterns embodied in the Lucas concept are sketched in Figure 1.15. Note that both swirling air and primary air jets are used to produce the desired flow reversal. As already noted, each mode of air injection is capable of achieving flow recirculation in its own right, but if both are used, and if a proper choice is made of swirl vane angle and the size, number, and axial location of the primary air holes, then the two separate flow recirculations created by the two separate modes of air injection will merge and blend in such a manner that each one complements and strengthens the other. The result is a strong and stable primary-zone airflow pattern that can provide wide stability limits, good ignition performance, and freedom from the type of flow instabilities that often give rise to combustion pulsations and noise. The Lucas company had a strong influence on British combustor design, and the basic aerodynamic features shown in Figure 1.15 can be found in the combustors designed for many British engines, including the Rolls Royce Nene, Derwent, Dart, Proteus, Avon, Conway, and Tyne.

1.8 Intermediate Zone

If the primary-zone temperature is higher than around 2000 K, dissociation reactions will result in the appearance of significant concentrations of carbon monoxide (CO) and hydrogen (H_2) in the efflux gases. Should these gases pass directly to the dilution zone and be rapidly cooled by the addition of massive amounts of air, the gas composition would be "frozen," and CO, which is both a pollutant and a source of combustion inefficiency, would be discharged from the combustor unburned. Dropping the temperature to an intermediate level by the addition of small amounts of air encourages the burnout of soot and allows the combustion of CO and any other unburned hydrocarbons (UHC) to proceed to completion.

In early combustor designs, an intermediate zone was provided as a matter of course. As pressure ratios increased, and more air was required for combustion and liner-wall cooling, the amount of air available for the intermediate zone went down accordingly. By around 1970, the traditional form of intermediate zone had largely disappeared. However, the desirability of an intermediate zone remains; therefore, should the developments now being made in wall-cooling techniques allow some air to become available, consideration might be given to its possible reinstatement.

1.9 Dilution Zone

The role of the dilution zone is to admit the air remaining after the combustion and wall-cooling requirements have been met, and to provide an outlet stream with a temperature distribution that is acceptable to the turbine. This temperature distribution is usually described in terms of "pattern factor" or "temperature traverse quality."

The amount of air available for dilution is usually between 20 and 40% of the total combustor airflow. It is introduced into the hot gas stream through one or more rows of holes in the liner walls. The size and shape of these holes are selected to optimize the penetration of the air jets and their subsequent mixing with the main stream.

In theory, any given traverse quality can be achieved either by the use of a long dilution zone or by tolerating a high liner pressure-loss factor. In practice, however, it is found that mixedness initially improves greatly with an increase in mixing length and thereafter at a progressively slower rate. This is why the length/diameter ratios of dilution zones all tend to lie in a narrow range between 1.5 and 1.8.

For the very high turbine entry temperature (around 2000 K) associated with modern high-performance engines, an ideal pattern factor would be one that gives minimum temperature at the turbine blade root, where stresses are highest, and also at the turbine blade tip, to protect seal materials. Attainment of the desired temperature profile is paramount, owing to its major impact on the maximum allowable mean turbine entry temperature and hot-section durability. Due to the importance and severity of the problem, a large proportion of the total combustor development effort is devoted to achieving the desired pattern factor.

The locations of the three main zones described above, in relation to the various combustor components and the air admission holes, are shown in Figure 1.16. Note also in this figure the "snout," which is formed by cowls that project upstream from the dome. The region inside the snout acts as

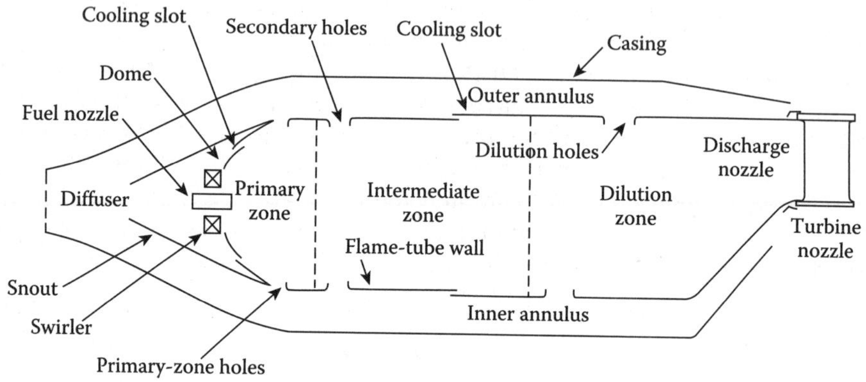

FIGURE 1.16
Main components of a conventional combustor.

a plenum chamber, providing a high uniform static pressure for feeding the air swirler, which is attached to the dome, the airblast atomizer, and the dome cooling airflows.

1.10 Fuel Preparation

The processes of liquid atomization and evaporation are of fundamental importance to the performance of a gas turbine combustion system. Normal liquid fuels are not sufficiently volatile to produce vapor in the amounts required for ignition and combustion unless they are atomized into a large number of droplets with a corresponding vastly increased surface area. The smaller the droplet size, the faster the rate of evaporation. The influence of drop size on ignition performance is of special importance, because large increases in ignition energy are needed to compensate for even a slight increase in mean drop size. Spray quality also affects stability limits, combustion efficiency, and pollutant emission levels.

1.10.1 Pressure-Swirl Atomizers

A common method of achieving atomization is by forcing the fuel under pressure through a specially designed orifice. Since the need to minimize combustor length, a spray cone angle of around 110° is customary. With the *simplex* atomizer, shown in Figure 1.17a, this is achieved by fitting a swirl chamber upstream of the discharge orifice. A major design problem is to achieve good atomization over a fuel flow range of around 40:1. If the atomizer discharge orifice is made small enough to ensure good atomization at

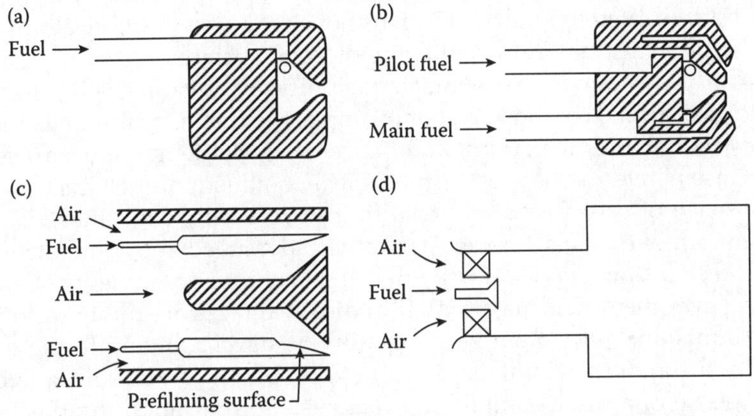

FIGURE 1.17
Trends in atomizer design: (a) simplex, (b) dual-orifice, (c) airblast, (d) premix-prevaporize.

low fuel flow rates, then the pressure required at high flows becomes excessive. On the other hand, if the orifice is made large, the fuel will not atomize satisfactorily at the low flow rates and low pressures associated with operation at high altitudes. A solution to this problem is provided by the *dual-orifice* atomizer, which incorporates two swirl chambers, one of which (the pilot) is located concentrically within the other (the main), as shown in Figure 1.17b. The orifices that feed fuel into the pilot swirl chamber are small in size, whereas the corresponding orifices for the main swirl chamber are much larger. At low fuel flows, all the fuel is supplied by the pilot and atomization quality is good because the delivery pressure, although not high, is adequate. As fuel flow is increased by increasing the fuel pressure, when a predetermined pressure is reached, a valve opens and fuel is also passed to the main atomizer. This arrangement allows satisfactory atomization to be achieved over a wide range of fuel flows without calling for excessive fuel pressures.

The principal advantages of pressure-swirl atomizers are good mechanical reliability and an ability to sustain combustion at very weak mixture strengths. Their drawbacks include potential plugging of the small passages and orifices by contaminants in the fuel and an innate tendency toward high soot formation at high-combustion pressures.

1.10.2 Airblast Atomizer

This atomizer employs a simple concept whereby fuel at low pressure is arranged to flow over a lip located in a high-velocity airstream. As the fuel flows over the lip it is atomized by the air, which then enters the combustion zone carrying the fuel droplets along with it. Minimum drop sizes are obtained by using designs that provide maximum physical contact between the air and the liquid. In particular, it is important to ensure that the liquid sheet formed at the atomizing lip is subjected to high-velocity air on both sides, as illustrated in Figure 1.17c. This not only gives optimum atomization, but also prevents fuel from depositing on solid surfaces.

The airblast atomizer has some very significant advantages in its application to gas turbine combustors. For example, the fuel distribution is dictated mainly by the airflow pattern, and hence the outlet temperature traverse is fairly insensitive to changes in fuel flow. Combustion is characterized by the absence of soot formation, resulting in relatively cool liner walls and a minimum of exhaust smoke. As another advantage, the component parts are protected from overheating by the air (at compressor outlet temperature) flowing over them. The major practical disadvantages are rather narrow stability limits and poor atomization quality at startup, owing to the low air velocity through the atomizer. Both these problems can be solved (albeit at the expense of a more complicated fuel system) by combining the airblast atomizer with a pilot pressure-swirl atomizer. By this means, the merits of the pressure-swirl atomizer at low fuel flows, namely, easy lightup and wide stability limits, are combined with all the virtues of airblast atomization (notably a soot-free exhaust) at high-fuel flow rates.

1.10.3 Gas Injection

Gaseous fuels, especially those of high-calorific value such as natural gas, present few problems from a combustion viewpoint. With low-heat-content (low Btu) gases, however, the fuel flow rate may comprise about one-fifth of the total combustor mass flow; this can lead to a mismatch between the compressor and the turbine, especially if the engine is intended for a multifuel application. Another problem with low Btu gases is their low burning rate, which may necessitate a larger combustion-zone volume, over and above the extra volume needed to accommodate the large volumetric flow of gaseous fuel. Achieving the required mixing rate in the combustion zone can also prove difficult. A mixing rate that is too high results in poor lean-blowout characteristics, whereas a mixing rate that is too low could give rise to rough combustion.

The methods used to inject gaseous fuels include plain orifices and slots, swirlers, and venturi nozzles.

1.11 Wall Cooling

The functions of the liner are to contain the combustion process and to facilitate the distribution of air to all the various combustor zones in the prescribed amounts. The liner must be structurally strong to withstand the buckling load created by the pressure differential across the liner wall. It must also have sufficient thermal resistance to withstand continuous and cyclic high-temperature operation. This is accomplished through the use of high-temperature, oxidant-resistant materials combined with the effective use of cooling air. On many combustors now in service, up to 20% of the total combustor air-mass flow is employed in liner-wall cooling. In practice, the liner-wall temperature is determined by the balance between (1) the heat it receives via radiation and convection from the hot gas, and (2) the heat transferred from it by convection to the annulus air and by radiation to the air casing.

The problem of liner-wall cooling has become increasingly severe as engine pressure ratios have increased (see Figure 1.18), but this is not due primarily to the higher pressure. In fact, an increase in pressure ratio is actually beneficial in reducing the specific surface area to be cooled. The difficulties arise from the increase in inlet air temperature that accompanies the higher pressure ratio. Higher inlet air temperature has a twofold adverse effect: (1) it raises the flame temperature, which, in turn, increases the rate of heat transfer to the liner wall, and (2) it reduces the effectiveness of the air as a coolant. As pressure ratios have increased over the years, turbine inlet temperatures have also had to rise accordingly (see Figure 1.19) in order to maximize fuel economy. This, too, has had a marked adverse effect on liner metal temperatures, especially at the rear end of the combustor. Further increases in the amount of air used in wall cooling (above the already high-current values) are not technically acceptable, because more air inserted along the liner

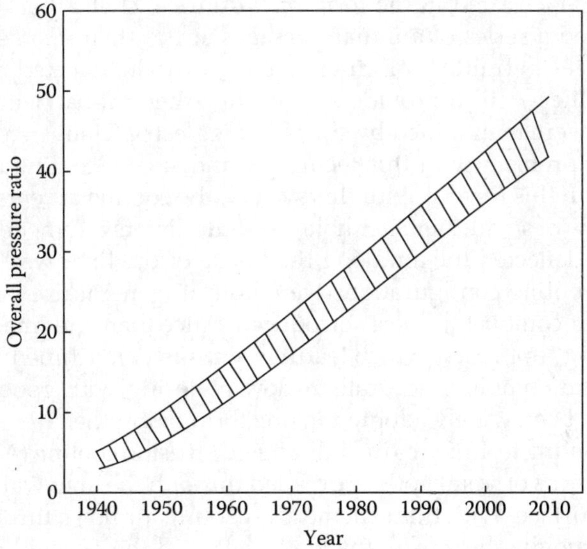

FIGURE 1.18
Historical trend of engine pressure ratio.

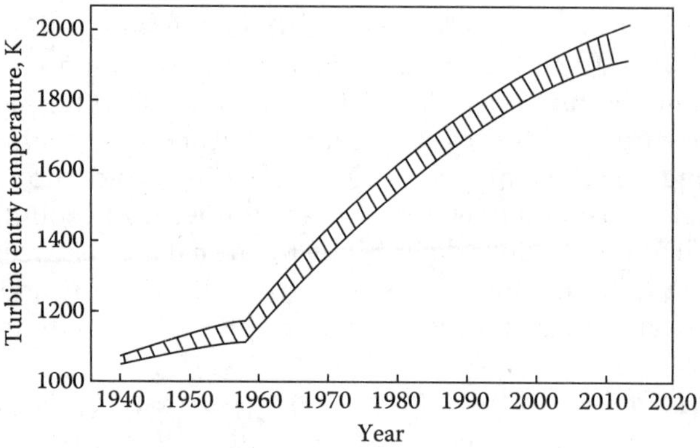

FIGURE 1.19
Historical trend of turbine entry temperature.

walls means that less is available for combustion and dilution. Moreover, it would worsen the radial temperature profile at the combustor outlet, thereby reducing the life of the turbine blades. Thus, the only practical alternative is to make more efficient use of the available cooling air or, better still, reduce the amount of air used in wall cooling.

1.11.1 Wall-Cooling Techniques

Many early gas turbine combustors used a louver cooling technique whereby the liner was fabricated in the form of cylindrical shells that, when assembled, provided a series of annular passages at the shell intersection points. These passages permitted a film of cooling air to be injected along the hot side of the liner wall to provide a protective thermal barrier. The annular gap heights were maintained by simple "wigglestrip" louvers. Air metering was a major problem with this technique and splash-cooling devices were much better in this regard. With this system, the cooling air entered the liner through a row of small-diameter holes, and the air jets impinged on a cooling skirt, which deflected the air along the inside of the liner wall. Wigglestrip and splash-cooling configurations were both in general use up to the time when annular combustors were introduced. Since then, the "machined-ring" or "rolled-ring" approach, which features accurately machined holes instead of louvers and combines accurate airflow metering with good mechanical strength, has been widely adopted in one form or another.

Modern cooling techniques include angled effusion cooling (AEC) whereby multiple patterns of small holes are drilled through the liner wall at a shallow angle to its surface. With this scheme, the cooling air flows through the liner wall, first removing heat from the wall itself, and then providing a thermal barrier between the wall and the hot combustion gases. AEC is perhaps the most promising contender among the various advanced combustor cooling techniques that are being actively developed for the new generation of industrial and aeronautical gas turbines. It is used extensively on the GE90 combustor, where it has reduced the normal cooling air requirement by 30%. The main drawback of AEC is an increase in liner weight of around 20%, which stems from the need for a thicker wall to achieve the required hole length and to provide buckling strength. (More detailed information on wall-cooling devices is contained in Chapter 8.)

With large industrial engines, where size and weight are of minor importance, it is practicable to line the combustor with refractory bricks to reduce the heat flux to the liner wall. Refractory bricks are clearly too heavy and cumbersome for application to aero- and most industrial engines. However, metallic tiles and thick thermal barrier coating offer an attractive solution. The V2500 engine is now in service with a tiled combustor and P&W is also using tiles on its radially staged combustor for the PW4000.

Using tiles effectively decouples the mechanical stresses, which are taken by the liner, from the thermal stresses, which are taken by the tiles. This method of construction has the advantage of tiles that can be cast from blade alloy materials having a much higher temperature capability (>100 K) than typical combustor alloys. Also, because the liner remains at a uniform low temperature, it can be made from relatively cheap alloys. The main drawback of tiled combustors is a substantial increase in weight.

An alternative to increasing the efficiency of cooling techniques is to spray a protective coating on the inner liner wall, and thermal barrier coatings are now used routinely to reduce liner-wall temperatures by up to 150 K. As it has for the past 60 years, the search continues for new liner materials that will allow operation at higher temperatures. Current production liners are typically fabricated from nickel- or cobalt-based alloys, such as Nimonic 263 or Hastelloy X. Candidates for liner and turbine blade materials now under investigation include carbon and carbon composites, ceramics, MX 4 (with TBC2), CMSX-4 and alloys of high-temperature materials, such as columbium. Techniques for the utilization of these materials are in varying stages of development; none is sufficiently advanced for routine application to production combustors.

1.12 Combustors for Low Emissions

A basic problem in combustor design is that of achieving easy ignition, wide burning range, high-combustion efficiency, and minimum pollutant emissions in a single, fixed combustion zone supplied with fuel from a single injection point. As some of these requirements conflict, the end result is inevitably a compromise of some kind. A good example of conflict in design is provided by the continuing requirement to reduce pollutant emissions. With conventional combustors, any modification that alleviates smoke and

NO_X will almost invariably increase the emissions of CO and UHC, and vice-versa.

One solution to this problem is to use some form of variable geometry to regulate the amount of air entering the primary combustion zone. At high pressures, large quantities of air are employed to minimize soot and NO_X formation. At low pressures, the primary airflow is partially blanked off, thereby raising the fuel/air ratio and reducing the velocity to give high-combustion efficiency (and, therefore, low emissions of CO and unburned UHC), as well as good lightup characteristics. Variable geometry has been used on a few large industrial engines, but the requirement for complex control and feedback mechanisms, which tend to increase cost and weight and reduce reliability, have so far ruled it out for small engines and, of course, for aeronautical applications.

Another alternative to attempting to achieve all the performance objectives in a single zone is to employ what is known as "staged" combustion. This may take the form of "axial" or "radial" staging, but in either case it uses two separate zones, each designed specifically to optimize certain aspects of combustion performance. The principle of axial staging is illustrated in Figure 1.20. It features a lightly loaded primary zone (Zone 1), operating at a fairly high equivalence ratio ϕ of around 0.8 (note that ϕ is the actual fuel/air ratio divided by the stoichiometric fuel/air ratio) to achieve high-combustion efficiency and to minimize the production of CO and UHC. Zone 1 provides all the temperature rise needed at low power conditions up to around idle speeds. At higher power levels, it acts as a pilot source of heat for the main combustion zone downstream (Zone 2), which is supplied with a premixed fuel–air mixture. When operating at full-load, the equivalence ratio

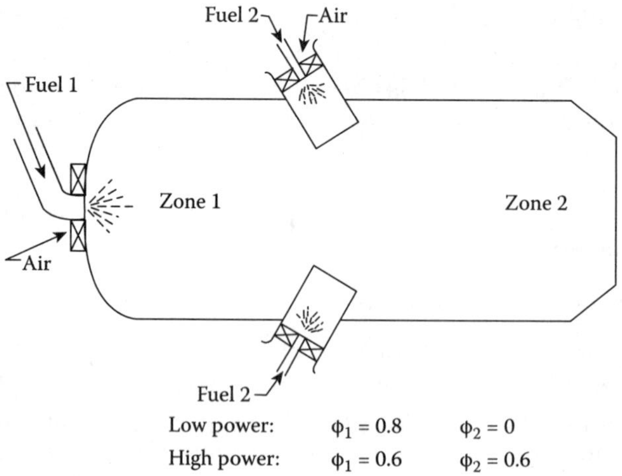

FIGURE 1.20
Principle of axial staging.

in both zones is kept low, at around 0.6, to minimize the emissions of NO_X and smoke.

Staged combustion is now widely used in industrial engines burning gaseous fuels, in both axial and radial configurations, to achieve low pollutant emissions without the need to resort to water or steam injection.

For liquid fuels, the lean premix prevaporize (LPP) combustor appears to have the most promise for ultralow NO_X combustion. The concept is shown schematically in Figure 1.17d. The design objective is to attain complete evaporation of the fuel and thorough mixing of fuel and air before combustion. By avoiding droplet burning, and by operating the reaction zone at a lean fuel/air ratio, NO_X emissions are drastically reduced because of the low flame temperature and the elimination of "hot spots" from the combustion zone. The main drawback of the LPP system is that the long time needed to fully vaporize and mix the fuel at low power conditions may result in the occurrence of autoignition or flashback in the fuel preparation duct at the high pressures and inlet temperatures associated with operation at maximum power. These problems may be overcome, at the expense of additional cost and complexity, through the use of staged combustion and/or variable geometry. Other concerns with LPP systems are those of durability, maintainability, and safety.

Another important contender in the ultralow NO_X emissions field is the rich-burn/quick-quench/lean-burn (RQL) combustor. This concept employs a fuel-rich primary zone in which NO_X formation rates are low because of the combined effects of low temperature and oxygen depletion. Downstream of the primary zone, the additional air required to complete the combustion process and reduce the gas temperature to the desired predilution zone level is injected in a manner that is designed to ensure uniform and rapid mixing with the primary-zone efflux. This mixing process must take place quickly, otherwise pockets of hot gas would survive long enough to produce appreciable amounts of NO_X. Thus, the design of a rapid and effective quick-quench mixing section is of decisive importance to the success of the RQL concept.

The device that appears to have the greatest potential of all for NO_X reduction, is the catalytic combustor. In this system, the fuel is first prevaporized and premixed with air at a very low equivalence ratio and the resulting homogeneous mixture is then passed through a catalytic reactor bed. The presence of the catalyst allows combustion to occur at very low fuel/air ratios that normally lie outside the lean flammability limit. In consequence, the reaction temperature is extremely low and NO_X formation is minimal. In most current designs, a thermal reaction zone is located downstream of the catalytic bed. Its function is to raise the gas temperature to the required turbine entry value and to reduce the concentrations of CO and UHC to acceptable levels.

The potential of catalytic reactors for very low pollutant emissions has been recognized for the past 25 years, but the harsh environment in a gas

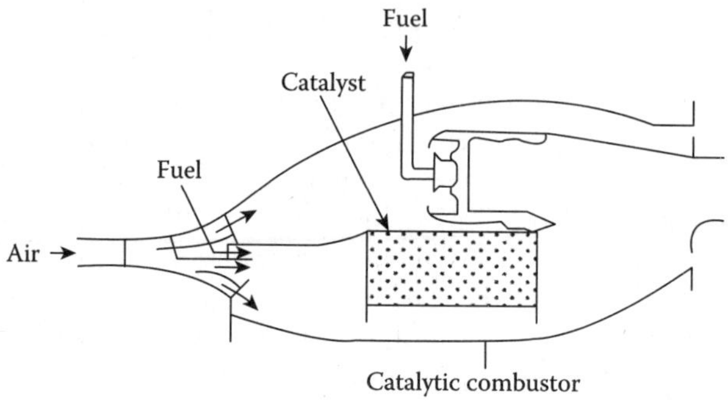

FIGURE 1.21
Combination of catalytic and conventional staged combustors.

turbine combustor and its wide range of operating conditions constitute a formidable barrier to the development of viable catalytic combustors for gas turbines. The long-term durability of catalyst materials is a major concern. Considerable progress on catalyst development continues to be made (see Chapter 9), but its application to aero engines is unlikely to happen until a large body of experience on stationary engines has been accumulated. When it does materialize, it will probably be in the form of a "radially staged," dual-annular combustor, as illustrated in Figure 1.21. The outer combustor is designed specifically for easy lightup and low emissions at engine idle conditions. At higher power settings, fuel is supplied premixed with air to the inner combustor containing the catalytic reactor. At maximum power conditions, this reactor provides most of the temperature rise needed to sustain the engine.

1.13 Combustors for Small Engines

On small engines, high shaft speeds necessitate close coupling of the compressor and turbine to alleviate shaft whirling problems. This requirement, especially when combined with the need for a low frontal area, has led to the almost universal use of annular reverse-flow or annular radial-axial combustors. A notable exception is the Allison T63 engine, which has a single tubular combustor mounted on the end of the engine to facilitate inspection and servicing.

An annular reverse-flow combustor is shown schematically in Figure 1.22. The main advantages of this layout, in addition to a very short shaft length, are efficient utilization of the available combustion volume and easy accessibility of the fuel injectors. Its main drawback is the high surface-to-volume

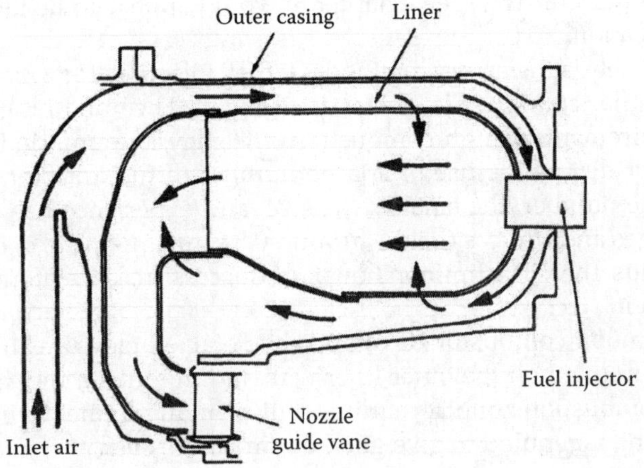

FIGURE 1.22
Reverse-flow annular combustor.

ratio of the liner, inherent to the reverse-flow concept, which adds to the problem of liner-wall cooling.

In reverse-flow annulars, the air that flows through holes in the outer liner wall approaches these holes from a direction which is opposite to that followed by the air entering the combustion zone through holes in the inner liner wall. Moreover, it is apparent from Figure 1.22 that the air in the inner annulus suffers a higher pressure loss (owing to its longer flow length) than the air in the outer annulus. For these reasons, it is impossible to balance the air jets emanating from the inner and outer liner walls in terms of initial angle, depth of penetration, and momentum. Consequently, the conventional double-vortex, primary-zone flow pattern is ruled out, and single-sided air admission, producing single-vortex flow recirculation in the primary zone, is generally used. The flow recirculation is created partly by air jets and partly by air that is introduced as a wall jet. This air serves to film-cool the liner dome before participating in primary combustion.

The main problem areas with small combustors are ignition, wall cooling, and fuel injection. The size and weight of ignition equipment are of special concern because on small engines they represent a larger proportion of the total engine size and weight than on large engines. Unfortunately, most small-engine applications call for a larger number of starts than large-engine applications, so that attempts to reduce the size and weight of ignition equipment can lead to lack of reliability and loss of performance.

Liner-wall cooling is especially difficult on small annular systems, in view of the relatively large surface area to be cooled. The situation is exacerbated by the low annulus velocities associated with centrifugal compressors, which result in low external convective cooling of the liner. New cooling methods that require only small air quantities per unit surface area of liner are clearly required. AEC (see Chapter 8) would appear to be ideally suited to this application.

No completely satisfactory method of fuel injection for small, straight-through annular chambers has been devised yet. The nub of the problem is that the requirements of high-combustion efficiency, low emissions, and good pattern factor dictate the use of a large number of fuel injectors. However, the larger the number, the smaller the size; and experience has shown that small passages and orifices (below around 0.5 mm) are prone to erosion and blockage. Thus, there is a limit on how far a successful, large atomizer can be scaled down in size.

A small annular combustor, developed by Solar, employs an airblast atomizer that is mounted on the outer liner wall and injects the fuel tangentially across the combustion zone. It requires only a small number of injectors per combustor and is reputed to give good atomization, even at startup.

Developments in compressors and air-cooled turbines are certain to lead to higher compression ratios and higher turbine inlet temperatures. More research is needed in the areas of wall cooling, fuel preparation and distribution, miniaturized ignition devices, and high-temperature materials, including ceramics, that will address the special needs of small annular combustors.

1.14 Industrial Chambers

Industrial engines are required to operate economically and reliably over long periods without attention. Compactness is no longer important and is only considered if the engine has to be constrained to fit into an existing building or if delivery is made difficult. Fuel economy and low pollutant emissions thus become the most important issues along with unit cost. In addition, accessibility for maintenance and minimal shut-down time will influence sales in a competitive market [3].

In order to meet these requirements, combustors in industrial engines tend to be much larger than their aeronautical counterparts. This results in longer residence times, which is clearly advantageous when burning poor quality fuels. Also, flow velocities are lower, hence pressure losses are smaller.

Most industrial units tend to fall into one of two categories:

1. Systems that are designed to burn gaseous fuels, heavy distillates, and residual oils, and depart fairly radically from aeronautical practice.
2. Systems that are essentially "industrialized" aero engines, or that follow aircraft practice closely. They usually burn gaseous and/or light to medium distillate fuels.

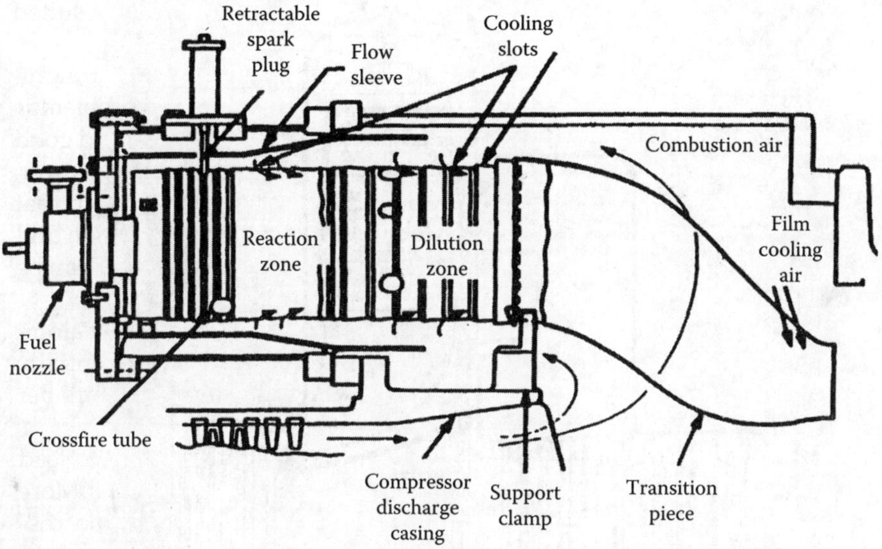

FIGURE 1.23
General Electric conventional industrial combustor.

One of the most successful industrial units in the first category is the 80 MW GE MS7001 gas turbine. Each machine has ten sets of combustion hardware, and each set comprises a casing, an end cover, a set of fuel nozzles, a flow sleeve, a combustion liner, and a transition piece. These components are indicated in Figure 1.23. The flow sleeve is an axisymmetric cylinder/cone that surrounds much of the combustion liner to aid in distributing the compressor air uniformly to all liners and to improve the external liner-wall cooling [3]. The conventional MS7001 combustion system has one fuel nozzle per combustor, but the more advanced dry low emissions (DLE) versions have multiple fuel nozzles per combustor (see Chapter 9).

Some manufacturers of industrial engines prefer to use a single, large combustor that is mounted outside the engine, as shown in Figure 1.24. This allows the combustor to be designed exclusively to meet the requirements of good combustion performance. It is also easier to design the outer casing of the unit to withstand the high-gas pressure. A further advantage of this approach is ease of inspection, maintenance, and repair, all of which can be accomplished without removing large casing components.

Two basic methods of liner construction are used:

1. An all-metal liner constructed of finned metal parts that are cooled by a combination of convection and film cooling.
2. A tube of nonalloy carbon steel that is lined with refractory bricks. This requires less cooling air than the all-metal type.

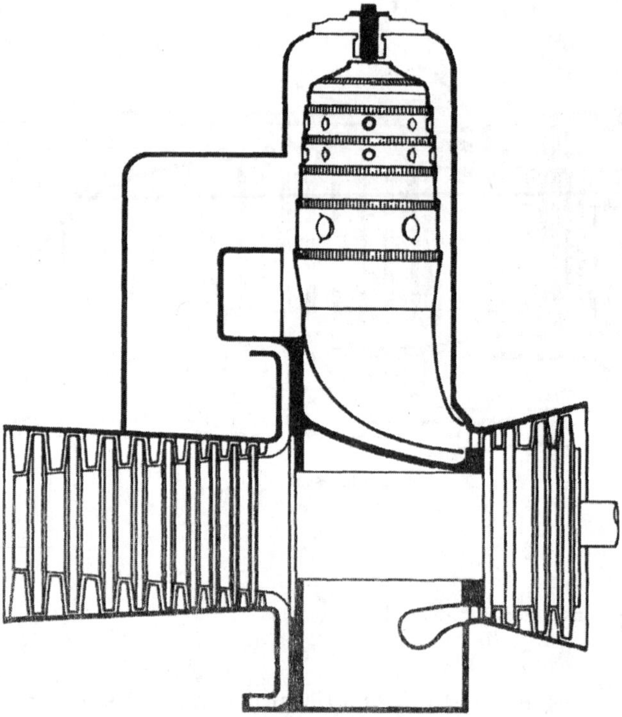

FIGURE 1.24
Industrial engine featuring single tubular combustor.

Multiple fuel injectors (burners) are generally preferred for these combustors because they provide a shorter flame and a more uniform distribution of temperature in the gases flowing into the dilution zone. The Siemens silo combustors are fitted with a number of "hybrid" burners that can operate on natural gas in either diffusion or premix modes to achieve low emissions over a wide operating range. Essentially, the system functions as a diffusion burner at low engine loads and as a premix burner in the upper load range. For their silo-type combustors, Siemens used the same fuel burner in engines of different power ratings, the number of burners being changed to accommodate variations in engine size. However, in their new annular combustors, the number of burners is kept constant at 24 to ensure a good pattern factor. The drawback to this approach is that the burners must be scaled with respect to the machine size, although the basic design remains the same. The Siemens hybrid burner is now fully established as a low-emissions system for large engines in the 150 MW class, but it has also been applied by MAN GHH to its THM 1304 engine. This 9 MW class gas turbine features two tubular combustion chambers mounted on top of the engine casing [4].

The ABB company has developed a conical premix burner module, called the EV-burner, which can operate satisfactorily on both gaseous and liquid fuels and has demonstrated good performance in a wide range of low-NO_X combustion applications [5,6]. The silo combustor for the ABB GT11N gas turbine is equipped with thirty-seven of these burners, all of which operate in a pure premix mode [5]. For part-load operation, fuel is supplied to only a fraction of the total number of burners. The same technology has also been used in the design of annular combustors. The ABB GT10 (23 MW) combustor features a single row of 18 EV burners, while the heavy-duty ABB GT13E2 gas turbine (>150 MW) has 72 EV burners that are arranged in two staggered circumferential rows.

1.14.1 Aeroderivative Engines

The notion of modifying aero engines to serve as industrial or marine engines is by no means new. One early example is the Allison 501 engine, which is basically this company's T56 aero engine, but adapted to burn DF2 fuel instead of aviation kerosine. Initially, this engine was fitted with six tubular (can) combustors, but the modern 501-K family of engines feature a can-annular configuration consisting of six tubular cans contained within an annular casing. The DLE version of this combustor for burning natural gas, as described by Razdan et al. [7,8], employs a dual-mode combustion approach to meet its emission goals without resorting to water or steam injection. Many other engine companies followed this same route of converting aero engines into power sources for a wide variety of industrial and transport applications. For example, Rolls Royce produced industrialized versions of its Avon, Tyne, and Spey aero engines. For the combustor, this "industrialization" process mainly involved changes to the fuel injector, sometimes to provide multifuel capability, but also to facilitate the injection of water or steam for NO_X reduction. It was also customary to modify the primary-zone airflow pattern, often by the addition of more air, to exploit the absence of a high-altitude relight requirement and to reduce soot formation and smoke. As emissions regulations became increasingly severe, such simple modifications to an existing aero combustor no longer sufficed and more sophisticated approaches were needed. Today's industrial DLE combustors take full advantage of the benefits to be gained from fuel staging and fuel–air premixing in achieving their emissions targets. The aero-derived GE LM6000 and RR211 DLE industrial engines both employ staged combustion of premixed gaseous fuel–air mixtures. Another interesting feature of these two engines is that both were derived from successful high-performance aero engines by simply replacing the

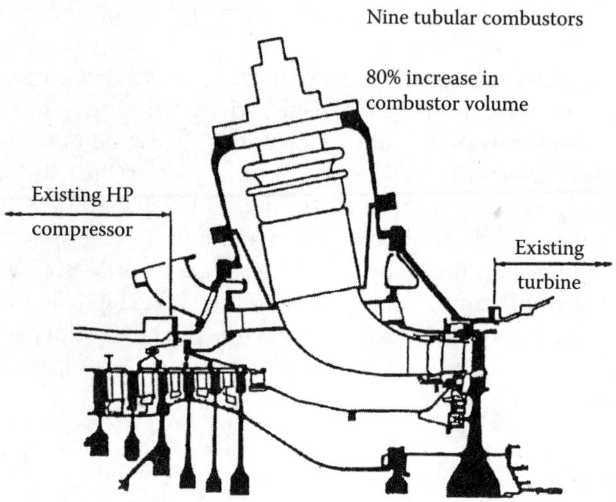

FIGURE 1.25
Industrial RB211 DLE combustor. (Courtesy of Rolls Royce plc.)

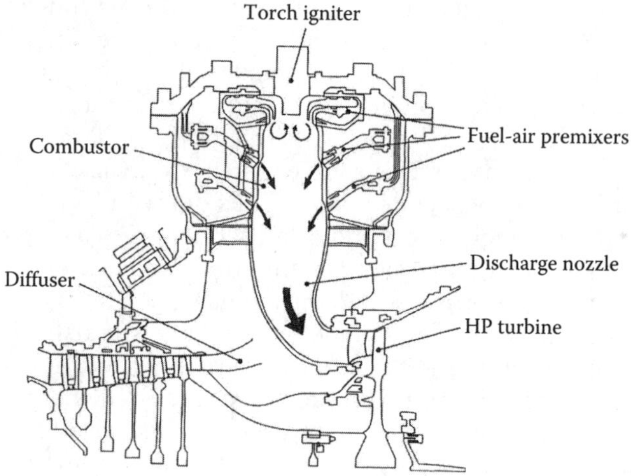

FIGURE 1.26
Industrial Trent DLE combustor. (Courtesy of Rolls Royce plc.)

existing aero combustors with DLE combustors of the same length, as illustrated for the RB211 in Figure 1.25. The Rolls Royce Industrial Trent is among the most recent aeroderivative engines. It employs three separate stages of premixed fuel–air injection, as shown in Figure 1.26.

References

1. Odgers, J., and Kretschmer, D., "Basic Considerations," *Northern Research and Engineering Corporation Report* No. 1344–1, 1980.
2. Adkins, R. C., "Selection of Pre-Combustor Diffuser," Lecture Notes RCA/1/93, Cranfield University, UK, 1993.
3. Washam, R. M., "Dry Low NO_X Combustion System for Utility Gas Turbine," ASME Paper 83-JPGC-GT-13, 1983.
4. Bonzani, F., Di Meglio, A., Pollarolo, G., Prade, B., Lauer, G., and Hoffmann, S., "Test Results of the V64.3A Gas Turbine Premix Burner," presented at Power-Gen Europe '97, Madrid, June 1997.
5. Sattelmayer, T., Felchin, M. P., Haumann, J., Hellat, J., and Styner, D., "Second Generation Low-Emission Combustors for ABB Gas Turbines: Burner Development and Tests at Atmospheric Pressure," *ASME Journal of Engineering for Gas Turbines and Power*, Vol. 114, No. 1, pp. 118–125, 1992.
6. Aigner, M., and Muller, G., "Second Generation Low-Emission Combustors for ABB Gas Turbines: Field Measurements with GT11N-EV," *ASME Journal of Engineering for Gas Turbines and Power*, Vol. 115, No. 3, pp. 533–536, 1993.
7. McLeroy, J. T., Smith, D. A., and Razdan, M. K., "Development and Engine Testing of a Dry Low Emissions Combustor for Allison 501-K Industrial Gas Turbine Engines," ASME Paper 95-GT-335, 1995.
8. Razdan, M. K., Bach, C. S., and Bautista, P. J., "Field Experience of a Dry Low Emissions Combustion System for Allison 501-K Series of Engines," ASME Paper 97/AA/13, 1997.

Bibliography

Franz, A., "The Development of the Jumo 004 Turbojet Engine," in W. Boyne and D. Lopez, eds., *The Jet Age: 40 Years of Jet Aviation*, Smithsonian Institution, Washington, DC, 1979.

Hawthorne, W. R., "Reflections on United Kingdom Aircraft Gas Turbine Industry," R. Tom Sawyer Award Lecture, *ASME Journal of Engineering for Gas Turbines and Power*, Vol. 116, pp. 495–510, 1994.

Meher-Homji, C. B., "The Development of the Junkers Jumo 004B—The World's First Production Turbojet," ASME Paper 96-GT-457, 1996.

Meher-Homji, C. B., "The Development of the Whittle Turbojet, ASME Paper 97-GT-528, 1997.

von Ohain, H., "The Origins and Future Possibilities of Airbreathing Jet Propulsion Systems," *Proceedings of Eighth International Symposium on Airbreathing Engines*, ISABE 87-7001, pp. 14–25, 1987.

Olsen, W. T., "Combustion Chamber Developments," in W. R. Hawthorne and W. T. Olsen, eds., *High Speed Aerodynamics and Jet Propulsion, Vol XI, Design and Performance of Gas Turbine Power Plants*, Oxford University Press, London, pp. 289–350, 1964.

Whittle, F., "The Birth of the Jet Engine in Britain," in W. Boyne and D. Lopez, eds., *The Jet Age: 40 Years of Jet Aviation*, Smithsonian Institution, Washington, DC, 1979.

2

Combustion Fundamentals

2.1 Introduction

The subject of combustion embraces a wide variety of processes and phenomena. Even a brief summary of the vast amount of material that has been published on combustion science and technology would be well beyond the scope and intention of this chapter. Instead, attention is focused on a few key aspects of combustion that are considered to be most relevant to the gas turbine and are not covered in the remaining chapters of this book.

Combustion is perhaps described most simply as an exothermic reaction of a fuel and an oxidant. In gas turbine applications, the fuel may be gaseous or liquid, but the oxidant is always air. Combustion occurs in many forms, not all of which are accompanied by flame or luminescence. Two important regimes of combustion can be distinguished [1].

2.1.1 Deflagration

This is a fast process that requires less than 1 ms for 80% completion. It is characterized by the presence of a flame that propagates through the unburned mixture. A flame may be defined as a rapid chemical change occurring in a very thin fluid layer, involving steep gradients of temperature and species concentrations, and accompanied by luminescence. From a macroscopic viewpoint, the flame front can be viewed as an interface between the burned gases and the unburned mixture. Compared with the unburned mixture, the burned gases are much higher in volume and temperature, and much lower in density. Deflagration waves in hydrocarbon fuel-air mixtures normally propagate at velocities below 1 m/s. All the flame processes that occur in gas turbine combustors fall within this category.

2.1.2 Detonation

The characteristic feature of detonation is a shock wave that is connected with and supported by a zone of chemical reaction. Detonation waves proceed at supersonic velocities, ranging from 1 to 4 km/s. They cannot occur in the conventional fuel–air mixtures employed in gas turbine combustors, but the possibility could arise in situations where oxygen injection is employed to facilitate ignition and engine acceleration. Currently, pulse detonation engines are of interest to the military and these engine combustors employ detonation waves.

2.2 Classification of Flames

Most fundamental studies of flame combustion are performed using gaseous or prevaporized fuels. Furthermore, although a flame (i.e., a combustion wave) can propagate through a static gas mixture, it is usual to stabilize the flame at a fixed point and supply it with a continuous flow of combustible mixture. Under these conditions, flames can be divided into two main classes— *premixed flames* and *diffusion flames*—depending on whether the fuel and air are mixed before combustion, or mixed by diffusion in the flame zone. Depending on the prevailing flow velocities, both types of flame can be further classified as *laminar* or *turbulent*.

A further complication arises in practical systems burning liquid fuels: if the fuel is not completely vaporized before entering the flame zone, heterogeneous spray combustion may take place. This process, involving diffusion flame burning of individual evaporating fuel droplets, may be superimposed on a premixed turbulent flame zone. However, if both reactants are in the same physical state, the combustion process is described as homogeneous.

The candle provides a simple example of a diffusion flame. Fuel vapor rises from the wick and can burn in the neighborhood of the wick only to the extent that it can mix with the oxygen in the air. For this type of flame, the rate of mixing between the fuel and the oxidant often limits the overall rate of combustion. Only for laminar premixed gaseous flames is the combustion process determined largely by flame chemistry, local heat and mass transfer, and definable macroscopic system parameters (pressure, temperature, and air/fuel ratio [AFR]), with gas-dynamic and gross heat and mass transfer processes having little effect.

With premixed gases, a combustible mixture is available from the outset. Once the flame has been initiated at some point in the mixture (by means of a hot surface, an electric spark, or some other ignition source), it will propagate throughout the entire volume of combustible mixture. The speed at which it propagates and the factors affecting its rate of propagation are of special interest to the designer of practical combustion systems. Turbulence is of prime importance because most flowing fuel–air mixtures are turbulent and turbulence is known to enhance flame speeds considerably.

2.3 Physics or Chemistry?

The subject of combustion embraces both physics and chemistry. In the present context, physics is taken to include heat transfer, mass transfer, thermodynamics, gas dynamics, and fluid dynamics. In many practical combustion devices, physical processes are much more limiting to combustion performance than chemical processes.

In general, chemical processes are important mainly for their influence on pollutant emissions and, in aircraft combustors, on lean lightoff and lean blowout limits at high altitudes. However, at most operating conditions, the main interest lies not so much on the limits of combustion as on the structure, heat-release rates, combustion products, and radiation properties of high-temperature flames. The release of energy by chemical reaction is, of course, an essential step in the overall combustion process, but in high-temperature flames it occurs so quickly in relation to the other processes involved that it can usually be disregarded. Also, heavier hydrocarbons quickly breakdown into simple, lighter radicals (e.g., methyl radicals), thereby simplifying the analysis of chemical processes.

For diffusion flames, the rate of interdiffusion of air and fuel and, for larger flames, the rate of large-scale mixing, are the rate-controlling steps. The aerodynamics of the system, which include turbulence levels and rates of entrainment of air and combustion products, are most important in determining flame size and stability.

2.4 Flammability Limits

Not all fuel–air mixtures will burn or explode; flames can propagate through fuel–air mixtures only within certain limits of composition. If small amounts of combustible fuel gas or vapor are gradually added to air, a point will be reached at which the mixture just becomes flammable. The percentage of fuel gas at this point is called the *lower flammable limit, weak limit,* or *lean limit.* If more fuel is added, another point will eventually be reached at which the mixture will no longer burn. The percentage of fuel gas at this point is called the *upper flammable limit* or *rich limit.* For many fuels, the weak limit corresponds to an equivalence ratio of around 0.5, and the rich limit to an equivalence ratio of around 3.

An increase in pressure above atmospheric usually widens the flammability limit of gases and vapors. This is especially true of hydrocarbon–air mixtures. Most of the widening occurs at the rich end of the range. In the practically important range of pressures from 10 kPa to 5 MPa, the weak flammability limit is not strongly pressure dependent.

The flammability range is also widened by an increase in temperature, but the effect is usually less than that of pressure.

For liquid fuels, the formation of combustible mixtures is only possible within definite temperature limits. The lower temperature limit is taken as the minimum temperature at which the fuel's vapor pressure is sufficient to form the weak-limit volume concentration of vapor in air. On being cooled below this temperature, the mixture becomes too weak for flammability. The upper temperature limit corresponds to the rich-limit concentration, and a subsequent increase in temperature enriches the mixture to a condition of nonflammability.

The lowest temperature at which a flammable mixture can be formed above the liquid phase is called the *flash point* when quoted for atmospheric pressure. The ease with which enough vapor is formed to produce a flammable mixture depends on the vapor pressure of the fuel. Highly volatile fuels produce high vapor pressures that give low flash points.

2.5 Global Reaction-Rate Theory

Although the combustion of a hydrocarbon fuel is an extremely complex process, it may be analyzed on the assumption that combustion can be fully described by a single global reaction in which fuel and air react at a certain rate to form combustion products. It is further assumed that the fuel and air entering the combustion zone are instantaneously mixed with all the other material within the zone, and that burned products leave the combustion zone with temperature and composition identical to those within the zone.

In a conventional combustion chamber, the essential requirement of low pressure loss prohibits sufficiently rapid mixing to ensure a truly homogeneous combustion zone. This condition has been closely approached in the "stirred reactor" of Longwell and Weiss [2], in which intimate mixing between fresh mixture and burned products is accomplished by imparting considerable energy to the inflowing jets, at the expense of appreciable pressure loss.

According to Longwell et al. [3], the rate of reaction between fuel and air may be expressed by the material balance equation

$$\eta_c \phi \dot{m}_A = C_{cf} V T^{0.5} \exp\left(-\frac{E}{RT}\right) \rho^n x_f^m x_o^{n-m}. \tag{2.1}$$

The following assumptions are made:

1. The consumed material forms a mixture of CO_2, CO, H_2, and H_2O in water–gas equilibrium at temperature T.
2. Fractions $(1-\eta_c)$ of the original fuel and $(1-\eta_c\phi)$ of the original oxygen remain and are regarded as the only reactants. In lean mixtures, the fuel forms only CO_2 and H_2O when consumed.

Longwell and Weiss have derived the chemical equations for octane burning in air. The corresponding equations for kerosine are discussed next.

2.5.1 Weak Mixtures

For weak mixtures ($\phi < 1$), the equations may be written

$$\phi C_{12}H_{24} + 18O_2 + 67.68N_2 = 12\eta_c\phi(CO_2 + H_2O) + (1-\eta_c)\phi C_{12}H_{24}$$
$$+18(1-\eta_c\phi)O_2 + 67.68N_2.$$

Hence

$$x_f = \frac{(1-\eta_c)\phi}{85.68 + \phi + 5\eta_c\phi}. \tag{2.2}$$

and

$$x_o = \frac{18(1-\eta_c\phi)}{85.68 + \phi + 5\eta_c\phi}. \tag{2.3}$$

In their original paper, Longwell et al. [3] used values for m and n of 1 and 2, respectively, corresponding to a second-order reaction. Inserting these values into Equation 2.1 and substituting Equations 2.2 and 2.3 into Equation 2.1 gives

$$\frac{\dot{m}_A}{VP^2} \alpha \frac{1}{T^{1.5}\exp(E/RT)} \frac{(1-\eta_c)(1-\eta_c\phi)}{\eta_c}. \tag{2.4}$$

Generally, it has been found that experimental data are best correlated using a value for n that is slightly less than 2. Longwell and Weiss [2] subsequently modified their value of n to 1.8, and others have confirmed the pressure dependence to be of this order [4,5]. Using values for m and n of 0.75 and 1.75, respectively, has the advantage of consistency with the burning-velocity parameter (see Chapter 5). Equation 2.4 then becomes

$$\frac{\dot{m}_A}{VP^{1.75}} \alpha \frac{1}{T^{1.25}\exp(E/RT)} \frac{1}{\phi^{0.25}} \frac{(1-\eta_c)^{0.75}(1-\eta_c\phi)}{\eta_c}. \tag{2.5}$$

2.5.2 Rich Mixtures

For rich kerosine mixtures ($\phi > 1$),

$$\phi C_{12}H_{24} + 18O_2 + 67.68N_2 = 18(1-\eta_c\phi)O_2 + (1-\eta_c)\phi C_{12}H_{24}$$
$$+12\eta_c\phi(CO_2 \text{ or } CO)$$
$$+12\eta_c\phi(H_2 \text{ or } H_2O) + 67.68N_2.$$

Following the same procedure as that employed for weak mixtures leads, for $n = 2$ and $m = 1$, to

$$\frac{\dot{m}_A}{VP^2} \alpha \frac{\phi}{T^{1.5}\exp(E/RT)} \frac{(1-\eta_c)^2}{\eta_c}, \tag{2.6}$$

while for $n = 1.75$ and $m = 0.75$, we have

$$\frac{\dot{m}_A}{VP^{1.75}} \alpha \frac{\phi^{0.75}}{T^{1.25}\exp(E/RT)} \frac{(1-\eta_c)^{1.75}}{\eta_c}. \tag{2.7}$$

The manner in which the heat-release rate varies with the fraction of the fuel burned η_c is illustrated in Figure 2.1. At low levels of η_c, heat-release rates are low because the temperature is low. As combustion proceeds, the temperature rises, thereby increasing the rate of heat release until a maximum is reached at a level of η_c that varies between 0.7 and 0.9, depending on the equivalence ratio of the mixture and its initial temperature. Beyond this point, any further increase in reaction rate, resulting from the continuing rise in temperature, is more than offset by the reduction caused by the fall in concentration of the oxygen and unburned fuel. Thus, the heat-release rate falls off, becoming zero at the maximum temperature, which also correspond to 100% combustion efficiency.

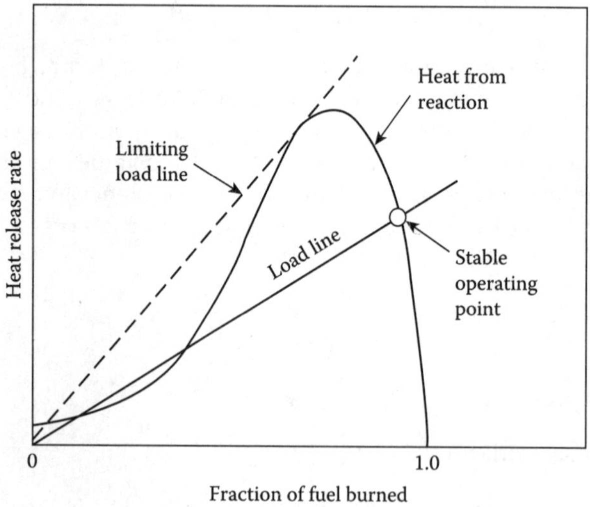

FIGURE 2.1
Mechanism of flame blowout.

The load line in Figure 2.1 represents the amount of heat required to raise the unburned mixture to the reaction temperature. The point at which it intersects the heat-release curve represents the operating point of the combustor. As the throughput is increased, the slope of this line increases until, finally, it no longer intersects the heat-release curve, and the flame blows out.

2.6 Laminar Premixed Flames

The burning velocity of a flame, i.e., the rate at which a plane combustion wave will propagate through a gaseous flammable mixture, is determined partly by the rate of chemical reaction in the thin flame zone, and by heat and mass transfer from the flame to the unburned gas. The key processes involved have been described by Gaydon and Wolfhard [6]. Conductive and radiative heating of the unburned gas serve to initiate reaction by a thermal mechanism, while back diffusion of active species from the flame zone can initiate reaction by a diffusion mechanism. The burning velocity of a flame is therefore affected by flame radiation and hence by flame temperature, by local gas properties such as viscosity and diffusion coefficient, and by the imposed variables of pressure, temperature, and AFR. The burning velocity may be defined as the velocity with which a plane flame front moves in a direction normal to its surface through the adjacent unburned gas. It is a fundamental property of a combustible mixture and is important practically, both in the stabilization of flames and in determining rates of heat release.

It is found in practice that for any fuel, the burning velocity has a reproducible constant value when the imposed variables are fixed. It is also of interest to note that the burning velocities of stoichiometric mixtures of many hydrocarbon fuels burning with air approach a single common value of about 0.43 m/s at normal atmospheric temperature and pressure. This is probably because most complex fuels are largely pyrolyzed to methane, other one- or two-carbon-atom hydrocarbons, and hydrogen before entering into the flame reaction zone. Hence, the gas composition entering the flame zone is substantially independent of the original fuel.

2.6.1 Factors Influencing Laminar Flame Speed

The most important factors governing the laminar burning velocity are equivalence ratio (fuel/air ratio), temperature, and pressure.

2.6.1.1 Equivalence Ratio

The variation of flame speed with mixture strength roughly follows that of flame temperature. In almost all cases, the maximum value occurs at an equivalence ratio of between 1.05 and 1.10. Notable exceptions to this general rule are hydrogen and carbon monoxide. Their laminar burning velocities reach a maximum at an equivalence ratio of around 2.

2.6.1.2 Initial Temperature

Dugger and Heimel [7] investigated the effect of initial mixture temperature on maximum burning velocity for mixtures with air of methane, propane, and ethylene, over temperatures ranging from 141 to 617 K. Their results showed that flame speed increases with an increase in temperature. The experimental data were correlated by the following empirical equations:

$$\text{Methane}: S_L = 0.08 + 1.6 \times 10^{-6} T_o^{2.11}$$

$$\text{Propane}: S_L = 0.10 + 3.42 \times 10^{-6} T_o^{2.0}$$

$$\text{Ethylene}: S_L = 0.10 + 25.9 \times 10^{-6} T_o^{1.74}$$

2.6.1.3 Pressure

Flame theory suggests that pressure is an important parameter whose effect may be related to the reaction order by an expression of the form

$$S_L \alpha P^{(n-2)/2}.$$

Thus, for a bimolecular reaction ($n = 2$), burning velocity should be independent of pressure. For the slow-burning fuels ($S_L < 0.6$ m/s) employed in gas turbines, such as natural gas and vaporized kerosine, the observed pressure dependence can be expressed as a simple law

$$S_L \alpha P^{-x},$$

where x varies from 0.1 to 0.5 [8–10].

2.7 Laminar Diffusion Flames

For laminar flames in premixed systems, chemical reaction rates are rate controlling. Even with non-premixed systems, if the mixing occurs rapidly compared with the chemical reactions, combustion rates may be considered solely in terms of homogeneous processes. However, there are some systems in which mixing is slow compared with chemical reaction rates, so that mixing time controls the burning rate. This is true for so-called "diffusion flames," in which the fuel and oxidant come together in a reaction zone through molecular and turbulent diffusion. The fuel may be in the form of a gaseous jet or a liquid or solid surface. Thus, there are two categories within diffusion-controlled combustion, according to the initial physical state of the fuel and/or oxidant. If both the fuel and the oxidant are initially gaseous, then the flame is referred to as a *diffusion flame*. If both the fuel and oxidant are initially in different physical states, i.e., liquid and gas or solid and gas, although the system is still diffusion controlled, the process is usually called *heterogeneous* combustion. Examples in this category include hydrocarbon-droplet and coal combustion.

2.8 Turbulent Premixed Flames

Although it has long been recognized that flame speeds can be appreciably increased by turbulence, as evidenced by the very high burning rates achieved in both piston and gas turbine engines, the manner and extent of this influence are still not fully resolved. The first contribution to the understanding of turbulent flames was made by Damkohler [11], who visualized a turbulent flame as being essentially the same in structure as a laminar flame. He attributed the observed increase in burning rate to the effect of turbulence in wrinkling the flame front, thereby increasing its specific surface area and hence also its ability to consume fresh mixture. Damkohler proposed the following equation for large-scale turbulence:

$$S_T = S_L + u', \tag{2.8}$$

where S_T is the turbulent flame speed, S_L is the laminar burning velocity, and u' is the RMS value of fluctuating velocity.

In due course, several more theories embodying the wrinkled-flame concept emerged, differing from Damkohler's theory and from each other mainly in the methods employed to relate turbulence properties to the increase in specific surface of the flame. Schelkin's [12] approach led to a relationship of the form

$$S_T = S_L\left[1 + B\left(\frac{u'}{S_L}\right)^2\right]^{0.5}, \tag{2.9}$$

in which B is a constant of the order of unity. At high velocities, Equations 2.8 and 2.9 become

$$S_T = u'. \tag{2.10}$$

Ballal and Lefebvre [13] carried out a series of experiments on enclosed, premixed turbulent flames. A number of different turbulence-promoting grids were located, in turn, at the upstream end of the test section, to create in the combustion zone conditions in which the separate effects of turbulence intensity and scale on burning velocity and flame structure could be determined. The scales encountered in turbulent flow range in size from the Kolmogoroff scale η, which represents the size of the smallest eddies in the flow, to the integral scale L, which represents the size of the largest eddies.

For conditions of low turbulence ($u' < 2S_L$), it was found that turbulence does not roughen the flame, which retains a smooth laminar appearance. However, burning velocity is increased owing to the effect of turbulence in wrinkling the flame, thereby extending its surface area, as first noted by Damkohler [11]. The ratio of turbulent to laminar flame speeds is given by

$$\left(\frac{S_T}{S_L}\right)^2 = 1 + 0.03\left(\frac{u'L}{S_L\delta_L}\right)^2. \tag{2.11}$$

At very high levels of turbulence, the turbulent eddies are too small to wrinkle the flame. Nevertheless, high burning rates are achieved, owing to the very large total area of flame surface created at the interfaces between the multitudinous small eddies and the combustion products in which they are enveloped. In this region, the concept of a continuous, coherent flame surface is no longer realistic, and the combustion zone may be regarded as a fairly thick matrix of burned gases interspersed with eddies of unburned mixture.

The ratio of turbulent to laminar flame speeds in this region of high turbulence is given by

$$\frac{S_T}{S_L} = 0.5\frac{u'\delta_L}{S_L\eta}. \tag{2.12}$$

The structure of the flame at the two extremes of low and high turbulence intensity is illustrated in Figure 2.2. The top photograph of this figure shows that when turbulence is low, the flame surface comprises an agglomeration of round swellings that gradually grow in size as the flame expands downstream. The laceration and disruption of the flame surface at conditions of high turbulence intensity are illustrated in the bottom photograph of Figure 2.2.

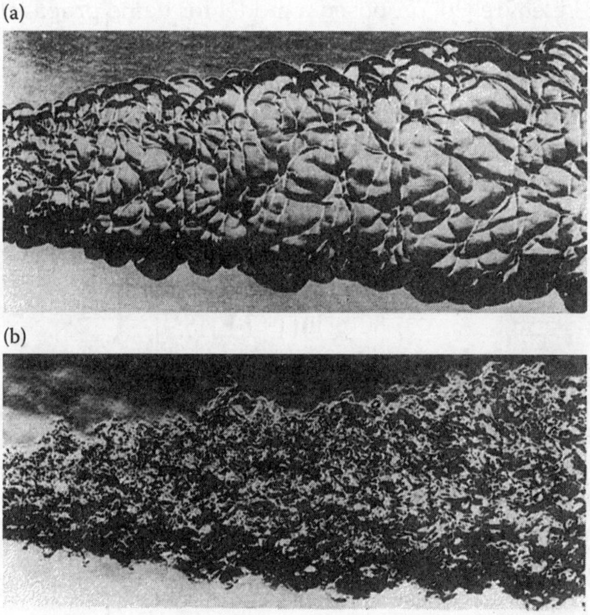

FIGURE 2.2
Stoichiometric propane-air flames under conditions of low and high turbulence. (a) Upper photograph, $u' = 3.1$ m/s. (b) Lower photograph, $u' = 30.5$ m/s. (Reprinted from Lefebvre, A.H. and Reid, R., *Combustion and Flame*, 10(4), 355–66, 1966. With permission from Elsevier, Inc.)

2.9 Flame Propagation in Heterogeneous Mixtures of Fuel Drops, Fuel Vapor, and Air

Comparatively few studies have been made of flame propagation through heterogeneous fuel–air mixtures, the earliest published work in this area being the classic treatise of Burgoyne and Cohen [15]. Subsequent studies include those of Cekalin [16], Mizutani and Nishimoto [17], Mizutani and Nakajima [18], Polymeropoulos and Das [19], Ballal and Lefebvre [20], Polymeropoulos [21], and Myers and Lefebvre [22]. The paucity of literature on this subject is not altogether surprising in view of the formidable experimental difficulties involved. Foremost among these is the creation of a uniform and reproducible multidroplet mist. Allied to this is the problem of accurate measurement of mean drop size, drop-size distribution, overall equivalence ratio, and concentration of fuel vapor. Another difficult task is the measurement of the rate of flame propagation through the mixture, where serious errors can arise due to the upward buoyancy of the burned gases and the downward settling velocity of the fuel drops relative to the flame. These effects are especially significant for slow-burning mixtures because they are of the same order of magnitude as the laminar flame speed.

Ballal and Lefebvre [20] proposed a model for flame propagation through quiescent combustible mixtures in which the fuel is present in the form of a multidroplet mist or vapor, or both. The basis of the model is that, under normal steady-state conditions, the rate of flame propagation through a fuel mist is always such that the quench time of the reaction zone is just equal to the sum of the evaporation and chemical reaction times. This model yields the following expression for flame speed

$$S = \alpha_g \left[\frac{C_3^3 (1-f) \rho_F D_{32}^2}{8 C_2 \rho_g \ln(1+B)} + \frac{\alpha_g^2}{S_L^2} \right]^{-0.5}, \qquad (2.13)$$

where α_g is the thermal diffusivity of fully vaporized fuel–air mixture, and f is the fraction of fuel initially present as vapor. C_2 and C_3 are drop-size distribution parameters; $C_2 = D_{20}/D_{32}$ and $C_3 = D_{30}/D_{32}$. Unless the distribution of drop sizes in the spray is known, values of C_2 and C_3 must be determined experimentally. Suitable values of C_2 and C_3 for pressure-swirl and airblast atomizers are 0.41 and 0.56, respectively [20].

In the above expression, the mass transfer number B provides a measure of the volatility of the fuel. Replacing B with the evaporation constant λ allows Equation 2.13 to be simplified to

$$S = \left[\frac{C_3^2 (1-f) D_{32}^2}{C_2 \alpha_g \lambda} + \frac{1}{S_L^2} \right]^{-0.5}. \qquad (2.14)$$

For a monodisperse spray of fuel drops and air, this equation reduces to

$$S = \left[\frac{D^2}{\alpha_A \lambda} + \frac{1}{S_L^2} \right]^{-0.5}. \qquad (2.15)$$

In the above equations, S_L is expressed essentially as the sum of two terms. The first characterizes the evaporation rate, and thus depends on fuel volatility, mean drop size, and vapor fraction. The second characterizes the chemical reaction rate. When the evaporation time is longer than the chemical reaction time, flame speed is enhanced by increases in gas density, fuel volatility, vapor concentration, and reduction in mean drop size. If conditions are such that chemical reaction rates are limiting to flame speed, the latter reverts to the normal burning velocity for the fully evaporated mixture. However, if the reaction time is small in comparison to the time required for evaporation, the equations predict that flame speed is inversely proportional to mean drop size. This theoretical finding is fully confirmed in the experimental investigation of flame propagation in heterogeneous mixtures of fuel drops and air carried out by Myers and Lefebvre [22]. Figures 2.3 and 2.4 are typical of the results obtained. Figure 2.3 shows flame speed plotted against the reciprocal of mean drop size for several different fuel/air ratios. It demonstrates, over wide ranges of mean drop size, a straight-line relationship between S_T and SMD^{-1}, indicating that evaporation rates are controlling the flame speed. In theory, the straight portion of the lines drawn in Figure 2.3 should pass through the origin, i.e., flame speed should become zero for infinite fuel drop size. In practice, the lines tend to intercept the abscissa at a finite value of SMD. Thus, although the theory suggests that flame speed should reduce gradually to zero, corresponding to infinitely large drops, the results for JP 7 and other hydrocarbon fuels indicate that, in practice, there is always a maximum mean drop size above which flame propagation is impossible. For the fuels and test conditions examined by Myers and Lefebvre, this practical limit on mean drop size is around 400 μm.

Generally, it is found that flame speed increases with reduction in mean drop size until a critical value is reached. For drop sizes smaller than the critical value, which is around 60–70 μm for kerosine-type fuels, the curves flatten out, indicating that for finely atomized sprays, flame speeds are much less dependent on evaporation rates, and are governed primarily by chemical reaction rates.

The influence of mean drop size on flame speeds is shown more directly in Figure 2.4. This figure also illustrates the beneficial effect on flame speed

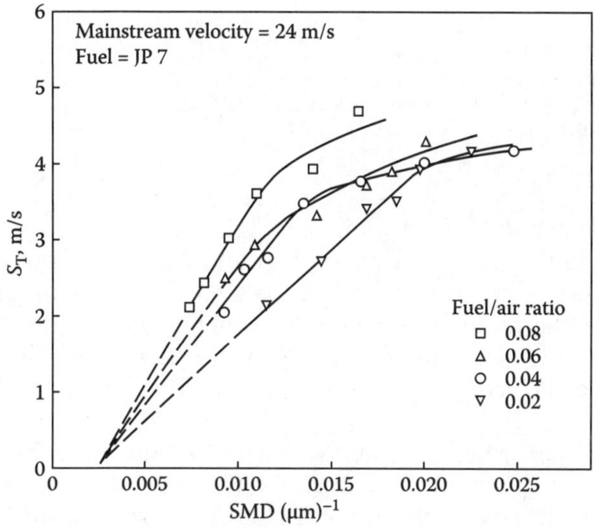

FIGURE 2.3
Influence of fuel/air ratio and mean drop size on flame speed for a mainstream velocity of 24 m/s. (Reprinted from Myers, G.D. and Lefebvre, A.H., *Combustion and Flame*, 66(2), 193–210, 1986. Copyright by The Combustion Institute.)

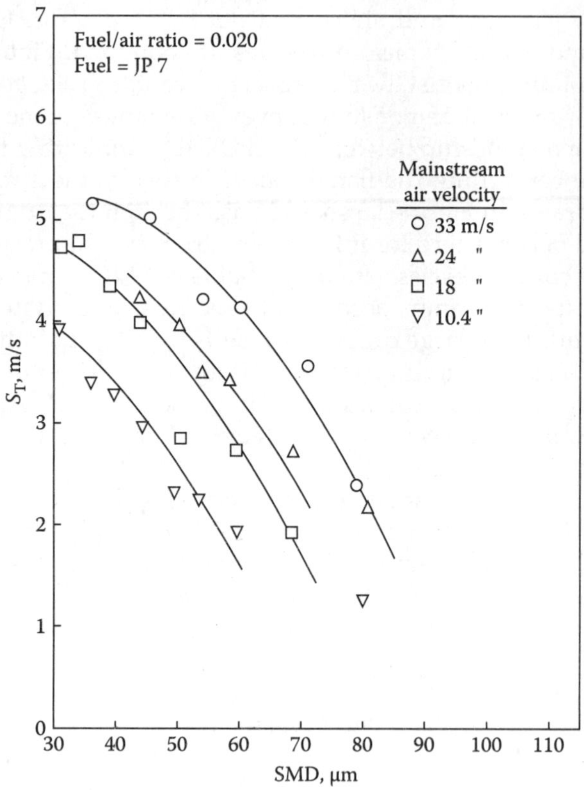

FIGURE 2.4
Influence of mainstream velocity and mean drop size on flame speed. (Reprinted from Myers, G.D. and Lefebvre, A.H., *Combustion and Flame*, 66(2), 193–210, 1986. Copyright by The Combustion Institute.)

of an increase in flow velocity. This benefit derives mainly from the increase in turbulence intensity that accompanies an increase in flow velocity. It is well established that turbulence promotes flame speeds in gaseous fuel–air mixtures (see previous section). With heterogeneous mixtures, it has the added advantage of enhancing fuel evaporation rates. The net effect is that flame speeds increase with an increase in flow velocity, as illustrated in Figure 2.4.

Equations 2.13 through 2.15 apply strictly to flame propagation through slow-moving or quiescent fuel mists of the type studied by Ballal and Lefebvre [20]. However, they are still valid for turbulent mixtures of fuel drops and air, provided that S_L is replaced by S_T, and λ_{eff} which takes account of the role of turbulence in enhancing evaporation rates, is used instead of λ.

2.10 Droplet and Spray Evaporation

The evaporation of fuel droplets in a spray involves simultaneous heat and mass transfer processes, in which the heat for evaporation is transferred to the drop surface by conduction and convection from the surrounding air or gas, and vapor is transferred by convection and diffusion back into the gas stream. The overall rate of evaporation depends on the pressure, temperature, and transport properties of the gas; the temperature, volatility, and diameter of the drops in the spray; and the velocity of the drops relative to that of the surrounding gas.

If a single-component fuel drop is suddenly immersed in gas at high temperature, it starts to heat up exactly like any other cold body when placed in a hot environment [23]. Figure 2.5 shows how the temperature of a fuel drop varies during its lifetime. Starting from its initial value T_F, the fuel temperature increases until eventually it reaches its steady-state value, T_{st}. This point denotes the end of the heat-up period, and from then on the drop temperature remains constant at T_{st}, until evaporation is complete. Thus, the total drop evaporation time can be subdivided into two main components, one for the heat-up period and another for the steady-state phase.

During the first phase of the evaporation process, almost all of the heat supplied to the drop serves merely to raise its temperature. Little or no mass transfer from the drop occurs during this stage, which corresponds to the horizontal portions of the curves drawn in Figure 2.6. As the fuel temperature rises, fuel vapor is formed at the drop surface and part of the heat

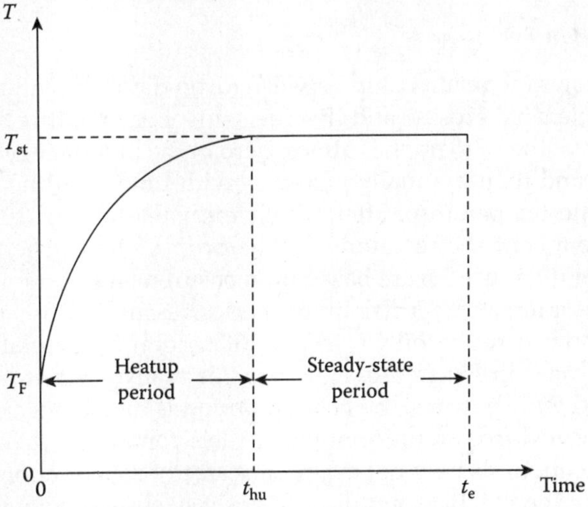

FIGURE 2.5
Variation of fuel temperature during drop lifetime.

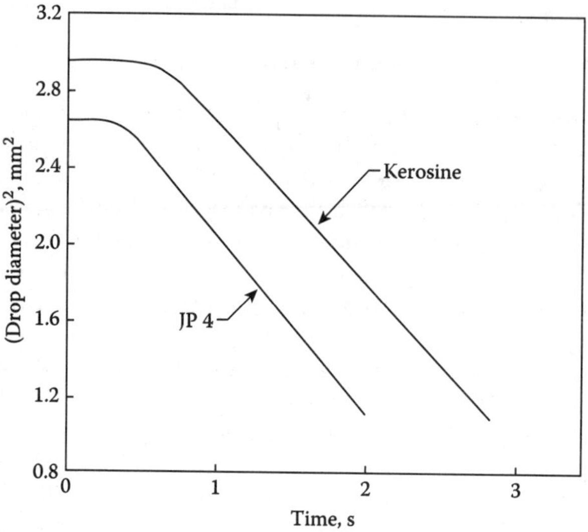

FIGURE 2.6
Evaporation rate curves for kerosine and JP 4.

transferred to the drop is now used to furnish the heat of vaporization of the fuel. Eventually, the drop attains its steady-state temperature and the heat supplied to the drop is used solely as heat of vaporization. This condition corresponds to the straight lines drawn in Figure 2.6.

2.10.1 Heat-Up Period

Figure 2.6 shows the relationship between (drop diameter)^2 and evaporation time for droplets of kerosine and JP 4 fuels. Inspection of this figure reveals that the slope of the D^2/t graph is almost zero in the first stage of the evaporation process and then gradually increases with time until the drop attains its steady-state temperature, after which the value of D^2/t remains fairly constant throughout the remainder of the drop lifetime. The vaporization curves drawn in Figure 2.6 are based on measurements carried out in air at 2000 K temperature and normal atmospheric pressure. At this pressure, the heat-up period constitutes only a very small portion of the total evaporation time, as indicated in Figure 2.6. However, for many fuels at high ambient pressures and temperatures, the heat-up period is much longer; so much so that drops formed from multicomponent fuels containing several different petroleum compounds may not experience steady-state evaporation during their lifetime. The practical significance of these observations is that actual drop and spray evaporation rates can be appreciably lower than the experimental values quoted in the literature, most of which were measured during steady-state evaporation at normal atmospheric pressure.

2.10.2 Evaporation Constant

One of the first theoretical approaches to the problem of droplet evaporation was made by Godsave [24], who derived the rate of evaporation of a single drop as

$$m_F = (\pi/4)\rho_F \lambda D, \tag{2.16}$$

where

$$\lambda = d(D)^2/dt. \tag{2.17}$$

Note that λ corresponds to the slope of the lines drawn in Figure 2.6.

The average rate of evaporation during the drop lifetime is obtained from Equation 2.16 as

$$\dot{m}_F = (\pi/6)\rho_F \lambda D_o. \tag{2.18}$$

The drop lifetime is also readily obtained by assuming λ is constant and integrating Equation 2.17 gives

$$t_e = D_o^2/\lambda. \tag{2.19}$$

Following Spalding, the evaporation rate of a single drop can also be expressed in terms of a mass transfer number B [25,26]. We have

$$\dot{m}_F = 2\pi D(k/c_p)_g \ \ln\ (1+B), \tag{2.20}$$

whereas drop lifetime is given by

$$t_e = \rho_F D_o^2/8(k/c_p)_g \ln(1+B). \tag{2.21}$$

A drawback to using B instead of λ for calculating m_F and t_e is that the accuracy of the results is very dependent on the choice of values of k and c_p [27].

Values of λ may be used to determine the transfer number B (and vice versa). Equating (2.16) and (2.20) gives

$$\lambda = 8(k/c_p)_g \ln(1+B)/\rho_F. \tag{2.22}$$

2.10.3 Convective Effects

In most continuous flow combustors, the fuel is sprayed into air or gas flowing at high velocity. Where relative motion exists between the droplets and the surrounding gas, the rate of evaporation is enhanced by forced convection. This effect can be accommodated by multiplying the evaporation rate calculated for quiescent conditions by the correction factor

$$1+0.22\mathrm{Re}_D^{0.5},$$

where Re_D, the drop Reynolds number, is typically around 5.

2.10.4 Effective Evaporation Constant

From a practical viewpoint, it would be very convenient if Godsave's evaporation constant, which corresponds to steady-state evaporation in quiescent air, could be modified to take into account both the adverse effect of the heat-up period and the beneficial effect of forced convection. To accomplish this, Chin and Lefebvre [28] defined an effective evaporation constant as

$$\lambda_{\mathrm{eff}} = D_o^2/t_e, \tag{2.23}$$

where t_e is the total time required to evaporate the fuel drop, including both convective and transient heat-up effects, and D_o is the initial drop diameter.

Calculated values of λ_{eff} for an ambient air pressure of 100 kPa are plotted in Figure 2.7. Similar graphs for higher levels of pressure may be found in Chapter 5. Figure 2.7 shows plots of λ_{eff} vs. T_{bn}, the normal boiling point, for various values of UD_o at three levels of ambient temperature, namely, 500, 1200, and 2000 K. While recognizing that no single fuel property can fully describe the evaporation characteristics of any given fuel, the normal boiling point has much to commend it for this purpose, because it is directly related to fuel volatility and vapor pressure. It also has the virtue of being quoted in fuel specifications. Figure 2.7 shows that λ_{eff} increases with increases in ambient temperature, velocity, and drop size, and diminishes with an increase in normal boiling temperature.

For any given conditions of pressure, temperature, and relative velocity, the lifetime of a fuel drop of any given size is obtained from Equation 2.23 as

$$t_e = D_o^2/\lambda_{\mathrm{eff}}, \tag{2.24}$$

while the average rate of fuel evaporation is given by

$$\dot{m}_F = (\pi/6)\rho_F\lambda_{\mathrm{eff}}D_o. \tag{2.25}$$

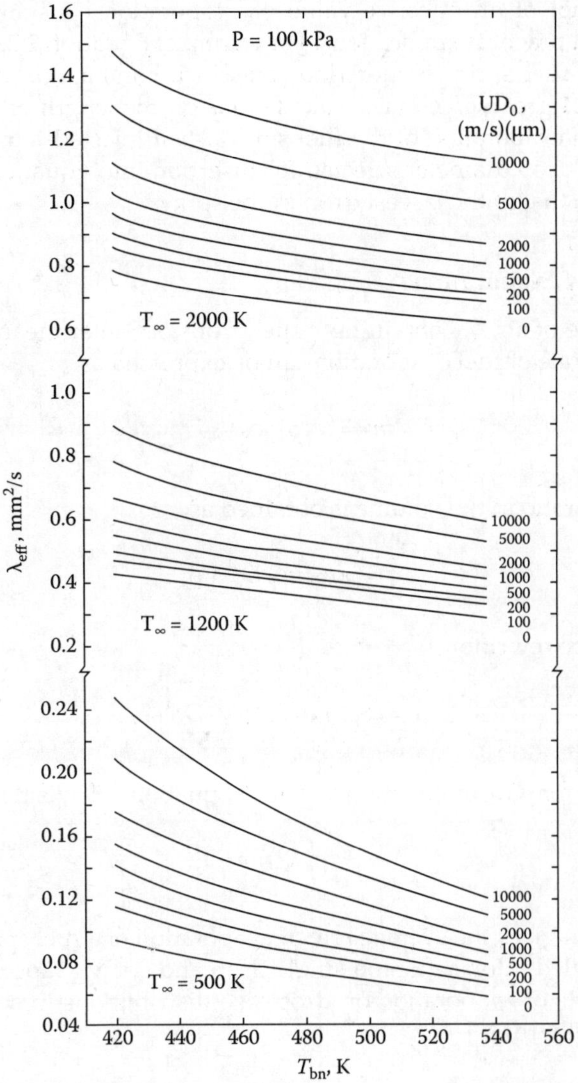

FIGURE 2.7
Variation of effective evaporation constant with normal boiling point at normal atmospheric pressure. (From Chin, J.S. and Lefebvre, A.H., *Proceedings of the 20th Automotive Technology Development Contractor Coordination Meeting*, 325–31 1982.)

The velocity term U in Figure 2.7 denotes the relative velocity between the drop and the surrounding gas. Small droplets rapidly attain the same velocity as the surrounding gas, after which they are susceptible only to the fluctuating component of velocity, u'. For gas turbine combustors, where the value of u' is usually high enough to affect evaporation rates, U in Figure 2.7 should be replaced with u'.

The concept of an effective value of evaporation constant concept has many useful practical applications. For example, Equation 2.29 may be used to calculate fuel spray evaporation rates in a combustion zone, whereas Equation 2.24 greatly facilitates calculations on the length of duct required for complete evaporation of the fuel spray when injected into a ducted airstream. The drop diameter selected for insertion into Equation 2.24 should, of course, be that of the largest drop in the spray.

2.10.5 Spray Evaporation

For a volume of air, V, containing n fuel drops of Sauter mean diameter D_o, the average rate of fuel evaporation can be expressed as

$$\dot{m}_F = (\pi/6) n \rho_F \lambda_{eff} D_o. \quad (2.26)$$

The fuel/air ratio in the volume is obtained as

$$q = (\pi/6) n \rho_F D_o^3 / V \rho_A, \quad (2.27)$$

which may be rewritten as

$$n = (6/\pi)(\rho_A/\rho_F)(V/D_o^3) q. \quad (2.28)$$

Substituting for n from Equation 2.28 into Equation 2.26 yields

$$\dot{m}_F = \rho_A \lambda_{eff} V q / D_o^2. \quad (2.29)$$

This equation gives the average rate of evaporation of a fuel spray.

More detailed information on single drop and spray evaporation, including the effects of evaporation on drop-size distributions in sprays, may be found in Lefebvre [27].

2.10.6 Some Recent Developments

Progress in the modeling of droplet vaporization up to 1994 has been reviewed by Peng and Aggarwal [29]. Their review includes the methodologies currently available for representing droplet motion and vaporization history in two-phase flow computations. More recent work includes a numerical study of two-component droplet evaporation by Stengele et al. [30]. This work is of special interest for gas turbine applications because it features gas temperatures of 800 and 2000 K and a range of pressures from 1 to 40 bars.

As a further advance on the λ_{eff} concept [28], Chin has developed more sophisticated models for the evaporation of multicomponent fuel drops that employ variable finite mass and thermal diffusivity. This work has culminated in a practical engineering calculation method for commercial gas turbine fuels. The procedures employed by Chin are too detailed for inclusion here, thus for further information, reference should be made to the original publications [31–33].

2.11 Ignition Theory

Most ignition theories are based on the idea that the transient ignition source, usually an electric spark, must supply sufficient energy to the combustible mixture to create a volume of hot gas that just satisfies the necessary and sufficient condition for propagation, namely, that the rate of heat generation just exceeds the rate of heat loss.

The work of Lewis et al. [34–36] did much to clarify and improve knowledge of spark ignition in quiescent mixtures. The first major contribution to ignition theory for flowing mixtures was made by Swett [37], who studied the influence on ignition energy of variations in pressure, velocity, equivalence ratio, and turbulence. Swett's theory is based on the ideas that (1) only a portion of the discharge length is important in the ignition process and (2) heat loss by thermal conduction is negligible compared with heat loss by eddy diffusion. Both of these ideas were fully confirmed in subsequent experiments carried out by Ballal and Lefebvre on ignition in flowing mixtures [38,39]. Unfortunately, Swett's treatment of turbulence is very limited and much of his experimental data are suspect for reasons discussed in Reference [39].

2.11.1 Gaseous Mixtures

Ballal and Lefebvre [38] analyzed the processes governing the rate of heat generation in an incipient spark kernel and the rate of heat loss by thermal conduction and turbulent diffusion. They conclude that, for the spark kernel to survive and propagate unaided throughout a gaseous mixture, its minimum dimension should always exceed the *quenching distance* as expressed by

$$d_q = \frac{10k}{c_p \rho_o (S_L - 0.16u')}, \tag{2.30}$$

for low-turbulence ($u' < 2S_L$) and

$$d_q = \frac{10k}{c_p \rho_o (S_T - 0.63u')}, \tag{2.31}$$

for highly turbulent mixtures ($u' \gg 2S_L$).

The minimum ignition energy E_{min} is defined as the amount of energy needed to heat to its adiabatic flame temperature, the smallest volume of gas whose minimum dimension is equal to the quenching distance. Clearly, the smallest volume that satisfies this criterion is a sphere of diameter d_q, so

$$E_{min} = c_p \rho_o \Delta T_{ad} (\pi/6) d_q^3. \quad (2.32)$$

Substituting d_q from Equations 2.30 and 2.31 into Equation 2.32 leads, respectively to

$$E_{min} = 5.24 \Delta T \frac{[k(S_L - 0.16u')^{-1}]^3}{(c_p \rho_o)^2}, \quad (2.33)$$

and

$$E_{min} = 5.24 \Delta T \frac{\left[k\left(S_T - 0.63u'\right)^{-1}\right]^3}{\left(c_p \rho_o\right)^2}. \quad (2.34)$$

The result of plotting measured values of E_{min} that were obtained over wide ranges of pressure, velocity, turbulence intensity, and mixture composition for both methane and propane fuels against values of d_q as calculated from Equations 2.30 and 2.31 is shown in Figure 2.8. The straight line drawn through the data points has a slope of 3.0, thus confirming the cubic relationship between E_{min} and d_q, as expressed in Equation 2.32.

The theory predicts that d_q (and hence E_{min}) increases with an increase in turbulence intensity; this is borne out by tests performed on propane–air mixtures, the results of which are shown in Figure 2.9. The effect of pressure on d_q is illustrated in Figure 2.10. Inspection of the data points indicates that quenching distance is roughly inversely proportional to pressure, as predicted by Equation 2.30. This corresponds to a pressure dependence for minimum energy of $E_{min} \alpha P^{-2}$.

Figure 2.10 also illustrates the beneficial effect (from an ignition standpoint) of replacing some or all of the nitrogen in the air with oxygen.

2.11.2 Heterogeneous Mixtures

All the evidence obtained in the studies of Subba Rao, Rao, and Lefebvre [40,41] on the ignition of flowing mixtures of fuel drops and air (see Chapter 5) serves to suggest that passage of the spark creates a kernel in which high gas temperatures are attained, partly from the energy supplied in the spark, but also from the heat liberated by the evaporation and rapid combustion of the smallest fuel drops. This initial high temperature then

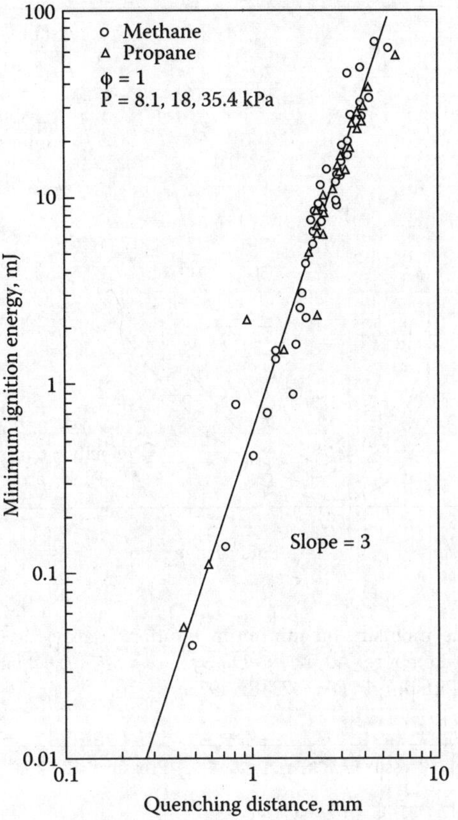

FIGURE 2.8
Relationship between E_{min} and d_q for both quiescent and flowing mixtures. $\phi = 1.0$. (From Ballal, D.R. and Lefebvre, A.H., *Proceedings of the Royal Society, London Series A*, 357 (1689), 163–81, 1977.)

falls as heat is lost by diffusion to the fresh mixture in contact with the outside surface of the kernel, and to the remaining fuel drops undergoing evaporation within the kernel. The key factor governing ignition is whether these droplets can evaporate and generate heat quickly enough to counter the heat loss from the kernel to its surroundings before it has shrunk below its minimum critical size, which corresponds to the quenching distance for the mixture.

When ignition is successful, combustion of the fuel vapor continues to produce heat that diffuses outward from the kernel to raise the temperature and initiate combustion in the surrounding unburned mixture. The flame then spreads rapidly to all regions where the air and fuel are in combustible proportions.

Analysis of all the experimental data obtained in these investigations [40,41] led to the conclusion that the sole criterion for the successful ignition of mixtures of fuel drops and air is a sufficiency of fuel vapor in the ignition

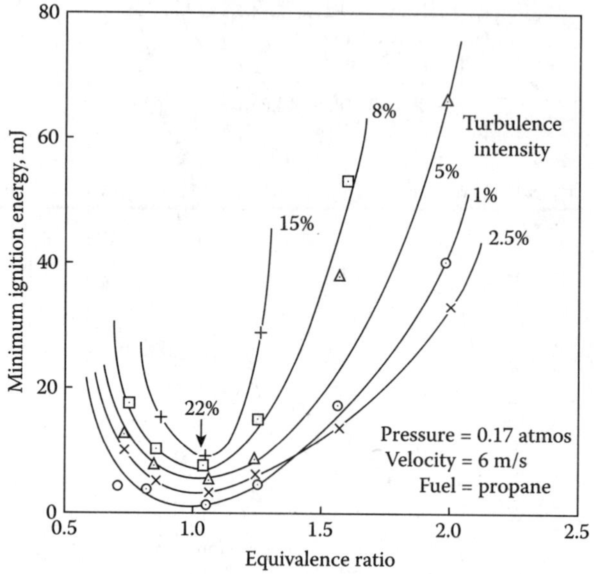

FIGURE 2.9
Influence of turbulence intensity on minimum ignition energy for propane–air mixtures. (From Ballal, D.R. and Lefebvre, A.H., *Fifteenth Symposium (International) on Combustion*, The Combustion Institute, Pittsburgh, PA, 1473–81, 1975.)

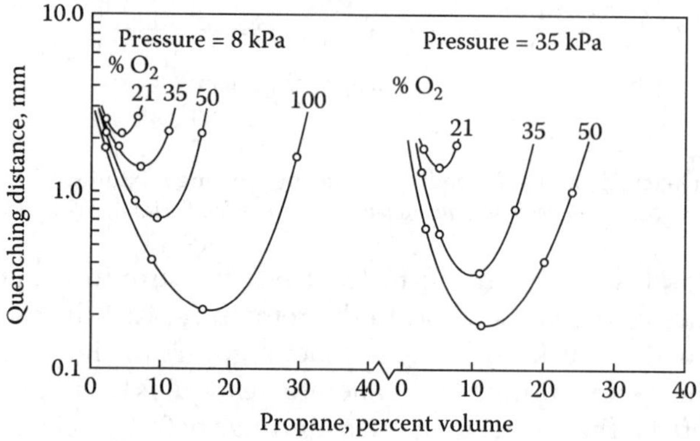

FIGURE 2.10
Influence of pressure and oxygen concentration on quenching distance for propane. Velocity = 15.3 m/s. (From Ballal, D.R. and Lefebvre, A.H., *Proceedings of the Royal Society, London Series A*, 357 (1689), 163–81, 1977.)

zone. Ignition will automatically ensue if passage of the spark creates sufficient thermal energy to produce the requisite amount of fuel vapor. The basic argument is that over wide ranges of operating conditions, the chemical reaction time is so short in comparison with the time required to produce an adequate amount of fuel vapor in the ignition zone that for all practical purposes it can be neglected [42]. This is in marked contrast to the ignition process in homogeneous mixtures, which is totally dominated by chemical reaction rates.

These considerations led to the development by Ballal and Lefebvre [42,43] of a theoretical model for the prediction of quenching distance and minimum ignition energy in liquid fuel sprays. The model is based on the assumption that chemical reaction rates are infinitely fast and that the onset of ignition is limited solely by the rate of fuel evaporation.

The process of ignition is envisaged as occurring in the following manner. Passage of the spark creates a small, roughly spherical, volume of air (henceforth referred to as the spark kernel) whose temperature is sufficiently high to initiate rapid evaporation of the fuel drops contained within the volume. Reaction rates and mixing times are assumed infinitely fast, so any fuel vapor created within the spark kernel is instantly transformed into combustion products at the stoichiometric flame temperature. If the rate of heat release by combustion exceeds the rate of heat loss by thermal conduction at the surface of the inflamed volume, then the spark kernel grows in size to fill the entire combustion volume. If, however, the rate of heat release is lower than the rate of heat loss, the temperature within the spark kernel falls steadily until fuel evaporation ceases altogether.

Thus, of crucial importance is the spark-kernel size for which the rate of heat loss at the kernel's surface is just balanced by the rate of heat release, due to the instantaneous combustion of fuel vapor, throughout its volume. As with homogeneous mixtures, this concept leads to the definition of *quenching distance* as the critical size that the inflamed volume must attain to propagate unaided; the amount of energy required from an external source to attain this critical size is termed the *minimum ignition energy*.

Analysis of the relevant heat-transfer and evaporation processes [42,43] yields the following expression for the quenching distance of quiescent or slow-moving multidroplet mists:

$$d_q = \left[\frac{\rho_F D^2}{\rho_A \phi \ln(1+B_{st})} \right]^{0.5}. \tag{2.35}$$

It should be noted that this equation was derived through consideration of the basic mechanisms of heat generation within the kernel and heat loss from its surface, and it contains no experimental or arbitrary constants. It is valid for monodisperse sprays only. However, for polydisperse sprays of the type provided by most practical atomizing devices, it can be shown [44] that the quenching distance is given by

$$d_q = \left[\frac{C_3^3 \rho_F D_{32}^2}{C_2 \rho_A \phi \ln(1+B_{st})} \right]^{0.5}. \tag{2.36}$$

Equations 2.35 and 2.36 provide simple dimensionless relationships between the quenching distance and the fuel drop size in the spray. Essentially, they state that quenching distance is directly proportional to drop size and is inversely proportional to the square root of gas pressure. An increase in ϕ and a reduction in ρ_F both reduce d_q because they promote evaporation by increasing the surface area of the fuel. Similarly, an increase in B also accelerates evaporation, thereby decreasing d_q.

Values of E_{min} may be obtained for quiescent or low-turbulence mixtures by inserting the calculated values of d_q from Equation 2.35 or Equation 2.36 into the expression

$$E_{min} = c_{pA}\rho_A \Delta T_{st} (\pi/6) d_q^3. \tag{2.37}$$

The results of such calculations are shown as solid curves in Figures 2.11 and 2.12. The very satisfactory level of agreement between theory and experiment, as demonstrated in these figures, suggests that the model can predict

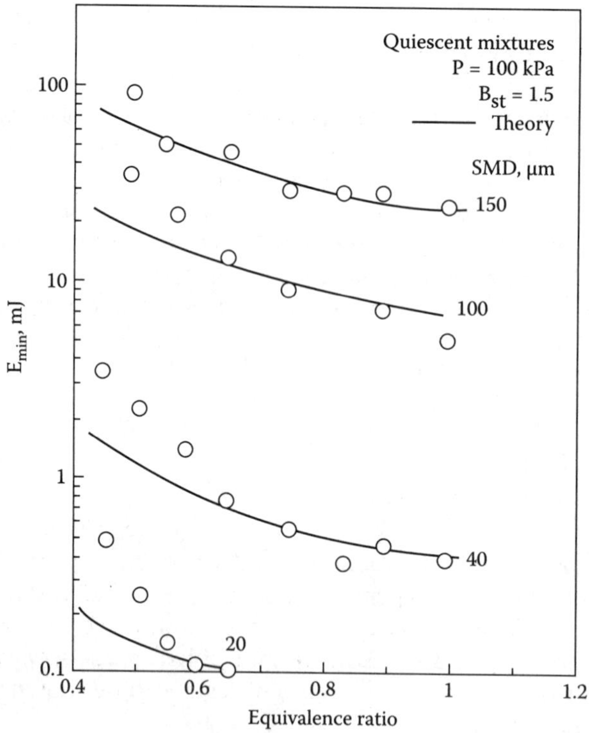

FIGURE 2.11
Minimum ignition energies of quiescent heavy fuel oil and air mixtures for various mean drop sizes. $P = 100$ kPa, $T_A = 290$ K. (From Ballal, D.R. and Lefebvre, A.H., *Proceedings of the Royal Society, London Series A*, 364 (1717), 277–94, 1978.)

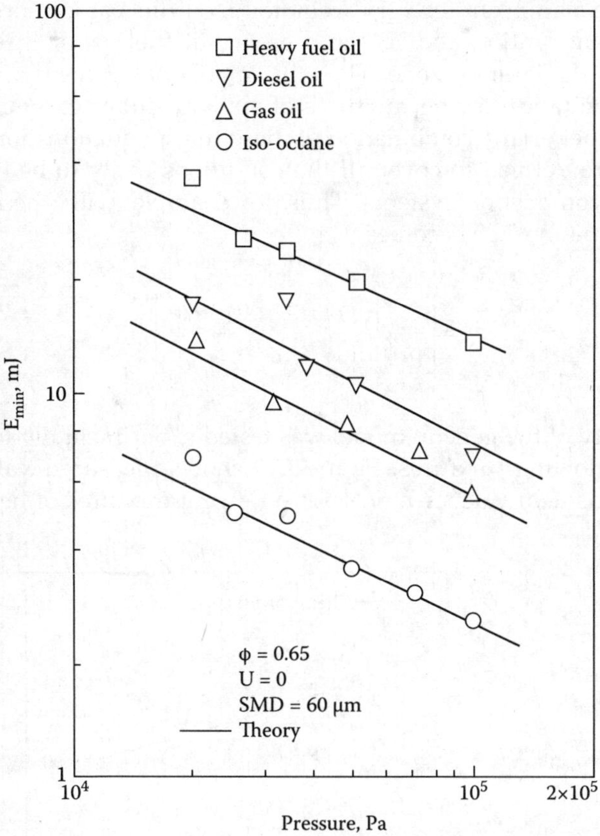

FIGURE 2.12
Effect of air pressure on minimum ignition energy. $\phi = 0.65$, SMD = 60 μm, $T_A = 290$ K, $U = 0$. (From Ballal, D.R. and Lefebvre, A.H., *Proceedings of the Royal Society, London Series A*, 364 (1717), 277–94, 1978.)

with fair accuracy the effect of variations in fuel volatility, mean drop size, and air pressure on minimum ignition energy. It also supports the basic assumption of the model, namely, that over a wide range of test conditions fuel evaporation is the rate-controlling step.

Although the above equations for d_q in heterogeneous fuel–air mixtures were derived for quiescent mixtures, they may be applied to flowing mixtures in combustion systems without much loss of accuracy. This is because, except for the very largest drops, most of the fuel spray is airborne, and the relative velocity between the fuel drops and the surrounding air or gas is too small to appreciably enhance either the rate of fuel evaporation or the rate of heat loss from the spark kernel.

In a later study, Ballal and Lefebvre [45] extended the model described above to include (1) the effects of finite chemical reaction rates, which are known to be significant for very well atomized fuels at low pressures and low equivalence ratios; and (2) the presence of fuel vapor in the mixture flowing into the ignition zone. Thus, the model has general application to both quiescent and flowing mixtures of air with either gaseous, liquid, or evaporated fuel or any combination of these fuels. Equations for quenching distance were derived to cover all the conditions likely to be encountered in practical combustion systems. Thus, for example, with chemical effects included, Equation 2.35 becomes

$$d_q = \left[\frac{\rho_F D_{32}^2}{\rho_A \phi \ln(1+B_{st})} + \left(\frac{10\alpha}{S_L} \right)^2 \right]^{0.5}. \qquad (2.38)$$

The validity of the general model was tested experimentally for both quiescent and flowing mixtures. Figure 2.13 shows measured values of E_{min} plotted against equivalence ratio ϕ for quiescent mixtures of heavy fuel oil

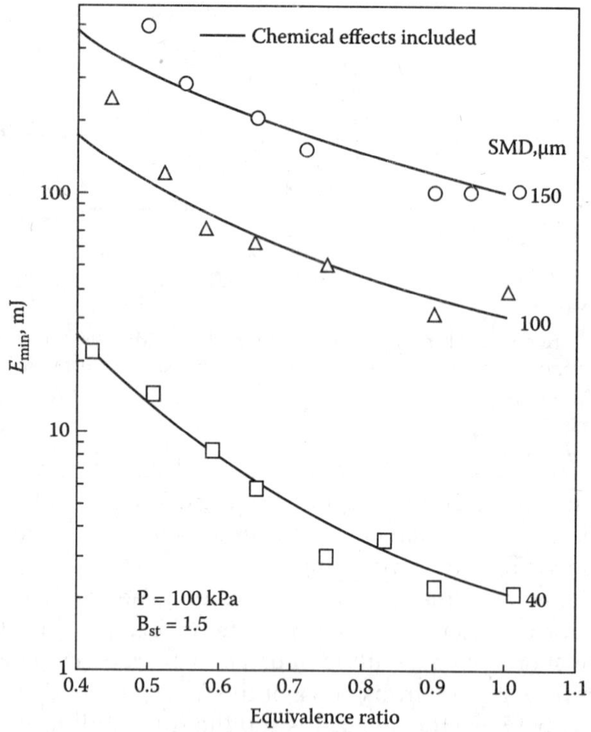

FIGURE 2.13
Improvements in correlation resulting from inclusion of chemical effects. Heavy fuel oil, $U = 0$. (From Ballal, D.R. and Lefebvre, A.H., *Eighteenth Symposium (International) on Combustion*, The Combustion Institute, Pittsburgh, PA, 1737–46, 1981.)

and air. The figure shows that the general model provides a good fit to the data over a range of SMDs from 40 to 150 μm.

Figure 2.14 shows the influence of pressure on E_{min} for four different fuels with a mean drop size of 60 μm when sprayed into a flowing airstream. Again, the excellent correlation achieved by including both chemical and evaporation effects is apparent in this figure.

In general, where evaporative effects are dominant, Equations 2.36 and 2.37 indicate for both quiescent and flowing mixtures that $E_{min}\ \alpha P^{-0.5}$. However, when chemical effects govern, Equations 2.30 and 2.32 show that $E_{min} \alpha P^{-2.0}$ for the extreme case where flame speed is independent of pressure. Thus, for both stagnant and flowing heterogeneous mixtures, the pressure exponent of E_{min} always lies between −0.5 and −2.0 and depends on the relative

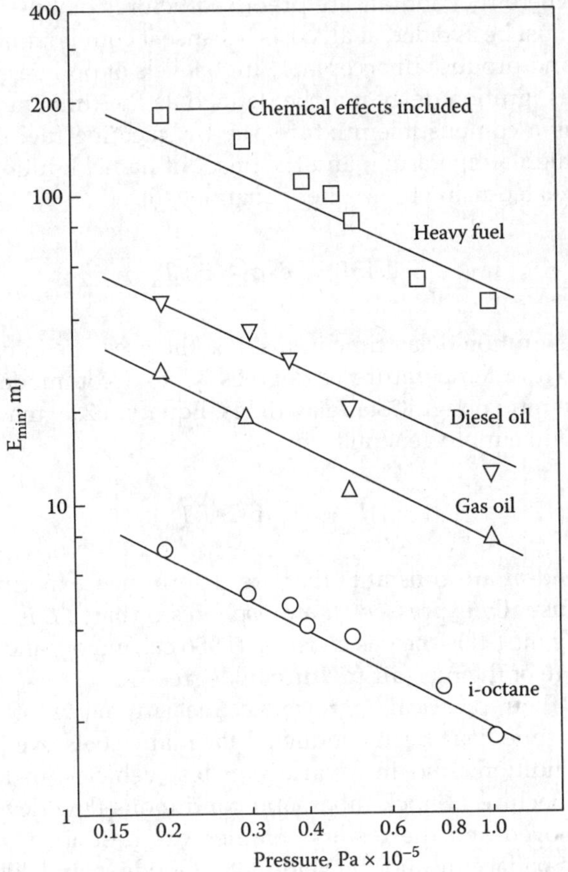

FIGURE 2.14
Improvements in correlation resulting from inclusion of chemical effects. $U = 15$ m/s, $\phi = 0.65$, SMD = 60 μm. (From Ballal, D.R. and Lefebvre, A.H., *Eighteenth Symposium (International) on Combustion*, The Combustion Institute, Pittsburgh, PA, 1737–46, 1981.)

importance of evaporative and chemical effects. In general, any change that enhances the role of reaction kinetics, such as a reduction in pressure and/or equivalence ratio, tends to increase the dependency of E_{min} on pressure.

2.12 Spontaneous Ignition

Spontaneous ignition, or autoignition, is a process whereby a combustible mixture undergoes a chemical reaction that leads to the rapid evolution of heat in the absence of any concentrated source of ignition, such as a flame or spark. In the lean-premix combustor, and other types of low-emissions combustors where fuel and air are premixed before combustion, spontaneous ignition must be avoided at all costs because it could damage combustor components and produce unacceptably high levels of pollutant emissions.

Spontaneous ignition delay may be defined as the time interval between the creation of a combustible mixture, say by injecting fuel into a flowing airstream at high temperature, and the onset of flame. Ignition delay times are often correlated using the Wolfer equation [46].

$$t_i = 0.43P^{-1.19}\exp(4650/T_m), \tag{2.39}$$

where t_i is the ignition delay time in ms, P is the pressure in bars, and T_m is the initial mixture temperature in degrees K. To accommodate the effects of equivalence ratio on ignition delay times, Equation 2.39 may be modified and expressed in a more general form as

$$t_i = AP^{-n}\phi^{-m}\exp(E/RT_m), \tag{2.40}$$

where A, n, and m are constants that are determined experimentally, P is the pressure (usually expressed in atmospheres or bars), E is the activation energy in cal/g mol, R is the gas constant (1.986 cal/g mol), and T_m is the initial temperature of the fuel–air mixture in degrees K.

In view of their practical importance, measurements of spontaneous ignition delay time have been conducted for many fuels over wide ranges of ambient conditions and in a variety of test vehicles, including rapid-compression machines, shock tubes, and continuous flow devices. The test methods employed and the results obtained are described in reviews by Mullins [47], Spadaccini and Te Velde [48], Goodger and Eissa [49], and Lundberg [50].

Freeman, Cowell, and Lefebvre [51,52] used a continuous flow apparatus to measure autoignition delay times. Twenty-five equispaced fuel-injection points ensured rapid mixing of gaseous fuel or fuel vapor with heated air at entry to the test section. The concept is shown schematically in Figure 2.15, where the ignition delay time is defined as the length L divided by the gas velocity U. This method has the advantage that when spontaneous ignition occurs, it does so under conditions that closely simulate those prevailing in the premixing passages of advanced combustors. Some of the results obtained for propane– and methane–air mixtures are shown in Figures 2.16 through 2.18.

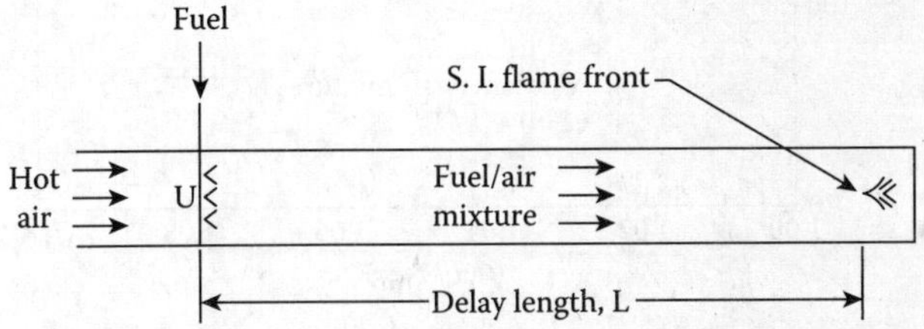

FIGURE 2.15
Basis of ignition delay time measurement technique. (Reprinted from Freeman, G. and Lefebvre, A.H., *Combustion and Flame*, 58(2), 153–62, 1984. With permission from Elsevier, Inc.)

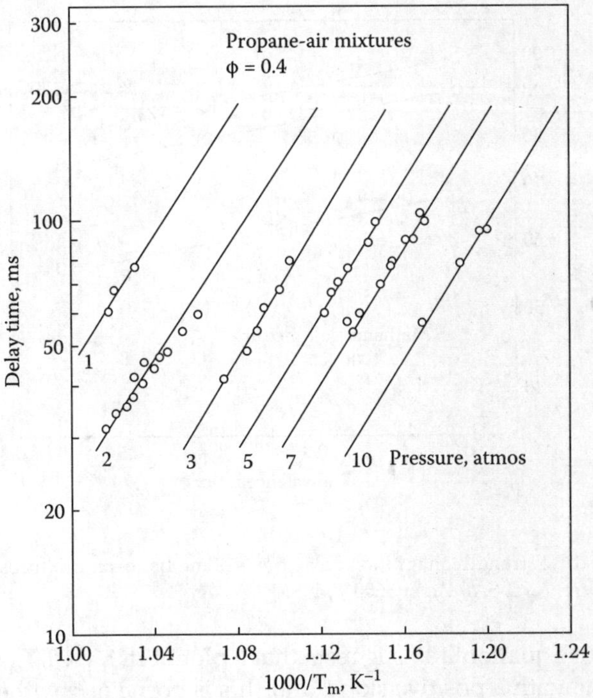

FIGURE 2.16
Influence of pressure and temperature on ignition delay times of propane-air mixtures. (From Cowell, L.H. and Lefebvre, A.H., SAE Paper 860068, 1986. With permission.)

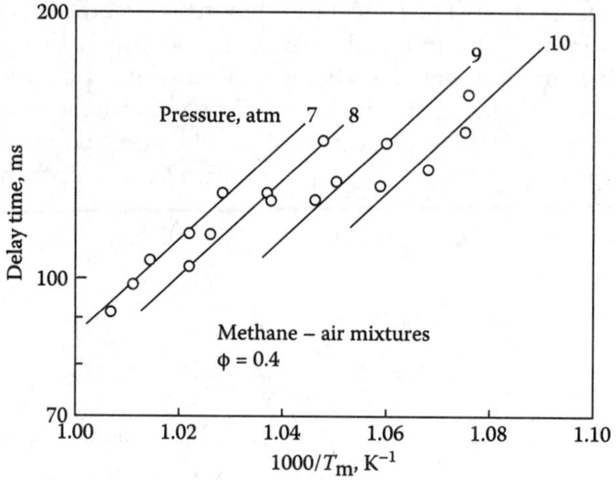

FIGURE 2.17
Influence of pressure and temperature on ignition delay times of methane–air mixtures. (From Cowell, L.H. and Lefebvre, A.H. SAE Paper 860068, 1986. With permission.)

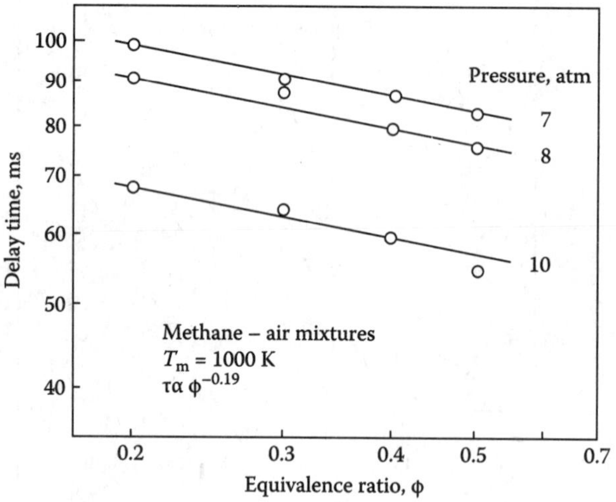

FIGURE 2.18
Influence of mixture strength on ignition delay times of methane–air mixtures. (From Cowell, L.H. and Lefebvre, A.H., SAE Paper 860068, 1986.)

The form of Equation 2.40 suggests that a plot of ln t_i vs. $1/T_m$ should yield a straight line with a positive slope, and this is borne out by the results presented in Figures 2.16 and 2.17. The values of E given by the slopes of the lines in these figures are 38.2 kcal/g mol for propane and 25.0 kcal/g mol for methane. For kerosine (Jet A), the value of E was found to be 29.6 kcal/g mol.

TABLE 2.1
Experimental values of constants in Equation 2.40

Fuel	E (kcal/g mol)	n	m
Propane	38.2	1.2	0.30
Methane	25.0	1.0	0.19
Kerosine	29.6	1.0	0.37

The influence of pressure on t_i is of great practical interest in view of the continuing trend toward engines of higher pressure ratio. Its importance is apparent from inspection of the experimental data plotted in Figures 2.16 and 2.17, which show a pronounced effect of pressure on t_i. Analysis of these and other data [53] led to values for n in Equation 2.40 of 1.2 for propane and 1.0 for both methane and kerosine (see Table 2.1).

There appears to be little agreement between different workers in regard to the influence of equivalence ratio on ignition delay time. Mullins [47] observed no effect, whereas Ducourneau [54] and Spadaccini and Te Velde [48] both found strong effects. Lefebvre et al. [53] examined the influence of ϕ on t_i for several fuels and found in all cases that delay times were reduced by an increase in equivalence ratio. Figure 2.18 is typical of the results obtained. It shows for methane that $t_i \alpha \phi^{-0.19}$. For propane and aviation kerosine, the measured values for m were higher at 0.30 and 0.37, respectively.

The explanation for the marked lack of consistency between different workers in regard to the influence of ϕ on t_i probably lies in the mode of fuel injection. With liquid fuels, there is always the potential for stoichiometric combustion in regions close to the evaporating spray. Thus, measured ignition delay times may be close to those for stoichiometric mixtures, even though the average equivalence ratio of the mixture differs appreciably from the stoichiometric value. Just how close will depend on the drop-size distribution in the spray, because this governs the initial rate of fuel evaporation and also the length of time that stoichiometric "streaks" of fuel–air mixture can survive. The number of fuel-injection points is also important. In this context, it is of interest to note that Tacina [55] obtained much more consistent autoignition data with a single orifice injector than with a 41-hole injector, which ostensibly should have provided a more uniform fuel–air mixture. Presumably, this was because with a single injector the rate of fuel–air mixing was so slow that the bulk of the prereactions leading up to the onset of ignition took place in near-stoichiometric mixtures, regardless of the average equivalence ratio. With gaseous fuels, the inconsistencies associated with slow fuel evaporation are no longer present, but the measured ignition delay times are still very dependent on the time required for the fuel and air to form a combustible mixture. As with liquid fuels, the longer the mixing time, the closer the measured ignition delay times will approach stoichiometric values.

Another probable reason for the conflicting evidence of the effect of equivalence ratio on ignition delay time is that in continuous flow experiments the fuel is almost invariably at a much lower temperature than the hot airstream into which it is injected. This has the advantage of closely simulating the actual engine situation but, from a fundamental viewpoint, it has the drawback that any change in ϕ must also change the temperature in the initial fuel–air mixing zone(s). As ignition delay time is exponentially dependent on temperature, the effect of a small change in temperature in these localized mixing zone(s) could be very pronounced and could largely offset the effect of the corresponding change in ϕ on ignition delay time. Note that the effect of this change in temperature on t_i, will always be such as to oppose the change in t_i caused by the change in ϕ. The net result is that measurements of t_i, carried out in continuous flow devices will always underpredict the effect of a change in ϕ on t_i by an amount that depends on the difference in temperature between the hot airstream and the injected fuel gas or vapor.

Most analyses and equations for ignition delay time ignore the effects of fuel vaporization, which is reasonable under conditions where the fuel evaporation time is appreciably shorter than the mixing and reaction times. However, it is important to bear in mind that spontaneous ignition delay times are affected by both physical and chemical processes. For liquid fuels, the physical delay is the time required to heat and vaporize the fuel drops and to mix the fuel vapor in flammable proportions with the surrounding air. The chemical delay is the time interval between the formation of a flammable mixture and the appearance of flame. Thus, the physical processes are important in the early stages of spontaneous ignition, while in the later stages the chemical processes become over-riding.

Rao and Lefebvre [56] have proposed a model for spontaneous ignition that takes both chemical and physical effects into account and has general application to both homogeneous and heterogeneous mixtures, including situations where both fuel drops and fuel vapor are initially present. This model leads to an equation where the ignition delay time, t_i, is derived as the sum of the times required for evaporation and chemical reaction. Calculated values of t_i from this equation show that fuel evaporation times are negligibly small in comparison with chemical reaction times for well atomized, highly volatile fuels, especially at conditions of low pressure and temperature. They also show that fuel evaporation times become increasingly significant with increases in pressure and temperature.

When calculating ignition delay times for liquid fuels, it is customary to disregard the fuel evaporation time. As discussed above, this is permissible for well-atomized volatile fuels injected into airstreams at relatively low pressures. However, it is important to recognize that for certain ultralow NO_x combustors (e.g., catalytic or lean premixed prevaporized (LPP)) when operating at high-power conditions corresponding to high air pressures and temperatures, the fuel evaporation time could be so long in relation to the chemical delay time that spontaneous ignition of the initial fuel vapor might occur before the remainder of the fuel spray has had time to evaporate. Very fine fuel atomization and rapid vapor–air mixing will be required to combat this problem.

In summary, autoignition data are very apparatus-dependent and, in particular, very fuel injector-dependent. Considerable caution should be exercised in comparing and selecting autoignition data and in no circumstances should experimentally derived equations for t_i, be extrapolated to pressures and temperatures outside the range of their experimental verification. Such extrapolations could lead to erroneous results because differences in reaction routes may occur over different levels of temperature and pressure. When liquid fuels are injected into air at high pressures and temperatures, very fine atomization is needed to promote rapid vaporization, thereby reducing the risk of spontaneous ignition in the fuel preparation zone.

2.13 Flashback

An intrinsic feature of all premixed-fuel combustion systems is a tendency toward flashback. Flashback occurs when the flame travels upstream from the combustion zone into the premixing sections of the combustor. This upstream propagation of flame takes place whenever the flame speed exceeds the approach flow velocity.

Three main types of flashback have been identified: (1) flashback occurring in the free stream, (2) flashback occurring through the low-velocity flow in the boundary layer along the walls of the premixing section, and (3) flashback driven by combustion-induced vortex breakdown (CIVB) in swirl-stabilized combustors. Any mechanism may involve homogeneous and/or heterogeneous reactions.

The most obvious free-stream mechanism would be the occurrence of flashback due to a flow reversal in the bulk flow through the combustor. This flow reversal could be a result of compressor surge or combustion instability. Flashback can also occur in the absence of flow reversal if the turbulent flame speed through the gas in a premixing section is greater than the local bulk velocity. Lean combustion tends to reduce flame speeds, but other factors associated with the engine cycle, such as high temperatures, pressures, and turbulence levels, and preignition reactions in the gas due to appreciable residence times at high temperature levels, cause increased flame speed. Therefore, flame speeds may be sufficiently high to necessitate increasing the minimum allowable velocity in premix-prevaporize sections to fairly high levels to avoid disturbances to the combustion process.

The boundary-layer mechanism involves flashback through retarded flow in a boundary layer. Important relevant parameters include the wall temperature and temperature distribution, and the boundary-layer structure, turbulence properties, and thickness. For more detailed information on flashback, reference should be made to Plee and Mellor [57].

The CIVB flashback has been studied by Sattelmayer and coworkers [58,59]. This phenomenon is characterized by an upstream propagation of the flame on the burner axis, with propagation speeds far beyond the turbulent flame speed, against high axial velocity flow. Especially, burner systems without central bluff body and purely swirl-stabilized flame stabilization are prone to CIVB. This type of flashback was successfully correlated with a chemical time scale describing local quenching of the flame.

2.14 Stoichiometry

Complete combustion of a hydrocarbon fuel requires sufficient air to convert the fuel completely to carbon dioxide and water vapor. Since 23% by mass (21% by volume) of oxygen in the air participates in combustion, the stoichiometric AFR can be calculated from the reaction equation. For example, one mole of C_7H_{16} requires 11 moles of oxygen for complete combustion:

$$C_7H_{16} + 11O_2 = 7CO_2 + 8H_2O.$$

Hence, by substituting the appropriate atomic weights (C = 12, H = 1, O = 16), we obtain

$$100\,\text{g} + 352\,\text{g} = 308\,\text{g} + 144\,\text{g}.$$

Thus, 1 g of fuel requires 3.52 g of oxygen or 3.52 × 100/23 = 15.3 g of air. This shows that for C_7H_{16}, the stoichiometric AFR is 15.3. Sometimes, this quantity is expressed in a reciprocal form, i.e., as the stoichiometric fuel/air ratio. For C_7H_{16}, the stoichiometric fuel/air ratio is 1/15.3 = 0.06535. For kerosine ($C_{12}H_{24}$), the stoichiometric fuel/air ratio is 0.0676.

Stoichiometric mixtures, by definition, contain sufficient oxygen for complete combustion; thus, operation at the stoichiometric AFR will release all the latent heat of combustion of the fuel. If the fuel is burned at a numerically larger AFR, the mixture is referred to as *lean* or *weak*. Combustion at an AFR lower than the stoichiometric value implies a deficiency of oxygen; hence, combustion is incomplete, and partially burned fuel, principally in the form of carbon monoxide and unburned hydrocarbons, will escape from the combustion zone.

In comparing the combustion characteristics of different fuels, it is sometimes convenient to express the mixture strength in terms of an equivalence ratio ϕ. The equivalence ratio is the actual fuel/air ratio divided by the stoichiometric fuel/air ratio. Thus, for all fuels, $\phi = 1$ denotes a stoichiometric mixture. Also, for all fuels, a value of $\phi < 1$ indicates a lean mixture, whereas a value of $\phi > 1$ indicates a rich mixture.

2.15 Adiabatic Flame Temperature

Flame temperature is perhaps the most important property in combustion because it has a controlling effect on the rate of chemical reaction. The term "flame temperature" may imply a measured value or a calculated one. If the latter, it is usually the *adiabatic* flame temperature. This is the temperature that the flame would attain if the net energy liberated by the chemical reaction that converts the fresh mixture into combustion products were fully utilized in heating those products. In practice, heat is lost from the flame by radiation and convection, so the adiabatic flame temperature is rarely achieved. Nevertheless, it plays an important role in the determination of combustion efficiency and in heat-transfer calculations. In high-temperature flames, say above 1800 K, dissociation of combustion products occurs to a significant extent and absorbs much heat. At low temperatures, combustion of a stoichiometric or lean fuel–air mixture would be expected to give only CO_2 and H_2O; however, at higher temperatures, these products are themselves unstable and partly revert to simpler molecular and atomic species and radicals, principally CO, H_2, O, H, and OH. The energy absorbed in dissociation is considerable, and its effect is to substantially reduce the maximum flame temperature.

2.15.1 Factors Influencing the Adiabatic Flame Temperature

The factors of prime importance to adiabatic flame temperature are fuel/air ratio, initial temperature and pressure, and vitiation of the inlet air by products of combustion.

2.15.1.1 Fuel/Air Ratio

The variation of adiabatic temperature rise with change in fuel/air ratio is illustrated in Figure 2.19. The departure from linearity as the flame temperature rises is due partly to the increase in specific heat of the combustion products with an increase in temperature and, at the highest temperatures (>1800 K), to the effects of dissociation.

2.15.1.2 Initial Air Temperature

An increase in initial air temperature will always increase the flame temperature. However, the extent of this increase diminishes with an increase in

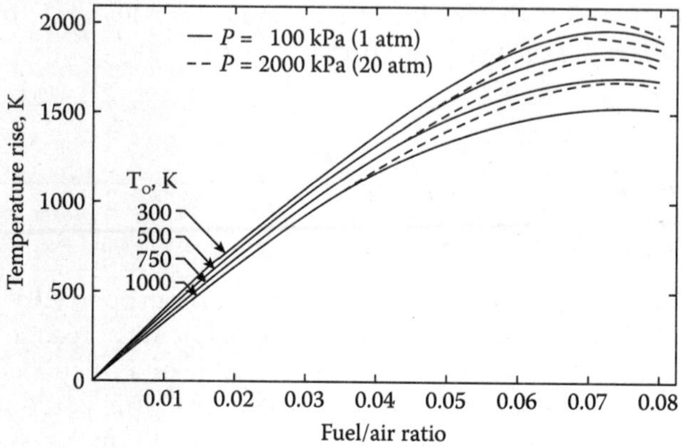

FIGURE 2.19
Adiabatic temperature rise curves for kerosine (JP 5) fuel. Lower specific energy = 43.08 MJ/kg (18,520 Btu/lb).

flame temperature. For near-stoichiometric mixtures, only about one half of an increase in initial air temperature is translated into an increase in flame temperature.

2.15.1.3 Pressure

For a constant inlet air temperature, an increase in pressure yields a higher flame temperature. This effect can be explained by examining the form of the CO_2 and H_2O dissociation equations:

$$CO_2 = CO + 0.5O_2,$$

$$H_2O = H_2 + 0.5O_2.$$

These reactions are endothermic, i.e., they absorb heat. They also lead to an increase in the number of moles (volume) of the combustion products. By opposing this increase in volume, an increase in pressure results in less dissociation and hence a higher flame temperature.

2.15.1.4 Inlet-Air Vitiation

Tests of combustion systems are sometimes carried out using vitiated air, i.e., high inlet-temperature requirements are met through precombustion, which results in abnormally high CO_2 and H_2O concentrations and lower O_2 concentrations. Their combined effect is to lower the maximum flame temperature. If the actual flame temperature is to be simulated, then oxygen must be added to replace that lost in the precombustion process.

Nomenclature

B	mass-transfer number
C	concentration
C_{cf}	collision factor
C_1	D_{10}/D_{32}
C_2	D_{20}/D_{32}
C_3	D_{30}/D_{32}
c_p	specific heat at constant pressure
D	drop diameter
D_{32}	Sauter mean diameter
D_{10}	mean diameter
D_{20}	mean surface diameter
D_{30}	mean volume diameter
d_q	quenching distance
E	activation energy
E_{min}	minimum ignition energy
f	fraction of total fuel in vapor form
k	thermal conductivity
L	integral (large) scale of turbulence
$\dot{m}$	mass flow rate
m	exponent of fuel concentration in Equation 2.40
n	reaction order, or pressure exponent in Equation 2.40
P	pressure
R	universal gas constant
S	burning velocity
S_L	laminar burning velocity
S_T	turbulent burning velocity
T	temperature
ΔT	temperature rise
T_f	flame temperature
T_i	ignition temperature
T_m	mixture temperature
t	time
t_e	evaporation time
t_i	ignition delay time
U	velocity
u'	RMS value of fluctuating velocity
V	volume
x_f	fuel concentration
x_o	oxygen concentration
α	thermal diffusivity ($k/c_p\rho$)
δ	thickness of reaction zone
δ_L	laminar flame thickness

λ	evaporation constant
η	Kolmogoroff scale of turbulence
η_c	combustion efficiency, or fraction of fuel burned
υ	kinematic viscosity
ρ	density
ϕ	equivalence ratio

Subscripts

A	air
ad	adiabatic value
F	fuel
g	gas
o	initial value, or fresh mixture value
st	stoichiometric value, or steady-state value

References

1. Hinde, P. T., "Fundamentals of Combustion," SME Lecture Supplement PL 1076, Cranfield University, Bedford, UK, 1972.
2. Longwell, J. P., and Weiss, M. A., "High Temperature Reaction Rates in Hydrocarbon Combustion," *Journal of Industrial and Engineering Chemistry*, Vol. 47, No. 8, pp. 1634–43, 1955.
3. Longwell, J. P., Frost, E. E., and Weiss, M. A., "Flame Stability in Bluff-Body Recirculation Zones," *Journal of Industrial and Engineering Chemistry*, Vol. 45, No. 8, pp. 1629–33, 1953.
4. Blichner, O., "A Fluid Dynamic Study of a Spherical and a Cylindrical Stirred Reactor," *Eighth Symposium (International) on Combustion*, pp. 995–1002, Williams and Wilkins, Baltimore, MD, 1962.
5. Clarke, A. E., Odgers, J., and Ryan, P., "Further Studies of Combustion Phenomena in a Spherical Combustor," *Eighth Symposium (International) on Combustion*, pp. 982–94, Williams and Wilkins, Baltimore, MD, 1962.
6. Gaydon, A. G., and Wolfhard, H. G., *Flames: Their Structure, Radiation and Properties*, 3rd ed., Chapman and Hall, London, 1970.
7. Dugger, G. L., and Heimel, S., "Flame Speeds of Methane-Air, Propane-Air, and Ethylene-Air Mixtures at Low Initial Temperatures," NACA TN 2624, 1952.
8. Sharma, S. P., Agrawal, D. D., and Gupta, C. P., "The Pressure and Temperature Dependence of Burning Velocity in a Spherical Combustion Bomb," *Eighteenth Symposium (International) on Combustion*, pp. 493–501, The Combustion Institute, Pittsburgh, PA, 1981.
9. Gulder, O. L., "Laminar Burning Velocities of Methanol, Ethanol and Isooctane-Air Mixtures," *Nineteenth Symposium (International) on Combustion*, pp. 275–81, The Combustion Institute, Pittsburgh, PA, 1982.

10. Okajima, S., Iinuma, K., Yamaguchi, Y., and Kumagai, S., "Measurements of Slow Burning Velocities and Their Pressure Dependence Using a Zero-Gravity Method," *Twentieth Symposium (International) on Combustion*, pp. 1951–6, The Combustion Institute, Pittsburgh, PA, 1984.
11. Damkohler G., "The Effects of Turbulence on the Flame Velocity in Gas Mixtures," NACA TM 1112, 1947.
12. Schelkin, K. I., *Soviet Physics – Technical Physics*, Vol. 13, Nos. 9, 10, 1943. English translation, NACA TM 1110, 1947.
13. Ballal, D. R., and Lefebvre, A. H., "The Structure and Propagation of Turbulent Flames," *Proceedings of the Royal Society, London Series A*, Vol. 344, pp. 217–34, 1975.
14. Lefebvre, A. H., and Reid, R., "The Influence of Turbulence on the Structure and Propagation of Enclosed Flames," *Combustion and Flame*, Vol. 10, No. 4, pp. 355–66, 1966.
15. Burgoyne, J. H., and Cohen, L., "The Effect of Drop Size on Flame Propagation in Liquid Aerosols," *Proceedings of the Royal Society, London Series A*, Vol. 225, pp. 375–92, 1954.
16. Cekalin, E. K., "Propagation of Flame in Turbulent Flow of Two-Phase Fuel-Air Mixture," *Eighth Symposium (International) on Combustion*, pp. 1125–29, Williams and Wilkins, Baltimore, MD, 1962.
17. Mizutani, Y., and Nishimoto, T, "Turbulent Flame Velocities in Premixed Sprays: pt. I, Experimental Study," *Combustion Science and Technology*, Vol. 6, pp. 1–10, 1972.
18. Mizutani, Y, and Nakajima, A., "Combustion of Fuel Vapor-Drop-Air Systems: pt. I, Open Burner Flames, pt. II, Spherical Flames in a Vessel," *Combustion and Flame*, Vol. 20, pp. 343–57, 1973.
19. Polymeropoulos, C. E., and Das, S., "The Effect of Droplet Size on the Burning Velocity of Kerosine-Air Sprays," *Combustion and Flame*, Vol. 25, pp. 247–57, 1975.
20. Ballal, D. R., and Lefebvre, A. H., "Flame Propagation in Heterogeneous Mixtures of Fuel Droplets, Fuel Vapor and Air," *Eighteenth Symposium (International) on Combustion*, pp. 321–28, The Combustion Institute, Pittsburgh, PA, 1980.
21. Polymeropoulos, C. E., "Flame Propagation in Aerosols of Fuel Droplets, Fuel Vapor, and Air," *Combustion Science and Technology*, Vol. 40, pp. 217–32, 1984.
22. Myers, G. D., and Lefebvre, A. H., "Flame Propagation in Heterogeneous Mixtures of Fuel Drops and Air," *Combustion and Flame*, Vol. 66, No. 2, pp. 193–210, 1986.
23. Faeth, G. M., "Current Status of Droplet and Liquid Combustion," *Progress in Energy and Combustion Science*, Vol. 3, pp. 191–224, 1977.
24. Godsave, G. A. E., "Studies of the Combustion of Drops in a Fuel Spray—The Burning of Single Drops of Fuel," *Fourth Symposium (International) on Combustion*, pp. 818–830, Williams and Wilkins, Baltimore, MD, 1953.
25. Spalding, D. B., *Some Fundamentals of Combustion*, Academic Press, New York; Butterworths, London, 1955.
26. Kanury, A. M., *Introduction to Combustion Phenomena*, Gordon & Breach, New York, 1975.
27. Lefebvre, A. H., *Atomization and Sprays*, Taylor & Francis, Philadelphia, PA, 1989.
28. Chin, J. S., and Lefebvre, A. H., "Effective Values of Evaporation Constant for Hydrocarbon Fuel Drops," *Proceedings of the 20th Automotive Technology Development Contractor Coordination Meeting*, pp. 325–31, Society of Automotive Engineers, Warrendale, PA, 1982.

29. Peng, F., and Aggarwal, S. K., "A Review of Droplet Dynamics and Vaporization Modeling for Engineering Calculations," ASME Paper 94-GT-215, 1994.
30. Stengele, J., Bauer, H. J., and Wittig, S., "Numerical Study of Bicomponent Droplet Vaporization in a High Pressure Environment," ASME Paper 96-GT-442, 1996.
31. Chin, J. S., "An Engineering Calculation Method for Multi-Component Stagnant Droplet Evaporation with Finite Diffusivity," ASME Paper 94-GT-440, 1994.
32. Chin, J. S., "Advanced Droplet Evaporation Model for Turbine Fuels," AIAA Paper 95-0493, 1995.
33. Chin, J. S., "An Engineering Calculation Method for Turbine Fuel Droplet Evaporation at Critical Conditions with Finite Liquid Diffusivity," AIAA Paper 95-0494, 1995.
34. Lewis, B., and von Elbe, G., *Combustion, Flames and Explosions of Gases*, 2nd ed., Academic Press, New York, 1961.
35. Calcote, H. F., Gregory, C. A., Barnett, C. M., and Gilmer, R. B., "Spark Ignition: Effect of Molecular Structure," *Journal of Industrial and Engineering Chemistry*, Vol. 44, No. 11, pp. 2656–62, 1952.
36. Linnett, J. W., *Discussion, in Selected Combustion Problems,* S. S. Penner and G. Ducarme, eds., Vol. II, p. 139, Butterworth, London, 1956.
37. Swett, C. C, "Spark Ignition of Flowing Gases Using Long-Duration Discharges," *Sixth Symposium (International) on Combustion*, pp. 523–32, Reinhold, New York, 1957.
38. Ballal, D. R., and Lefebvre, A. H., "Ignition and Flame Quenching in Flowing Gaseous Mixtures," *Proceedings of the Royal Society, London Series A*, Vol. 357, No. 1689, pp. 163–81, 1977.
39. Ballal, D. R., and Lefebvre, A. H., "The Influence of Flow Parameters on Minimum Ignition Energy and Quenching Distance," *Fifteenth Symposium (International) on Combustion*, pp. 1473–81, The Combustion Institute, Pittsburgh, PA, 1975.
40. Subba Rao, H. N., and Lefebvre, A. H., "Ignition of Kerosine Fuel Sprays in a Flowing Air Stream," *Combustion Science and Technology*, Vol. 8, pp. 95–100, 1973.
41. Rao, K. V. L., and Lefebvre, A. H., "Minimum Ignition Energies in Flowing Kerosine-Air Mixtures," *Combustion and Flame*, Vol. 27, No. 1, pp. 1–20, 1976.
42. Lefebvre, A. H., "An Evaporation Model for Quenching Distance and Minimum Ignition Energy in Liquid Fuel Sprays," presented at Fall Meeting of the Combustion Institute (Eastern Section), 1977.
43. Ballal, D. R., and Lefebvre, A. H., "Ignition and Flame Quenching of Quiescent Fuel Mists," *Proceedings of the Royal Society, London Series A*, Vol. 364, No. 1717, pp. 277–94, 1978.
44. Ballal, D. R., and Lefebvre, A. H., "Ignition and Flame Quenching of Flowing Heterogeneous Fuel-Air Mixtures," *Combustion and Flame*, Vol. 35, No. 2, pp. 155–68, 1979.
45. Ballal, D. R., and Lefebvre, A. H., "General Model of Spark Ignition for Gaseous and Liquid Fuel/Air Mixtures," *Eighteenth Symposium (International) on Combustion*, pp. 1737–46, The Combustion Institute, Pittsburgh, PA, 1981.
46. Wolfer, H. H., "Der Zundverzug im Dieselmotor," *V.D.I. Forschungsh*, Vol. 392, pp. 15–24, 1938.
47. Mullins, B. P., "Spontaneous Ignition of Liquid Fuels," AGARDograph 4, 1955.
48. Spadaccini, L. J., and Te Velde, J. A., "Autoignition Characteristics of Aircraft-Type Fuels," NASA CR 159886, 1980.

49. Goodger, E. M., and Eissa, A. F. M., "Spontaneous Ignition Research; Review of Experimental Data," *Journal of the Institute of Energy*, Vol. 84, pp. 84–89, 1987.
50. Lundberg, K., Ulstein Turbine Technical Report UTU94003, 1994.
51. Freeman, G., and Lefebvre, A. H., "Spontaneous Ignition Characteristics of Gaseous Hydrocarbon-Air Mixtures," *Combustion and Flame*, Vol. 58, No. 2, pp. 153–62, 1984.
52. Cowell, L. H., and Lefebvre, A. H., "Influence of Pressure on Autoignition Characteristics of Gaseous Hydrocarbon-Air Mixtures," SAE Paper 860068, 1986.
53. Lefebvre, A. H., Freeman, W., and Cowell, L., "Spontaneous Ignition Delay Characteristics of Hydrocarbon Fuel/Air Mixtures," NASA CR-175064, 1986.
54. Ducourneau, F, "Inflammation Spontanee de Melanges Riches Air-Kerosine," *Entropie*, Vol. 10, No. 59, pp. 11–18, 1974.
55. Tacina, R. R., "Autoignition in a Premixing-Prevaporizing Fuel Duct Using Three Different Fuel Injection Systems at Inlet Air Temperatures to 1250K," NASA TM-83938, 1983.
56. Rao, K. V. L., and Lefebvre, A. H., "Spontaneous Ignition Delay Times of Hydrocarbon Fuel-Air Mixtures," *First International Specialist Combustion Symposium*, pp. 325–30, The Combustion Institute, Pittsburgh, PA, 1981.
57. Plee, S. L., and Mellor, A. M., "Review of Flashback Reported in Prevaporizing Premixing Combustors," *Combustion and Flame*, Vol. 32, pp. 193–203, 1978.
58. Kiesewetter, F., Konle, M., and Sattelmayer, T. "Analysis of Combustion Induced Vortex Breakdown Driven Flashback in a Premix Burner with Cylindrical Mixing Zone," *ASME Journal of Engineering for Gas Turbines and Power*, Vol. 129, pp. 929–36, 2007.
59. Konle, M. and Sattelmayer, T., "Time Scale Model for the Prediction of the Onset of Flame Flashback Driven by Combustion Induced Vortex Breakdown (CIVB), " *ASME Journal of Engineering for Gas Turbines and Power*, Vol. 132, 041503, 2010.

Bibliography

Beer, J. M., and Chigier, N. A., *Combustion Aerodynamics*, Wiley, New York, 1972.

Goodger, E. M., *Combustion Calculations*, Macmillan, London, 1987.

Gulder, O. L., "Flame Temperature Estimation of Conventional and Future Jet Fuels," *Journal of Engineering for Gas Turbines and Power*, Vol. 108, pp. 376–80, 1986.

Lewis, B., and von Elbe, G., *Combustion, Flames and Explosions of Gases*, Academic, New York, 1961.

Kuo, K. K., *Principles of Combustion*, John Wiley, New York, 1986.

Peters, J. E., and Hammond, D. C, "Introduction to Combustion for Gas Turbines," in A. M. Mellor, ed., *Design of Modern Gas Turbine Combustors*, pp. 38–79, Academic Press, San Diego, CA, 1990.

Williams, F. A., *Combustion Theory*, Addison-Wesley, Reading, MA, 1965.

3

Diffusers

3.1 Introduction

In axial-flow compressors, the stage pressure rise is very dependent on the axial flow velocity. To achieve the design pressure ratio in the minimum number of stages, a high axial velocity is essential; in many aircraft engines, compressor outlet velocities may reach 170 m/s or higher. It is, of course, impractical to attempt to burn fuels in air flowing at such high velocities. Quite apart from the formidable combustion problems involved, the fundamental pressure loss would be excessive. For example, for an air velocity of 170 m/s and a combustor temperature ratio of 2.5, the pressure loss incurred in combustion would be approximately 25% of the pressure rise achieved in the compressor. Thus, before combustion can proceed, the air velocity must be greatly reduced, usually to about one-fifth of the compressor outlet velocity. This reduction in velocity is accomplished by fitting a diffuser between the compressor outlet and the upstream end of the liner.

In its simplest form, a diffuser is merely a diverging passage in which the flow is decelerated and the reduction in velocity head is converted to a rise in static pressure. The efficiency of this conversion process is of considerable importance because any losses that occur are manifested as a fall in total pressure across the diffuser. In long diffusers of low divergence angle, the pressure loss is high due to skin friction along the walls, as shown in Figure 3.1. Such diffusers are, in any case, impractical because of their extreme length. On all aircraft engines, and also on many industrial engines, length is crucial, and it is essential, therefore, that diffusion is accomplished in the shortest possible distance. With an increase in divergence angle, both diffuser length and friction losses are reduced, but stall losses arising from boundary-layer separation become more significant. Clearly, for any given area ratio, there is an optimum angle of divergence at which the pressure loss is a minimum. Usually this angle lies between 6° and 12°.

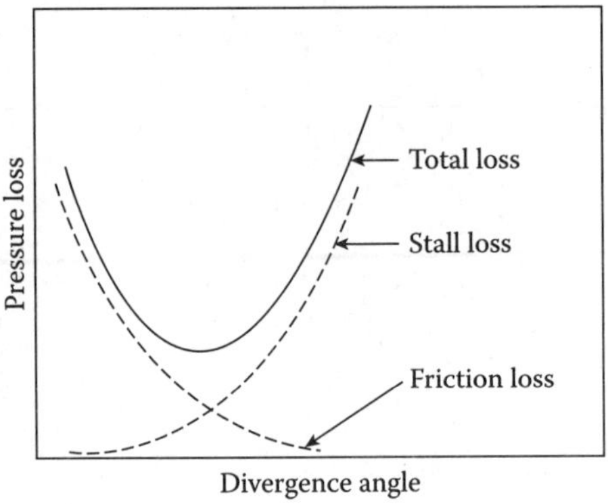

FIGURE 3.1
Influence of divergence angle on pressure loss.

From a designer's viewpoint, an ideal diffuser is one that achieves the required velocity reduction in the shortest possible length, with minimum loss in total pressure, and with uniform and stable flow conditions at its outlet. Sufficient experimental data are now available to design such a diffuser, provided that the inlet velocity profile is symmetrical and not too peaked. Unfortunately, on many engines, the compressor outlet velocity profile is both peaked and asymmetric and is also subject to appreciable variation with changes in engine operating conditions. Under these circumstances, stable flow conditions cannot always be achieved, with the result that some engines are plagued by various deficiencies, such as a lack of consistency in the temperature distribution at the combustor exit and an increase in exhaust gas pollutants.

There is no lack of reliable experimental data on the performance of conventional conical diffusers. Available data on two-dimensional and annular diffusers are less comprehensive, and nearly all these data are summarized in a few important papers. Unfortunately, the performance charts presented in these papers are for boundary-layer-type inlet flows, developed in approach sections, which differ appreciably from the compressor-generated flows encountered in combustor diffusers. Moreover, in comparison with conventional diffusers, there are a number of additional geometric parameters that strongly affect the performance of combustor diffusers, such as the size and shape of the liner and its position relative to the diffuser exit. This complex interaction between the liner and diffuser explains why there are no general performance charts for combustor diffusers comparable to those for conventional diffusers.

At the present time, there is no completely general and accurate method for predicting combustor-diffuser performance. However, much useful progress has been achieved with numerical modeling techniques, which can now successfully predict the gross features of flow fields in combustor diffusers.

3.2 Diffuser Geometry

The geometry of straight-walled diffusers may be defined in terms of three geometric parameters, as shown in Figure 3.2. Area ratio, *AR*, is an obvious choice as a major parameter because it is directly related to the primary function of the diffuser in achieving a prescribed reduction in velocity. Some form of nondimensional length is a logical selection for another because, as pointed out by Sovran and Klomp [1], in combination with the area ratio, such a length defines the overall pressure gradient; the principal factor in boundary-layer development. Usually, either the wall length, *L*, or the axial length, *N*, is used as a characteristic length; it is expressed in nondimensional form by dividing by a representative inlet dimension.

A third parameter is the divergence angle, 2θ, which is not an independent variable, but is related to the other parameters by

$$AR = 1 + 2\frac{L}{W_1}\sin\theta, \tag{3.1}$$

for two-dimensional units, and

$$AR = 1 + 2\frac{L}{R_1}\sin\theta + \left(\frac{L}{R_1}\sin\theta\right)^2, \tag{3.2}$$

for conical units.

Sovran and Klomp [1] recommend the use of $L/\Delta R_1$ as the characteristic dimension for annular diffusers, where *L* is the average wall length, and ΔR_1 is the annulus height at the diffuser inlet. This gives an expression for area ratio that is similar to the expression for conical units when the inlet radius ratio approaches zero, and similar to the expression for two-dimensional units when it approaches unity. Thus, the performance characteristics of all three types of diffuser may be plotted on a single set of coordinate axes as, for example, in Figure 3.3.

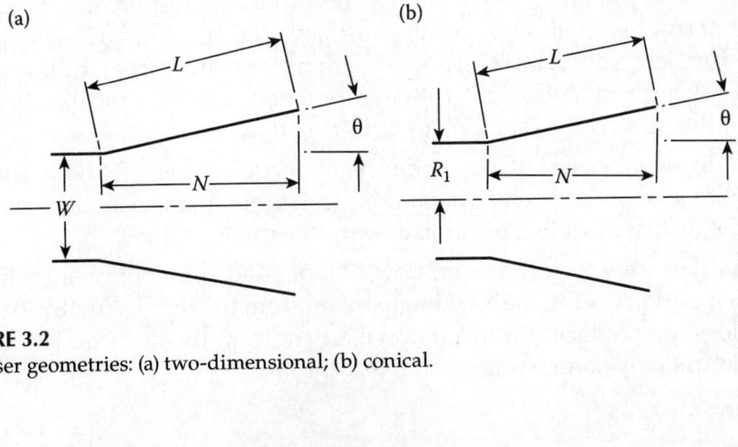

FIGURE 3.2
Diffuser geometries: (a) two-dimensional; (b) conical.

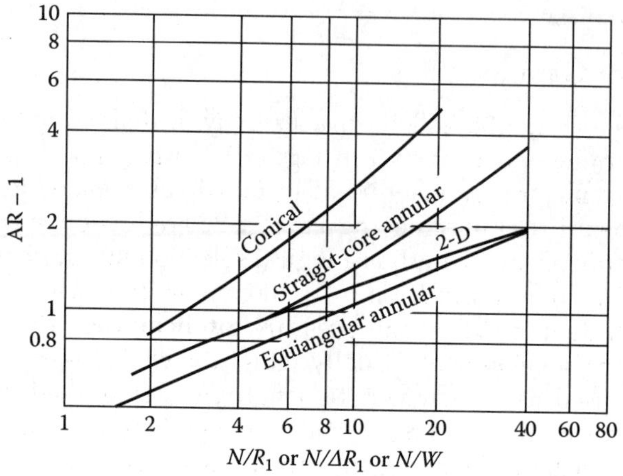

FIGURE 3.3
Lines of first stall. (From Howard, J.H.G., Thornton-Trump, A.B., and Henseler, H.J., "Performance and Flow Regimes for Annular Diffusers," ASME Paper 67-WA/FE-21, 1967. With permission.)

3.3 Flow Regimes

The first systematic study of flow patterns in diffusers was carried out by Kline et al. [2]. Tests were conducted on two-dimensional diffusers with straight walls, and the inlet flow conditions, the wall length, and the throat width were held constant. They observed that as the divergence angle is progressively increased from zero, a number of different flow regimes are found, which can be described as follows:

1. No "appreciable" stall, with the main flow well behaved and apparently unseparated.
2. Transitory stall, whereby eddies are formed that run along the diffuser, some in close proximity to the wall. These eddies assist the diffusion process by transporting lethargic air away from the boundary layer and replacing it with more energetic air from the main core of the flow. This is a region of pulsating flow.
3. Fully developed stall, where the major portion of the diffuser is filled with a large triangular-shaped recirculation region, extending from the diffuser exit to a position close to the diffuser throat.
4. Jet flow, in which the main flow is separated from both walls. The separation begins slightly downstream from the throat, and the flow does not reattach until well downstream from the diffuser. Jet flow occurs only at high angles of divergence.

Howard et al. [3] used wool-tuft techniques to examine the nature of stall in annular diffusers. The lines of occurrence of first appreciable stall are shown in Figure 3.3, where they are compared with data obtained on conical diffusers by McDonald and Fox [4] and on two-dimensional diffusers by Reneau et al. [5]. The superior performance of conical diffusers is explained on the grounds that flow separation is delayed due to the absence of corners [5]. It is also clear from Figure 3.3 that stall may exist in annular diffusers under less severe geometric conditions than in conical diffusers.

3.4 Performance Criteria

The function of a diffuser is to reduce velocity and to convert kinetic energy or dynamic pressure into a rise in static pressure, as shown schematically in Figure 3.4. To assess the efficiency of conversion, it is necessary to define the quantity of available dynamic pressure. This is usually based on a mean velocity, u, which is obtained directly from the continuity equation as

$$u = \dot{m}/\rho_A. \tag{3.3}$$

The dynamic pressure is then obtained as

$$q = \rho u^2/2. \tag{3.4}$$

The pressure loss in the diffuser is defined as

$$\Delta P_{diff} = P_1 - P_2, \tag{3.5}$$

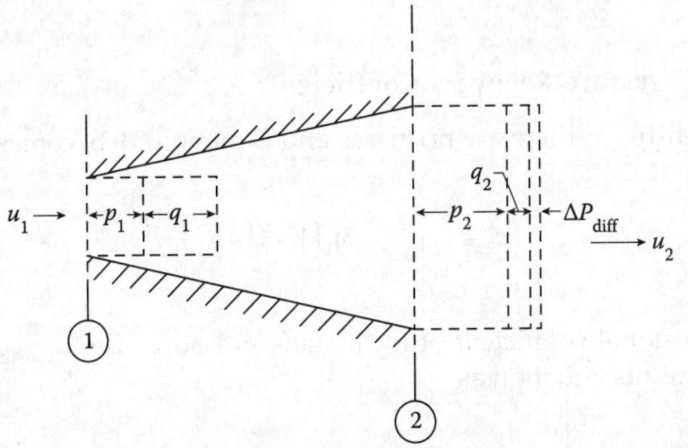

FIGURE 3.4
Energy conversion in a diffuser.

where ΔP_{diff} includes both the internal energy loss and the effects of redistribution of velocity between inlet and outlet.

For one-dimensional incompressible flow we have

Continuity:

$$\dot{m} = \rho A_1 u_1 = \rho A_2 u_2, \tag{3.6}$$

hence

$$A_2/A_1 = u_1/u_2 = AR. \tag{3.7}$$

Bernoulli:

$$p_1 + q_1 = p_2 + q_2 + \Delta P_{diff}. \tag{3.8}$$

From Equations 3.7 and 3.8, the rise in static pressure is given by

$$p_2 - p_1 = q_1[1 - 1/AR^2] - \Delta P_{diff}. \tag{3.9}$$

Several useful parameters for expressing diffuser performance can be derived from this equation.

3.4.1 Pressure-Recovery Coefficient

The pressure-recovery coefficient C_p is calculated as

$$C_p = (p_2 - p_1)/q_1. \tag{3.10}$$

3.4.2 Ideal Pressure-Recovery Coefficient

In an ideal diffuser, there are no losses and Equation 3.9 becomes

$$(p_2 - p_1)_{ideal} = q_1[1 - 1/AR^2]. \tag{3.11}$$

A nondimensional coefficient of ideal static pressure rise, C_{pideal}, is derived directly from this equation as

$$C_{P_{ideal}} = (p_2 - p_1)_{ideal}/q_1 = [1 - 1/AR^2]. \tag{3.12}$$

This equation shows that C_{pideal} is dependent solely on area ratio to which it is related by a law of diminishing returns.

3.4.3 Overall Effectiveness

This is the ratio η of the actual pressure rise to the maximum theoretically obtainable, i.e.,

$$\eta = C_{P_{measured}} / C_{P_{ideal}}, \tag{3.13}$$

or

$$\eta = (p_2 - p_1)/q_1 = [1 - 1/AR^2]. \tag{3.14}$$

Thus, η is related to C_p by the equation

$$\eta = C_p / [1 - 1/AR^2]. \tag{3.15}$$

Typically η varies between 0.5 and 0.9, depending on the geometry and flow conditions [6].

3.4.4 Loss Coefficient

This is usually defined as

$$\lambda = (\bar{P}_1 - \bar{P}_2)/\bar{q}_1, \tag{3.16}$$

where the overbar denotes a mass-flow weighted value derived from a detailed traverse across the duct.

The value of λ depends largely on the type of diffuser employed. Typical values of λ for combustor diffusers range from around 0.15 for "aerodynamically clean" faired diffusers to around 0.45 for dump diffusers of high liner/depth ratio (D_L/h_1) containing a normal complement of support struts and fuel injectors. For vortex-controlled diffusers (VCD), reported values of λ range from 0.05 to 0.15.

Some researchers also measure local values of temperature and static pressure and so are able to calculate local values of density, which are then used to obtain accurate mass flow distributions across the duct. This procedure is not normally justified except when the static pressure variations in a cross section are large, for example, just downstream of the compressor outlet guide vanes.

3.4.5 Kinetic-Energy Coefficient

With nonuniform flows, the kinetic energy flux is greater than it would be for the same flow rate under uniform flow conditions. To take account of this, a velocity profile energy coefficient α is defined as

$$\alpha = \frac{\int \frac{1}{2} u^2 \rho u dA}{\frac{1}{2} \bar{u}^2 \dot{m}}. \tag{3.17}$$

The value of α varies from 1.0 for completely uniform flow, up to around 2.0 for flow on the point of separation. For fully developed turbulent flow, its value is about 1.05.

The kinetic-energy coefficient may be incorporated into Equation 3.8 to give

$$p_1 + \alpha_1 q_1 = p_2 + \alpha_2 q_2 + \Delta P_{\text{diff}}. \tag{3.18}$$

The relevant performance parameters then become

$$C_{P_{\text{ideal}}} = \left[1 - \frac{\alpha_2}{\alpha_1} \cdot \frac{1}{AR^2}\right]. \tag{3.19}$$

$$C_p = \frac{p_2 - p_1}{\alpha_1 \bar{q}_1}. \tag{3.20}$$

$$\eta = \frac{p_2 - p_1}{\bar{q}_1\left(\alpha_1 - \alpha_2/AR^2\right)}. \tag{3.21}$$

$$\lambda = C_{P_{\text{ideal}}} - C_p. \tag{3.22}$$

3.5 Performance

Although considerable progress has been made, no method exists for accurately predicting the quantitative performance of a diffuser with arbitrary shape and flow. In practice, this means that the designer usually resorts to model tests carried out on each particular configuration. However, it is possible to predict with reasonable accuracy the performance of an important range of diffusers that Cocanower et al. [7] have described as "Class A" and that have the following characteristics:

1. The flow is subsonic, but not necessarily incompressible
2. The inlet Reynolds number is greater than 2.5×10^4, so that problems of transition from laminar to turbulent flow are avoided
3. The inlet velocity profile is symmetrical
4. Flow within the diffuser is essentially unstalled
5. The diffuser itself is symmetrical and nonturning
6. The diffuser is either of two-dimensional, conical, or annular geometry

3.5.1 Conical Diffusers

A performance chart for conical diffusers is provided by Figure 3.5, in which the pressure-recovery coefficient is plotted as a function of area ratio, *AR*, and nondimensional length, N/R_1. It was constructed by Sovran and Klomp [1] from the experimental data of Cockrell and Markland [8] for one particular value of boundary-layer thickness.

Figure 3.5 includes two lines that are useful for design purposes. One (line C_p^*) is the locus of points that define the diffuser area ratio producing maximum pressure recovery in a prescribed nondimensional length. The other (line C_p^{**}) is the locus of points that define the nondimensional length producing maximum pressure recovery at a prescribed area ratio. Although the divergence angle varies along the C_p^* line, it is very nearly equal to 6° all along the C_p^{**} line.

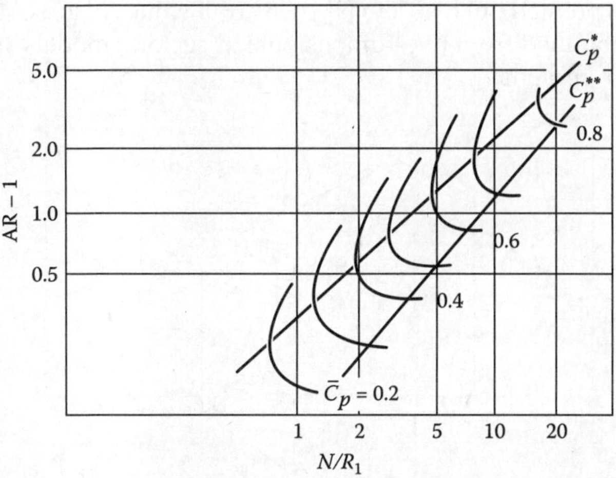

FIGURE 3.5
Performance chart for conical diffusers at 2% inlet blockage. (From Sovran, G. and Klomp, E.D., *Fluid Mechanics of Internal Flow*, Elsevier, Amsterdam, 270–319, 1967. With permission.)

3.5.2 Two-Dimensional Diffusers

Perhaps the best known study on two-dimensional diffusers of the type illustrated in Figure 3.6 is that of Reneau et al. [5], who carried out a series of tests on two-dimensional diffusers with divergence angles 2θ ranging from 5° to 30°, corresponding to variations in N/W_1 from 1.5 to 25. Figure 3.7 shows contour plots of overall effectiveness as a function of area ratio and nondimensional length for an inlet blockage of 1.5%. Similar plots, constructed for other inlet conditions, reveal that effectiveness is diminished by an increase in inlet boundary-layer thickness.

Sovran and Klomp [1] examined all the data obtained by Reneau et al. and made the interesting observation that the location of the C_p^* line is constant and independent of inlet blockage. This fortuitous result means that although pressure recovery is affected by inlet boundary-layer thickness, for a wide range of area ratios the optimum geometry can be chosen without regard for inlet boundary-layer conditions.

Two-dimensional diffusers have been widely used in combustor development because they offer three useful advantages:

1. Flow visualization is easy to apply and helps to identify and avoid regions of flow separation
2. Geometrical changes are relatively simple, which facilitates experimental studies on the influence of geometric parameters on flow patterns and performance
3. Airflow requirements are appreciably less than for a fully annular system

However, as pointed out by Klein [9], measured values of losses and pressure recoveries obtained from two-dimensional or sectoral models are inaccurate for two main reasons:

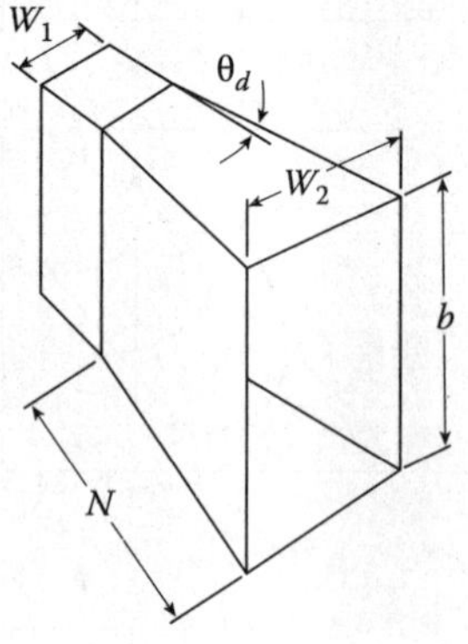

FIGURE 3.6
Two-dimensional diffuser.

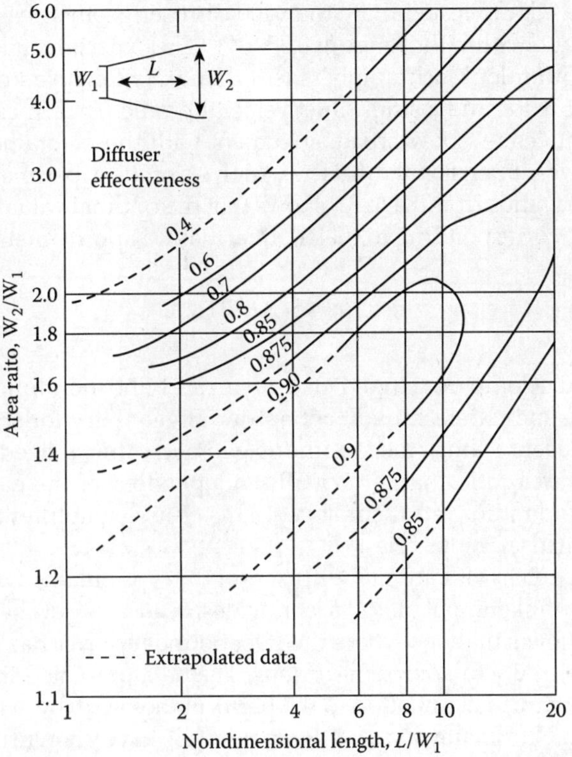

FIGURE 3.7
Two-dimensional-diffuser data at 1.5% inlet blockage. (From Reneau, L.R., Johnston, J.P., and Kline, S.J., *Journal of Basic Engineering* 95, 141–50, 1967. With permission from ASME.)

1. The boundary layers growing along the side walls and the secondary flows created by them have considerable influence on the main flow. Little and Manners [10] have shown that these effects cannot be ignored either in two-dimensional models, even for aspect ratios higher than 10, or in 90° sectors.
2. In two-dimensional models, the radial velocity profile diffuses only in the direction of the flow, whereas in a fully annular model it also diffuses in the circumferential direction. This creates additional mixing and, therefore, higher pressure losses.

3.5.3 Annular Diffusers

Diffusers of the annular type are widely used in aircraft engine combustors. Their characteristics have been studied by Sovran and Klomp [1], Howard et al. [3], Klein [9], Kunz [11], Adkins et al. [12,13], Takehira et al. [14], and Stevens and Williams [15].

Sovran and Klomp tested more than 100 diffuser geometries, all having an inlet radius ratio within the range of 0.55–0.70. The experimental program was conducted with inlet Mach numbers less than 0.30, Reynolds numbers from 4.8×10^5 to 8.5×10^5, and a single inlet velocity profile. Their measurements of diffuser effectiveness show broad agreement with those obtained by Reneau et al. [5] for two-dimensional diffusers. Analysis of the experimental data over a range of area ratios from 1.4 to 3.0 shows that the optimal value of nondimensional length, corresponding to the C_p^* line, may be approximated as

$$L/h_1 = 5(AR - 1). \tag{3.23}$$

Sovran and Klomp concluded that area ratio and nondimensional wall length are the main determinants of optimum geometry for straight-walled annular diffusers. Thus, the optimum geometry for a fixed wall length occurs at an area ratio that is virtually independent of the combination of wall angles and radius ratios employed. This clearly simplifies the treatment of annular-diffuser geometries.

Performance charts for annular diffusers of the type obtained by Sovran and Klomp [1] and Takehira et al. [14] are in widespread use, but their application is limited to flow situations where the inlet boundary layer has been allowed to develop naturally in approach sections. The data presented in these charts tend to be pessimistic in relation to the performance actually achieved on the engine [9]. Lakshminarayana and Reynolds [16] have shown that when an annular diffuser is sited downstream of a compressor rotor, there is a redistribution of radial turbulence energy that delays separation and improves the radial velocity profile. These improvements allow larger area ratios to be used than those of the optimum geometries given by standard performance charts.

3.6 Effect of Inlet Flow Conditions

From the very earliest studies on diffusers, dating back almost a century, it was realized that inlet conditions have an important influence on the subsequent development of flow within the diffuser. Most of these early studies were confined to conventional diffuser geometries, but more recent work has shown that the performance of gas turbine diffusers is also strongly affected by inlet flow conditions. For example, Stevens et al. [17,18] and Klein [19] have demonstrated the influence of the wakes produced by compressor outlet blade vanes on diffuser performance, whereas Lohmann et al. [20] and Carrotte et al. [21] have examined the effects of inlet swirl on flow conditions within and downstream of various prediffuser configurations. In addition, many combustors contain struts, either within the prediffuser or between its exit plane and the liner dome, which are there mainly for structural reasons, but also to accommodate supply lines to the bearing compartment. The presence of these struts in the compressor efflux creates more wakes in the flow.

Another important parameter influencing flow stability and diffuser performance is the radial distribution of velocity at the compressor exit. On many aircraft engines, the compressor outlet velocity is both peaked and asymmetric and, therefore, subject to appreciable variation with changes in aircraft speed and altitude. Radial velocity profiles can also vary between development and production engines and even among engines from the same production line. Usually, the velocity profile at the diffuser inlet peaks toward the outer diffuser wall. In any case, the propensity for separation is enhanced at the low-velocity wall and suppressed at the high-velocity wall [9]. Faired diffusers are especially susceptible to the adverse effects on flow stability arising from variations in inlet radial velocity profile and this is the main reason why they are no longer favored for application to modern engine combustors.

Although the radial velocity profile is of paramount importance, many other inlet flow parameters can have a significant effect on both diffuser and overall combustor performance. They include Reynolds number, Mach number, turbulence, and swirl.

3.6.1 Reynolds Number

The influence of Reynolds number is most pronounced when the inlet boundary layer is not fully developed. Under these conditions, an increase in Reynolds number improves performance by reducing the boundary-layer thickness and increasing the turbulence level. At Reynolds numbers larger than 3×10^5 at the diffuser inlet, the performance of conical diffusers is insensitive to variations in Reynolds number [22]. For annular diffusers, Reynolds number has little or no effect on performance for $Re > 5 \times 10^4$ (based on the hydraulic mean diameter at inlet). Only a limited amount of data is available for dump diffusers, but Hestermann et al. [23] found no effect of Reynolds number on performance for values of Re in the range from 9.2×10^4 to 1.6×10^5 for both small and large dump gaps.

Typical values of Reynolds number in gas turbines are in the order of several millions and turbulence levels are relatively high. Therefore, combustor diffusers are unlikely to be Reynolds-number sensitive. However, as pointed out by Adkins [24], care must be taken during component testing when air densities may be close to normal ambient and Reynolds numbers are much lower than on the engine. Measurements of performance carried out at such conditions could give pessimistic results.

3.6.2 Mach Number

The flow characteristics and performance of diffusers are insensitive to Mach number when it is below around 0.3. Between this value and 0.6, performance can actually improve slightly because much of the pressure recovery occurs nearer the diffuser inlet, thereby providing more opportunity for settling of the flow in the downstream portion of the diffuser. Above a Mach number of 0.6, the pressure gradient near the inlet of a straight-walled diffuser becomes excessive and the performance begins to deteriorate. Large-scale separation can occur when the Mach number approaches 0.7, and the performance falls off dramatically. As combustor diffusers always operate at Mach numbers below 0.4, these compressibility effects have little practical significance.

3.6.3 Turbulence

Moore and Kline [25] were among the first to examine the influence of turbulence on diffuser performance. Their tests on two-dimensional configurations showed that turbulence has little effect on the line of first stall; however, the divergence angle at which fully developed stall occurs is significantly increased, as shown in Figure 3.8. Later work by Hoffmann et al. [26–28] on two-dimensional diffusers confirmed these early findings. Their results show that increasing the turbulence intensity up to 3.5% can improve performance markedly. For a diffuser N/W_1 ratio of 15 and a divergence angle of 20°, an increase in turbulence level raised the pressure recovery coefficient from 0.58 to 0.71.

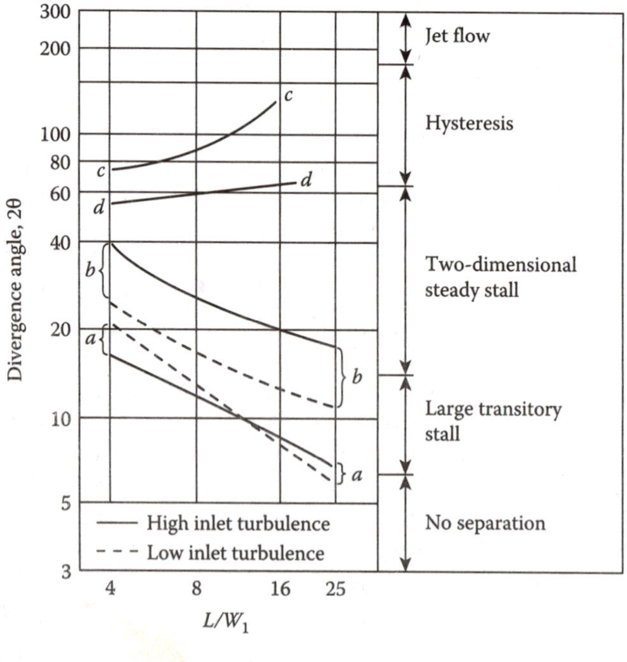

FIGURE 3.8
Regions of two-dimensional plane-wall diffuser flow. (From Moore, C.A. and Kline, S.J., "Some Effects of Vanes and of Turbulence in Two-Dimensional Wide-Angle Subsonic Diffusers," NACA TN 4080, 1958. With permission from NASA.)

Similar performance improvements from raising turbulence levels are also observed in annular combustors. For example, Stevens and Williams [15] investigated the performance of an annular diffuser with constant inner diameter and found that increasing the turbulence intensity improved flow stability at the diffuser exit. Recent work by Hestermann and Wittig [23] has shown that combustor dump diffusers also exhibit this same tendency.

In general, by delaying flow separation in the prediffuser and flattening the velocity profile at exit, an increase in turbulence level enhances prediffuser performance and also improves flow conditions into the inner and outer annuli.

Compressor-generated turbulence seems to be especially beneficial to diffuser performance. It produces improvements that are even larger than those achieved by grid- or spoiler-generated turbulence of the same intensity. This is attributed to the special type of turbulence created by the compressor, which has a large component in the radial direction [16].

3.6.4 Swirl

Some degree of swirl is normally present in the efflux from compressors. It is generally regarded as undesirable due to its deleterious effect on temperature pattern factor at the turbine inlet, although it has been suggested that in wide-angled diffusers, swirl can sometimes prove advantageous by inhibiting flow separation. Comparatively little evidence is available on the effects of swirl on diffuser performance, but Lohmann et al. [20] have shown that inlet swirl angles can increase as the flow is diffused due to the decrease in axial velocity and radius that is experienced by the flow passing to the inner annulus. Quantitative evidence in support of this finding has been provided by Carrotte et al. [21], who carried out an experimental investigation on several modern dump diffuser configurations with inlet conditions being generated by an axial-flow compressor. For all the geometries tested, their results show how the presence of a small amount of inlet swirl (≈3°), typical of that in a gas turbine engine, can result in large swirl angles (≈15°) being generated further downstream, especially in the flow within the inner annulus.

3.7 Design Considerations

3.7.1 Faired Diffusers

Many gas turbine combustors now in service employ a "faired" type of diffuser, in which the main objective is to achieve a gradual reduction in flow velocity without inducing flow separation (stall) and with minimum loss in pressure. A typical faired diffuser for an annular combustor is shown in Figure 3.9. Note in this figure that a section is located at the combustor inlet where the flow velocity is maintained sensibly constant at the compressor outlet value. This section provides a settling length whose primary function is to dissipate the large radial and circumferential flow distortions that exist in the compressor efflux due to the wakes of the compressor outlet guide vanes and to the secondary vortices of the last row of rotor blades. To alleviate the adverse effects of these flow distortions on diffuser and combustor performance, Lefebvre [29] recommends that the settling length should be twice the chord width of the exit blade or vane.

The need for a settling length has not found universal acceptance. Based on the results of his own research, and that of other workers, Klein [9] has argued that a settling length is not required because, although the mixing losses associated with the decay of flow distortions create high losses in the prediffuser, they also generate a fairly uniform velocity profile at the diffuser outlet, which results in lower losses and more stable flow conditions in the regions downstream. According to Klein, inlet flow distortions decay more rapidly when their level is high. Thus, a settling length offers no advantage because the resulting weaker flow distortions entering the prediffuser will decay at a slower rate.

These observations of Klein are of considerable practical interest and importance, but they are relevant only to flow in a single channel. If the flow is divided, either by vanes or the combustor snout, instabilities could arise due to unequal diffusion down the separate channels. Further information on the role of the settling length may be found in Klein [9].

A faired diffuser incorporates at least three separate diffusion regions, as illustrated in Figure 3.9. The first region is located immediately downstream of the settling length (if any), and its purpose is to achieve some reduction in velocity, typically around 35%, before the air reaches the "snout," which is formed by two cowlings that extend upstream from the liner dome. At the snout, most of the air is subjected to a change in direction as it enters the diffusing passages, which are created in the space between the cowlings and the combustor casing. It is important that this change in flow direction be carried

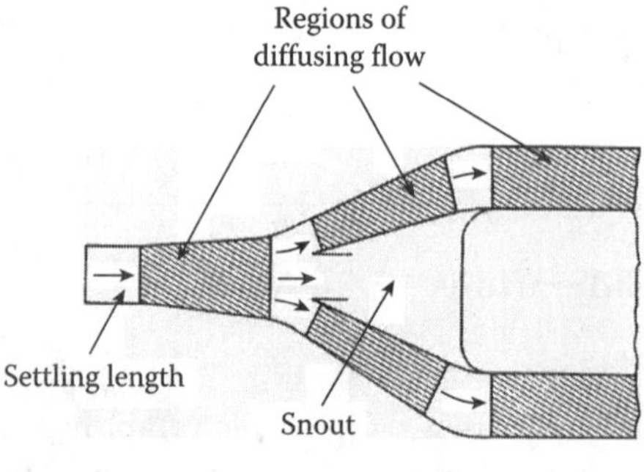

FIGURE 3.9
Faired diffuser.

out at constant velocity to reduce the possibility of breakaway at this point. Some rounding of the snout lip can be helpful, especially if the compressor-outlet velocity profile is subject to variation with changes in engine operating conditions. However, care must be taken to ensure that it creates no sudden local increase in divergence angle; otherwise, flow separation may be initiated. Downstream of the flow split produced at the snout is a second diffusion region. From this region, air flows around a bend at constant velocity into the third diffusion region, which is formed by the annular space between the liner and the casing. Beyond this region, further diffusion takes place as more air flows into the liner, thereby reducing the flow velocity in the annulus.

Many faired diffusers incorporate a fourth diffusion region that affects only the center portion of the air entering the snout. In early combustor designs, the snout air comprised around 10% of the total combustor airflow. Today, due to the large and increasing air requirements of airblast atomizers and other fuel preparation devices for low pollutant emissions, this figure is appreciably higher. It is customary to fit a short annular diffuser at the entry to the snout, as illustrated in Figure 3.10. This allows the snout air to be further decelerated before being "dumped" into the dome region. The primary purpose of this additional diffusion zone is to provide a higher and more uniform pressure in the air supply to the atomizers, air swirlers, and dome-cooling slots.

Clearly, a basic requirement of faired diffusers is that all diffusion regions must perform without stall. Lines of first stall for conventional conical, two-dimensional, and annular configurations are shown in Figure 3.3. Pressure losses in each diffuser section may be calculated using performance charts of the type presented in References [1–9,30,31] and illustrated in Figures 3.5 and 3.7.

The faired diffuser comprises a number of subdiffusers whose passage heights are quite small. In consequence, their dimensions, and hence

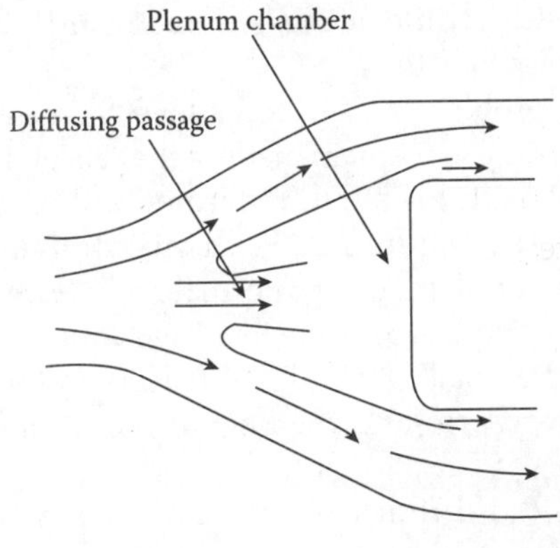

FIGURE 3.10
Illustration of pressure-balancing slots.

also their effective area ratios, are especially subject to manufacturing variations, differential thermal expansion between the liner and casing, and the thermal distortion that occurs with changes in engine operating conditions. Another drawback is that the distribution of air between the inner and outer annuli, which governs the flow distribution throughout the liner, is very sensitive to variations in inlet velocity profile. The effect is minor when the inlet velocity is low, because the fraction of total pressure that is tied up in kinetic energy is so small that the static pressure drop across the combustor effectively controls its airflow distribution. In consequence, the airflow pattern tends to remain constant, regardless of variations in inlet velocity profile. However, when the inlet velocity is high, a significant proportion of the total pressure is in the form of dynamic pressure, and the static pressure drop across the combustor is correspondingly lower. Under these conditions, it is the distribution of dynamic pressure in the inlet airstream that controls the airflow distribution throughout the combustor. For a flat or symmetrical velocity profile, this would present no special problems because the extra velocity could readily be accommodated by an increase in diffuser pressure loss. Unfortunately, an increase in compressor outlet velocity is almost invariably accompanied by a deterioration in velocity profile, to the extent that often makes it virtually impossible to achieve a balanced airflow distribution inside the liner.

To balance the aerodynamic flow pattern within the liner, it is necessary to attain symmetry in terms of the mass flows of the opposing jets, the penetration of these jets, and their momentum. If the air entering the combustion chamber has a flat velocity profile, it is possible to achieve matching, or symmetry, on all three counts. However, if the velocity profile is distorted, i.e., peaked either to the outside or the inside of the diffuser centerline, then one can achieve symmetry only with two of the three parameters. The designer can choose which two to balance by suitable alterations to passage areas and by modifications to liner hole sizes, but the fact remains that a distorted velocity profile at the combustor inlet always results in an unbalanced liner airflow pattern.

To some extent, the problems described above can be alleviated by arranging for the snout diffuser to accept an airflow that exceeds its normal requirements. The excess air then flows into the inner and outer passages through short slots, as illustrated in Figure 3.10. With this configuration, the snout acts as a plenum chamber, supplementing the airflow in both liner passages—but especially in whichever passage is deficient in air due to a shift in inlet velocity profile. Another drawback to faired diffusers is the risk of air leakage between the burner feed arm and the hole in the combustor snout through which it is inserted.

In summary, the faired diffuser has the great advantage of low pressure loss, some one-third less than that of the dump diffuser. This important attribute is, however, more than offset by the following drawbacks that virtually prohibit its application to modern aircraft engines:

1. Excessive length
2. Performance and flow stability are very sensitive to changes in inlet velocity profile
3. Performance is very susceptible to thermal distortions and manufacturing tolerances

3.7.2 Dump Diffusers

The annular dump diffuser concept is illustrated in Figure 3.11. The unit consists of a short conventional diffuser in which the air velocity is reduced to about 60% of its initial value. In complete disregard for conventional diffuser design principles, the air is then "dumped" and left to divide and flow around the liner dome before entering the two annuli that surround the liner. The standing vortices created by projection of the prediffuser walls into the dump region help to maintain a uniform and stable division of flow around the liner. Due to the sudden expansion at the prediffuser outlet, the dump configuration has an inherently higher pressure loss than the faired diffuser, typically by around 50%. However, this penalty is more than compensated by substantial savings in length and weight. These attributes are especially advantageous for aircraft applications. The dump diffuser has the further important advantage of producing a stable flow pattern, which is fairly insensitive to manufacturing tolerances, differential thermal expansions between the liner and combustor casing, and variations in inlet velocity profile.

In an early study on dump diffusers, Fishenden and Stevens [32] examined the influence on performance and stability of diffuser geometry, distance

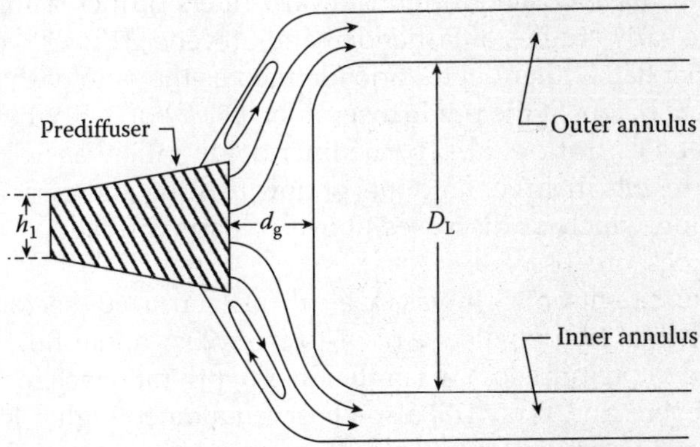

FIGURE 3.11
Dump diffuser.

between prediffuser outlet and liner dome (dump gap), and the division of flow between the inner and outer annuli. They found that nearly all the static pressure rise occurs in the prediffuser, and most of the loss in total pressure occurs in the dump and settling regions. It was also found that the presence of the liner has a beneficial effect on the performance and stability of the flow in the prediffuser. In particular, it tends to suppress flow separation, which allows divergence angles and area ratios to be made larger than those of conventional annular diffusers of the same relative length. Hestermann et al. [23] also found that flow separation in the prediffuser could be suppressed by reducing the dump gap. The larger the divergence angle, the smaller the dump gap must be in order to prevent flow separation. For a divergence angle of 22°, they found that flow separation could only be suppressed by reducing d_g/h_1 to below unity. However, as Carrotte et al. [21] have shown, if the dump gap is made too small, the overall dump diffuser pressure loss will increase due to excessive local flow acceleration and an increase in flow turning and curvature around the liner dome. These considerations suggest that for any given diffuser configuration there will be a certain value of dump gap ratio (d_g/h_1), depending on the amount of prediffusion and the liner depth ratio, at which the overall diffuser pressure loss is a minimum. This is confirmed in the experimental data of Honami and Marioka [33]. These data also show that the optimum value of d_g/h_1 decreases with an increase in the prediffuser divergence angle.

3.7.2.1 Influence of Liner Depth Ratio

For aircraft applications, the constant desire to reduce engine length and weight has increased the demand for shorter length diffusers. At the same time, the requirements of lower lean blowout limits and better relight capability at high altitudes, along with the continuing pressure to reduce pollutant emissions, have created a trend toward liners of increasing depth. For example, the early studies of Fishenden and Stevens [32] were conducted on a diffuser that had a liner depth ratio (defined as the ratio of the liner depth to the passage height at the prediffuser inlet, D_L/h_1) of 3–5, whereas in more recent work by Carrotte et al. [21], the liner depth ratio was 5.5. An increase in this ratio results in larger amounts of flow turning and curvature around the liner dome, which, as discussed above, impairs diffuser performance by raising the pressure loss.

The measurements of Srinivasan et al. [34] on a 60°-sector dump diffuser model that incorporated a cowl to pass 20% dome flow indicated a 60% increase in pressure loss when the liner depth ratio was increased from 3.1 to 4.1. Stevens and Wray [18] also observed a much higher loss than had been measured in previous work by Fishenden and Stevens [32], which they attributed to an increase in liner depth ratio from 3.5 to 5.5.

In an attempt to clarify the influence of liner depth on diffuser performance, Klein [9] examined measured values of loss coefficient, λ, from several sources, all relating to dump diffusers of similar design and the same dump gap ratio of unity. The results of this study indicate a general increase in λ of around 60% as the liner depth ratio is raised from 3.5 to 5.5. It is clear, therefore, that liner depth ratio should be regarded as an important geometric parameter affecting diffuser performance.

3.7.3 Splitter Vanes

The notion of using splitter vanes to create a multiple passage diffuser of large total divergence angle is by no means new, and was demonstrated successfully by Cochran and Kline [35] as early as 1958. In their design, the vanes were arranged symmetrically, as shown in Figure 3.12, and their length and divergence angle were chosen so that each vane passage operated near the line of first stall. By this means, near-optimum pressure recoveries and unstalled flow were obtained for total divergence angles of up to 42°. The vanes were also found to improve the outlet velocity profile, making it more uniform. Subsequent tests by Miller [36] on three-vane diffusers demonstrated satisfactory performance up to a total divergence angle of 50°.

The combustor prediffuser of the NASA/GE "Energy Efficient Engine" employs a single splitter vane to achieve the desired large area ratio of 1.8 and to direct the flow toward the two dome regions as well as to the inner and outer annuli. The flow split between the two diffusing passages is such that the inner duct passes 52% of the total air mass flow and the outer duct 48%. The presence of the splitter vane allows a length reduction of 50% relative to a single annular diffuser of the same area ratio. The overall pressure loss coefficient for the entire combustor inlet is 0.30, which corresponds to 35% of the total combustor pressure loss [9,37].

The General Electric LM6000 dry low-emissions combustor, shown in Figure 9.30, requires almost twice the volume of the conventional combustor it replaces in order to meet its emissions goals. The use of three splitter vanes

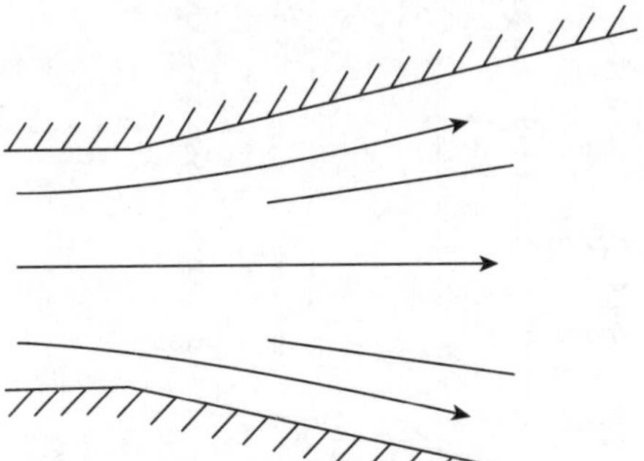

FIGURE 3.12
Diffuser incorporating splitter vanes.

to create four annular diffuser passages allows this large increase in volume to be achieved without any increase in overall combustor length.

3.7.4 Vortex-Controlled Diffuser

The notion of achieving rapid and efficient diffusion by boundary-layer bleed at the throat of a sudden expansion has been studied by Adkins [38]. A tubular version of his VCD is sketched in Figure 3.13. The basic mechanism of vortex control is not yet fully understood, but Adkins has proposed a model that is best described by reference to Figure 3.14. According to this model, the application of external suction causes the static pressure inside the vortex chamber to fall below that of the main stream. In consequence, stream *a*, which is being drawn into the vortex, experiences considerable acceleration. On the other hand, stream *b*, which flows down the diffuser, is flowing into a region of greater static pressure and therefore decelerates. A shearing action, produced by the velocity differential between the streams, then results in the creation of an extremely turbulent layer that inhibits flow separation.

The experiments indicate that diffusion is achieved in a very short length (clearly advantageous for aircraft applications) and the effectiveness is almost

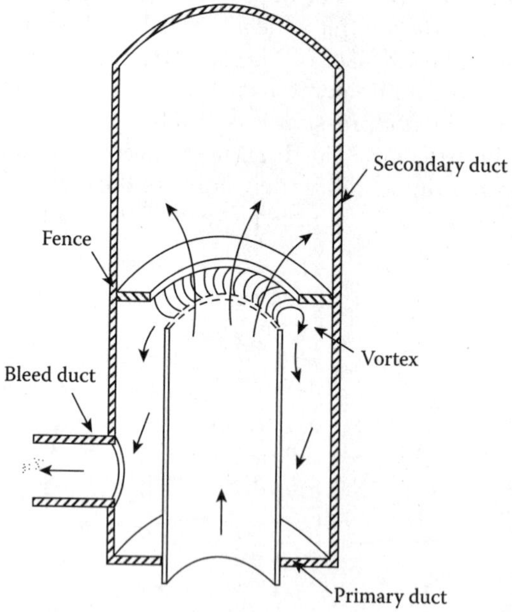

FIGURE 3.13
Vortex-controlled diffuser of tubular configuration. (From Adkins, R.C., *Journal of Fluids Engineering*, 297–302, 1975. With permission.)

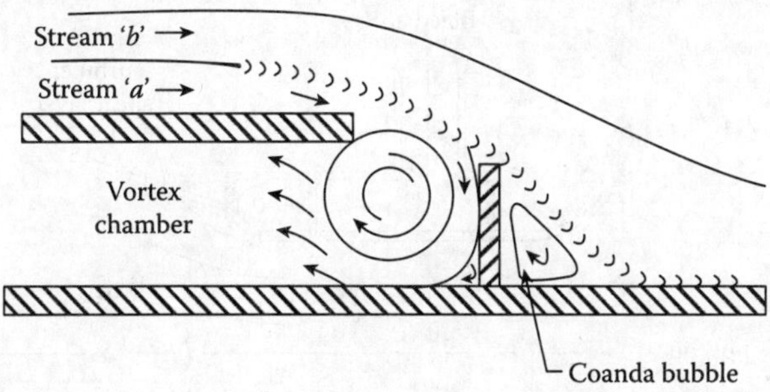

FIGURE 3.14
Flow mechanisms of vortex control. (From Adkins, R.C., Matharu, D.S., and Yost, J.D., *Journal of Engineering for Power*, 103, 229–36, 1981. With permission.)

equal to the theoretical optimal value. Tests have also confirmed that the technique can be applied to conical, annular, and two-dimensional in configurations. The air-bleed requirements depend mainly on area ratio; most combustor applications, the bleed amounts to about 4% of the mainstream flow. As with most diffusers, under some conditions the flow can become unstable. Adkins and Yost [39] have described the instabilities that occur when the optimum area ratio is exceeded and have derived empirical equations for predicting the optimum area ratio, the level of pressure recovery, and the bleed-air requirement, all of which are governed by the degree of nonuniformity of the inlet velocity profile.

Some features of the VCD are not yet fully resolved, such as the design of the suction slot and the location of the vortex retaining fence. Also, the trade-off between suction rate and diffuser length, as it affects static-pressure rise, total-pressure loss, and exit flow stability, warrants further study. However, sufficient work has already been done to demonstrate its considerable potential for application to gas turbines, and especially to high-temperature engines in which bleed air from the combustor is required to cool the hot sections downstream. It offers significant savings in engine length and weight, combined with an increase in available liner pressure drop, which could improve almost all aspects of combustion performance.

3.7.5 Hybrid Diffuser

The main drawback of the VCD is its high bleed-off requirement. To surmount this problem, Adkins et al. [39–41] developed a hybrid concept that combines a VCD with a conventional wide-angled postdiffuser located at its exit. Hybrid systems of this type have been demonstrated successfully by Juhasz and Smith [42,43] and Verdouw [44]. Figure 3.15 shows a hybrid concept of Adkins et al. [40]. In this design, the vortex-controlled step accounts for only a small increase in cross-sectional area, thus requiring only a minimal bleed-off. The turbulent layer generated by the step is then used to

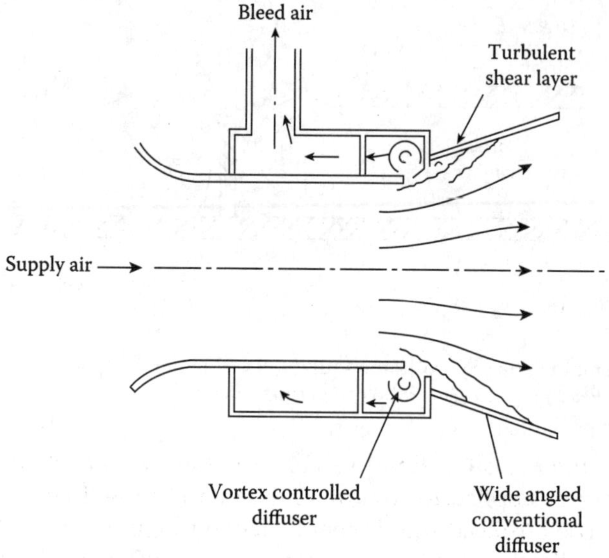

FIGURE 3.15
Hybrid diffuser arrangement. (From Adkins, R.C., Matharu, D.S., and Yost, J.D., *Journal of Engineering for Power*, 103, 229–36, 1981. With permission.)

inhibit flow separation from a relatively wide-angled conventional diffuser that contains a much greater part of the overall increase in area. The results obtained for an overall area ratio of 2 and an inlet Mach number of 0.25 are shown in Figure 3.16. This figure also contains corresponding performance data for a conventional conical diffuser of the same area ratio. The superior performance of the hybrid arrangement is very evident from this figure.

Another impressive feature of the hybrid diffuser is its ability to yield high pressure recovery, even without the application of bleed. This is demonstrated in Figure 3.16 where, at zero bleed, the hybrid diffuser is shown to produce pressure recoveries far greater than the equivalent conventional conical diffuser. Thus, the benefits of the new diffuser can be demonstrated either as an increase in pressure recovery of at least 25% over that of a conventional diffuser of the same length, or as the attainment of the same recovery in a much shorter length. For example, to achieve a value for C_p of 0.52, a conventional diffuser needs twice the length of a hybrid diffuser.

The ability to achieve high pressure recoveries in a short length with small amounts of bleed would appear to make the hybrid diffuser very attractive for gas turbine applications. Unfortunately, it cannot be used on most engines because the pressure of the bleed air is too low for turbine blade cooling.

A solution to this problem is to fit a short prediffuser to the upstream end of the VCD, as shown in Figure 3.17. Only a small area ratio of around 1.3–1.4 is needed to raise the bleed air pressure to a level that is sufficient for turbine cooling [44,45]. The main diffusion takes place downstream in either a VCD

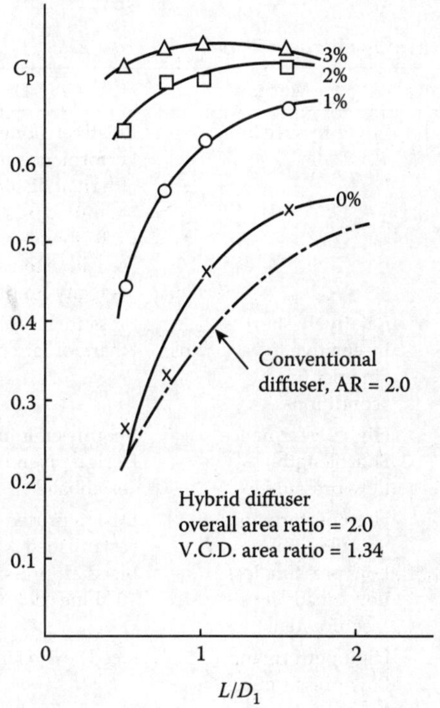

FIGURE 3.16
Comparison of hybrid and conventional diffuser performance. (From Adkins, R.C., Matharu, D.S., and Yost, J.D., *Journal of Engineering for Power*, 103, 229–36, 1981. With permission.)

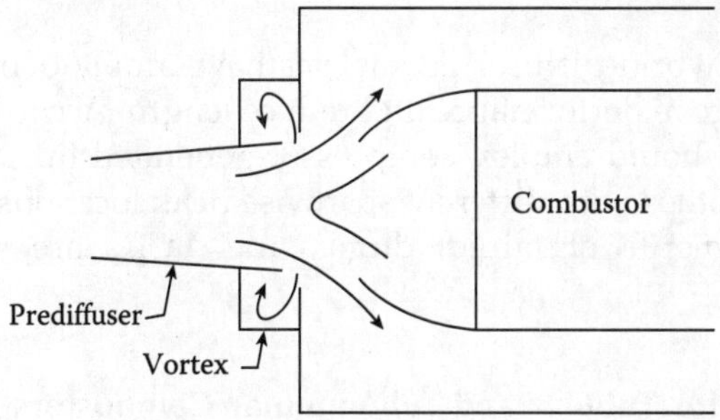

FIGURE 3.17
VCD fitted with prediffuser. (From Adkins, R.C., "Tests on a Vortex-Controlled Diffuser Combined with a Pre-Diffuser and Simulated Combustor," Cranfield SME Report No. C1331-D2, Cranfield University, November 1975.)

or hybrid diffuser. The main drawback to a prediffuser is the additional length required.

The results obtained from various VCD and hybrid diffusers, with and without prediffusers, have been reviewed and summarized by Klein [9]. More detailed performance data may be found in the publications of Adkins et al. [39–41].

TABLE 3.1
Relative Merits of Various Diffuser Types

Diffuser Type	Merits	Drawbacks
Aerodynamic or faired	Low pressure loss	Relatively long Performance susceptible to thermal distortion and manufacturing tolerances Performance and stability sensitive to variations in inlet velocity profile
Dump	Relatively short Insensitive to variations in inlet flow conditions	Pressure loss about 50% higher than for faired type
Vortex-controlled	High performance Short length Low pressure loss	Requires minimum of 4% air bleed Design procedures not fully established
Hybrid	High performance Short length Low pressure loss Low bleed air requirement	Design procedures not fully established Bleed air pressure too low for turbine cooling
Hybrid with prediffuser	High performance Low pressure loss Low bleed air requirement High bleed air pressure	Needs extra length

In summary, hybrid-diffuser geometries have amply demonstrated their potential to improve performance and reduce length. According to Klein [9], future research should employ compressor-generated inlet conditions and inlet velocity profiles with different spanwise peak locations.

The relative merits of faired, dump, and VCDs are summarized in Table 3.1.

3.7.6 Diffusers for Tubular and Tuboannular Combustors

With tubular combustors, the diffuser also serves as a transition piece, converting from an annular segment at the compressor outlet to a circular section at the combustor casing. The geometry is almost always complex, and pressure losses tend to be high. Normally, a hollow snout is fitted to the liner dome. The snout should be designed to ensure smooth deceleration of the flow into the annulus.

For tuboannular (or can-annular) combustors, the normal arrangement includes a settling length that terminates in a conventional annular diffuser of modest area ratio, say about 1.5. This is followed by a dump chamber if length is at a premium, or by a snout-combustor configuration if further diffusion is required to minimize the loss in total pressure.

Potentially, the tuboannular configuration has many advantages over the annular one. According to Adkins and Binks [46] they include:

1. A superior aerodynamic flow pattern inside the combustor liner. Here, the air jets flow radially through the liner wall to meet the radially expanding fuel spray, thereby ensuring good matching between the air and fuel flows. As a result, the combustion process is more homogeneous, giving smaller quantities of exhaust gas pollutants and a more uniform temperature distribution, both around the combustor liner and at turbine inlet. It follows that cooling air can be used more efficiently.
2. A smaller diameter liner gives improved mechanical rigidity without the need for expensive cooling devices, such as machined rings. Stress levels are lower, and these allow either a higher liner temperature or a longer life.
3. Combustor development can be accomplished by tests on single cans, thereby using only a fraction of the total engine airflow.

In practice, these potential advantages are rarely achieved because of the unsatisfactory condition of the flow between the combustor casing and the liners. The large diameter of the liners, relative to the duct height at the compressor exit, dictates that the divergence angle of the diffuser greatly exceeds the maximum permissible angle for good aerodynamic stability. This causes the flow to adhere to only one of the diffuser walls—usually the outer wall—with the result that most of the compressor efflux is directed to the radially outboard regions of the combustor. Some of this air flows radially inward through the spaces between liners and then flows upstream along the inner casing to create a large recirculatory flow pattern. The end result is that some of the air entering the liner is supplied from the normal downstream direction, whereas the remaining air is supplied from the reverse direction. This destroys the symmetry of the liner airflow distribution, which leads to excessive exhaust smoke and other problems associated with poor combustion performance.

A method for greatly improving the aerodynamics of tuboannular combustors by the use of a novel diffusion system termed the "wedge diffuser" has been described by Adkins and Binks [46].

3.7.7 Testing of Diffusers

It is important that the initial diffuser design be tested while the engine is still at an early stage of development. Water flow-visualization rigs are ideally suited for this purpose and, when properly used, will clearly reveal any irregularities in the flow. It is important to simulate the compressor discharge conditions as closely as possible, in terms of velocity profile and angle of swirl. To this end, it is helpful to incorporate a set of compressor-outlet guide vanes at the diffuser inlet. The presence of swirl in the compressor efflux precludes the use of sectors for most test purposes, which means that flow visualization studies on diffusers for annular combustors require full-scale, fully annular models.

As discussed below, computational fluid dynamics (CFD) simulations can also provide valuable guidance in the diffuser design and development stages.

3.8 Numerical Simulations

Adkins et al. [12,13] have developed relatively simple calculation methods for designing optimum annular diffusers. Potential flow computations have also been used successfully; one example being in the design of a two-passage diffuser for the NASA/GE "Energy Efficient Engine." The main advantage of these methods is simplicity, but they are not suited to the complex geometries and flow conditions of some modern combustor diffusers.

Advances in computer technology have led to the increasingly widespread use of CFD for calculating flow fields throughout the combustor, including the diffuser. Diffuser calculations do not have the problems of two-phase flows and chemical reactions, but they do involve regions of adverse pressure gradients, developing boundary layers, flow recirculation, and strong streamline curvature. Other complications arise from the complex geometry of diffusers and the presence in the flow of burner feed arms and liner support struts.

Klein [9] has reviewed the merits and drawbacks of various CFD simulations as published in the literature, most of which are based on the k-ε model of turbulence. Shyy [47,48] has also compared various numerical schemes employed in a CFD application to a dump diffuser configuration. Special importance is attached to the generation of the computational grid. The orthogonal Cartesian or cylindrical grids used in the simulation of flows in simple geometries cannot be applied to combustor diffusers because they cannot predict the flow behavior near the walls. This means that a boundary-fitted curvilinear, non-orthogonal coordinate system must be used. Also, the grid distribution must include very fine meshes in regions where large gradients of the flow properties normal to the flow direction could exist. Any available experimental evidence on the flow fields in such regions could clearly provide useful guidance in the selection of mesh shape and the number of grid nodes [47–49].

The k-ε model is well established, relatively cheap, and easy to use. Its drawback in diffuser applications is that it predicts the flow to remain attached in situations where experiments indicate separation [50]. Another shortcoming of the model arises in flows containing high strain rates produced by strong curvature. Their effects are known to be poorly predicted by eddy viscosity models [51]. An alternative to the use of such models is to derive transport equations for the Reynolds stresses themselves. This approach adds appreciably to the cost and complexity of the computations, but it is far superior to the k-ε model for predicting flows in regions of strong curvature.

Jones and Manners [52] used the k-ε model to compute the flow in a faired diffuser that had previously been subjected to extensive experimental study by Stevens et al. [53,54]. The model was found to overpredict the velocity near the concave wall. Velocity profiles were poorly predicted in many regions of the flow. The overall pressure recovery was also grossly overpredicted. These discrepancies between computation and experiment were attributed to deficiencies of the k-ε model. Many other workers, including Shyy [48], Koutmos and McGuirk [49], Mayer and Kneeling [55], and Ando et al. [56], have tested the accuracy of the k-ε model by comparing static pressure recoveries and/or velocity profiles for a number of different geometries, including split and dump diffusers. The k-ε model gave erroneous results in all cases.

Despite their shortcomings, CFD methods have been shown to predict the gross features of diffuser flows quite well. This is illustrated in Figure 3.18

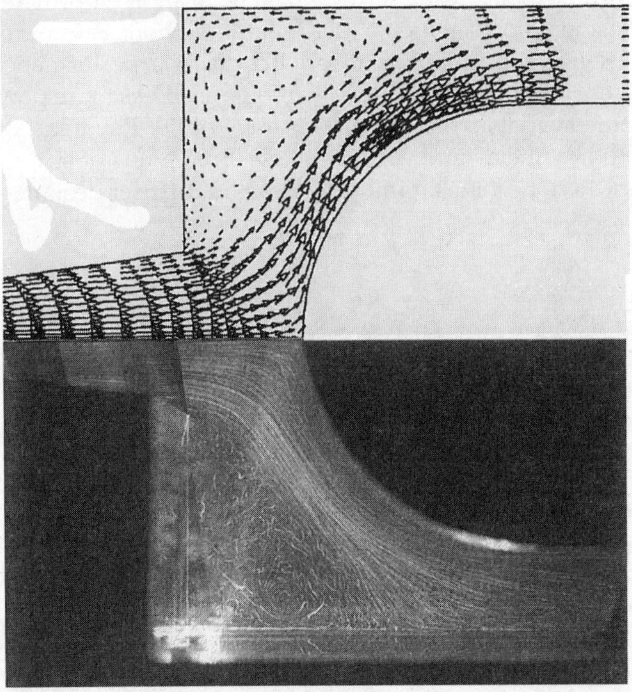

FIGURE 3.18
Comparison of computed and visualized flow fields in a dump diffuser. (From Hestermann, R., Kim, S., Ben Khaled, A., and Witting, S., *Journal of Engineering for Gas Turbines and Power*, 117(4), 686–94, 1995. With permission.)

from Reference [23], which shows a comparison between computed (k-ε model) and visualized flow fields in a two-dimensional dump diffuser. Flow visualization was achieved using a hydrogen bubble technique on a water table. The level of agreement between the flow-visualization method and the numerical simulation, as illustrated in Figure 3.18, is clearly very satisfactory. Hestermann et al. [23] and Zhiben and Guangshi [57] obtained good agreement between computed and measured velocity profiles at the exits of annular and two-dimensional prediffusers, respectively. Furthermore, Karki et al. [58], using the high Reynolds number form of the k-ε model, obtained useful insight into the highly three-dimensional nature of combustor diffuser flows. Their results showed significant asymmetric effects in the flow due to the presence of support struts and fuel nozzles.

The conclusion to be drawn from these various studies is that the k-ε model is incapable of predicting pressure recoveries with any degree of accuracy, but can provide a good representation of velocity profiles in the absence of strong curvature.

Many other examples to illustrate the value of numerical methods in predicting the gross features of diffuser flow fields have been presented by Klein [9]. Developments in CFD methods have now reached a stage at which they are an important tool in diffuser design. They can even provide details of three-dimensional flows, which experimental methods cannot reveal. Their main limitation is their inability to predict diffuser performance, which is due to deficiencies in the k-ε models currently used. As more powerful computers become available, the greater accuracy of the Reynolds stress model in predicting flow fields and pressure losses in regions of strong curvature should enable it to become an important tool in diffuser design.

Nomenclature

A	geometrical area
AR	area ratio (A_2/A_1)
C_p	pressure recovery coefficient
C_p^*	maximum pressure recovery coefficient in a prescribed length
C_p^{**}	maximum pressure recovery coefficient in a prescribed area ratio
D_L	liner depth
D_L/h_1	liner depth ratio
d_g	dump gap width
d_g/h_1	dump gap ratio
h	annulus height
L	wall length
$\dot{m}$	mass flow rate

N axial length
P total pressure
ΔP loss in total pressure across diffuser
p static pressure
q dynamic pressure ($\rho u^2/2$)
R radius, conical diffusers
R_o outer radius
R_i inner radius
ΔR annulus height (R_o–R_i)
u velocity
W width (distance between divergent walls) of two-dimensional diffusers
ρ density
η overall effectiveness
θ half-divergence angle
λ loss coefficient

Subscripts

1. diffuser inlet plane
2. diffuser outlet plane

Other

— mean value, mass-flow derived

References

1. Sovran, G., and Klomp, E. D., "Experimentally Determined Optimum Geometries for Rectilinear Diffusers with Rectangular, Conical or Annular Cross Section," in G. Sovran, ed., *Fluid Mechanics of Internal Flow*, Elsevier, Amsterdam, pp. 270–319, 1967.
2. Kline, S. J., Abbott, D. E., and Fox, R. W., "Optimum Design of Straight-Walled Diffusers," *Journal of Basic Engineering*, Vol. 81, pp. 321–31, 1959.
3. Howard, J. H. G., Thornton-Trump, A. B., and Henseler, H. J., "Performance and Flow Regimes for Annular Diffusers," ASME Paper 67-WA/FE-21, 1967.
4. McDonald, A. T., and Fox, R. W., "An Experimental Investigation of Incompressible Flow in Conical Diffusers," *International Journal of Mechanical Science*, Vol. 8, pp. 125–39, 1966.
5. Reneau, L. R., Johnston, J. P., and Kline, S. J., "Performance and Design of Straight, Two-Dimensional Diffusers," *Journal of Basic Engineering*, Vol. 95, pp. 141–50, 1967.
6. Cockrell, D. J., and King, A. L., "A Review of the Literature on Subsonic Fluid Flow through Diffusers," The British Hydromechanics Research Association, TN 902, 1967.

7. Cocanower, A. B., Kline, S. J., and Johnston, J. P., "A Unified Method for Predicting the Performance of Subsonic Diffusers of Several Geometries," Stanford University PD-10, 1965.
8. Cockrell, D. J., and Markland, E., "Effects of Inlet Conditions on Incompressible Flow through Conical Diffusers," *Journal of the Royal Aeronautical Society*, Vol. 66, pp. 51–2, 1962.
9. Klein, A., "Characteristics of Combustor Diffusers," *Progress in Aerospace Science*, Vol. 31, pp. 171–271, 1995.
10. Little, A. R., and Manners, A. P., "Predictions of the Pressure Losses in 2D and 3D Model Dump Diffusers," ASME Paper 93-GT-184, 1993.
11. Kunz, H. R., "Turbulent Boundary-Layer Growth in Annular Diffusers," *Transactions of the ASME*, set. D, Vol. 87, p. 535, 1965.
12. Adkins, R. C., "A Simple Method for Designing Optimum Annular Diffusers," ASME Paper 83-GT-42, 1983.
13. Adkins, R. C., and Wardle, M. H., "A Method for the Design of Optimum Annular Diffusers of Canted Configuration," *Journal of Engineering for Gas Turbines and Power*, Vol. 114, pp. 8–12, 1990.
14. Takehira, A., Tanaka, M., Kawashima, T., and Hanabusa, H., "An Experimental Study of the Annular Diffusers in Axial-Flow Compressors and Turbines," *Proceedings of the Joint Gas Turbine Congress*, Japan Society of Mechanical Engineers, Tokyo, Japan, 319–28, 1977.
15. Stevens, S. J., and Williams, G. J., "The Influence of Inlet Conditions on the Performance of Annular Diffusers," *Journal of Fluids Engineering*, Vol. 102, pp. 357–63, 1980.
16. Lakshminarayana, B., and Reynolds, B., "Turbulence Characteristics in the Near Wake of a Compressor Rotor Blade," *AIAA Journal*, Vol. 18, 1354–62, 1980.
17. Stevens, S. J., Harasgama, S. P., and Wray, A. P., "The Influence of Blade Wakes on the Performance of Combustor Shortened Prediffusers," *AIAA Journal of Aircraft*, Vol. 21, No. 9, pp. 641–8, 1984.
18. Stevens, S. J., and Wray, A. P., "The Influence of Blade Wakes on the Performance of Outwardly Curved Combustor Prediffusers," AIAA Paper 85-1291, 1985.
19. Klein, A., "The Relation Between Losses and Entry-Flow Conditions in Short Dump Diffusers of Combustors," *Zeitschrift Flugwissenschaft und Weltraumforsch*, December 1988.
20. Lohmann, R. P., Markowski, S. J., and Brookman, E. T., "Swirling Flow Through Annular Diffusers With Conical Walls," *Journal of Fluids Engineering*, Vol. 101, pp. 224–29, 1979.
21. Carrotte, J. F., Bailey, D. W., and Frodsham, C. W., "Detailed Measurements on a Modern Dump Diffuser," *Journal of Engineering for Gas Turbines and Power*, Vol. 117, No. 4, pp. 678–85, 1995.
22. Klein, A., "Review: Effects of Inlet Conditions on Conical-Diffuser Performance," *Journal of Fluids Engineering*, Vol. 103, pp. 250–57, 1981.
23. Hestermann, R., Kim, S., Ben Khaled, A., and Wittig, S., "Flow Field and Performance Characteristics of Combustor Diffusers: A Basic Study," *Journal of Engineering for Gas Turbines and Power*, Vol. 117, No. 4, pp. 686–94, 1995.
24. Adkins, R. C., private communication, 1995.
25. Moore, C. A., and Kline, S.J., "Some Effects of Vanes and of Turbulence in Two-Dimensional Wide-Angle Subsonic Diffusers," NACA TN 4080, 1958.

26. Hoffmann, J. A., "Effects of Free-Stream Turbulence on Diffuser Performance," *Journal of Fluids Engineering*, Vol. 103, pp. 385–90, 1981.
27. Hoffmann, J. A., and Gonzalez, G., "Inlet Free-Stream Turbulence Effects on Diffuser Performance," NASA-CR-166516, 1983.
28. Hoffmann, J. A., and Gonzalez, G., "Effects of Small-Scale, High Intensity Inlet Turbulence on Flow in a Two-Dimensional Diffuser," *Journal of Fluids Engineering*, Vol. 106, pp. 121–24, 1984.
29. Lefebvre, A. H., Gas Turbine Combustion, Taylor and Francis, Philadelphia, PA, 1983.
30. "Performance of Circular Annular Diffusers in Incompressible Flow," Engineering Science Data Unit, London, Item No. 75026, 1976.
31. "Introduction to Design and Performance Data for Diffusers," *Engineering Science Data Unit*, London, Item No. 76027, 1976.
32. Fishenden, C. R., and Stevens, S. J., "The Performance of Annular Combustor Dump Diffusers," *Journal of Aircraft*, Vol. 10, pp. 60–7, 1977.
33. Honami, S., and Morioka, T., "Flow Behavior in a Dump Diffuser with Distorted Flow at the Inlet," ASME Paper 90-GT-90, 1990.
34. Srinivasan, R., Freeman, G., Grahmann, J., and Coleman, E. "Parametric Evaluation of the Aerodynamic Performance of an Annular Combustor-Diffuser System," AIAA Paper 90-2163, 1990.
35. Cochran, D. L., and Kline, S. J., "The Use of Short Flat Vanes as a Means for Producing Efficient Wide-Angle Two-Dimensional Subsonic Diffusers," NACA TN 4309, 1958.
36. Miller, D. S., "Internal Flow Systems," *BHRA Fluid Engineering*, Cranfield, UK, 1978.
37. Sabla, P. E., Taylor, J. R., and Gaunter, D. J., "Design and Development of the Combustor Diffuser for NASA/GE Energy Efficient Engine," *Journal of Energy*, Vol. 6, pp. 275–82, 1982.
38. Adkins, R. C., "A Short Diffuser with Low Pressure Loss," *Journal of Fluids Engineering*, Vol. 97, pp. 297–302, 1975.
39. Adkins, R. C., and Yost, J. O., "A Combined Diffuser Arrangement," presented at the International Joint Gas Turbine Congress and Exhibition, Haifa, Israel, 1979.
40. Adkins, R. C., Matharu, D. S., and Yost, J. O., "The Hybrid Diffuser," *Journal of Engineering for Power*, Vol. 103, pp. 229–36, 1981.
41. Adkins, R. C., and Yost, J. O., "A Compact Diffuser System for Annular Combustors," *International Journal of Turbo and Jet Engines*, Vol. 3, pp. 257–67, 1986.
42. Juhasz, A. J., and Smith, J. M., "Performance of a Short Annular Dump Diffuser Using Suction-Stabilized Vortex Flow Control," NASA TM X-3535, 1977.
43. Smith, J. M., and Juhasz, A. J., "Performance of High Area Ratio Annular Dump Diffuser Using Suction-Stabilized Vortices at Inlet Mach Numbers to 0.41," NASA TP-1194, 1978.
44. Verdouw, A. J., "Performance of the Vortex-Controlled Diffuser (VCD) in an Annular Swirl-Can Combustor Flowpath," in A. H. Lefebvre, ed., *Gas Turbine Design Problems*, Hemisphere, Washington, DC, pp. 12–25, 1980.
45. Adkins, R. C., "Tests on a Vortex-Controlled Diffuser Combined with a Pre-Diffuser and Simulated Combustor," Cranfield SME Report No. C1331-D2, Cranfield University, November 1975.

46. Adkins, R. C., and Binks, D., "A Configuration to Improve the Aerodynamics and Scope of Can-Annular Combustors," ASME Paper 83-GT-37, 1983.
47. Shyy, W., "A Numerical Study of Annular Dump Diffuser Flows," *Computer Methods in Applied Mechanics and Engineering*, Vol. 53, pp. 47–65, 1985.
48. Shyy, W., "A Further Assessment of Numerical Annular Dump Diffuser Flow Calculations," AIAA Paper 85-1440, 1985.
49. Koutmos, P., and McGuirk, J. J., "Numerical Calculations of the Flow in Annular Combustor Dump Diffuser Geometries," *Proceedings of the Institution of Mechanical Engineers*, Vol. 203, *Part C: Journal of Mechanical Engineering Science*, pp. 319–31, 1989.
50. Rodi, W., and Scheuerer, G., "Scrutinizing the k-ε Turbulence Model Under Adverse Pressure Gradient Conditions," *Journal of Fluids Engineering*, Vol. 108, pp. 174–79, 1986.
51. Bradshaw, P., "Effects of Streamline Curvature on Turbulent Flow," *AGARDograph* AG 169, 1973.
52. Jones, W. P., and Manners, A., "The Calculation of the Flow Through a Two-dimensional Faired Diffuser," in F. Durst et al., eds., *Turbulent Shear Flows*, Vol. 6, Springer-Verlag, Berlin, pp. 18–31, 1989.
53. Stevens, S. J., and Eccleston, B., "The Performance of Combustion Chamber Annular Diffusers," Dept. of Transport Technology, University of Loughborough, Rep. TT 69 R 02, 1969.
54. Stevens, S. J., and Fry, P., "Measurements of the Boundary-Layer Growth in Annular Diffusers," *Journal of Aircraft*, Vol. 10, pp. 73–80, 1973.
55. Mayer, D. W., and Kneeling, W. D., "Evaluation of Two Flow Analyses for Subsonic Diffuser Design," AIAA Paper 92-0273, 1992.
56. Ando, Y., Kawai, M., Sato, Y., and Toh, H., "Development of a Numerical Method for the Prediction of Turbulent Flows in Dump Diffusers," *IHI Engineering Review*, Vol. 20, pp. 95–100, 1987.
57. Zhiben, H., and Guangshi, H., "Numerical Study of a Dump Diffuser Flow Field in a Short Annular Combustor," *International Journal of Turbo and Jet-Engines*, Vol. 6, pp. 73–82, 1989.
58. Karki, K. C., Oechsle, V. L., and Mongia, H. C., "A Computational Procedure for Diffuser-Combustor Flow Interaction Analysis," *Journal of Engineering for Gas Turbines and Power*, Vol. 114, pp. 1–7, 1992.

4

Aerodynamics

4.1 Introduction

Aerodynamic processes play a vital role in the design and performance of gas turbine combustion systems. When good aerodynamic design is allied to a matching fuel-injection system, a trouble-free combustor requiring only nominal development is virtually assured.

Many types of combustors, differing widely in size, concept, and method of fuel injection, have been designed. However, close inspection reveals that many aerodynamic features are common to all systems. In the diffuser and annulus, the main objectives are to reduce the flow velocity and distribute the air in prescribed amounts to all combustor zones, while maintaining uniform flow conditions with no parasitic losses or flow recirculation of any kind. Within the combustion liner itself, attention is focused on the attainment of large-scale flow recirculation for flame stabilization, effective dilution of the combustion products, and efficient use of cooling air along the liner walls.

Mixing processes are of paramount importance in the combustion and dilution zones. In the primary zone, good mixing is essential for high burning rates and to minimize soot and nitric oxide formation, whereas the attainment of a satisfactory temperature distribution (pattern factor) in the exhaust gases is very dependent on the degree of mixing between air and combustion products in the dilution zone. A primary objective of combustor design is to achieve satisfactory mixing within the liner and a stable flow pattern throughout the entire combustor, with no parasitic losses and with minimal length and pressure loss.

Successful aerodynamic design demands knowledge of flow recirculation, jet penetration and mixing, and discharge coefficients for all types of air admission holes, including air swirlers. The purpose of this chapter is to review existing knowledge in these areas and to develop relationships among combustor size, pressure loss, and pattern factor, thus providing a rational basis for good aerodynamic design.

4.2 Reference Quantities

A number of flow parameters have been defined to facilitate the analysis of combustor flow characteristics and to allow comparison of the aerodynamic performance of different combustor designs. These parameters include the reference velocity, U_{ref}, which is the mean velocity across the plane of maximum cross-sectional area of the casing in the absence of a liner; i.e.,

$$U_{ref} = \frac{m_3}{\rho_3 A_{ref}}.$$

Also,

$$q_{ref} = \frac{\rho_3 U_{ref}^2}{2},$$

and

$$M_{ref} = \frac{U_{ref}}{(\gamma R T_3)^{0.5}}.$$

4.3 Pressure-Loss Parameters

Two dimensionless pressure-loss parameters are important in combustor design. One is the ratio of the total pressure drop across the combustor to the inlet total pressure ($\Delta P_{3-4}/P_3$), and the other is the ratio of the total pressure drop across the combustor to the reference dynamic pressure ($\Delta P_{3-4}/q_{ref}$). The two parameters are related by the equation

$$\frac{\Delta P_{3-4}}{P_3} = \frac{\Delta P_{3-4}}{q_{ref}} \frac{R}{2} \left(\frac{\dot{m}_3 T_3^{0.5}}{A_{ref} P_3} \right)^2. \tag{4.1}$$

The left-hand side of Equation 4.1 is normally referred to as the *overall pressure loss* and is usually quoted as a percentage. Values range from 4 to 8%. Normally, it does not include the *hot loss*, i.e., the fundamental loss in pressure due to combustion.

The term $\Delta P_{3-4}/q_{ref}$ is the so-called *pressure-loss factor*. It is of prime importance to the combustion engineer, since it denotes the flow resistance introduced into the airstream between the compressor outlet and the turbine inlet. Aerodynamically, it may be regarded as equivalent to a "drag coefficient." Unlike the overall pressure loss, which depends on operating conditions, the pressure-loss factor is a fixed property of the combustion chamber. It represents the sum of two separate sources of pressure loss: (1) pressure drop in the diffuser and (2) pressure drop across the liner. Thus,

$$\frac{\Delta P_{3-4}}{q_{\text{ref}}} = \frac{\Delta P_{\text{diff}}}{q_{\text{ref}}} + \frac{\Delta P_{\text{L}}}{q_{\text{ref}}}. \tag{4.2}$$

It is important to keep ΔP_{diff} to a minimum, since any pressure loss incurred in the diffuser makes no contribution to combustion. In practice, there is little the combustion engineer can do to minimize diffuser pressure loss other than observe the recognized principles of diffuser design. It is equally important to minimize the liner pressure-loss factor, although in this case there is an important difference in that a high liner pressure drop is beneficial to the combustion and dilution processes. It gives high injection air velocities, steep penetration angles, and a high level of turbulence, which promotes good mixing and can result in a shorter liner.

The liner pressure-loss factor is determined essentially by the total effective hole area in the liner $A_{\text{h.eff}}$. Thus,

$$\frac{\Delta P_{\text{L}}}{\rho_3} = \frac{U_{\text{j}}^2}{2}, \tag{4.3}$$

or

$$\frac{\Delta P_{\text{L}}}{P_3} = \frac{R}{2}\left(\frac{\dot{m}_3 T_3^{0.5}}{A_{\text{h.eff}} P_3}\right)^2. \tag{4.4}$$

Substituting the right side of Equation 4.4 into Equation 4.1 gives

$$\frac{\Delta P_{\text{L}}}{q_{\text{ref}}} = \left(\frac{A_{\text{ref}}}{A_{\text{h.eff}}}\right)^2. \tag{4.5}$$

Thus, the total effective area of the holes in the liner is governed by the casing reference area and the available pressure drop across it. Rearranging Equation 4.5 gives

$$A_{\text{h.eff}} = \frac{A_{\text{ref}}}{\left(\Delta P_{3-4} / q_{\text{ref}} - \Delta P_{\text{diff}} / q_{\text{ref}}\right)^{0.5}}. \tag{4.6}$$

The effective flow area of the liner may be calculated from the expression

$$A_{\text{h.eff}} = \sum_{i=1}^{i=n} C_{\text{D.i}} A_{\text{h.i}}, \tag{4.7}$$

where $C_{\text{D.i}} A_{\text{h.i}}$ is the effective flow area of ith hole and n is the total number of holes.

The quantity $(R/2)\ (\dot{m}_3 T_3^{0.5} / A_{ref} P_3)^2$ in Equation 4.1 is effectively a measure of combustor reference velocity, since it can be rewritten as $U_{ref}^2 / 2RT_3$. As $\dot{m}_3 T_3^{0.5} / P_3$ is fixed by the compressor design, the only control over $\dot{m}_3 T_3^{0.5} / A_{ref} P_3$ afforded to the combustion engineer is in the selection of the maximum casing area, A_{ref}. Thus, Equation 4.1 confronts the designer with a serious dilemma. For low fuel consumption, the overall pressure loss of the chamber (the left side of Equation 4.1) must be low. Typically, a one percentage point increase in pressure loss can produce either a half percent reduction in thrust or around a quarter of a percent increase in specific fuel consumption. However, if the chamber is to be small and have adequate mixing, both terms on the right side of the equation must be large. These conflicting requirements can be optimized only in the context of specific engine applications; e.g., for a lift engine, a high overall pressure loss, and hence a high fuel consumption, may be tolerated in return for a small engine (low value of A_{ref}). On a long-range cruise engine, low fuel consumption is of paramount importance, and a large-diameter, low-pressure-loss combustor would be advantageous.

Some typical values of "cold" pressure loss in practical chambers are given in Table 4.1. The third column on the table shows that the annular chamber has the lowest pressure-loss factor. This may seem to conflict with the fact that a high pressure-loss factor is necessary for good mixing. However, in annular chambers, most of the losses arise across the liner air-inlet holes, where they contribute to mixing, and there are fewer losses in the diffuser and inlet ducting. The resultant advantage of the annular chamber shows up clearly in the rightmost column. Although the overall pressure loss is the same as that for a tubular chamber (second column), the term $\dot{m}_3 T_3^{0.5} / A_{ref} P_3$ is much larger, implying a lower value of A_{ref} and hence a smaller chamber for any given aerodynamic loading.

As indicated in Table 4.1, the value of the pressure-loss factor to be used in determining the reference area can vary from around 20 for a straight-through annular combustor to almost 40 for a tubular combustor. In practice, the actual value of $\Delta P_{3-4}/q_{ref}$ is governed by the various performance requirements that dictate the pressure drop across the liner, such as pattern factor and pollutant emissions, and by the compressor outlet velocity and the type of diffuser employed, e.g., dump, faired, or hybrid.

TABLE 4.1

Pressure Losses in Combustion Chambers

Type of Chamber	$\frac{\Delta P_{3-4}}{P_3}$	$\frac{\Delta P_{3-4}}{q_{ref}}$	$\frac{m_3 T_3^{0.5}}{A_{ref} P_3}$
Tubular	0.07	37	0.0036
Tuboannular	0.06	28	0.0039
Annular	0.06	20	0.0046

The values of overall pressure loss listed in Table 4.1 represent the cold loss only, i.e., the losses arising from turbulence and friction that can be measured with reasonable accuracy from cold-flow tests. Under burning conditions, these losses are augmented by the fundamental loss due to combustion. For uniform mixtures flowing at low Mach number through a constant-area duct, this may be derived from momentum considerations as

$$\frac{\Delta P_{hot}}{q_{ref}} = \frac{\rho_3}{\rho_4} - 1, \tag{4.8}$$

where ρ_3 and ρ_4 are the air densities at inlet and outlet air temperatures of T_3 and T_4, respectively. Hence,

$$\frac{\Delta P_{hot}}{q_{ref}} \cong \frac{T_4}{T_3} - 1. \tag{4.9}$$

Analyses of the factors governing $\Delta P_{hot}/q_{ref}$ in practical chambers have indicated relationships of the form

$$\frac{\Delta P_{hot}}{q_{ref}} = K_1\left(\frac{T_4}{T_3} - K_2\right). \tag{4.10}$$

Actual values of K_1 and K_2, as determined experimentally, are contained in Reference [1]. For chambers of moderate temperature rise, ΔP_{hot} usually lies between 0.5 and 1.0% of P_3.

4.4 Relationship between Size and Pressure Loss

For straight-through combustors, the optimal cross-sectional area of the casing A_{ref} is determined from considerations of overall pressure loss and combustion loading, as discussed in Chapter 5. However, for most industrial combustors and some aircraft combustors, the casing area needed to meet combustion requirements is so low as to give an unacceptably high pressure loss. Under these conditions, the overall pressure loss dictates the casing size, and A_{ref} is obtained from Equation 4.1 as

$$A_{ref} = \left[\frac{R}{2}\left(\frac{\dot{m}_3 T_3^{0.5}}{P_3}\right)^2 \frac{\Delta P_{3-4}}{q_{ref}}\left(\frac{\Delta P_{3-4}}{P_3}\right)^{-1}\right]^{0.5}. \tag{4.11}$$

At first sight, it might appear advantageous to make the liner cross-sectional area as large as possible, since this results in lower velocities and longer residence times within the liner, both of which are highly beneficial to ignition, stability, and combustion efficiency. Unfortunately, for any given casing area, an increase in liner diameter can be obtained only at the expense of a reduction in annulus area. This raises the annulus velocity and lowers the annulus static pressure, thereby reducing the static pressure drop across the liner holes. This is undesirable, since a high static pressure drop is needed to ensure that the air jets entering the liner have adequate penetration and sufficient turbulence intensity to promote rapid mixing with the combustion products. These considerations suggest that a satisfactory criterion for mixing performance would be the ratio of the static pressure drop across the liner Δp_L to the dynamic pressure of the flow in the combustion zone q_{pz}. If the ratio of liner cross-sectional area to casing cross-sectional area is denoted by k, then the optimal value of k, which gives the highest value of $\Delta p_L/q_{pz}$, can be derived as [2]

$$k_{opt} = 1 - \left[\frac{(1-\dot{m}_{sn})^2 - \lambda}{\Delta P_{3-4}/q_{ref} - \lambda r^2} \right]^{1/3}. \tag{4.12}$$

Hence

$$A_L = k_{opt} A_{ref}. \tag{4.13}$$

Equations 4.11 through 4.13 are valid for all "straight-through" tubular, tuboannular, and annular chambers.

4.5 Flow in the Annulus

Flow conditions in the annulus have a substantial effect on the airflow pattern within the liner and influence the level and distribution of liner wall temperatures. The mean velocity in the annulus is governed by the combustor reference velocity and the ratio of liner area to casing area. In practice, appreciable variations in annulus velocity occur because of changes in inlet velocity profile and as air is drawn into the liner through rows of holes and cooling slots.

Although a high annulus velocity augments the convective cooling of the liner walls, low velocities are generally preferred because they provide the following benefits:

- Minimum variation in annulus velocity and static pressure, ensuring that all the liner holes in the same row pass the same airflow

- Higher hole discharge coefficients
- Steeper angles of jet penetration
- Lower skin-friction loss
- Lower "sudden-expansion" losses downstream of liner holes and cooling slots

In most combustors, the critical areas are at the upstream end of the liner and in the vicinity of the dilution holes. At the upstream end, the air issuing from the diffuser sometimes has a thick boundary layer that prohibits the use of static-pressure-fed cooling slots. Even with total-head cooling slots, problems are caused by the flow separation that often occurs as a result of the bend formed at the junction of the diffuser and annulus. As the flow proceeds along the annulus, its velocity profile gradually improves as air is drawn off through the various apertures in the liner; however, if the air is allowed to flow without restriction into the space downstream of the dilution holes, flow disturbances can arise that cause air to recirculate upstream within the annulus in an intermittent and random manner. Tuboannular combustors are especially prone to this phenomenon, which, in severe cases, causes some of the liner holes to receive air from different directions. This produces an internal flow pattern that is not only distorted, but also varies in an irregular manner with time.

One method of alleviating this problem is by fitting a "backstop" just downstream of the dilution holes. This is simply an annular plate that is designed to fit between the inner and outer combustor casings. It is pierced with large, round holes to accommodate the liners, and provision is made for some air to flow through the plate to cool the hot sections downstream. Plates of this type can be very effective in combating large, random recirculations in the annulus flow.

On annular liners, the backstop is often in the form of a continuous dam, located immediately downstream of the dilution holes. Typically, the dam blocks off about two-thirds of the annulus area. Another method of controlling the flow in the dilution-hole region is by tapering the liner walls outward to prevent excessive diffusion in the annulus passage adjacent to these holes.

If the pitch of the dilution holes is greater than the annulus height, a vortex can form in the flow entering the hole; this changes the penetration and mixing characteristics of the dilution air jet. The strength of the vortex depends on the ratio of the annulus area, as measured in the plane of the holes, to the hole area. A high value for this ratio inhibits vortex formation, thus providing another argument in support of designing liners with adequate pressure drop.

Vortex formation, which can occur on both tubular and annular liners, may be eliminated or subdued by fitting a longitudinal splitter plate across each dilution hole. The plate, which may be attached to either the liner or the outer casing, is especially effective when used in conjunction with a spectacle plate, or dam, as illustrated in Figure 4.1.

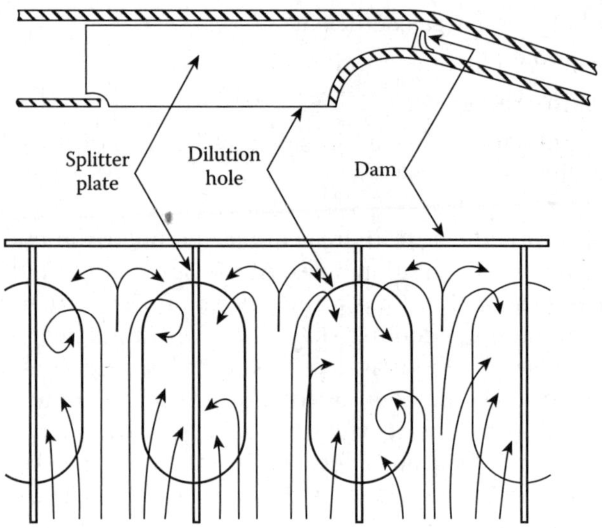

FIGURE 4.1
Flow control through dilution holes.

4.6 Flow through Liner Holes

The flow through a liner hole depends not only on its size and the pressure drop across it, but also on the duct geometry and flow conditions in the vicinity of the hole, which can strongly influence its effective flow area.

4.6.1 Discharge Coefficient

The basic equation for flow through a hole may be expressed as

$$\dot{m}_h = C_D A_{h.geom}[2\rho_3(P_1 - p_j)]^{0.5}, \tag{4.14}$$

where P_1 is the total pressure upstream of the hole and p_j is the static pressure downstream of the hole.

It has been observed that some deflection of the flow streamlines occurs in the vicinity of the liner holes, to an extent that depends on the geometry of the system, the approach velocity, and the pressure drop across the liner [3]. Thus, in practice, the coefficient of discharge of liner holes is affected by

- Type (e.g., plain or plunged)
- Shape (e.g., circular or rectangular)
- Ratio of hole spacing to annulus height
- Liner pressure drop

- Distribution of static pressure around the hole inside the liner
- Presence of swirl in the upstream flow
- Local annulus air velocity

From a detailed analysis of the factors governing the flow of a jet through a liner wall, Kaddah [4] concluded that, for incompressible, nonswirling flow, the coefficient of discharge for plain circular, oval, and rectangular holes is given by

$$C_D = \frac{1.25(K-1)}{[4K^2 - K(2-\alpha)^2]^{0.5}}, \tag{4.15}$$

where α is the ratio of hole mass flow rate to annulus mass flow rate $(\dot{m}_h / \dot{m}_{an})$, and K is the ratio of the jet dynamic pressure to the annulus dynamic pressure upstream of the holes.

The very satisfactory level of agreement between the predictions of Equation 4.15 and experimental data are illustrated in Figure 4.2.

In subsequent experimental measurements of the discharge coefficients of plunged holes, Freeman [5] found that Equation 4.15 still gave very satisfactory correlation of the data, provided that the value of the constant was raised from 1.25 to 1.65, as shown in Figure 4.3. Thus, for plunged holes

$$C_D = \frac{1.65(K-1)}{[4K^2 - K(2-\alpha)^2]^{0.5}}. \tag{4.16}$$

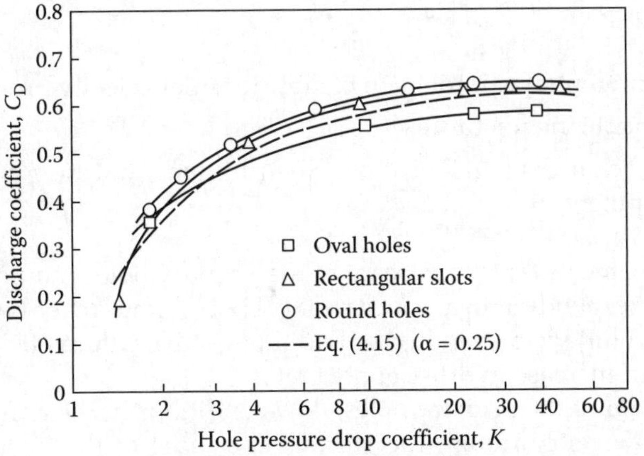

FIGURE 4.2
Influence of hole shape on discharge coefficient. (From Kaddah, K.S., College of Aeronautics MSc Thesis, Cranfield, UK, 1964.)

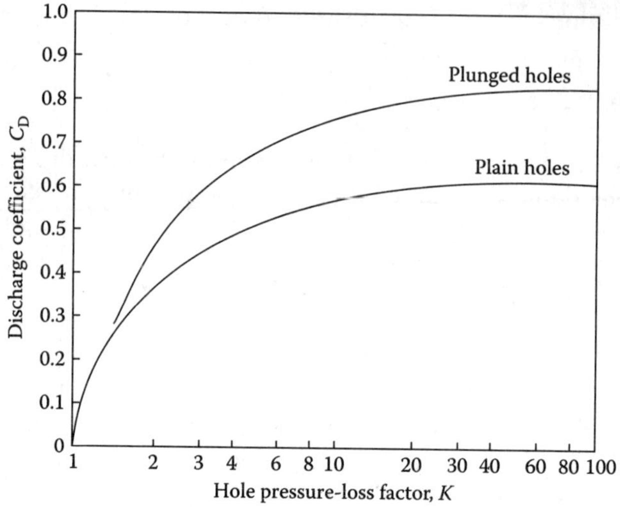

FIGURE 4.3
Influence of hole type and pressure-drop coefficient on discharge coefficient. (From Freeman, B.C., College of Aeronautics MSc Thesis, Cranfield, UK, 1965.)

The flow through scoops and louvers has not yet been subjected to theoretical analyses of the type conducted for plain and plunged holes. Experimental data for these and other types of hole configurations are contained in References [6–10].

It is apparent in Figure 4.3 that a high value of K not only provides a high value of C_D, but also ensures that C_D is fairly insensitive to small changes in K. This is an important asset because, in practice, local values of K can change appreciably around the liner annuli due to a number of factors, including:

- Circumferential variations in combustor inlet velocity profile
- Manufacturing tolerances
- Small changes in airflow distribution during combustor development

Ideally, the value of K just upstream of the primary holes should not be less than six, corresponding to a velocity ratio, U_h/U_{an} of just over two. A higher value of K would yield even higher and more stable values of C_D, but at the expense of an increase in diffuser area ratio.

Downstream of the primary holes, flow conditions in the annuli improve as air is drawn off through the various apertures in the liner wall. This means that if the annular feed passages are sized to provide a satisfactory value for K at the primary holes, then, for all the holes further downstream, higher values of K and C_D are virtually assured.

4.6.2 Initial Jet Angle

It is clear from Figure 4.4 that any reduction in initial jet angle must reduce the effective hole area. Thus, one would expect the initial angle θ to be related to C_D, and such a relationship has been established [11] as follows:

$$\sin^2\theta = C_D / C_{D\infty}, \tag{4.17}$$

where $C_{D\infty}$ is the asymptotic value of C_D as K tends to infinity. This relationship is plotted alongside Kaddah's [4] experimental data in Figure 4.5.

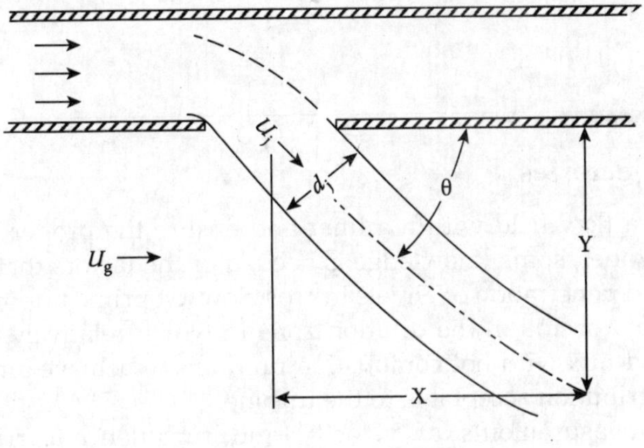

FIGURE 4.4
Representation of flow through liner hole.

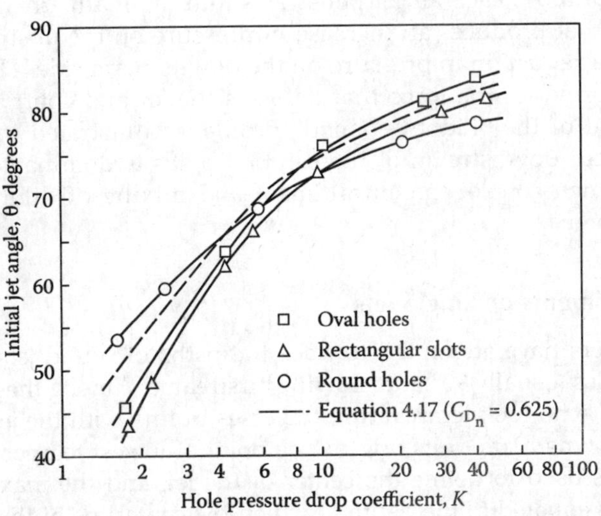

FIGURE 4.5
Variation in initial jet angle with pressure-drop coefficient for various hole shapes. (From Kaddah, K.S., College of Aeronautics MSc Thesis, Cranfield, UK, 1964.)

In practice, if the value of K for any row of holes is too small to provide the desired initial jet angle, some steepening of this angle can be achieved by attaching short chutes to each hole. Such chutes were fitted to Rolls Royce Spey combustors some 30 years ago as a means of improving primary air recirculation and reducing smoke. This company now uses chutes on its more modern RB211 and Trent combustors for both primary and intermediate holes.

Drawbacks to the use of chutes include additional cost and weight. Also, some form of cooling is usually required in order to protect them from overheating by the hot combustion gases in which they are immersed.

4.7 Jet Trajectories

To establish a flow field within a liner and ensure the proper distribution of air to all zones, some knowledge is needed of the factors that govern the trajectory and penetration of air jets in crossflow. Of prime importance is the jet mixing that occurs in the dilution zone in which relatively cold air jets penetrate and mix with hot combustion products to achieve an outlet temperature distribution acceptable to the turbine.

Detailed measurements of velocity and turbulence carried out by Carrotte and Stevens [12] have indicated the structure of a dilution jet and the mixing processes that develop as the jet interacts with the surrounding gas stream. As each air jet penetrates into the main stream, it creates a blockage that produces an increase in pressure on the upstream side of the jet and a reduction in pressure on the downstream side. This pressure difference provides the force that deforms the jet and contributes to the development of the "kidney-shaped" profile, as illustrated in Figure 4.6. The flow field downstream of the dilution holes is dominated by vortex systems, which control the entrainment and mixing of dilution air and mainstream gas.

4.7.1 Experiments on Single Jets

Many workers have attempted to trace the paths of round jets injected at various angles (usually 90°) into flowing airstreams. Among the earliest was Norster [13], who took temperature traverses in line with the jet at various distances downstream of its origin. The point of lowest temperature in the traverse was used to define the center of the jet, and the maximum penetration was equated to the depth at which the centerline of the jet became asymptotic to the mainstream flow.

Norster's data are especially relevant to gas turbine combustors, because his test conditions were chosen to simulate those in an actual dilution

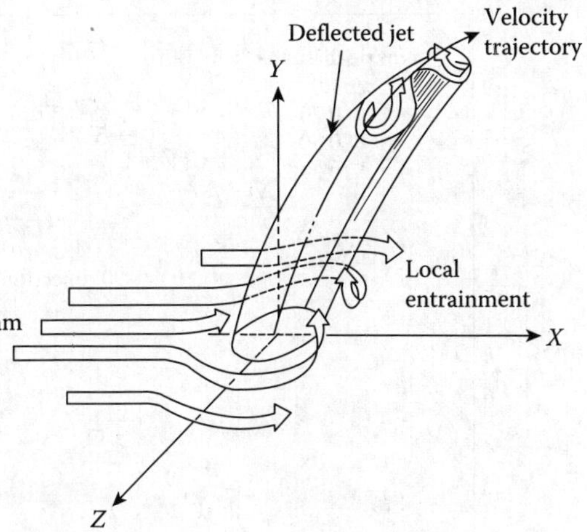

FIGURE 4.6
Representation of a jet in crossflow.

zone. From analysis of these and other data, Lefebvre [14] concluded that the trajectory of a jet in crossflow, as illustrated in Figure 4.4, is adequately described by the expression

$$Y / d_j = 0.82 J^{0.5} (X / d_j)^{0.33}, \tag{4.18}$$

where

$$J = (\rho_j U_j^2) / (\rho_g U_g^2).$$

If a single jet is injected into a crossflow at an angle θ, where θ is less than 90°, its trajectory is readily obtained by multiplying the value of Y/d_j for 90° by sin θ.

Equation 4.18 implies that jet penetration increases continually with an increase in downstream distance. In practice, the jet may attain its maximum penetration within a fairly short distance downstream of its origin. Thus, Equation 4.18 and other similar equations found in the literature, are useful only for describing the initial portion of the jet trajectory and lose their validity as the jet centerline becomes asymptotic to the crossflow.

Of more practical interest is the maximum penetration achieved by the jet. For a single round jet injected into a circular duct, Norster [13] found that the maximum penetration is given by

$$Y_{max} = 1.15 d_j J^{0.5} \sin\theta. \tag{4.19}$$

The data on which this equation is based are shown in Figure 4.7.

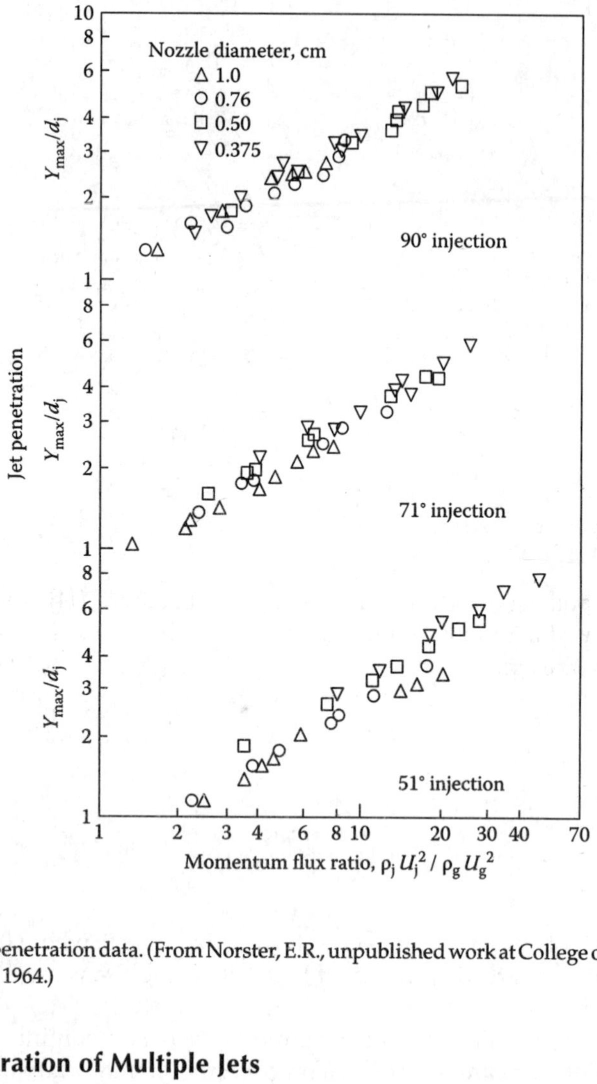

FIGURE 4.7
Maximum jet penetration data. (From Norster, E.R., unpublished work at College of Aeronautics, Cranfield, UK, 1964.)

4.7.2 Penetration of Multiple Jets

The first systematic study of the penetration of multiple jet configurations was carried out by Sridhara and Norster [15,16]. For a circular duct, Sridhara [15] found that the penetration of multiple jets was lower than that for a single jet. He attributed this to the blockage effect of the jets in producing a local increase in mainstream velocity. From analysis of these data, Norster [16] recommended the following equation for estimating the maximum penetration of round air jets into a tubular liner:

$$Y_{max} = 1.25 d_j J^{0.5} \dot{m}_g / (\dot{m}_g + \dot{m}_j). \quad (4.20)$$

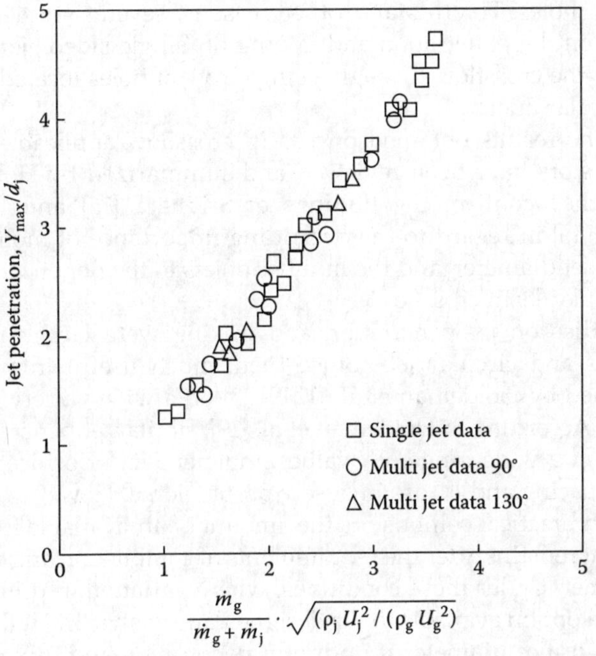

FIGURE 4.8
Multiple-jet penetration correlation. (From Sridhara, K., College of Aeronautics MSc Thesis 20/187, Cranfield, UK, 1967. [See also National Aeronautics Laboratory Report TN 30, Bangalore, India, 1970].)

The level of agreement between the predictions of this equation and measured values of Y_{max}/d_j is shown in Figure 4.8.

Sridhara's investigation was confined to circular ducts, but a considerable amount of numerical and experimental work on the penetration of multiple jets has been carried out using rectangular ducts to simulate flow conditions in annular combustors. Most annular combustors feature two rows of dilution holes in the same axial plane, one row on the inner liner and the other on the outer liner. Usually, the number of holes in each row is the same. In most designs, the opposing holes are arranged to be "in-line" so that the jets impinge on each other, but in other designs, the holes are staggered circumferentially to allow the jets to penetrate past each other.

In almost all annular combustors, the aerodynamic quality of the air approaching the outer row of dilution holes is impaired by the presence of various obstacles in the outer annulus, such as fuel-nozzle feed arms, liner support pins, and igniters, with the result that the flow approaching the outer dilution holes contains numerous small eddies and other flow perturbations that have an adverse effect on the uniformity of the flow entering the dilution holes. For this and other reasons, several studies have been carried out on the penetration and mixing of "single-sided" jets, i.e., jets of air entering the crossflow through a single row of holes located in one wall of a rectangular duct.

Most of the results obtained on jets in crossflow applicable to gas turbine combustors have been reviewed and summarized by Holdeman [17]. Generally, they confirm the findings of Sridhara [15] and Norster and Lefebvre [16,18] in regard to the overriding importance of momentum-flux ratio, initial jet diameter, and the number of jets to the penetration and mixing of multiple jets in crossflow.

Early studies on jet penetration and mixing were largely confined to round holes and circular jets. Since then, much useful information has been obtained by Holdeman et al. [17,19–23] on a wide variety of hole configurations. According to Holdeman et al. [19], similar jet penetration can be obtained over a wide range of J values, independent of orifice diameter, if the orifice spacing and $J^{0.5}$ are inversely proportional. However, it should be noted that in practical combustors the amount of air available for dilution is usually what remains after the requirements of combustion and wall cooling have been met. Under these conditions, where variation in dilution airflow rate is not an option available to the designer, any change in J will necessitate a change in orifice diameter if optimum penetration and mixing are to be maintained.

Kamotani and Greber [24] used smoke photographs to investigate the detailed features of jet interaction in single- and multiple-row jets. They observed that when two closely spaced jets are arranged parallel to the crossflow, the rear jet, being in the wake of the front jet, remains almost undeflected until it meets the front jet, whereupon the two jets quickly combine. The penetration of this combined jet is slightly greater than that of a single jet injected from a hole having a cross-sectional area equal to the sum of the two separate holes. Holdeman [17] also compared the penetrations of a single row of round holes and several equal-area double-row circular hole configurations at intermediate momentum-flux ratios ($J = 26$) and found the average penetration to be nearly the same for all configurations.

In a separate series of experiments, Kamotani and Greber [24] studied the effects of an opposite wall on the characteristics of turbulent jets injected into a crossflow. They found that an opposite wall has relatively little effect on a single jet unless J is large enough to cause the jet to impinge on the wall. They also conducted experiments in which two jets impinged on each other, to compare behavior in this situation with behavior when there is interaction with an opposite wall. Their measurements showed that the trajectories for these two situations are virtually indistinguishable from each other. Thus, as far as velocity trajectories are concerned, the plane of symmetry between two opposing jets can be considered equivalent to a wall. However, for this to be true, it is very important that the velocities of the two opposing jets be matched quite closely.

4.8 Jet Mixing

Jet-mixing processes play an important role in achieving satisfactory combustion performance. In the primary zone, good mixing promotes efficient combustion and minimum pollutant formation. In the intermediate zone, rapid mixing of the injected air with the hot gases from the primary zone is needed to accelerate soot oxidation and to convert any dissociated species into normal products of combustion. Finally, the attainment of a satisfactory pattern factor at the combustor exit is dependent on thorough mixing of air and combustion products in the dilution zone.

In general, the rate of mixing between the air jets and the hot gases contained within the liner is mainly influenced by the following factors:

- The size and (to a much lesser extent) the shape of the hole through which the jet issues
- The initial angle of jet penetration
- The momentum-flux ratio, J
- The presence of other jets, both adjacent and opposed
- The length of the jet mixing path
- The proximity of walls
- The inlet velocity and temperature profiles of the jet and the hot gases.

Equal significance should not be attached to all of these parameters. In practice, the key factors governing mixing rates are momentum-flux ratio, length of mixing path, and the number, size, and initial angle of the jets.

Most of the early work on jet mixing was carried out in cylindrical ducts designed to simulate tubular combustion liners. Subsequent investigations, notably those of Holdeman and co-workers [17,19–23,25–30] have been directed toward jet mixing in rectangular ducts, using both single-sided and double-sided air injection. The results and conclusions of these various studies carried out before 1993 have been summarized by Holdeman [17].

4.8.1 Cylindrical Ducts

Sridhara [15] investigated the mixing of air jets injected into hot gas streams under conditions that allowed the temperature and velocity of the hot and cold streams, the injection-hole diameter, the angle of injection, and the mixing length to be accurately controlled and varied over a wide range. Figure 4.9 is typical of the results obtained using a circular duct of 17 cm diameter to contain a hot gas stream into which the air jets were injected at an initial angle of 90°. Temperature distributions were measured at a plane

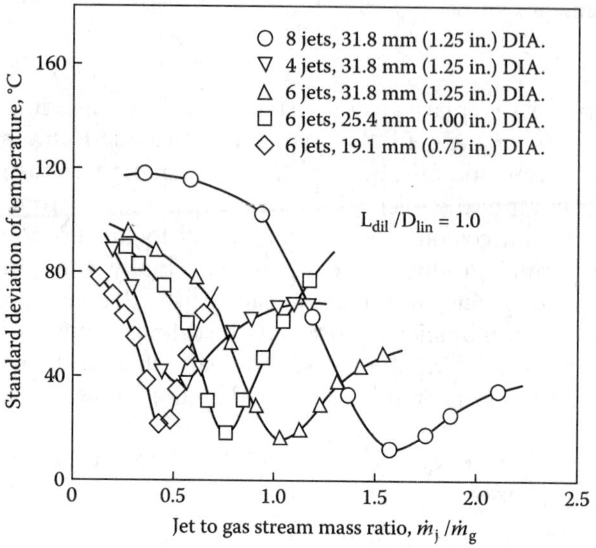

FIGURE 4.9
Influence of dilution-hole geometry on temperature distributions. (From Sridhara, K., College of Aeronautics MSc Thesis 20/187, Cranfield, UK, 1967. [See also National Aeronautics Laboratory Report TN 30, Bangalore, India, 1970].)

17 cm downstream of the air-injection holes. The figure shows that for all configurations, an increase in jet flow improves the level of mixing up to an optimum value, after which any further increase causes the temperature distribution to deteriorate. Presumably, this is because under-penetrating jets fail to reach the hot gas core at the liner axis, whereas over-penetrating jets collide to produce a cold core at the liner axis.

The main conclusion to be drawn from the Cranfield experiments [13,15] is that jet-to-mainstream momentum-flux ratio J, and jet diameter d_j, are the most significant factors affecting jet penetration and mixing in a cylindrical combustion liner. The data acquired in these investigations was subsequently used by Lefebvre and Norster [18] to derive a method for determining the optimum number and size of air injection holes for achieving the most uniform mixing between the air jets and the hot gas stream.

Talpallikar et al. [30] used a CFD code to numerically analyze variations in J and slot aspect ratio (SAR) on jet mixing in a rich quick-quench lean combustor (RQL) combustor. The value of J was varied parametrically from 16 to 64 by variation in jet velocity from 120 to 240 m/s. Their results showed that SAR has a significant effect on jet penetration and mixing performance. It was concluded that conventional correlations for optimum mixing effectiveness for round holes may not be applicable for slots.

Hatch et al. [31] conducted a series of parametric studies to determine the influence of geometry and flow variations on mixing patterns in a cylindrical duct. The quality of mixing was assessed from temperature measurements carried out downstream of a row of cold jets injected into a heated cross stream. The geometries investigated included round holes and slanted slots, i.e., slots oriented at various angles with respect to the mainstream flow. No attempt was made to optimize the number of orifices for each value of J, and eight equispaced holes were used throughout the test program, which covered a range of J values from 25 to 80. The jet to mainstream mass ratio ($\dot{m}_j / \dot{m}_g$) was kept constant at 2.2. These values of J and $\dot{m}_j / \dot{m}_g$ would be considered exceptionally high for a conventional combustion liner, but they are appropriate for RQL combustors, which pose formidable challenges in jet mixing in a confined crossflow. One important consequence of high $\dot{m}_j / \dot{m}_g$ values is that they necessitate the use of slots instead of round holes around the liner perimeter.

The results of Hatch et al. [31] confirmed the importance of J to penetration and mixing in a cylindrical combustion liner. They also found that increasing the aspect ratio of slanted slots reduces jet penetration to the center and enhances mixing along the walls. At low and intermediate values of J, the 4:1 aspect ratio slots gave better mixing than the 8:1 slots at all downstream locations. At the highest value of J tested, the higher aspect ratio slots exhibited better mixing. For slanted slots, it was found that the optimum value of J for best mixing varied with slot angle.

Zhu et al. [32] also carried out a numerical study on the penetration and mixing of radial jets in a "necked-down" cylindrical duct designed to simulate the mixing section of an RQL combustor. Two types of jet slots were considered: rectangular straight slots aligned in the streamwise direction, and slanted slots. The parameters investigated included wide variations in J (2–64), n (2–12), and SAR (1–4), while maintaining a constant value of $\dot{m}_j / \dot{m}_g$. Their results confirmed that J has the most prominent effect on mixing performance and showed that the optimum configuration changes with the number of orifices and the SAR. Another interesting conclusion from this study is that modifying the flow by necking down or introducing swirl merely serves to raise the pressure loss of the system without enhancing mixing performance.

4.8.2 Rectangular Ducts

Mixing studies in rectangular ducts provide useful guidance in the design of dilution zones for annular combustors without the complications of radius effects. For this reason, the mixing of multiple air jets in a confined rectangular crossflow has been extensively treated in the literature. The early studies of Holdeman, Walker, and Kors [19,20,25] identified the main flow and geometric variables that characterize the mixing process. More recent studies, in particular those of Holdeman and Srinivasan et al. (see, e.g., references [22,23,26,27,33–37]), have extended the available experimental data and yielded useful empirical correlations on the mixing of multiple jets in crossflow. The geometric variables encompassed in these studies include area convergence, noncircular orifices, double rows of holes, and opposed rows of jets, both in-line and staggered.

4.8.2.1 Influence of Density Ratio

Holdeman, Walker, and Kors [19] suggested that density ratio did not need to be considered independently from momentum-flux ratio. This was confirmed over a broad range of density ratios in the experiments by Srinivasan et al. [26]. Another interesting conclusion from these experiments is that circumferential nonuniformities mix much more rapidly with increasing downstream distance than do radial nonuniformities.

The main results from the various numerical and experimental studies carried out so far are summarized below. More detailed information may be obtained from the quoted references.

- Momentum-flux ratio and orifice spacing [17] or size [18] are the main factors governing mixing performance.
- Jet penetration remains largely unchanged when the product of d_j and $J^{0.5}$ is kept constant [18].
- The temperature distributions obtained using square orifices are almost identical to those observed from equal-area circular orifices at all downstream locations [17,33,34,38].
- The penetration and mixing of 45° slanted slots are less than for streamlined or bluff slots or equal-area circular orifices. Moreover, the temperature distributions for slanted slots are skewed and shifted laterally with respect to the injection centerplane [17,34].
- Double rows of jets have penetrations and temperature distributions similar to those from a single row of equally spaced, equal-area, circular orifices [17,23,25,27].
- Flow area convergence, especially injection-wall convergence, significantly improves mixing performance [17].
- For opposed rows of jets with the orifice centers in-line, the optimum ratio of orifice spacing to duct height is one-half of the optimum value for single-sided injection at the same momentum-flux ratio [17,22,39].
- For opposed rows of jets with the orifice centers staggered, the optimum ratio of orifice spacing to duct height is double the optimum value for single-sided injection at the same momentum-flux ratio [17,22,36].
- In-line configurations have better initial mixing than staggered configurations at their respective optimum values of orifice spacing-to-duct height ratio [28].
- Density ratio does not need to be considered independently from momentum-flux ratio [19,26].

4.8.3 Annular Ducts

Very little information is available in the open literature on jet penetration and mixing in annular ducts, despite their enormous practical importance to the design of annular gas turbine combustors, due mainly to the very high cost of conducting systematic experimental research on annular configurations. Holdeman [17] has reported on the results of numerical calculations on the mixing of jets in an annular duct whose inside radius was made equal to the duct height. The orifice geometry was opposed rows of in-line holes with $J = 6.6$. Similar penetration and mixing was achieved by specifying the hole spacing for the annular duct to be equal to that in a rectangular duct at the radius that divides the annulus into equal areas.

The results and correlations of temperature distributions obtained in the Cranfield and NASA test programs cannot be applied directly to the dilution zones of practical combustors owing to the effects of liner-cooling airflow, the nonuniform temperature distribution of the hot gases entering the dilution zone, and the convergence of the flow area in the dilution zone. Nevertheless, they provide useful guidance in liner design, as discussed below.

4.9 Temperature Traverse Quality

One of the most important and, at the same time, most difficult problems in the design and development of gas turbine combustion chambers is that of achieving a satisfactory and consistent distribution of temperature in the efflux gases discharging into the turbine. In the past, experience has played a major role in the determination of dilution-zone geometry, and trial-and-error methods have, of necessity, been employed in developing the temperature traverse quality of individual combustor designs to a satisfactory standard.

The temperature attained by an elemental volume of gas at the chamber outlet is dependent on its history from the time it emerges from the compressor. During its passage through the combustor, its temperature and composition change rapidly under the influence of various combustion, heat-transfer, and mixing processes, none of which are perfectly understood. The final mixing process, for example, is affected in a complicated manner by the dimensions, geometry, and pressure drop of the liner, the size, shape, and discharge coefficients of the liner holes, the airflow distribution to various zones of the chamber, and the temperature distribution of the hot gases entering the dilution zone. For any given combustor, the latter is strongly influenced by fuel spray characteristics, such as drop size, spray angle, and spray penetration, because these control the pattern of burning and hence the distribution of temperature in the primary-zone efflux. It is known that spray characteristics are strongly influenced by pressure, especially with atomizers of the simplex- or dual-orifice type. It is to be expected, therefore, that the temperature traverse will also vary with pressure, although the extent of this variation varies from one chamber to another, depending on design and, in particular, on length. Thus, it is highly desirable that rig work on the improvement of temperature traverse quality should be carried out at the maximum engine pressure, because this corresponds to maximum heat-transfer rates to nozzle guide vanes and turbine blades.

The most important temperature parameters are those that affect the power output of the engine and the life and durability of the hot sections downstream of the combustor. As far as overall engine performance is concerned, the most important temperature is the turbine inlet temperature, T_4, which is the mass-flow-weighted mean of all the exit temperatures recorded for one standard of liner. Since the nozzle guide vanes are fixed relative to the combustor, they must be designed to withstand the maximum temperature found in the traverse. Thus, the parameter of most relevance to the design of nozzle guide vanes is the overall temperature distribution factor, which highlights this maximum temperature. This *pattern factor* is normally defined as

$$\text{Pattern factor} = \frac{T_{\text{max}} - T_4}{T_4 - T_3}, \tag{4.21}$$

where T_{max} is the maximum recorded temperature, T_3 is the inlet air temperature, and T_4 is the mean exit temperature.

The temperatures of most significance to the turbine blades are those that constitute the average radial profile. They are obtained by adding together the temperature measurements around each radius of the liner and then dividing by the number of locations at each radius, i.e., by calculating the arithmetic mean at each radius. A typical radial temperature profile is shown in Figure 4.10. The expression used to describe the radial temperature distribution factor, also known as the *profile factor*, is

$$\text{Profile factor} = \frac{T_{\text{mr}} - T_4}{T_4 - T_3}, \tag{4.22}$$

where T_{mr} is the maximum circumferential mean temperature.

The pattern factor and profile factor, as defined above, are best suited for situations where a perfectly uniform exit-temperature distribution would be considered ideal. However, in modern high-performance engines, which employ extensive air cooling of both nozzle guide vanes and turbine blades, the desired average radial distribution of temperature at the combustor exit plane is far from flat; instead, it usually has a profile that peaks above the midheight of the blade, as illustrated in Figure 4.10. The objective is to provide

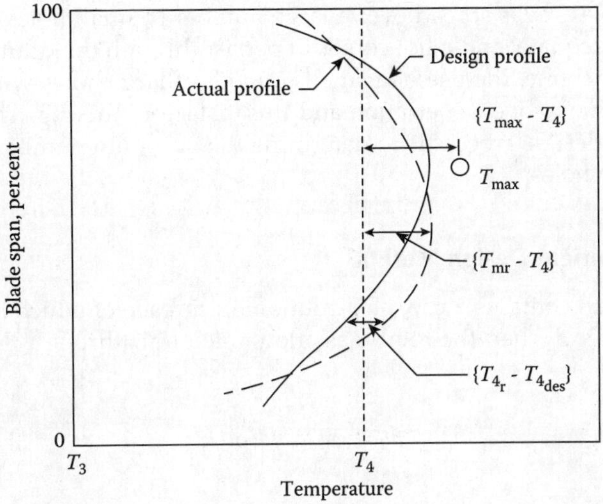

FIGURE 4.10
Explanation of terms in exit-temperature profile parameters.

lower temperatures at the turbine blade root, where mechanical stresses are highest, and at the tip of the blade, which is the most difficult to cool [40]. A parameter that takes the design profile into account is the *turbine profile factor,* which is defined as

$$\text{Turbine profile factor} = \frac{(T_{4.r} - T_{4.des})_{max}}{T_4 - T_3}, \tag{4.23}$$

where $(T_{4.r} - T_{4.des})_{max}$ is the maximum temperature difference between the average temperature at any given radius around the circumference and the design temperature for that same radius.

4.10 Dilution Zone Design

At this stage in the design process, the amount of air available for dilution purposes will have been established, along with estimates of liner diameter and liner pressure-loss factor. The principal dilution-zone design variables are the number and size of the air-admission holes and the zone length. To ensure a satisfactory temperature profile at the chamber outlet, there must be adequate penetration of the dilution air jets, coupled with the correct number of jets to form sufficient localized mixing regions. The penetration of a round jet is a function of its diameter (see Equation 4.20). If the total dilution-hole area is spread over a large number of small holes, penetration will be inadequate, and a hot core will persist through the dilution zone. At the other extreme, the use of a small number of large holes will result in a cold core, due to overpenetration and unsatisfactory mixing. Thus, the first step in the design process is to determine the optimum number and size of the dilution holes.

4.10.1 Cranfield Design Method

If a liner wall contains a row of n dilution holes, each of which has an effective diameter d_j, then the total mass flow rate of air through these holes is given by

$$\dot{m}_j = (\pi / 4) n d_j^2 \rho_3 U_j. \tag{4.24}$$

Now

$$U_j = (2\Delta P_L / \rho_3)^{0.5}. \tag{4.25}$$

Hence

$$\dot{m}_j = (\pi / 4) n d_j^2 (2\Delta P_L \rho_3)^{0.5},$$

and

$$n d_j^2 = 15.25\, \dot{m}_j (P_3 \Delta P_L / T_3)^{-0.5}. \tag{4.26}$$

This equation may be used along with Equation 4.20 to determine optimum values of n and d_j for both tubular and annular combustors. For tubular combustors, Y_{max} in Equation 4.20 is made equal to $0.33D_L$. This allows d_j to be calculated from Equation 4.20 and this optimum value of d_j is then substituted into Equation 4.26 to determine the optimum number of dilution holes. The actual geometric diameter of the holes is then given by

$$d_h = d_j / C_D^{0.5}, \tag{4.27}$$

where C_D is obtained from Figure 4.2 or 4.3, or from References [6–10].

For annular combustors, the procedure is exactly the same except that Y_{max} in Equation 4.20 is made equal to $0.40D_L$.

For both tubular and annular combustors, the length of the dilution zone should be around $1.5D_L$. Shorter lengths lead to inadequate mixing, whereas longer lengths do not improve the pattern factor significantly because the additional wall-cooling air required reduces the amount available for dilution.

4.10.2 NASA Design Method

Holdeman et al. [23,34,37] analyzed experimental and CFD data from a number of NASA-Lewis-sponsored studies to obtain the following expression for calculating the optimum number of dilution holes for best mixing:

$$n_{opt} = \pi(2J)^{0.5} / C, \tag{4.28}$$

where C is an experimentally derived constant.

According to Holdeman et al. [17,23], the optimum value of C for cylindrical ducts and rectangular ducts featuring single-sided air injection is 2.5. For rectangular ducts having opposed rows of jets, C_{opt} is 1.25 for in-line injection and 5.0 for staggered injection.

Most of the data on which Equation 4.28 is based were obtained using round dilution holes and low values of $\dot{m}_j / \dot{m}_g$ (usually less than unity). Srinivasan et al. [33] found that the optimum value of C for round holes in a cylindrical duct applied equally well to square holes of the same area, a result that was later confirmed by Zhu et al. for higher values of $\dot{m}_j / \dot{m}_g$ [32]. However, Zhu et al. noted that the value of C_{opt} for round holes needs to be modified for rectangular or slanted slots.

Bain et al. [28] also employed a high value of $\dot{m}_j / \dot{m}_g$ (2.0) in their study on axially opposed rows of staggered and in-line jets injected into a rectangular crossflow. Their results generally substantiate Equation 4.28, but they recommend that C should be increased by a factor of 1.8 for two-sided, in-line configurations.

4.10.3 Comparison of Cranfield and NASA Design Methods

The main difference between these two approaches to dilution-zone design is that one stresses the importance of hole size, while the other places primary emphasis on hole spacing. Thus, for any given values of *J* and downstream distance, the Cranfield method leads to an optimum hole *size* and the spacing between adjacent holes is then chosen to provide the design value of $\dot{m}_j / \dot{m}_g$. On the other hand, the NASA procedure first identifies the optimum hole *spacing* and the hole size is then estimated to give the required ratio of $\dot{m}_j$ to $\dot{m}_g$.

Unfortunately, the two different approaches do not always lead to the same conclusions. For example, if *J* is increased by increasing U_j, then both methods make the same recommendations in regard to how the number, size, and spacing of the holes should be changed in order to retain optimum penetration and mixing. However, if *J* is increased by reducing U_g, then the two approaches give different results.

An interesting feature of the Cranfield method is that it takes account in a very direct manner (see Equation 4.20) the adverse effects on jet penetration and mixing arising from the aerodynamic blockage created by the presence of adjacent air jets. This could be a useful asset in situations where the ratio $\dot{m}_j / \dot{m}_g$ is exceptionally high as, for example, in the quick-quench zone of an RQL combustor (see Chapter 9).

4.11 Correlation of Pattern Factor Data

Two parameters of crucial importance to pattern factor are the liner length, which controls the time and distance that are available for mixing, and the pressure drop across the liner, which governs the penetration of the dilution jets and their rate of mixing with the products of combustion. From the analysis of experimental data on tubular, tuboannular, and annular combustors, it is found that

$$\frac{T_{max} - T_4}{T_4 - T_3} = f\left(\frac{L_L}{D_L} \times \frac{\Delta P_L}{q_{ref}}\right), \tag{4.29}$$

where $\Delta P_L/q_{ref}$ is the liner pressure-loss factor, L_L is the total liner length, and D_L is the liner diameter or height.

The correlation of data obtained for tubular and annular liners is shown in Figures 4.11 and 4.12, respectively. In connection with these figures, it should be noted that the correlation is based not on the *L/D* ratio of the dilution zone, but on that of the complete liner. This is found to provide a better fit to the data.

For tubular and tuboannular combustors, it is found that

$$\frac{T_{max} - T_4}{T_4 - T_3} = 1 - \exp\left(-0.070\frac{L_L}{D_L}\frac{\Delta P_L}{q_{ref}}\right)^{-1}, \tag{4.30}$$

while for annular combustors

$$\frac{T_{max} - T_4}{T_4 - T_3} = 1 - \exp\left(-0.050\frac{L_L}{D_L}\frac{\Delta P_L}{q_{ref}}\right)^{-1}. \tag{4.31}$$

Equations 4.30 and 4.31 embody the assumption that the total length of the liner is available for mixing. This is valid for most practical purposes because pattern factor is important to engine life only at the highest combustion pressures, where evaporation rates are so fast that the time and space

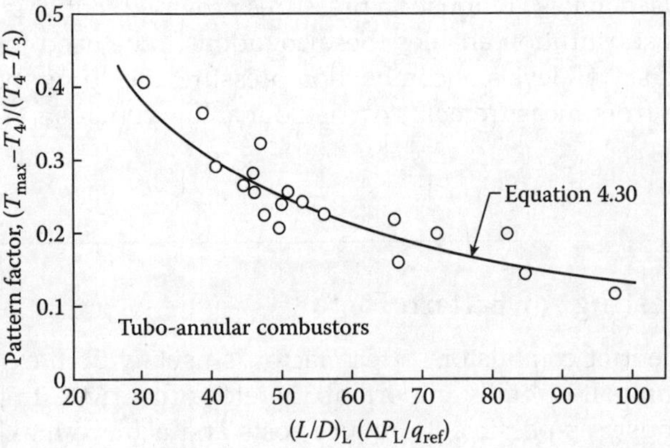

FIGURE 4.11
Pattern factor correlation for tubular and tuboannular combustors.

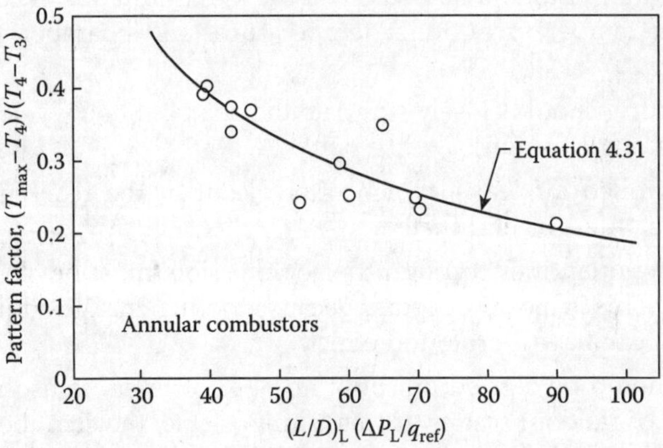

FIGURE 4.12
Pattern factor correlation for annular combustors.

occupied in fuel evaporation are negligibly small. However, at low combustion pressures, the liner length employed in fuel evaporation constitutes a significant fraction of the total line length and can no longer be ignored. It may be obtained as [41]

$$L_e = 0.33 \cdot 10^6 \dot{m}_A D_o^2 / \rho_g A_L \lambda_{eff}, \tag{4.32}$$

where λ_{eff} is a measure of fuel volatility as discussed in Chapter 2.

Equation 4.32 shows that the influence of fuel volatility on evaporation length (and hence on pattern factor) is relatively small at high combustion pressures, corresponding to high values of ρ_g, but becomes increasingly important with a reduction in combustion pressure. Equations 4.30 and 4.31

can still be used at low combustion pressures provided that L_L is replaced by (L_L–L_e). This substitution enables these equations to be used for predicting pattern factors at all levels of combustion pressure, including those at maximum power, from measurements carried out at low combustion pressures.

4.12 Rig Testing for Pattern Factor

The assessment of combustor pattern factor is usually both expensive and time consuming. It is vitally important, therefore, that rig testing be carried out as rigorously as possible, to reduce costs and achieve maximum credibility for the results obtained. The following general guidelines should be observed:

- Rig testing should be carried out at the maximum power conditions of inlet temperature, outlet temperature, and chamber reference velocity.
- The test rig should closely simulate the chamber entry velocity profile and angle of swirl.
- Uncontrolled air leaks, which often occur at the rear end of the chamber, must be eliminated.
- The flow quantities through the cooling slots must be closely controlled. This is most important because tooling changes during production can have a profound effect.
- Tests should be carried out on a number of liners to ascertain the degree of random scatter. If this is impossible, the chamber should be stripped and rebuilt a few times, and the pattern factor measured after each reassembly.
- The fuel injectors should be checked for flow number and spray patternation.

4.13 Swirler Aerodynamics

The primary-zone airflow pattern is of prime importance to flame stability. Many different types of airflow patterns are employed, but one feature common to all is the creation of a toroidal flow reversal that entrains and recirculates a portion of the hot combustion products to mix with the incoming air and fuel. These vortices are continually refreshed by air admitted through holes pierced in the liner walls, supplemented in most cases by air flowing through swirlers and flare-cooling slots, and by air employed in atomization.

One of the most effective ways of inducing flow recirculation in the primary zone is to fit a swirler in the dome around the fuel injector. Vortex breakdown is a well-known phenomenon in swirling flows; it causes recirculation in the core region when the amount of rotation imparted to the flow is high, as illustrated in Figure 4.13 [42]. This type of recirculation provides better mixing than is normally obtained by other means, such as bluff bodies, because swirl components produce strong shear regions, high turbulence, and rapid mixing rates. These characteristics of swirling flows have long been recognized and have been used in many practical combustion devices to control the stability and intensity of combustion and the size and shape of the flame region.

A comprehensive review of swirling flows and their practical applications is contained in the volume *Swirl Flows* by Gupta et al. [42]. Others have studied the flow characteristics of air swirlers, notably Beer and Chigier [43], Mathur and Maccallum [44], and Kilik [45]. The work of Kilik is of special interest because it directly addresses the types of swirlers of most relevance to the gas turbine. These and other experimental studies have provided valuable information on the flow characteristics and aerodynamic performance of swirlers. The manner in which this knowledge is employed in the design of swirlers for gas turbine combustors has been described by Dodds and Bahr [40].

Air swirlers are widely used in both tubular and annular combustors. The two main types of swirlers are *axial* and *radial*, as illustrated in Figure 4.14 [40]. They are often fitted as single swirlers, but sometimes as double swirlers

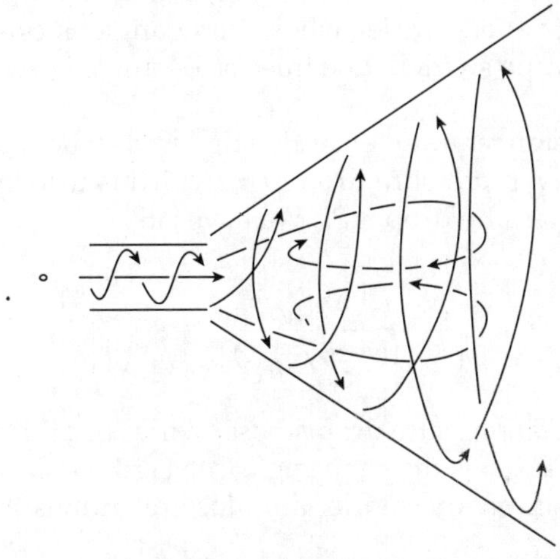

FIGURE 4.13
Flow recirculation induced by strong swirl. (From Gupta, A.K., Lilley, D.G., and Syred, N., *Swirl Flows*, Abacus Press, Tunbridge Wells, UK, 1984. With permission.)

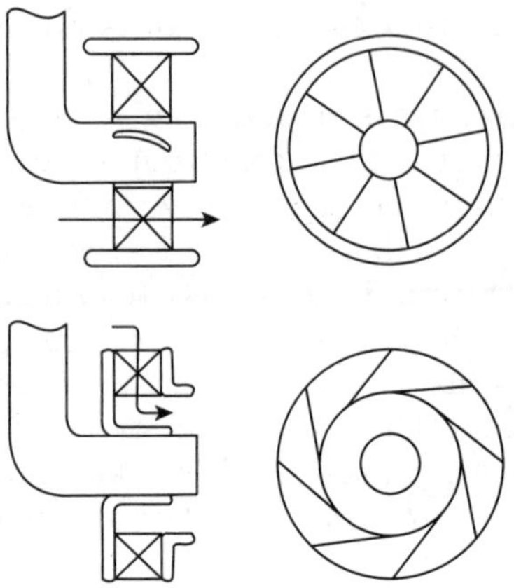

FIGURE 4.14
Two main swirler types. (From Dodds, W. J. and Bahr, D. W., *Design of Modern Gas Turbine Combustors*, Academic Press, San Diego, CA, 343–476, 1990. Copyright Elsevier. With permission.)

that are mounted concentrically and arranged to supply either co-rotating or counter-rotating airflows. Examples of modern combustors fitted with axial, radial, and double swirlers may be found in Chapter 9.

4.14 Axial Swirlers

The conventional notation for axial swirlers is indicated in Figure 4.15. This figure shows a flat-vaned swirler whose vane angle is constant and equal to θ. With curved-vane swirlers, the inlet blade angle is zero and the outlet angle is θ.

An important design requirement is that the swirler should pass the desired airflow rate for a given pressure drop ΔP_{sw}, which is usually assumed to be equal to the liner pressure drop, ΔP_L. We have [46]

$$\dot{m}_{sw} = \left\{ \frac{2\rho_3 \Delta P_{sw}}{K_{sw}\left[(\sec\theta / A_{sw})^2 - 1/A_L^2\right]} \right\}^{0.5}, \tag{4.33}$$

where ΔP_{sw} is the total pressure drop across swirler ($\cong \Delta P_L$), A_{sw} is the frontal area of the swirler, and θ is the vane angle (see Figure 4.15).

Note that A_{sw} is simply the swirler annulus area minus the area occupied by the vanes, i.e.,

$$A_{sw} = (\pi/4)\left(D_{sw}^2 - D_{hub}^2\right) - 0.5 n_v t_v \left(D_{sw} - D_{hub}\right). \tag{4.34}$$

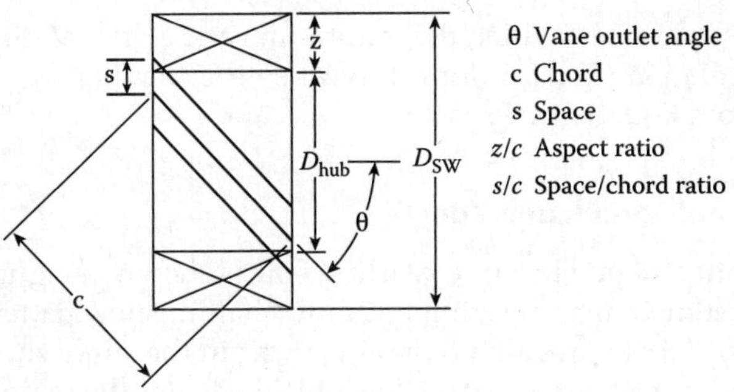

FIGURE 4.15
Notation for axial swirlers.

Typical ranges of values for the design variables in Equations 4.33 and 4.34 are [40,46]

Vane angle, θ	30°–60°
Vane thickness, t_v	0.7–1.5 mm
Number of vanes, n_v	8–16
ΔP_{sw}	3%–4% of P_3
K_{sw}	1.3 for flat vanes, and 1.15 for curved vanes

The hub diameter, D_{hub}, is determined by the need to provide space for a centrally mounted fuel injector. The outer swirler diameter, D_{sw}, is then obtained by substituting the calculated values of A_{sw} from Equation 4.33 into Equation 4.34.

4.14.1 Swirl Number

Beer and Chigier [43] have proposed the following nondimensional criterion to characterize the amount of rotation imparted to the axial flow:

$$S_N = 2G_m / (D_{sw}G_t), \tag{4.35}$$

where G_m is the axial flux of angular momentum and G_t is the axial thrust.

For values of swirl number less than around 0.4, no flow recirculation is obtained, and the swirl is described as *weak*. Most swirlers of practical interest operate under conditions of strong swirl (that is, $S_N > 0.6$).

Expressions for calculating swirl numbers for various types of swirl generators have been derived by Beer and Chigier [43]. For an annular swirler with constant vane angle θ, they give

$$S_N = \frac{2}{3}\frac{1-(D_{hub}/D_{sw})^3}{1-(D_{hub}/D_{sw})^2}\tan\theta. \tag{4.36}$$

Thus, for a simple axial swirler, the minimum vane angle required to obtain strong recirculation ($S_N > 0.6$) for a typical swirler having $D_{hub}/D_{sw} = 0.5$ is calculated from Equation 4.36 as 38°.

4.14.2 Size of Recirculation Zone

The recirculation region in a free swirling flow is shown in Figure 4.16. Since the flow is assumed to be axisymmetric, only half the flow pattern is considered. The recirculation region is contained within the curve *ACB*. The point B is called the *stagnation point*. The flow outside *ACB* is the main flow, which drives the recirculation along the solid curve *AB*. Conditions of zero axial velocity are represented by the dashed curve *AB*.

Typical axial and swirl velocity profiles are shown in Figure 4.17. All the velocity components decay in the downstream direction. After the stagnation point, the reverse axial velocities disappear, and further downstream the peak of the axial velocity profile shifts toward the centerline as the effect of swirl diminishes.

Factors governing the size of the recirculation zone have been studied by several workers. The most comprehensive investigation is that of Kilik [45], who examined the separate effects on recirculation-zone size of variations

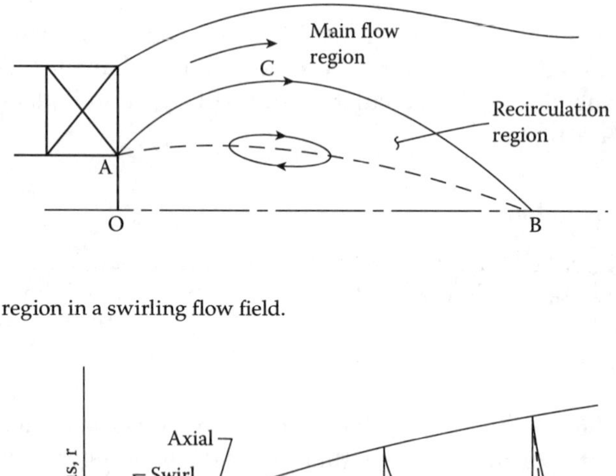

FIGURE 4.16
Recirculation region in a swirling flow field.

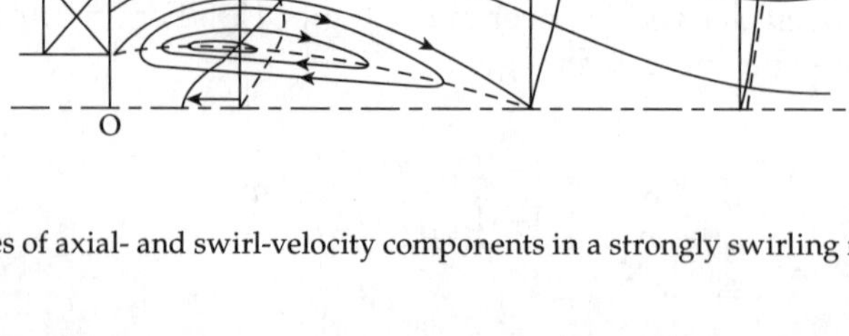

FIGURE 4.17
Typical profiles of axial- and swirl-velocity components in a strongly swirling flow.

in vane type (flat or curved), vane angle, vane aspect ratio, and space/chord ratio. His experimental data show that the size of the recirculation zone is increased by

1. An increase in vane angle
2. An increase in the number of vanes
3. A decrease in vane aspect ratio
4. Changing from flat to curved vanes

4.14.3 Flow Reversal

One of the primary functions of the swirler is to induce combustion products to flow upstream to meet and merge with the incoming fuel and air. For weak swirl, there is little or no flow recirculation, but when the swirl number is increased and reaches a critical value ($S_N > 0.4$), the static pressure in the central core just downstream of the swirler becomes low enough to create flow recirculation, as indicated in Figure 4.13. From velocity measurements carried out along the swirler axis for several swirler designs, Kilik [45] was able to ascertain the influence of the key geometric parameters on the reverse mass flow rate. His results showed that curved-vane swirlers induce larger reverse mass flows than the corresponding flat-vane swirlers, and that the reverse flow is increased by an increase in swirl number. The effect of swirl number on the maximum reverse mass flow rate is illustrated in Figure 4.18.

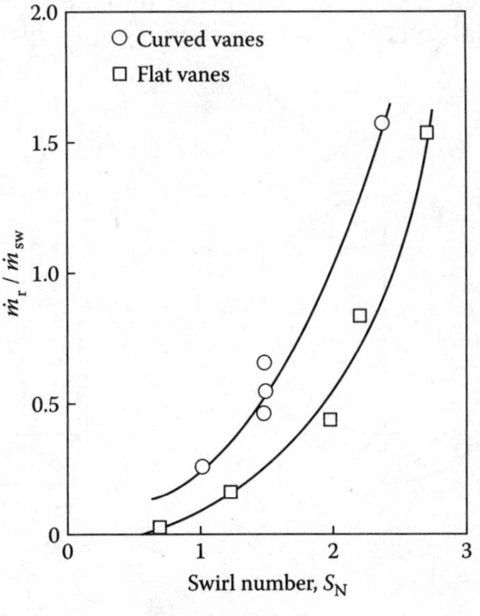

FIGURE 4.18
Influence of swirl number on maximum reverse mass flow.

This figure includes Kilik's data for curved vane swirlers and Mathur and Maccallum's [44] results for flat-vaned swirlers. It is of interest to note that under conditions of very strong swirl, corresponding to vane angles of around 65°, the reverse mass flow created by the swirler can actually exceed the swirler flow.

4.14.4 Influence of Swirler Exit Geometry

The geometry of the transition between the swirler exit and the combustor dome can have a strong influence on the primary-zone flow field and can also affect the trajectories of droplets injected into this flow field [40]. For example, a diverging passage downstream of the swirler exit increases both the size of the recirculation zone and the amount of air recirculated [43].

4.15 Radial Swirlers

Radial inflow swirlers are now widely used in both conventional and dry low emissions (DLE) combustors. Their flow characteristics have not been studied to the same extent as those of axial swirlers, but experience has shown that the flow fields generated by the two different swirler types are broadly the same. Thus, the design rules established for axial swirlers can provide useful guidance in the design of radial swirlers. The notation for conventional radial swirlers is indicated in Figure 4.19.

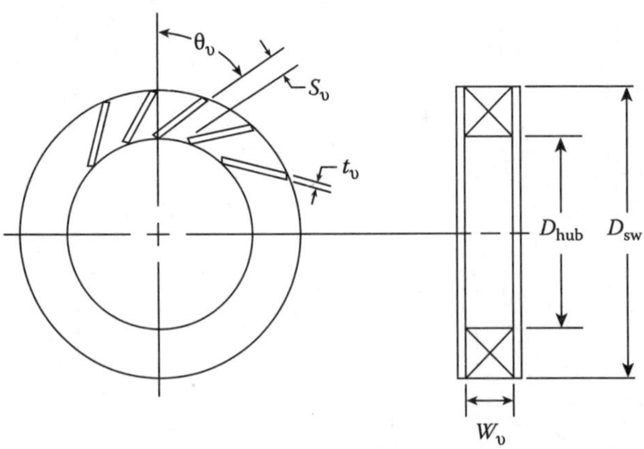

FIGURE 4.19
Notation for radial swirlers. (From Dodds, W. J. and Bahr, D. W., *Design of Modern Gas Turbine Combustors*, Academic Press, San Diego, CA, 343–476, 1990. Copyright Elsevier. With permission.)

The airflow into the swirler is determined by the effective flow area at the trailing edge of the vane. This can be calculated as

$$A_{sw} = n_{\upsilon} s_{\upsilon} w_{\upsilon} C_D, \tag{4.37}$$

where n_{υ} is the number of vanes, s_{υ} is the vane gap, w_{υ} is the vane width, and C_D is the vane discharge coefficient (see Figure 4.19). According to Dodds and Bahr [40], for preliminary design purposes an appropriate value for C_D is 0.7. The dimension w_{υ} can then be adjusted during combustor development to obtain the desired swirler airflow rate.

4.16 Flat Vanes Versus Curved Vanes

All the evidence presented in this section on swirling flows shows that curved vanes are more efficient aerodynamically than flat vanes. This is because they allow the incoming axial flow to gradually turn, which inhibits flow separation on the suction side of the vane. Thus, more complete turning and higher swirl- and radial-velocity components are generated at the swirler exit, which results in a larger recirculation zone and a higher reverse flow rate. However, these arguments in favor of curved vanes should not exclude the use of flat vanes in certain applications. One advantage of flat vanes is that they are cheap and easy to produce. Moreover, the flow striations associated with flat-vane swirlers, which are created by the stalled regions attached to each vane, tend to promote a more stable flame and reduce combustion noise. Another asset of the flat-vane axial swirler is that its exit velocity profile is less peaked and less biased radially outboard than that of the corresponding curved-vane swirler. In consequence, it provides better aeration of the main soot-forming zone, which is normally located just downstream of the fuel injector.

For these reasons, flat-vane swirlers are still preferred in some combustor configurations. However, when air swirlers are incorporated into airblast atomizers, curved vanes should always be used because the wakes produced downstream of flat vanes could adversely affect the quality of atomization.

Nomenclature

A area, m^2

$A_{h.geom}$ hole area, geometric, m^2

$A_{h.eff}$ total effective liner hole area, m^2

A_{an}	annulus area, m^2
A_r	area ratio ($A_{h \cdot geom}/A_{an}$)
C_D	discharge coefficient
D_{hub}	swirler hub diameter, m (see Figure 4.15)
D_L	liner diameter or height, m
D_o	Sauter mean diameter of fuel spray, m
D_{sw}	swirler diameter, m (see Figure 4.15)
d	diameter, m
d_s	diameter of secondary hole, m
J	momentum-flux ratio, $(\rho_j U_j^2)/(\rho_g U_g^2)$
K	hole pressure-drop coefficient, $(1 + \Delta p_L/q_{an})$
K_1, K_2	constants in Equation 4.10
k	ratio of liner to casing area
L	length, m
L_e	evaporation length, m
M	Mach number
$\dot{m}$	*m*air mass flow rate, kg/s
$\dot{m}_e$	entrained mass flow rate, kg/s
$\dot{m}_p$	ratio of primary-zone airflow to total chamber airflow
$\dot{m}_r$	maximum reverse mass flow rate, kg/s
m_{sn}	ratio of air entering snout to total chamber airflow
n	number of holes
P	total pressure, Pa
p	static pressure, Pa
q	dynamic pressure, Pa
R	gas constant, 286.9 Nm/(kg K)
r	ratio of casing area to combustor inlet area
S	length of jet path, m
S_N	swirl number
T	total temperature, K
U	velocity, m/s
X	distance downstream of hole, m
X	downstream distance, m
Y	jet penetration, m
α	hole bleed ratio ($\dot{m}_h / \dot{m}_{an}$)
γ	ratio of specific heats
Δ	difference
θ	initial jet angle, or swirler vane angle
λ	diffuser pressure-loss coefficient (see Chapter 3)
λ_{eff}	effective evaporation constant, m^2/s
ρ	density

Subscripts

0	initial value
3	combustor inlet plane

4	combustor outlet plane
an	annulus value
diff	diffuser
g	gas value
h	hole value
I	local value
j	jet value
L	liner value
max	maximum value
pz	primary-zone value
ref	reference value
sw	swirler value

References

1. Scull, W. E., and Mickelsen, W. R., "Flow and Mixing Processes in Combustion Processes," Chap. II, NACA Report 1300, 1957.
2. Lefebvre, A. H., and Norster, E. R., "The Design of Tubular Combustion Chambers for Optimum Mixing Performance," Technical Advances in Gas Turbine Design," pt. 3N, *Proceedings of the Institution of Mechanical Engineers*, Institution of Mechanical Engineers, London, UK, 1969.
3. Sullivan, D. A., "Gas Turbine Combustor Analysis," ASME Paper 74-GT-2, 1974.
4. Kaddah, K. S., "Discharge Coefficients and Jet Deflection Angles for Combustor Liner Air Entry Holes," College of Aeronautics MSc thesis, Cranfield, UK, 1964.
5. Freeman, B. C., "Discharge Coefficients of Combustion Chamber Dilution Holes," College of Aeronautics MSc thesis, Cranfield, UK, 1965.
6. Callaghan, E. E., and Bowden, D. T., "Investigation of Flow Coefficients of Circular, Square, and Elliptical Orifices at High Pressure Ratios," NACA TN 1947, 1949.
7. Dittrich, R. T., and Graves, C. C., "Discharge Coefficients for Combustor Liner Air Entry Holes I, Circular Holes," NACA TN 3663, 1956.
8. Dittrich, R. T., "Discharge Coefficients for Combustor Liner Air Entry Holes II, Flush Rectangular Holes, Step Louvers, and Scoops," NACA TN 3924, 1958.
9. Graves, C. C., and Grobman, J. S., "Theoretical Analysis of Total Pressure Loss and Airflow Distribution for Tubular Turbojet Combustors with Constant Annulus and Liner Cross-Sectional Areas," NACA Report 1373, 1958.
10. Grobman, J. S., "Comparison of Calculated and Experimental Total Pressure Loss and Airflow Distribution in Tubular Turbojet Combustors with Tapered Liners," NASA Memo 11-26-58E, 1959.
11. Fletcher, R. S., and Bastress, E.K., "The Design and Performance of Gas Turbine Combustion Chambers," Northern Research and Engineering Corporation, NREC Report 1082, 1964.
12. Carrotte, J. F., and Stevens, S. J., "The Influence of Dilution Hole Geometry on Jet Mixing," *Journal of Engineering for Gas Turbines and Power*, Vol. 112, No. 1, pp. 73–9, 1990.

13. Norster, E. R., "Jet Penetration and Mixing Studies," unpublished work at College of Aeronautics, Cranfield, UK, 1964.
14. Lefebvre, A. H., unpublished work, 1979.
15. Sridhara, K., "Gas Mixing in the Dilution Zone of a Combustion Chamber," College of Aeronautics MSc thesis 20/187, Cranfield, UK, 1967 (see also "Gas Mixing in the Dilution Zone of a Combustion Chamber," National Aeronautics Laboratory Report TN 30, Bangalore, India, 1970).
16. Norster, E. R., private communication.
17. Holdeman, J. D., "Mixing of Multiple Jets with a Confined Subsonic Crossflow," *Progress in Energy and Combustion Science*, Vol. 19, No. 1, pp. 31–70, 1993.
18. Lefebvre, A. H., and Norster, E. R., "A Design Method for the Dilution Zones of Gas Turbine Combustion Chambers," College of Aeronautics Note 169, Cranfield, UK, 1966.
19. Holdeman, J. D., Walker, R. E., and Kors, D. L., "Mixing of Multiple Dilution Jets with a Hot Primary Airstream for Gas Turbine Combustors," AIAA Paper 73-1249, 1973.
20. Holdeman, J. D., and Walker, R. E., "Mixing of a Row of Jets with a Confined Crossflow," *AIAA Journal*, Vol. 15, No. 2, pp. 243–9, 1977.
21. Holdeman, J. D., "Perspectives on the Mixing of a Row of Jets with a Confined Crossflow," AIAA Paper 83-1200, 1983.
22. Holdeman, J. D., Srinivasan, R., and Berenfeld, A., "Experiments in Dilution Jet Mixing," *AIAA Journal*, Vol. 22, No. 10, pp. 1436–43, 1984.
23. Holdeman, J. D., Srinivasan, R., Coleman, E. B., Meyers, G. D., and White, C. D., "Experiments in Dilution Jet Mixing—Effects of Multiple Rows and Noncircular Orifices," AIAA Paper 85-1104, 1985.
24. Kamotani, Y., and Greber, I., "Experiments on Confined Turbulent Jets in Cross Flow," NASA CR-2392, 1974.
25. Walker, R. E., and Kors, D. L., "Multiple Jet Study," Final Report, NASA CR 121217, 1973.
26. Srinivasan, R., Berenfeld, A., and Mongia, H. C., "Dilution Jet Mixing Program Phase I Report," NASA CR-168031, 1982.
27. Holdeman, J. D., and Srinivasan, R., "Perspectives on Dilution Jet Mixing," AIAA Paper 86-1611, 1986.
28. Bain, D. B., Smith, C. E., and Holdeman, J. D., "Mixing Analysis of Axially-Opposed Rows of Jets Injected into Confined Crossflow," *Journal of Propulsion and Power*, Vol. 11, No. 5, pp. 885–93, 1995.
29. Vranos, A., Liscinsky, D. S., True, B., and Holdeman, J. D., "Experimental Study of Cross-Stream Mixing in a Cylindrical Duct," AIAA Paper 91-2459, 1991.
30. Talpallikar, M. V., Smith, C. E., Lai, M. C., and Holdeman, J. D., "CFD Analysis of Jet Mixing in Low NO_x Flametube Combustors," *Journal of Engineering for Gas Turbines and Power*, Vol. 114, No. 2, pp. 416–24, 1992.
31. Hatch, M. S., Sowa, W. A., Samuelsen, G. A., and Holdeman, J. D., "Geometry and Flow Influences on Jet Mixing in a Cylindrical Duct," *Journal of Propulsion and Power*, Vol. 11, No. 3, pp. 393–402, 1995.
32. Zhu, G., Lai, M. C., and Lee, T., "Penetration and Mixing of Radial Jets in Neck-Down Cylindrical Crossflow," *Journal of Propulsion and Power,* Vol. 11, No. 2, pp. 252–60, 1995.
33. Srinivasan, R., Coleman, E., and Johnson, K., "Dilution Jet Mixing Program Phase II Report, NASA CR-174624, 1984.

34. Holdeman, J. D., Srinivasan, R., Coleman, E. B., Meyers, G. D., and White, C. D., "Effects of Multiple Rows and Non-Circular Orifices on Dilution Jet Mixing," *Journal of Propulsion and Power,* Vol. 3, No. 3, pp. 219–26, 1987.
35. Holdeman, J. D., and Srinivasan, R., "Modeling of Dilution Jet Flowfields," *Combustion Fund Research,* NASA CP-2309, pp. 175–87, 1984.
36. Holdeman, J. D., and Srinivasan, R., "Modeling Dilution Jet Flowfields," *Journal of Propulsion and Power,* Vol. 2, No. 1, pp. 4–10, 1986.
37. Holdeman, J. D., Srinivasan, R., and White, C. D., "An Empirical Model of the Effects of Curvature and Convergence on Dilution Jet Mixing," AIAA Paper 88-3180, 1988.
38. Liscinsky, D. S., True, B., and Holdeman, J. D., "Crossflow Mixing of Noncircular Jets," *Journal of Propulsion and Power,* Vol. 12, No. 2, pp. 225–30, 1996.
39. Wittig, S. L. K., Elbahar, O. M. F., Nott, B. E., and Kutz, R., "Temperature Profile Development in Turbulent Mixing of Coolant Jets with a Confined Hot Crossflow," *Journal of Engineering for Gas Turbines and Power,* Vol. 106, No. 1, pp. 193–7, 1984.
40. Dodds, W. J., and Bahr, D. W., "Combustion System Design," in A. M. Mellor, ed., *Design of Modern Gas Turbine Combustors,* pp. 343–476, Academic Press, San Diego, CA, 1990.
41. Chin, J. S., and Lefebvre, A. H., "Effective Values of Evaporation Constant for Hydrocarbon Fuel Drops," *Proceedings of the Twentieth Automotive Technology Development Contractor Coordination Meeting,* SAEP-120, pp. 325–32, Society of Automotive Engineers, Warrendale, PA, 1982.
42. Gupta, A. K., Lilley, D. G., and Syred, N., *Swirl Flows,* Abacus Press, Tunbridge Wells, UK, 1984.
43. Beer, J. M., and Chigier, N. A., *Combustion Aerodynamics,* Applied Science, London, 1972.
44. Mathur, M. L., and Maccallum, N. R. L., "Swirling Air Jets Issuing from Vane Swirlers, part I, Free Jets," *Journal of the Institute of Fuel,* Vol. 40, pp. 214–25, 1967.
45. Kilik, E., "The Influence of Swirler Design Parameters on the Aerodynamics of the Downstream Recirculation Region," PhD thesis, School of Mechanical Engineering, Cranfield Institute of Technology, UK, 1976.
46. Knight, M. A., and Walker, R. B., "The Component Pressure Losses in Combustion Chambers," Aeronautical Research Council R and M 2987, UK, 1957.

5

Combustion Performance

5.1 Introduction

Combustion chambers are required to burn stably over a wide range of operating conditions with levels of combustion efficiency close to 100%. Another important requirement is easy and reliable lightup during ground starting while the engine is being cranked up to its self-sustaining speed. The aircraft gas turbine has the additional requirement of rapid relighting of the combustor after a flameout in flight. Thus, the combustion performance parameters of prime importance to the gas turbine are combustion efficiency, stability, and ignition. The essential features of these three major topics are reviewed in this chapter.

5.2 Combustion Efficiency

Failure to achieve high levels of combustion efficiency is generally regarded as unacceptable, partly because combustion inefficiency represents a waste of fuel, but mainly because it is manifested in the form of pollutant emissions, such as unburned hydrocarbons and carbon monoxide. That is why current emissions regulations call for combustion efficiencies in excess of 99%. For the modern aircraft engine, combustion efficiency is effectively 100% at takeoff conditions. To avoid the production of "white" smoke, efficiency must exceed 96% and at no point in the operating cycle is it less than 90%. High combustion efficiency is necessary at this "off design" point because, with the engine windmilling, the pressure and temperature of the air flowing through the combustor are close to ambient values. At high altitudes, these are so low that the stability limits are very narrow. This means that when the engine control system attempts to compensate for combustion inefficiency by supplying more fuel to the combustor, this extra fuel may cause a "rich extinction" of the flame. Thus, an important design requirement for an aircraft combustor is that it be sized large enough to ensure an adequate level of combustion efficiency during engine restart at the highest altitude at which relight capability is required.

5.2.1 The Combustion Process

The primary purpose of combustion is to raise the temperature of the airflow by efficient burning of fuel. From a design viewpoint, an important requirement is a means of relating combustion efficiency to the operating variables of air pressure, temperature, and mass flow rate, and to the combustor dimensions. Unfortunately, the various processes taking place within the combustion zone are highly complex and a detailed theoretical treatment is precluded at this time. Until more information is available, suitable parameters for relating combustion performance to combustor dimensions and operating conditions can be derived only through the use of very simplified models to represent the combustion process. One such model starts from the well-established and widely accepted notion that the total time required to burn a liquid fuel is the sum of the times required for fuel evaporation, mixing of fuel vapor with air and combustion products, and chemical reaction. As the time available for combustion is inversely proportional to the airflow rate, the combustion efficiency may be expressed [1] as

$$\eta_c = f(\text{airflow rate})^{-1}\left(\frac{1}{\text{evaporation rate}}+\frac{1}{\text{mixing rate}}+\frac{1}{\text{reaction rate}}\right)^{-1}. \tag{5.1}$$

In practical combustion systems, the maximum rate of heat release under any given operating conditions may be governed by either evaporation, mixing, or chemical reaction, but rarely by all three at the same time. However, when the combustion process is in transition from one regime to another, two of the three key rates will participate in determining the overall combustion efficiency. Before exploring that situation, let us examine the separate effects on combustion efficiency of chemical reaction, mixing, and evaporation.

5.3 Reaction-Controlled Systems

The two most widely used approaches for describing combustion efficiency under conditions where the overall rate of heat release is limited by chemical kinetics are the *burning velocity* and *stirred reactor* models.

5.3.1 Burning Velocity Model

Here, the combustion zone is envisaged as being similar in structure to the flame brush produced on a Bunsen burner under turbulent flow conditions. Combustion performance is then described as a function of the ratio of turbulent burning velocity to the velocity of the fresh mixture entering the combustion zone. It is assumed that evaporation rates and mixing rates are both infinitely fast, and that all of the fuel that burns does so completely. Combustion inefficiency arises when some of the mixture succeeds in passing through the combustion zone without being entrained by a turbulent flame front (see Figure 5.1).

This model was used by Greenhough and Lefebvre [2] in deriving a parameter that was shown to correlate experimental data on combustion efficiency obtained over wide ranges of pressure, temperature, and airflow rate for various designs of combustion chamber. The model is described only briefly below; for further details, reference should be made to the original paper and to the subsequent development of the model [3].

Combustion efficiency is defined as:

$$\eta_c = (\text{heat released in combustion})/(\text{heat available in fuel}), \tag{5.2}$$

$$= \left(\rho_g A_f S_T c_p \Delta T\right) / \left(q \dot{m}_A H\right). \tag{5.3}$$

Now $c_p \Delta T = qH$, by definition; also the flame area, A_f, may be assumed to be proportional to the combustor reference area, A_{ref}. Thus, Equation 5.3 simplifies to:

$$\eta_c \propto S_T / U_{ref}. \tag{5.4}$$

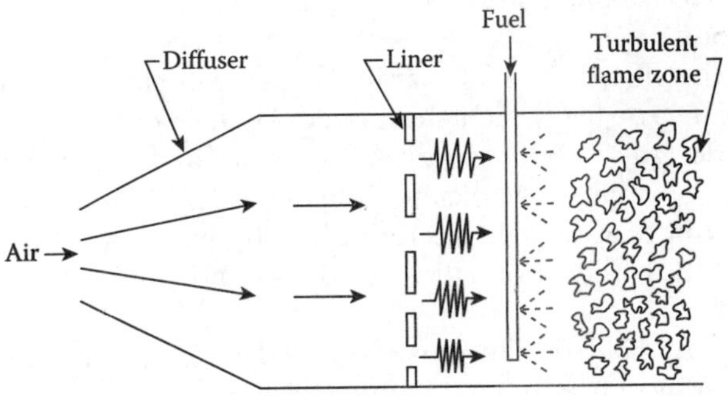

FIGURE 5.1
Burning velocity model for combustion efficiency.

If one expresses U_{ref} in terms of $\dot{m}_A$, P_3, and A_{ref} and describes S_T in terms of laminar burning velocity and turbulence intensity (which, in turn, is related to the liner pressure loss factor), Equation 5.4 becomes

$$\eta_c = \left[P_3 A_{ref} (P_3 D_{ref})^m \exp(T_3 / b) / \dot{m}_A \right] \left[\Delta P_L / q_{ref} \right]^{0.5m}. \quad (5.5)$$

Lefebvre and Halls [3] demonstrated that combustion efficiency data obtained during low pressure tests on several types of combustion chambers could be satisfactorily correlated by assigning values to m and b of 0.75 and 300, respectively. Substitution of these values into Equation 5.5, and neglecting the pressure loss term that varies little between one combustor and another, leads to the well-known θ parameter

$$\eta_\theta = f(\theta) = f\left[P_3^{1.75} A_{ref} D_{ref}^{0.75} \exp(T_3 / 300) / \dot{m}_A \right]. \quad (5.6)$$

Equation 5.6 has been applied with considerable success to the correlation of experimental data on combustion efficiency, and has proved very useful in reducing the amount of rig testing required to evaluate new combustor designs. As shown in Figure 5.2, only a few test points are needed to establish the complete performance curve for a chamber. Furthermore, it is possible to predict, with reasonable accuracy, combustion efficiencies at flow conditions that lie outside the capacity of the test facility—provided, of course, that at these extrapolated conditions the combustion performance is not limited by fuel evaporation or by any factor other than chemical reaction rates.

The main advantage of Equation 5.6 is that it provides a method of scaling combustor dimensions and operating conditions to common values so that any differences in performance that remain can be attributed directly to differences in design. This is a tremendous asset when one is attempting to select a design for a new combustion chamber from several existing designs, none of which is of the required size or has been tested at the relevant operating conditions.

The manner in which the θ parameter is used can be demonstrated by reference to Figure 5.3, which shows performance curves for two different combustor designs. Clearly, design A is superior to design B, because for any given value of combustion efficiency the θ parameter has a lower value. This means that under any given operating conditions of $\dot{m}_A$, P_3, and T_3, design A can equal the combustion efficiency of design B, and yet be made smaller in size.

Any new chamber design must be based to a large extent on previous experience. A most useful way in which past experience can be summarized is by the use of charts where combustion efficiency data from all known systems are correlated against all the relevant variables. Such a chart is shown in Figure 5.4, in which the hatched areas include experimental data obtained

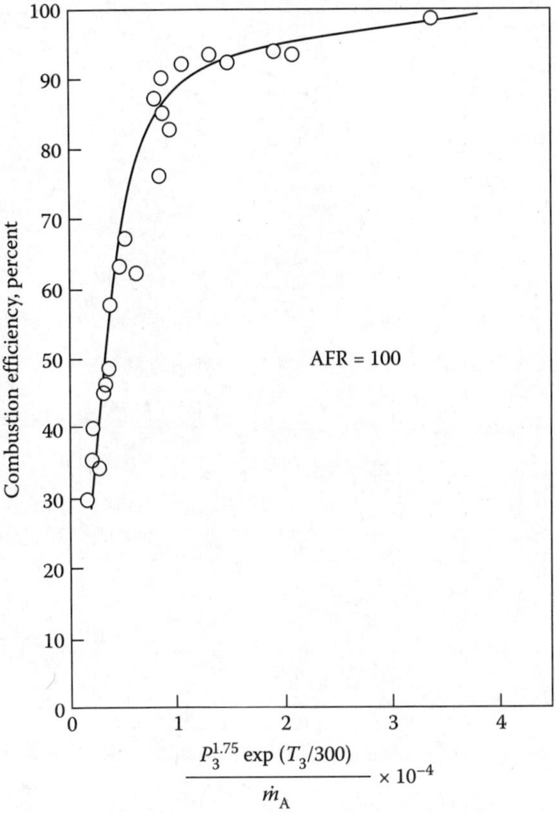

FIGURE 5.2
Correlation of combustion efficiency data for an aircraft combustor.

from a large number of multi-can, can-annular, and annular chambers. Figure 5.4 may be used to determine the size of chamber needed to meet any stipulated performance requirement. For all types of engines, the most arduous operating conditions are those at which the inlet pressure P_3, is a minimum. For aircraft engines, this usually corresponds to the engine windmilling after a flameout at high altitude.

When flameout occurs in flight, the engine rotational speed falls rapidly to its windmilling value. The relight sequence is first to use the ignition system to relight the combustor. When this has been accomplished, the next step is to accelerate the engine up to its normal rotational speed. This normally calls for a minimum combustion efficiency of around 80%. As previously discussed, a lower level of combustion efficiency could result in a rich extinction of the flame, whereas a higher level would lead to an unnecessarily large combustor.

Appropriate values of A_{ref} and D_{ref} may be obtained from Figure 5.4 by reading off a value of θ at a point along a horizontal line within the hatched

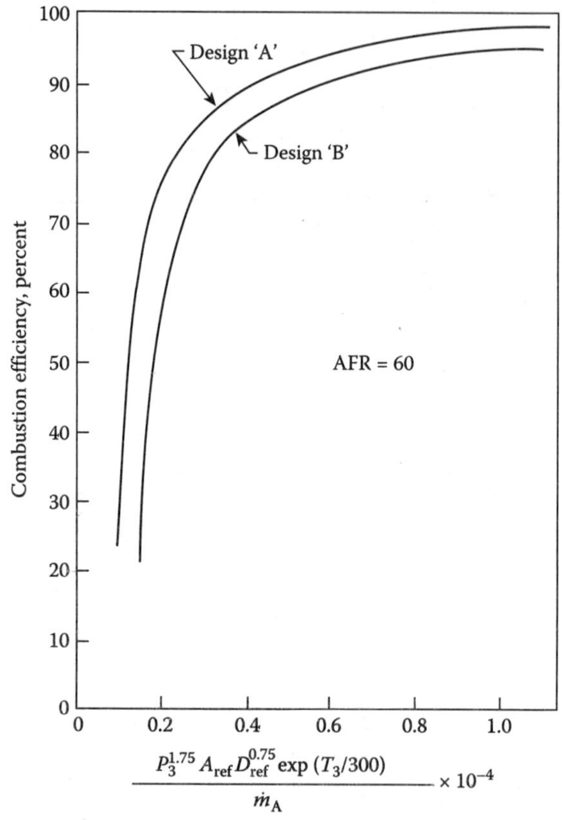

FIGURE 5.3
Combustion efficiency curves for two different combustion designs.

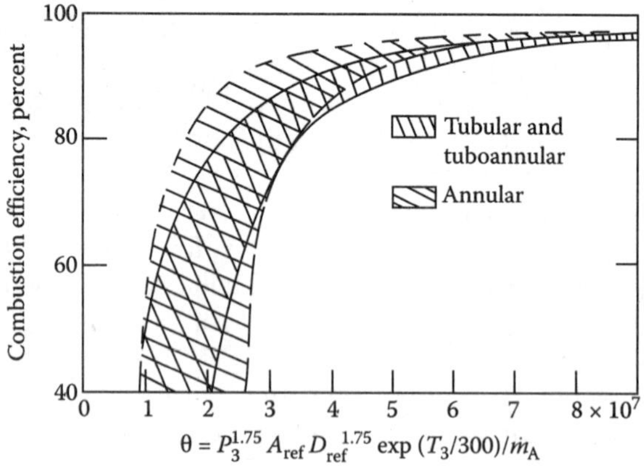

FIGURE 5.4
Design chart for conventional combustors.

area at 80% combustion efficiency, and then substituting into it the values of P_3, T_3, and $\dot{m}_A$ corresponding to the engine windmilling at the maximum guaranteed relight altitude. The actual point chosen within the hatched area will represent a balance between the conflicting needs of high performance, small combustor size, and low development cost.

A notable feature of the θ parameter is that it ignores the influence of drop size on combustion efficiency. The fact that it has been shown to work successfully over wide ranges of combustor types and operating conditions tends to confirm that drop sizes are indeed irrelevant to combustion efficiency, as the experimental study of Odgers et al. has demonstrated [4]. However, for fuels heavier than Jet A (JP5), the effects of atomization and evaporation cannot be ignored.

5.3.2 Stirred Reactor Model

Another simplified approach to the problem of providing a quantitative description of the combustion process in a gas turbine combustor is to regard the combustion zone as a perfectly stirred reactor into which fuel and air are fed at a constant rate and are instantaneously mixed with all the other material within the zone. Burned material leaves the zone at a constant rate, with temperature and composition identical to that within the zone.

The application of simple reaction rate theory to practical combustion systems seems to have been considered first by Childs [5], and then independently by Avery and Hart [6] and Bragg [7]. The approach is based on the notion that one limiting reaction governs the overall rate of combustion. Examination of the chemistry of the process shows that this is unlikely, and its justification lies entirely in the fact that it allows great simplification while explaining experimental results to a satisfactory degree.

A drawback to the stirred reactor approach is that heat-release rates are related to the reaction temperature. This is clearly a serious deficiency in any calculation aimed at determining combustion efficiency because the combustion efficiency must be known, a priori, in order to calculate the reaction temperature. However, Bragg [7] and Greenhough and Lefebvre [2] found that overall heat-release rates could be expressed in terms of inlet air temperature, as illustrated in Equation 5.6. We have

$$\eta_\theta = f\left[P_3^2 V_c \exp(T_3/300)/\dot{m}_A\right] \quad \text{for } n = 2, \tag{5.7}$$

or

$$\eta_\theta = f\left[P_3^{1.75} V_c \exp(T_3/300)/\dot{m}_A\right] \quad \text{for } n = 1.75. \tag{5.8}$$

(NOTE: The experiments of Longwell and Weiss [8] on stirred reactors indicate a pressure exponent, n, of 1.8, which is in close agreement with the value of 1.75, as obtained by Greenhough and Lefebvre [2] from analysis of combustion efficiency data.)

Inspection of Equations 5.6 and 5.8 shows that there is little difference between them. This is because the temperature dependence derived for Equation 5.6 has been assigned to Equation 5.8. For practical purposes, e.g., the correlation of experimental combustion efficiency data obtained from a given combustion system or from geometrically similar systems, the two parameters are effectively the same. It is in the comparison of combustion systems of basically different design that points of variance between the two parameters emerge. In the burning velocity model, the importance of the cross-sectional area of the burning zone is emphasized, as opposed to volume in the reaction rate analysis. This could be a useful advantage because it ensures that any new combustor designed on the basis of the θ parameter not only meets combustion efficiency requirements, but also has adequate combustion stability.

5.4 Mixing-Controlled Systems

If evaporation and chemical reaction rates are both infinitely fast, Equation 5.1 becomes

$$\eta_m = f(\text{mixing rate/air flow rate}). \tag{5.9}$$

The mixing rate between a turbulent air jet and the surrounding gas is given by the product of the eddy diffusivity, the mixing area, and the density gradient. If it is assumed that the eddy diffusivity is proportional to the product of a mixing length, l, and the turbulent velocity in the air jet, then

$$\begin{aligned} \text{mixing rate} &= (\text{eddy diffusivity})(\text{mixing area})(\text{density gradient}) \\ &= (lU_j)(l^2)(\rho / l) \\ &= \rho U_j l^2. \end{aligned} \tag{5.10}$$

Substituting in Equation 5.10 for $U_j \alpha (\Delta P_L/\rho)^{05}$ yields

$$\text{mixing rate} \propto (P_3 l^2 / T_3^{0.5})(\Delta P_L / P_3)^{0.5}. \tag{5.11}$$

Where mixing limits performance, combustion efficiency depends on the ratio of the mixing rate to the airflow rate. Thus, if one assumes that turbulence scale is proportional to combustor size, Equation 5.11 becomes

$$\eta_m = f\left(P_3 A_{ref} / \dot{m}_A T_3^{0.5}\right)\left(\Delta P_L / P_3\right)^{0.5}. \tag{5.12}$$

One important example of a system where mixing rates (and, under certain conditions, evaporation rates as well) limit performance is the combustion chamber of an industrial gas turbine. Such systems are not normally required to operate at subatmospheric pressures, thus the θ parameter and design charts based on it have no practical significance.

5.5 Evaporation-Controlled Systems

Consider the case in which the mixing and reaction rates are fast enough for fuel evaporation to be the rate-controlling step. In Chapter 2, the average rate of evaporation of a fuel spray is given as:

$$\dot{m}_F = \rho_A \lambda_{ref} V_c q / D_o^2, \tag{5.13}$$

where V_c is the combustion volume, q is the fuel/air ratio by mass, D_o is the initial Sauter mean diameter of the spray, and λ_{eff} is the effective evaporation constant. It is assumed that as the fuel evaporates, it instantly mixes and burns with the surrounding air. Thus, combustion efficiency is obtained as the ratio of the rate of fuel evaporation within the combustion zone to the rate of fuel supply, i.e.,

$$\eta_e = \dot{m}_F / q_{ov} \dot{m}_A = \dot{m}_F / f_c q_c \dot{m}_A, \tag{5.14}$$

where f_c is the fraction of the total combustor airflow ($\dot{m}_A$) employed in combustion, and q_c is the fuel/air ratio in the combustion zone. Substituting for $\dot{m}_F$ from Equation 5.13 into Equation 5.14 gives

$$\eta_e = \left(\lambda_{eff} \rho_g V_c\right) / \left(f_c \dot{m}_A D_o^2\right). \tag{5.15}$$

Equation 5.15 may be used for calculating or correlating combustion efficiencies in situations where fuel evaporation is known to be the rate-controlling step. For these conditions, it shows that combustion efficiency is improved by increases in fuel volatility, turbulence intensity, combustion volume, and gas pressure, and is impaired by increases in air mass flow rate and mean drop size.

It should be noted that the form of Equation 5.15 allows η_e to exceed unity. When this occurs, it simply means that the time required for fuel evaporation is less than the time available, so the fuel is fully vaporized within the primary recirculation zone. In these circumstances, η_e should be assigned a value of unity.

Equation 5.15 relates combustion efficiency to combustor dimensions (via V_c), combustor operating conditions (via $\dot{m}_A$, λ_{eff}, and ρ_g), fuel nozzle characteristics (via D_o), and fuel type (via D_o and λ_{eff}).

Values of λ_{eff} are shown plotted in Figures 5.5 and 5.6 for ambient pressures of 1 and 2 MPa, respectively. The manner in which these charts are used to

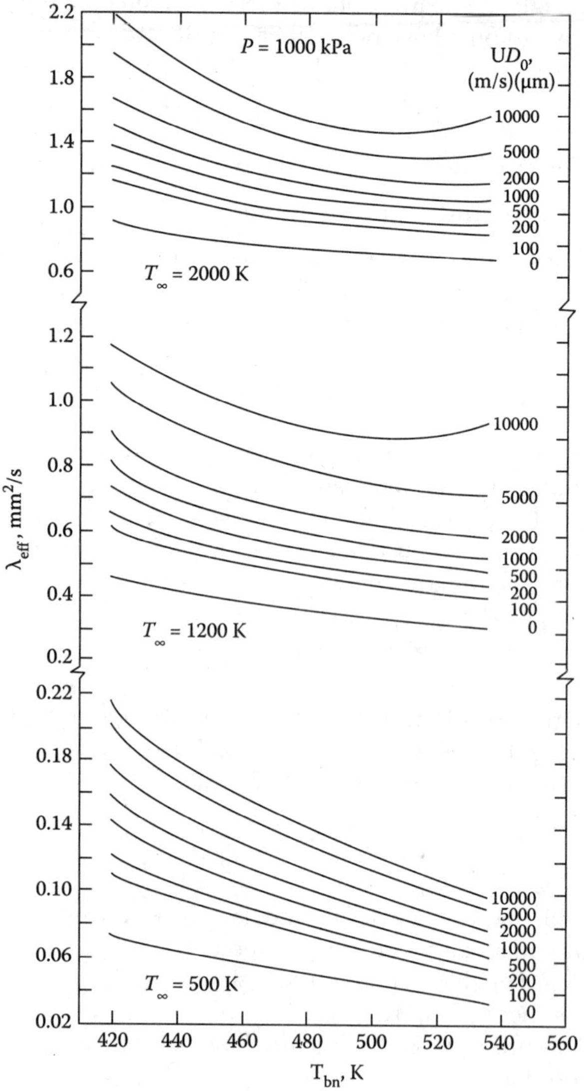

FIGURE 5.5
Variation of effective evaporation constant with normal boiling point for a pressure of 1 MPa. (From Chin, J.S. and Lefebvre, A.H., *Proceedings of 20th Automotive Technology Development Contractors Meeting*, P-120, 325–331, 1982.)

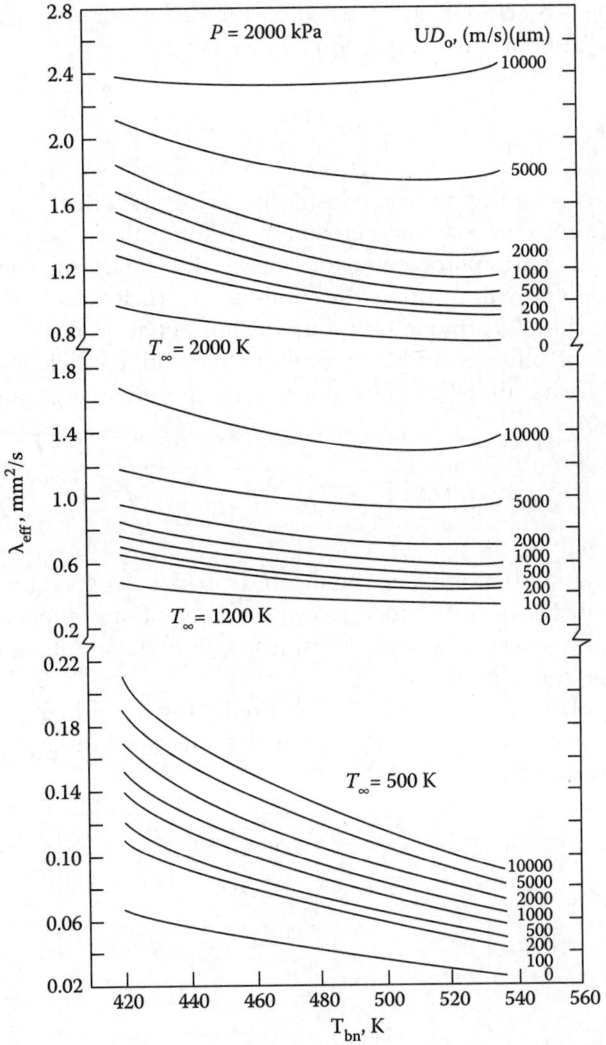

FIGURE 5.6
Variation of effective evaporation constant with normal boiling point for a pressure of 2 MPa. (From Chin, J.S. and Lefebvre, A.H., *Proceedings of 20th Automotive Technology Development Contractors Meeting*, P-120, 325–331, 1982.)

calculate values of λ_{eff} for a range of liquid hydrocarbon fuels at different levels of ambient air temperature is described in Chapter 2, which also contains similar data on λ_{eff} for an ambient pressure of 0.1 MPa (Figure 2.7).

A useful guide to the influence of fuel type on combustion efficiency may be obtained by defining a dimensionless efficiency parameter as the ratio of the combustion efficiency of any alternative fuel "a" to that of some baseline fuel "b," when both fuels are burned in the same combustor at the same operating conditions. From Equation 5.15 we have

$$\eta_{ca} / \eta_{cb} = \left(\lambda_{eff} / D_o^2\right)_a / \left(\lambda_{eff} / D_o^2\right) b. \tag{5.16}$$

This expression provides a means of assessing the effect on combustion efficiency of replacing aviation kerosine with some alternative fuel, without reference to any particular combustor or operating conditions. It may also be used more generally to compare the combustion efficiency characteristics of any two liquid fuels, provided, of course, that fuel evaporation is known to be the rate-controlling step. If the level of combustion efficiency of interest is high, say greater than 90%, it is more useful and more accurate to rewrite Equation 5.16 as

$$\left(2-\eta_{cb}\right) / \left(2-\eta_{ca}\right) = \left(\lambda_{eff} / D_o^2\right)_a / \left(\lambda_{eff} / D_o^2\right) b. \tag{5.17}$$

For pressure-swirl atomizers, the mean drop size depends on the surface tension and viscosity of the fuel. However, conventional fuels exhibit only slight differences in surface-tension values, so only the influence of viscosity on D_o need be considered.

The practical utility of Equation 5.17 for predicting the influence on combustion efficiency of a change in fuel type is demonstrated in Figure 5.7, which

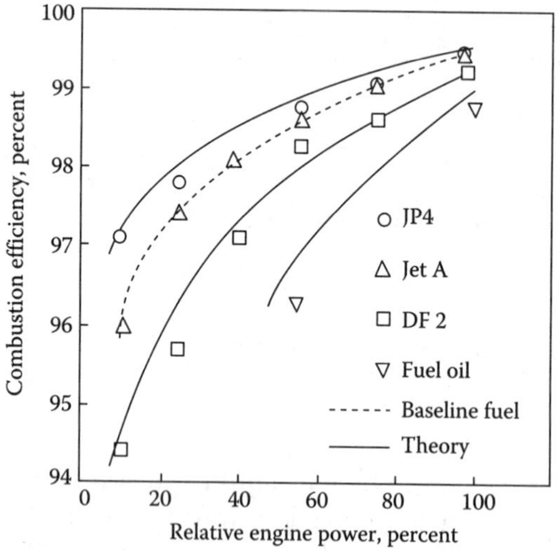

FIGURE 5.7
Influence of fuel type on combustion efficiency. (From Moses, C.A., "Studies of Fuel Volatility Effects on Turbine Combustor Performance," Joint Spring Meeting of Western and Central States Sections of the Combustion Institute, San Antonio, Texas, 1975. With permission.)

shows experimental data obtained by Moses [10] using a T63 combustor. With Jet A designated as the baseline fuel, and using values for λ_{eff} from Figure 5.5, the combustion efficiencies of the other fuels are readily calculated from Equation 5.17 by assuming that mean drop sizes are proportional to (fuel viscosity)$^{0.25}$ for the pressure-swirl atomizer employed in the T63 combustor. The results of the calculations are shown in Figure 5.7, where the dashed curve describes the combustion efficiency of the baseline fuel as determined experimentally. The solid curves represent the predictions of Equation 5.17 for the other fuels. The level of agreement between the predicted and measured values is clearly satisfactory.

5.6 Reaction- and Evaporation-Controlled Systems

In some situations, for example, with fuels of low volatility burning at low pressures, the rate of heat release may be limited by both chemical reaction and evaporation rates. Under these conditions, the combustion efficiency is obtained as the product of the evaporation efficiency, η_e, and the reaction efficiency, η_θ, i.e.,

$$\eta_c = \eta_e \times \eta_\theta. \tag{5.18}$$

The first term on the right hand side of Equation 5.18 represents the fraction of the fuel that is evaporated within the combustion zone. For $\eta_e > 1$, $\eta_c = \eta_\theta$, which denotes the fraction of fuel vapor that is converted into combustion products by chemical reaction.

From analysis of the available experimental data on combustion efficiency, the following expressions for η_θ and η_e have been derived [11].

$$\eta_\theta = 1 - \exp\left[0.022 P_3^{1.3} V_c \exp(T_c / 400) / f_c \dot{m}_A\right], \tag{5.19}$$

and

$$\eta_e = 1 - \exp\left[-36 \times 10^{-6} P_3 V_c \lambda_{eff} / T_c D_o^2 f_c \dot{m}_A\right], \tag{5.20}$$

where P_3 is the combustor inlet pressure in kPa, V_c is the combustion (predilution) volume in m^3, T_c is the combustion (predilution) temperature in degrees Kelvin, f_c is the fraction of air used in combustion, $\dot{m}_A$ is the total combustor airflow in kg/s, D_o is the Sauter mean diameter of the fuel spray in microns, and λ_{eff} is the effective evaporation constant in mm^2/s.

It should be noted in Equations 5.19. and 5.20 that the temperatures are expressed in terms of T_c rather than T_3. Either may be used (with suitable adjustment to the constants), but T_c is the adiabatic flame temperature in the combustion zone assuming complete combustion of the fuel. It is calculated from the expression

$$T_c = T_3 + \Delta T_c, \tag{5.21}$$

where ΔT_c is obtained from standard temperature rise charts for the fuel in question, using appropriate values of P_3, T_3, and $q_c (= q_{ov}/f_c)$.

In the late 1970s, the US Air Force, Army, and Navy, along with NASA and the engine manufacturers, initiated programs to determine the effects of anticipated future fuels on the life and performance of existing aircraft engines. As a result of these studies, data became available that yielded new and useful insights into the effects of fuel properties, combustor design, and engine operating conditions on all major aspects of combustion performance, including combustion efficiency [12–17]. Some of these data are shown in Figures 5.8 and 5.9, which demonstrate the satisfactory correlation of combustion efficiency data provided by Equation 5.18. Further examples to illustrate the application of this equation to the correlation of experimental data on combustion efficiency may be found in [11].

It should be noted that these investigations and correlations apply to petroleum-based conventional jet fuels. For synthetic and bio-based jet fuels, the validity of this approach and correlations are yet to be studied and demonstrated.

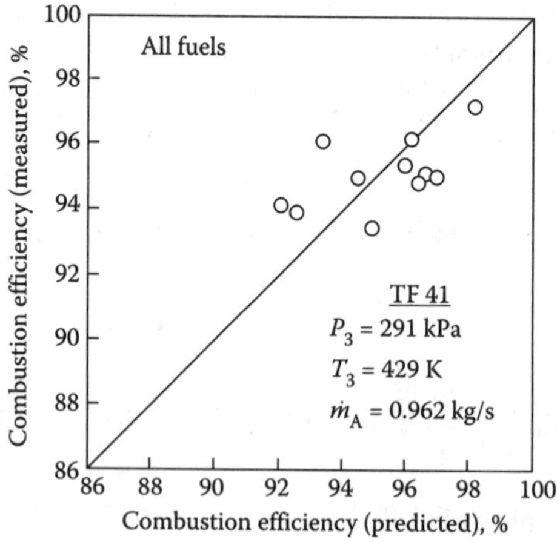

FIGURE 5.8
Comparison of measured and predicted values of combustion efficiency for a TF 41 combustor. (From Lefebvre, A.H., *Journal of Engineering for Gas Turbines and Power*, 107, 24–37, 1985. With permission.)

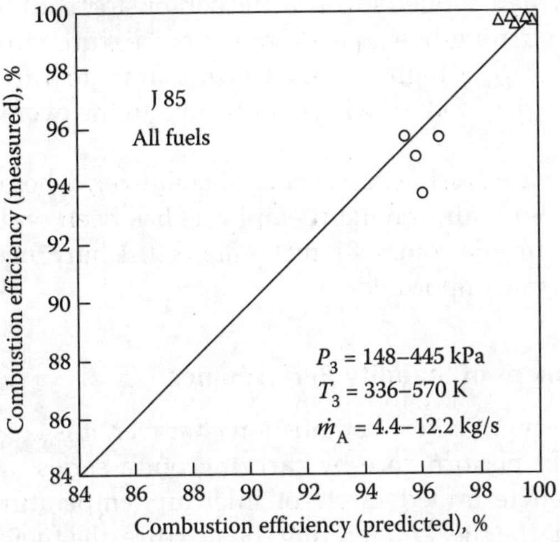

FIGURE 5.9
Comparison of measured and predicted values of combustion efficiency for a J 85 combustor. (From Lefebvre, A.H., *Journal of Engineering for Gas Turbines and Power*, 107, 24–37, 1985. With permission.)

5.7 Flame Stabilization

One of the primary requirements of a gas turbine combustor is that combustion must be maintained over a wide range of operating conditions. This is especially true for the aircraft combustion chamber, which must sometimes operate at low temperatures and pressures, and at fuel/air ratios that lie well outside the normal limits of flammability for hydrocarbon/air mixtures. Combustion must be initiated and sustained in highly turbulent airstreams, flowing at speeds many times greater than the normal burning velocity of the fuels employed. Moreover, the flame must cope with the various abnormal conditions that are sometimes encountered in flight, such as those created by the ingestion of tropical rain or ice.

The usual method of surmounting this problem is to create a sheltered zone of low velocity at the upstream end of the liner, where flame speeds are greatly enhanced by imparting a high level of turbulence to the primary air jets and by arranging for hot combustion products to recirculate and mix with the incoming air and fuel.

5.7.1 Definition of Stability Performance

In combustion parlance, the term "stability" is often used rather loosely to describe either the range of fuel/air ratios over which stable combustion can be achieved, or as a measure of the maximum air velocity the system can tolerate before flame extinction occurs. Thus, the description "good stability performance," when applied to a specific combustor, could mean either that it is capable of burning over a wide range of mixture strengths or that its blowout velocity, U_{BO}, is high. Clearly, it is important to differentiate between these two properties, both of which contribute to the overall stability of the system.

In general, with experimental forms of stabilizer, where the fuel is supplied premixed with air, the main emphasis has been on blowout velocity, whereas in gas turbine combustion chambers, the burning range is usually considered of prime importance.

5.7.2 Measurement of Stability Performance

In the development of a new combustion chamber, it is customary to determine its stability performance by carrying out a series of extinction tests at constant, predetermined levels of inlet air temperature and pressure. After turning on the fuel and igniting the mixture, the fuel flow is gradually reduced until flame extinction occurs. After noting the fuel and air flows at this "weak extinction" or "lean blowout" point, combustion is re-established and the fuel flow slowly increased until "rich extinction" occurs. This process is repeated at increasing levels of air mass flow until the complete stability loop can be drawn. Figure 5.10 illustrates the main features of a stability loop obtained by this technique. The region of stable burning is bounded by rich and weak limits that gradually converge with increasing mass flow rate until eventually a flow rate is reached beyond which combustion is unattainable at any fuel/air ratio. Of special interest with such curves are the "rich"

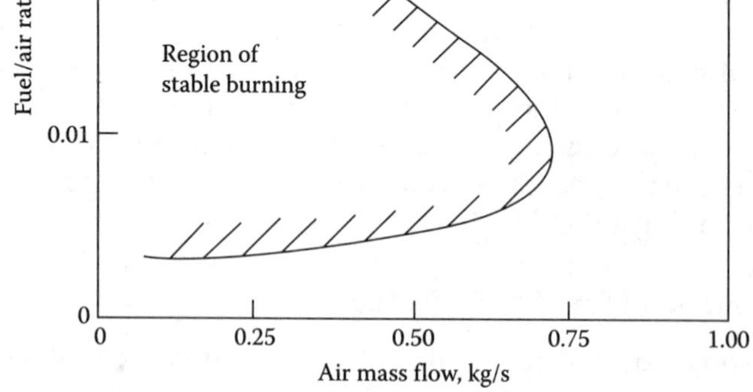

FIGURE 5.10
Typical combustion chamber stability loop.

and "weak" extinction points obtained at the air mass flows corresponding to the design value of chamber reference velocity.

The complete stability performance of an aero-engine combustor is obtained by carrying out sufficient extinction tests to allow a number of stability loops to be drawn at different levels of pressure, as illustrated in Figure 5.11. It is then a fairly straightforward procedure to translate these stability loops into performance charts illustrating the range of flight conditions over which stable combustion is possible, as illustrated in Figure 5.12. In practice, the risk of overheating the liner and rig ducting tends to restrict the number of rich extinction points that can be obtained, especially at high pressures. Moreover, even with large-scale test facilities, it is usually impossible to determine the peaks of the loops, owing to limitations on the amount of air that can be supplied at subatmospheric pressures. Fortunately, these are not serious drawbacks, as it is the lean blowout limit that is of primary interest and importance. With aircraft systems, which must provide for sudden changes in throttle setting, it is essential that the lean blowout limit should exceed 250 AFR at atmospheric pressure. In this context, the method of fuel injection is of paramount importance.

In conventional combustion chambers, good stability performance is usually attained without undue difficulty. However, in chamber designs featuring complete fuel evaporation and thorough mixing of fuel and air prior to combustion, current views on what is acceptable in terms of stability performance may have to be modified considerably in order to achieve worthwhile reductions in pollutant emissions. For this reason, it is essential

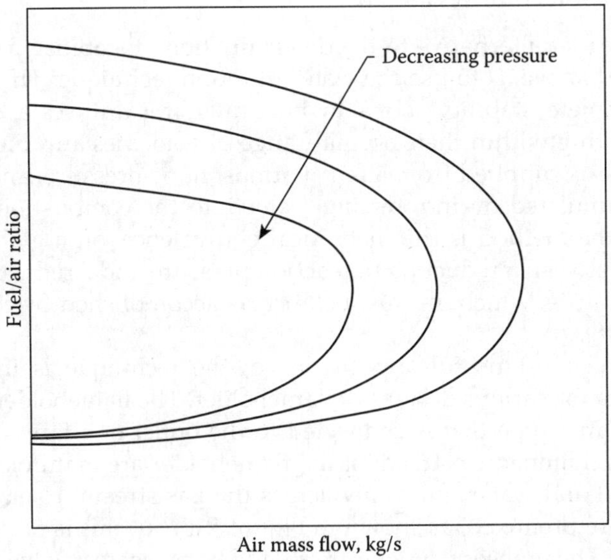

FIGURE 5.11
Influence of pressure on stability loops.

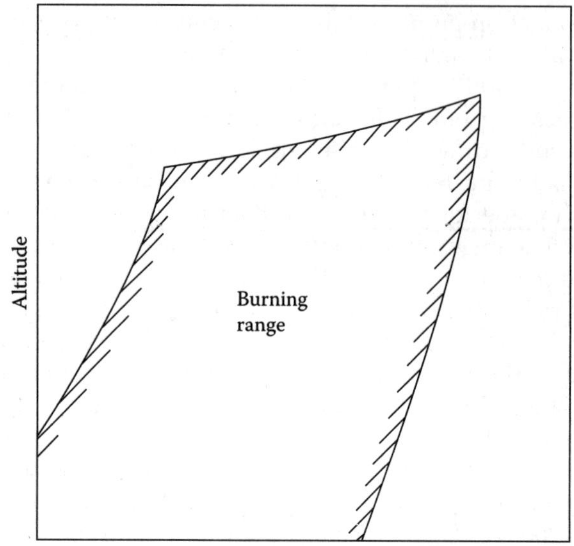

FIGURE 5.12
Stability performance of an aircraft combustor.

that the combustion engineer be fully aware of all the factors that govern lean blowout limits in practical combustion systems.

5.7.3 Water Injection Technique

A much cheaper alternative to the determination of stability performance, as described above, is to use the water injection technique. This technique allows complete stability loops to be drawn for full-scale combustors while operating within their normal range of velocities and fuel/air ratios [2,18–24]. Air is supplied from a fan at atmospheric pressure, and low pressures are simulated by introducing water into the combustion zone. The essence of the method is the theoretical equivalence, on a global reaction rate basis, between a reduction in reaction pressure and a reduction in reaction temperature, which, in this instance, is accomplished by the addition of water [2,18].

One of the most useful applications of the technique is in obtaining blowout data for various designs of flameholder. The flameholder under test is mounted in a pipe that is connected to the outlet of a fan via a preheat combustion chamber. Upstream of the flameholder are manifolds designed to inject fuel and water uniformly across the gas stream. Usually, the fuel and water are premixed, as shown in Figure 5.13, but this is not essential to the method. The preheat temperature should be preset to a value that is high enough to ensure that the fuel and water are fully vaporized upstream of the flameholder.

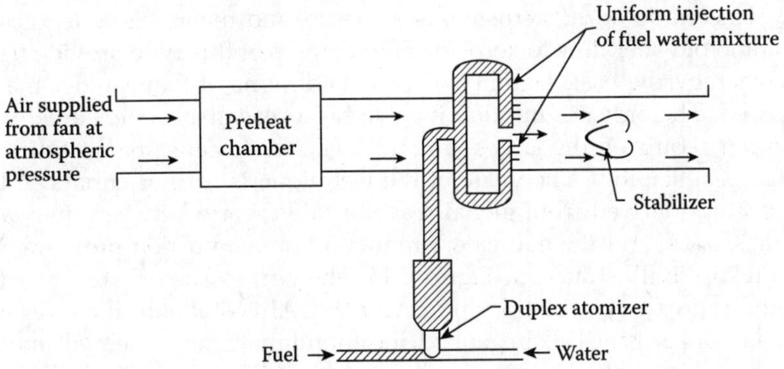

FIGURE 5.13
Basic rig requirements for stability tests using water injection technique.

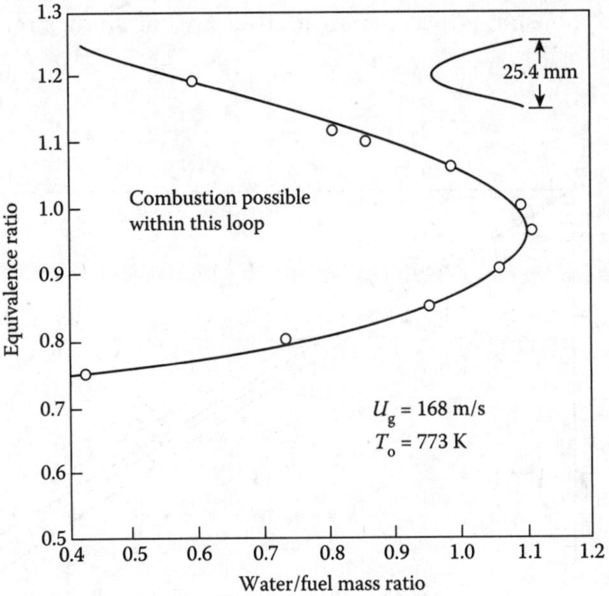

FIGURE 5.14
Typical stability loop obtained using water injection technique.

The test procedure is quite simple. The velocity and temperature of the gas flowing over the stabilizer are adjusted to the desired values; the fuel is turned on, and a flame is established in the recirculation zone downstream of the stabilizer. Water is then gradually mixed with the fuel in increasing amounts until extinction occurs. This process is repeated at a sufficient number of fuel flow rates for a complete stability loop to be drawn. A typical stability loop is shown in Figure 5.14, in which the ordinate represents the equivalence ratio of the kerosine/air mixture, and the abscissa denotes the mass ratio of water flow to kerosine flow. Curves of this type provide useful data whereby the basic stability of various designs of flameholder may be compared. The only assumption involved is a reasonable one, namely, that the gutter requiring the largest amount of water to cause flame extinction has the best stability. The value of the technique is further enhanced by a relationship (derived from global reaction rate theory) between the water/fuel mass ratio and the equivalent reduction in combustion pressure. This relationship is illustrated in Figure 5.15. The curves shown are for octane fuel and represent calculations by Taylor [19]. Almost identical curves were obtained for kerosine [18]. In general, it is found that injecting equal amounts of water and liquid fuel into the combustion zone is equivalent to halving the combustion pressure.

The simulation of low pressures by nitrogen gas dilution has also been used successfully by several workers, including Norster [25] and Sturgess et al. [26]. Nitrogen is clearly more costly than water, but it does not require a preheat combustor, which makes it attractive for studies on small-scale flameholders.

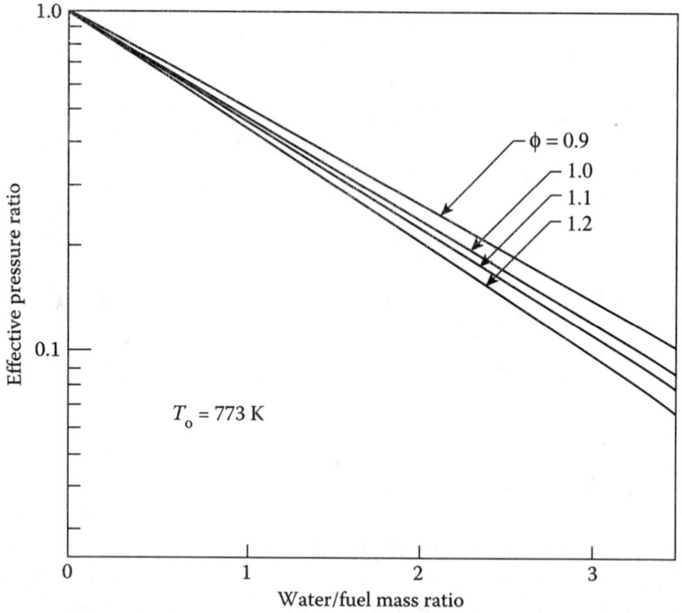

FIGURE 5.15
Relationship between water-fuel ratio and equivalent pressure ratio for isooctane fuel. (From Taylor, J.S., "Large-Scale Bluff Body Flame Stabilization," MSc thesis, School of Mechanical Engineering, Purdue University, 1980.)

5.8 Bluff-Body Flameholders

Bluff-body flameholders are widely used to stabilize flames in flowing combustible mixtures, and their many practical applications include ramjet and turbojet afterburner systems. The practical importance of bluff-body stabilizers has given rise to a large number of theoretical and experimental studies. Much of our present understanding of the flame stabilization process is due to the pioneering studies carried out in the 1950s by Longwell et al. [27], Zukowski and Marble [28], Barrère and Mestre [29], De Zubay [30], and Spalding [31]. More recent studies include those of Lefebvre et al. [20–24,32–34], whose work culminated in equations for predicting stability limits in terms of bluff-body dimensions, blockage ratio, and the pressure, temperature, velocity, turbulence properties, and equivalence ratio of the incoming fresh mixture. Plee and Mellor [35] have successfully correlated lean blowout data for bluff-body stabilized flames, using a characteristic-time model.

5.8.1 Experimental Findings on Bluff-Body Flame Stabilization

The flameholding properties of bluff-body stabilizers have been studied extensively for both homogeneous gaseous fuel–air mixtures and for combustion zones supplied with heterogeneous mixtures of fuel drops and air.

5.8.1.1 Homogeneous Mixtures

Ballal and Lefebvre [32] investigated the effects of inlet air temperature, pressure, velocity, and turbulence on the lean blowout performance of flameholders supplied with homogeneous mixtures of gaseous propane and air. The apparatus employed comprised a flameholder in the form of a hollow cone that was located at the center of a circular pipe with its apex pointing upstream. Fourteen geometrically similar conical baffles were manufactured to various sizes and used in conjunction with three different pipe diameters to allow the effects of baffle size and blockage ratio to be studied independently over wide ranges of operating conditions.

The influence of flameholder size on blowout limits is illustrated in Figure 5.16. In this figure, measured values of weak extinction equivalence ratio are shown plotted against flameholder diameter for two different levels of approach stream velocity and three different values of blockage, B_g. (Note that B_g is defined as the ratio of the flameholder cross-sectional area to that of the pipe.) The improvement in stability with an increase in flameholder diameter is attributed to the longer residence time of the reactants in the recirculation zone. Figures 5.17 and 5.18 contain similar data to illustrate the effects of inlet air temperature and approach velocity on lean blowout limits. All the experimental data obtained by Ballal and Lefebvre [32] are consistent

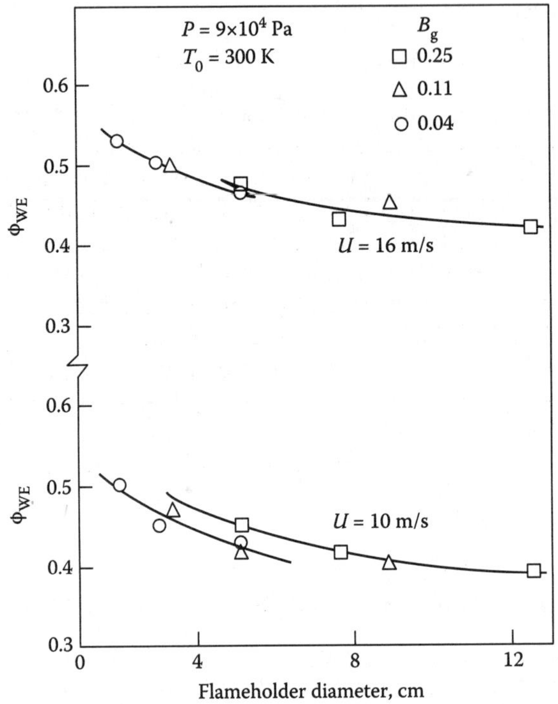

FIGURE 5.16
Influence of flameholder size on weak extinction limits for propane–air mixtures. (From Ballal, D.R. and Lefebvre, A.H., *Journal of Engineering for Power*, 101(3), 343–348, 1979. With permission.)

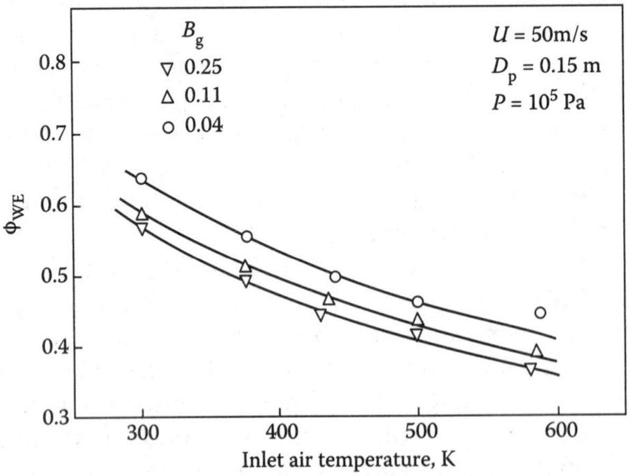

FIGURE 5.17
Influence of temperature on weak extinction limits for propane–air mixtures. (From Ballal, D.R. and Lefebvre, A.H., *Journal of Engineering for Power*, 101(3), 343–348, 1979. With permission.)

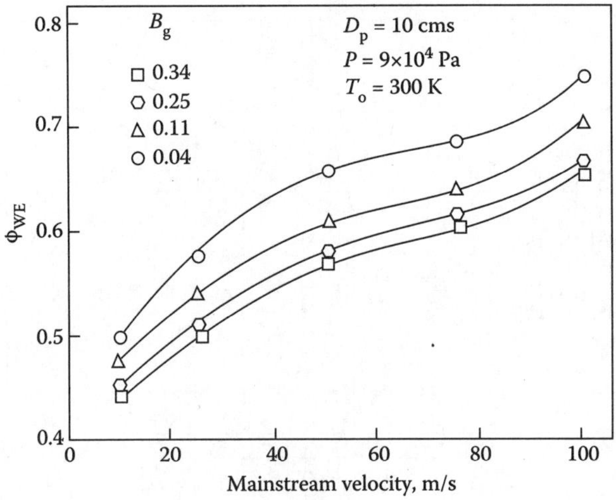

FIGURE 5.18
Influence of velocity on weak extinction limits for propane–air mixtures. (From Ballal, D.R. and Lefebvre, A.H., *Journal of Engineering for Power*, 101(3), 343–348, 1979. With permission.)

with the notion that lean blowout limits are improved (i.e., extended to lower values of equivalence ratio) by increases in pressure and temperature (via increase in reaction rates) and by an increase in flameholder size and/or reduction in approach velocity (via increase in residence time).

Baxter and Lefebvre [24] employed fully vaporized kerosine–air mixtures to examine the effects of various flow parameters on the lean blowout limits of Vee-gutter flameholders of various widths and included angles. A novel feature of their apparatus is that it includes both a preheat combustor and a heat exchanger in order to allow the effects of inlet air vitiation on stability limits to be examined independently from those of inlet gas temperature. The adverse effect on flame stability of increasing the degree of vitiation while maintaining a constant inlet temperature is illustrated in Figure 5.19. The data contained in this figure were obtained using two Vee-gutter flameholders, both of 60 degree included angle. The gutter widths and flow conditions were selected to represent those encountered in modern afterburner systems. An interesting feature of this figure is that the slopes of the lines drawn through the data points become increasingly steep with an increase in upstream vitiation, as indicated by the preheater equivalence ratio, ϕ_p. This is due to the rapid depletion of available oxygen with the increase in ϕ_p. The strong adverse effect of inlet air vitiation on weak extinction limits, as illustrated in Figure 5.19, is clearly of practical importance to the design of afterburner systems, where the hot gas stream approaching the flameholder array has experienced appreciable oxygen depletion because of prior combustion in the main combustor.

The shape of a bluff-body flameholder affects its stability characteristics through its influence on the size and shape of the wake region. The effect of

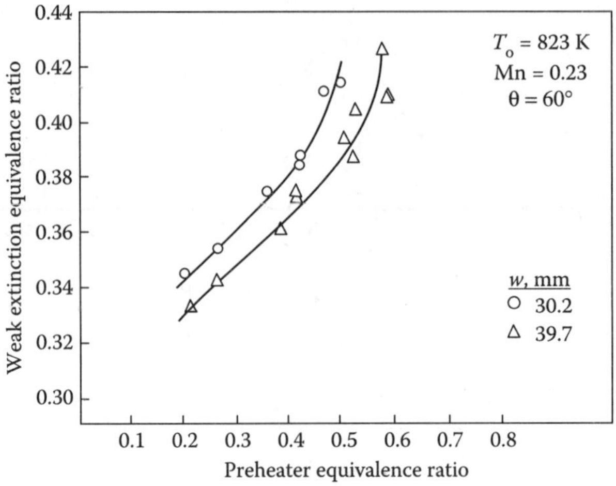

FIGURE 5.19
Influence of inlet-air vitiation on weak extinction limits. (From Baxter, M.R. and Lefebvre, A.H., *Journal of Engineering for Gas Turbines and Power*, 114(4), 776–782, 1992. With permission.)

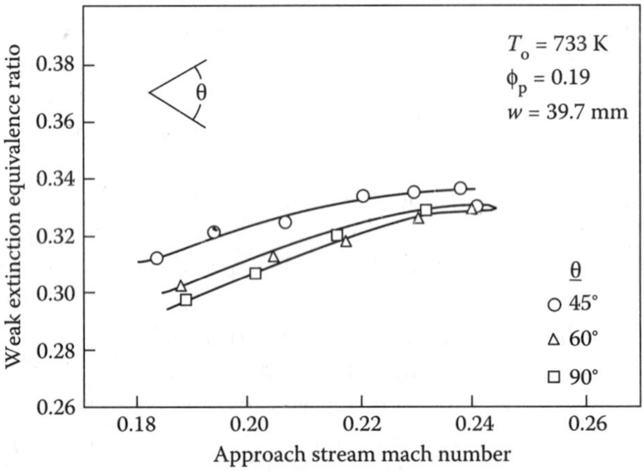

FIGURE 5.20
Influence of approach stream velocity and flameholder shape on weak extinction limits. (From Baxter, M.R. and Lefebvre, A.H., *Journal of Engineering for Gas Turbines and Power*, 114(4), 776–782, 1992. With permission.)

shape is illustrated in Figure 5.20, which shows lean blowout data for three gutters having different included angles, but the same projected width. These data confirm the results of Barrère and Mestre [29] in showing that the characteristic dimension of a bluff-body flameholder should not be its geometric width, *w*, but rather the maximum aerodynamic width of the wake created behind it, w_a. The ratio w_a/w increases with an increase in gutter included angle, thereby improving stability performance by enlarging the recirculation-zone volume.

Surprisingly perhaps, the influence of combustion pressure on lean blowout limits is quite small. Ballal and Lefebvre [32] observed only a slight increase in ϕ_{WE} as the inlet air pressure was reduced from 1.0 to 0.2 bars.

5.8.1.2 Heterogeneous Mixtures

Ballal and Lefebvre [33] examined the factors governing the lean blowout limits of bluff-body stabilized flames supplied with flowing mixtures of liquid fuel drops and air. Their test program included wide variations in inlet air pressure and velocity, and also covered wide ranges of fuel volatility and mean fuel drop size. The experimental apparatus was essentially the same as that used previously for homogeneous fuel/air mixtures, except that means were provided for spraying fuel drops directly into the combustion zone through a simplex pressure nozzle that was located inside the conical flameholder. Ten fuel nozzles of different flow numbers were used in order to cover wide ranges of air velocity, air pressure, and mean drop size. Some of the results obtained in this investigation are shown in Figures 5.21 through 5.23. They show that lean blowout limits are improved, i.e., extended to lower equivalence ratios, by increases in fuel volatility and air pressure, and by reductions in mainstream velocity and mean drop size. Figure 5.21 indicates a strong effect of pressure on ϕ_{WE}, which is in marked contrast to the small effect observed with homogeneous mixtures. This result is of special interest because it shows that the influence of pressure on the lean

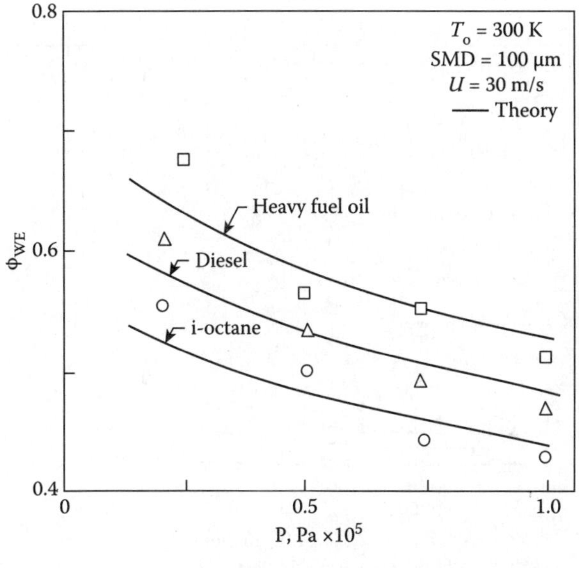

FIGURE 5.21
Influence of pressure on weak extinction limits. (From Ballal, D.R. and Lefebvre, A.H., *Journal of Engineering for Power*, 102(2), 416–421, 1980. With permission.)

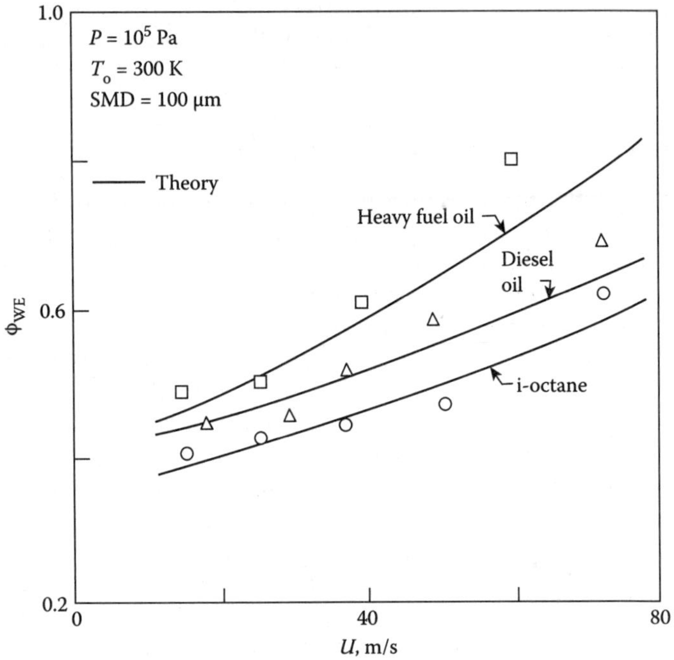

FIGURE 5.22
Influence of velocity on weak extinction limits. (From Ballal, D.R. and Lefebvre, A.H., *Journal of Engineering for Power*, 102(2), 416–421, 1980. With permission.)

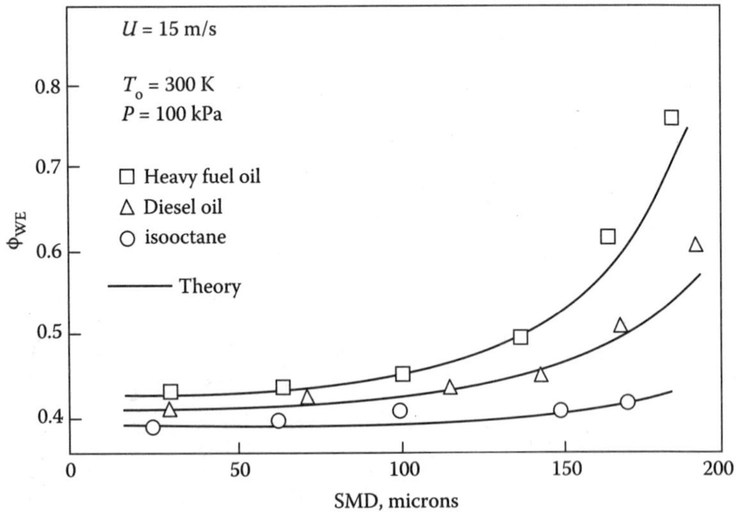

FIGURE 5.23
Influence of mean drop size on weak extinction limits. (From Ballal, D.R. and Lefebvre, A.H., *Journal of Engineering for Power*, 102(2), 416–421, 1980. With permission.)

blowout limits of heterogeneous mixtures is manifested primarily through fuel evaporation rates rather than chemical reaction rates. Also worthy of note is that in Figure 5.23 the curves of ϕ_{WE} vs. SMD become horizontal at low values of SMD. This denotes that drop sizes are small enough for evaporation to be complete within the reaction zone. Thus, over the range of SMDs, where ϕ_{WE} is independent of SMD, the combustion zone operates effectively as a homogeneous stirred reactor.

5.8.2 Summary of Experimental Findings

A considerable body of evidence has been accumulated on the factors governing flame stability, and several broad conclusions can be drawn. In general, stability limits are extended by:

1. A reduction in approach stream velocity
2. An increase in approach stream temperature
3. An increase in gas pressure
4. A reduction in turbulence intensity
5. Any change in equivalence ratio toward unity
6. An increase in flameholder size
7. An increase in flameholder base-drag coefficient
8. A reduction in flameholder blockage (for a constant flameholder size)

For liquid fuels, stability is further improved by:

1. An increase in fuel volatility
2. Finer atomization, i.e., reduction of mean drop size

5.9 Mechanisms of Flame Stabilization

Many theoretical studies have been carried out on bluff-body flame stabilization and several models have been proposed to account for the experimental observations on flame extinction. Most of these models tend to fall into two main categories. One of these, following Longwell et al. [27], views the wake region of a bluff-body essentially as a homogeneous chemical reactor. Flame extinction occurs when the time available for chemical reaction becomes less than the time required to generate sufficient heat to raise the fresh mixture up to its ignition temperature. The other category includes models in which attention is focused mainly on the shear layer surrounding the wake region [28,35]. According to Zukowski and Marble [28], ignition of the fresh mixture occurs in the shear layer when it is turbulently mixed with combustion products from the recirculation zone. The burning mixture then flows downstream through the shear layer, where it ignites neighboring mixture kernels. When it reaches the end of the wake region, some of the burning mixture continues to flow downstream, and the remainder is entrained into the recirculatory flow, which conveys it upstream to mix with and ignite the shear layer. A flame is anchored on the baffle through continuation of this process. Flame extinction occurs when the fresh mixture does not spend enough time in the shear layer to be ignited by the hot recirculation zone. Thus, the criterion for blowout is that the ignition delay time be equal to the residence time in the shear layer adjacent to the recirculation zone.

Which of these two basic approaches has the most fundamental significance and relevance to flame stabilization is uncertain but, fortunately, is of academic interest only as far as the development of a suitable correlation for weak extinction is concerned. This is because the time spent by the fresh mixture in the shear layer, and the residence time of the combustion products in the recirculation zone, are both proportional to the characteristic dimension of the flameholder, D_c. Since the material entering the recirculation zone at its downstream edge has already passed through the shear layer, it would seem more logical to define the residence time as the sum of the times spent in the shear layer and the recirculation zone. However, because this total time is also proportional to D_c, this assumption does not change the resulting correlation. This, of course, is why many workers, who appear to base their analyses on a seemingly different set of assumptions, all eventually arrive at the same general conclusion, namely, that the equivalence ratio at blowout is a function of U, P^x, D_c^y, and T_o^z or $\exp(T_o/z)$, where U is the velocity in the plane of the flameholder, P is pressure, and T_o is the inlet gas temperature. For information on the values x, y, and z, obtained experimentally before 1960, reference should be made to the survey papers of Longwell [36] and Herbert [37]. The purpose of this brief discussion is to point out that, regardless of how simple or sophisticated the assumptions employed in its derivation, the best correlating parameter is one that is based on sound principles and has the greatest ease and breadth of application.

The general approach adopted by Ballal and Lefebvre [32,34] for homogeneous fuel–air mixtures was to assume that flame blowout occurs when the rate of heat liberation in the combustion zone becomes insufficient to heat the incoming fresh mixture up to the required reaction temperature. With heterogeneous mixtures, an additional factor is the time required for fuel evaporation. For fuel sprays of low volatility and large mean drop size, this time is relatively long and is often the main factor limiting the overall rate of heat release. Thus, in the analysis of lean blowout limits, it is appropriate to consider homogeneous mixtures first and then to examine how the results obtained should be modified to take account of fuel evaporation.

5.9.1 Homogeneous Mixtures

Following Longwell et al. [27], Ballal and Lefebvre [32] viewed the reaction zone of a bluff-body flameholder as a homogeneous chemical reactor in which the temperature and chemical composition are constant throughout. Their proposed model is based on the notion that flame extinction occurs when the amount of heat needed to ignite the fresh mixture being entrained into the wake region just exceeds the amount of heat liberated by combustion in that zone. The rate of entrainment of fresh mixture into the wake region is assumed to be proportional to the product of gas density, surface area, and the velocity difference between the fresh mixture flowing over the wake region and the adjacent, co-flowing combustion products. This velocity difference is proportional to the velocity of the flow over the edge of the baffle, which is equal to $U/(1-B_g)$. If it is further assumed that for a conical flameholder of diameter D, the surface area available for the entrainment of fresh mixture is proportional to D^2, then

$$\dot{m} \propto \rho D^2 U/(1-B_g). \tag{5.22}$$

The maximum airflow rate corresponding to flame blowout is derived from global reaction rate considerations [32] as

$$\dot{m}_{max} = 1.93 V_c P^{1.25} \exp(T_o/150)\phi^{6.25}. \tag{5.23}$$

Equating Equations 5.22 and 5.23 and substituting for $\rho = P/\mathrm{R}T$ and $V_c \propto D_c^3$ leads to

$$\phi_{WE} \propto \left[\frac{U}{P^{0.25} T_o \exp(T_o/150) D_c (1-B_g)}\right]^{0.16}. \tag{5.24}$$

This equation shows that the weak extinction value of ϕ is affected mainly by temperature, to a lesser extent by velocity, and hardly at all by pressure. An increase in the characteristic dimension, D_c, of the flameholder always improves the weak extinction performance (i.e., reduces ϕ_{WE}), provided that the increase in D_c is not accompanied by an increase in the blockage ratio B_g. If this occurs, then the increase in D_c still improves the weak extinction performance, but only up to a certain value of B_g, beyond which the performance starts to decline. For a conical baffle mounted in a circular pipe, the critical value of blockage ratio is 33% (that is, $B_g = 0.33$). Equation 5.24 also indicates that, for a constant value of D_c, an increase in B_g caused, say, by a reduction in pipe diameter or by the introduction of more flameholders in the same plane, always has an adverse effect on weak extinction performance.

As mentioned earlier, the true characteristic dimension of a flameholder from a stability viewpoint is not its geometric size, but the corresponding "aerodynamic" value measured downstream of the baffle in the plane of maximum aerodynamic blockage. The ratio of the aerodynamic blockage B_a to the geometric blockage B_g depends on the forebody shape of the flameholder. The more streamlined the forebody shape, the lower the ratio of B_a to B_g [29,38].

The predictions of Equation 5.24 in regard to the influence of flameholder dimensions and operating conditions on weak extinction limits show good agreement with the results obtained by Ballal and Lefebvre over wide ranges of pressure, temperature, velocity, and stabilizer dimensions, as demonstrated in Figure 5.24. They also show good qualitative agreement with the experimental data obtained by other workers for baffle-stabilized flames [27–30,39,40].

5.9.2 Heterogeneous Mixtures

Equation 5.24 may also be used to predict the lean blowout limits of combustion systems supplied with heterogeneous fuel–air mixtures, provided that the rate of fuel evaporation is sufficiently high to ensure that all the fuel is fully vaporized within the primary combustion zone. If the fuel does not fully vaporize, then clearly the "effective" fuel/air ratio will be lower than the nominal value. However, if the fraction of fuel that is vaporized is

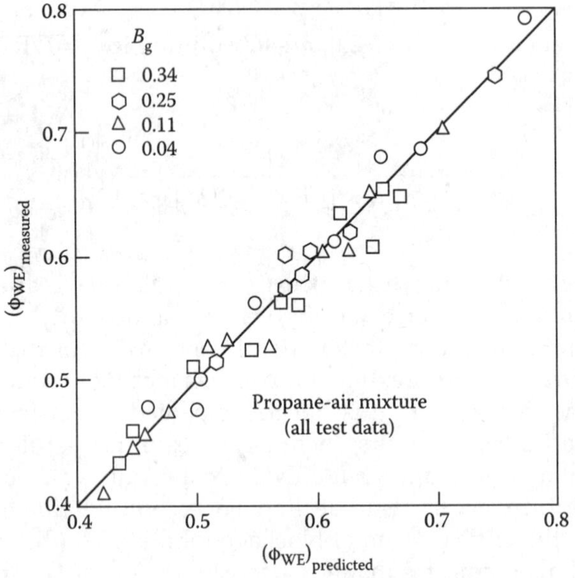

FIGURE 5.24
Comparison of measured and predicted values of weak extinction limits. (From Ballal, D.R. and Lefebvre, A.H., *Journal of Engineering for Power*, 101(3), 343–348, 1979. With permission.)

known, or can be calculated, it can be combined with Equation 5.24 to yield the fuel/air ratio at lean blowout, i.e.,

$$\phi_{WE}(\text{heterogeneous}) = \phi_{WE}(\text{homogeneous}) / f_f, \quad (5.25)$$

where f_f is the fraction of fuel that is vaporized within the primary zone.

From analysis of the factors governing the rate of evaporation of a fuel spray, it was found [9,33] that

$$f_f = 8\rho_g V_c \lambda_{eff} / f_{pz} \dot{m}_A D_o^2, \quad (5.26)$$

where f_{pz} is the fraction of the total airflow rate, $\dot{m}_A$, that enters the primary zone.

If the value of f_f determined from Equation 5.26 exceeds unity, this means that the time required for fuel evaporation is less than the time available, so the fuel is fully vaporized within the recirculation zone, f_f should then be assigned a value of 1.0, so that

$$\phi_{WE}(\text{heterogeneous}) = \phi_{WE}(\text{homogeneous}).$$

The validity of this approach to the determination of weak extinction limits for flames supplied with flowing heterogeneous fuel–air mixtures may be tested by comparing measured values of ϕ_{WE} with the corresponding predicted values from Equations 5.24 through 5.26. This is done in Figures 5.21 through 5.23, in which the full lines represent the predicted values from these equations [33]. Clearly, the level of agreement between the measured and predicted values is generally satisfactory, thus confirming the basic premise of the model.

5.10 Flame Stabilization in Combustion Chambers

The designer has very little control over the amount of fresh mixture that is entrained into the recirculation zone of a bluff-body flameholder. Usually, this amount represents only a very small fraction of the mainstream flow, a fraction that varies markedly with changes in air velocity and temperature [41]. With main combustors, however, air enters the recirculation zone through various apertures in the liner wall, and the designer can control the amount of air participating in primary combustion to within fairly close limits by proper selection of the number, size, and type of aperture.

Figure 5.25a shows the type of primary zone employed in most tubular combustors. The essential feature, as far as the stabilization process is

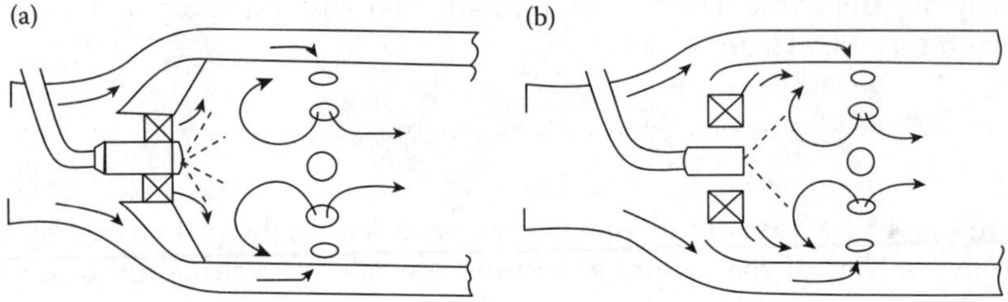

FIGURE 5.25
Typical primary-zone configurations.

concerned, is the toroidal flow reversal that is created and maintained by air entering through swirl vanes located around the fuel injector and through a single row of holes in the wall of the liner. In addition to its main role as the major heat-release zone of the chamber, an important function of the primary zone is to recirculate burned and burning gases to mix with the incoming air and fuel. By this means, a mechanism of continuous ignition is established, and combustion can be sustained over wide ranges of pressure, velocity, and fuel/air ratio.

Figure 5.25b shows a typical primary-zone configuration for an annular combustor in which the primary air feed is supplemented appreciably by film-cooling air in the dome region and, to an increasing extent, by the air employed in the fuel-preparation process.

Flame stabilization in gas turbine combustors has not been subjected to the same experimental and theoretical study as in bluff-body flameholders, but, as a general rule, maximum stability is achieved by injecting the primary air through a small number of large holes. This is because large holes produce large jets and large-scale flow recirculations that provide ample time for combustion. However, for a given air mass flow rate, an increase in hole size can be obtained only at the expense of a reduction in the number of holes. Although no firm guidelines have been laid down for the optimal number of liner holes for annular combustors, one opposing pair of holes per fuel injector should be regarded as the absolute minimum; twice that number would be preferable.

For aircraft engines, lean blowout limits are especially important when the aircraft is descending through inclement weather with the engine idling. At this flight condition, the combustor AFR is typically around 120, but a lean blowout AFR of around 250 is usually specified in order to provide a safety margin for engine to engine variations, fuel control tolerances, and the possible ingestion of water and/or ice.

5.10.1 Influence of Mode of Fuel Injection

One of the key factors governing lean blowout limits is the mode of fuel injection. The poor fuel distribution of pressure-swirl atomizers of the simplex or dual-orifice type ensures that some combustion takes place at mixture strengths that are appreciably richer than the average value. This means that, even when the nominal fuel/air ratio falls to well below the normal lean blowout limit, the flame can still survive due to the presence, within the burning zone, of pockets of near-stoichiometric mixtures. This is why pressure atomizers are noted for wide burning limits—in particular, for good lean blowout values (typically around 1000 AFR). By contrast, the airblast atomizer, which provides much better mixing of fuel and air, is characterized by fairly narrow burning limits (a typical lean blowout limit is around 250 AFR). Methods of overcoming the poor lean blowout performance of airblast atomizers include the use of piloting devices as employed, for example, in the *hybrid or piloted airblast* atomizer (see Chapter 6) and staged fuel injection (see Chapter 9). Nowhere is the problem of flame extinction more important than in the lean premix prevaporize combustor, which must, of necessity, always operate close to the lean blowout limit.

5.10.2 Correlation of Experimental Data

For gas turbine combustors, the lean blowout limit is usually expressed in terms of overall combustor fuel/air ratio rather than equivalence ratio, which is more common for bluff-body flameholders.

From an analysis of lean blowout data acquired from a large number of aircraft combustion chambers [12–17], the following equation for lean blowout fuel/air ratio, q_{LBO}, was derived [11]

$$q_{LBO} = \left[\frac{A}{V_{pz}}\right]\left[\frac{\dot{m}_A}{P_3^{1.3}\exp(T_3/300)}\right]\left[\frac{D_r^2}{\lambda_r H_r}\right], \quad (5.27)$$

where D_r is the mean drop size relative to that for JP4, H_r is the lower calorific value relative to that for JP4, and λ_r is the effective evaporation relative to that for JP4.

A is a constant whose value depends on the geometry and mixing characteristics of the combustion zone and also on the amount of air employed in primary combustion [11]. Having determined the value of *A* for any given combustor at any convenient test condition, Equation 5.27 may then be used to predict the lean blowout fuel/air ratio at any other operating condition.

The first term on the right hand side of Equation 5.27, which contains combustion volume, is the only term that can be varied at the discretion of the designer. The second term represents the combustor operating conditions. The third term embodies the relevant fuel-dependent properties of mean drop size, effective evaporation constant, and the heating value of the fuel.

For each combustor for which lean blowout data were available [11], a value of *A* was chosen for insertion into Equation 5.27 that gave the best fit to the experimental data. These values are listed in Table 5.1. The satisfactory level

TABLE 5.1

Values of A and B Employed in Equations 5.27 and 5.29

Engine	A	B
J 79-17A	0.042	0.20
J 79-17C	0.031	–
F 101	0.032	0.090
TF 41	0.013	0.63
TF 39	0.037	0.21
J 85	0.064	0.18
TF 33	0.025	0.27
F 100	0.023	0.17

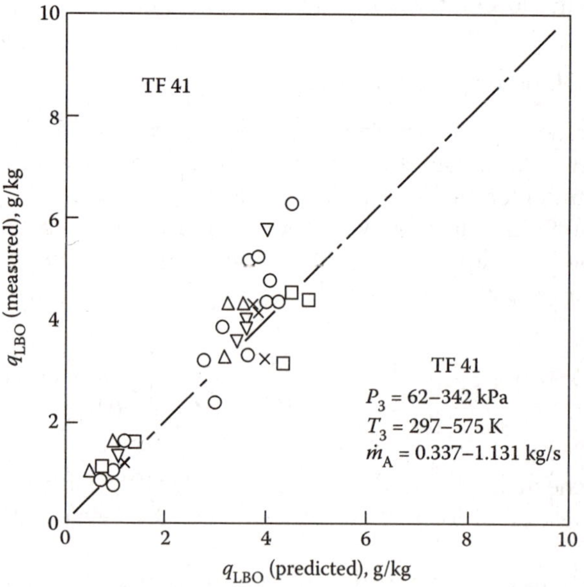

FIGURE 5.26
Comparison of measured and predicted values of q_{LBO} for a TF41 combustor. (From Lefebvre, A.H., *Journal of Engineering for Gas Turbines and Power*, 107, 24–37, 1985. With permission.)

of correlation achieved is illustrated for three combustors in Figures 5.26 through 5.28.

The especially low value of *A* shown in Table 5.1 for the TF41 combustor can be attributed to its excellent atomizing characteristics at low fuel flows, stemming from the use of an exceptionally low primary nozzle flow number. The high value of *A* obtained for the J85 combustor is attributed to a design modification made during its development that introduced additional air into the front end of the liner as a smoke-reduction measure.

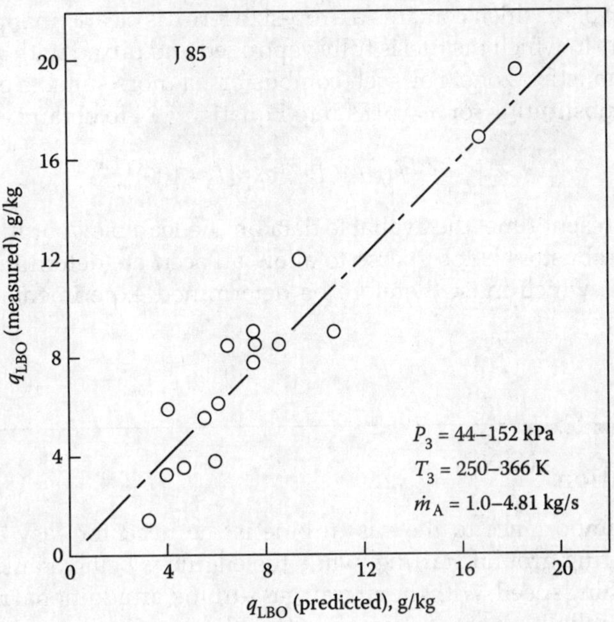

FIGURE 5.27
Comparison of measured and predicted values of q_{LBO} for a J 85 combustor. (From Lefebvre, A.H., *Journal of Engineering for Gas Turbines and Power*, 107, 24–37, 1985. With permission.)

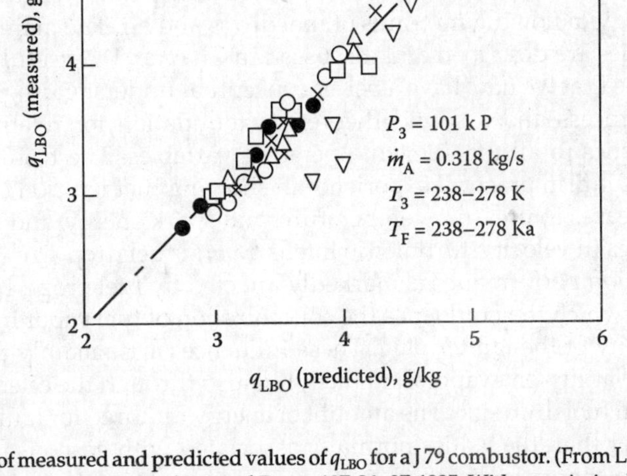

FIGURE 5.28
Comparison of measured and predicted values of q_{LBO} for a J 79 combustor. (From Lefebvre, A.H., *Journal of Engineering for Gas Turbines and Power*, 107, 24–37, 1985. With permission.)

Equation 5.27, which contains a drop-size term, is clearly inappropriate for combustors in which the fuel is fully vaporized and mixed with air upstream of the combustion zone. For such combustors, a more suitable expression is given by substituting for $\dot{m} = \rho UA$ into Equation 5.24 to obtain

$$q_{\mathrm{LBO}} = C\{\dot{m}_{\mathrm{A}} / V_{\mathrm{c}} P_3^{1.25} \exp(T_3 / 100)\}^{0.16}. \tag{5.28}$$

At the present time, the available data on the lean blowout limits of lean-premix combustors is too sparse to allow an accurate determination of the constant C, which must therefore be determined experimentally for each combustor.

5.11 Ignition

Of prime importance to the gas turbine is the need for easy and reliable lightup during ground starting, while the engine is being cranked up to its self-sustaining speed. With the aircraft gas turbine, an additional requirement is for rapid relighting of the chamber after a flameout in flight. Under adverse climatic conditions, or on takeoff from a wet runway, where there is a risk of ingestion of excessive amounts of water or ice, the ignition system must also be capable of continuous operation in order to ensure immediate relighting of the engine in the event of flame extinction.

The ignition of a combustible mixture may be accomplished by various means, but in the gas turbine it is usually effected by means of an electric spark. Large amounts of energy are needed to ignite the heterogeneous and highly turbulent mixtures flowing at velocities of the order of 25 m/s.

Since the early 1970s, many theoretical and experimental studies have been carried out on the influence of fuel and flow parameters on minimum spark energy in flowing mixtures of fuel drops and air. The main findings of these studies are described and discussed in Chapter 2. A useful outcome of this work is that we now have a better conceptual understanding of the basic ignition process and a sound theoretical foundation for relating ignition characteristics to all the relevant operating variables. The results obtained generally confirm practical experience in showing that ignition is made easier by increases in pressure, temperature, and spark energy, and is impaired by increases in velocity, turbulence intensity, and fuel drop size. With liquid fuels, ignition performance is markedly affected by fuel properties through the way in which they influence the concentration of fuel vapor in the immediate vicinity of the igniter plug. These influences arise mainly through the effect of volatility on evaporation rates, but also through the effect of viscosity on mean fuel drop size. The amount of energy required for ignition is very much larger than the values normally associated with gaseous fuels. Much of this extra energy is absorbed in the evaporation of fuel droplets, the actual amount depending on the distribution of fuel throughout the primary zone and on the quality of atomization.

The continuing trend toward engines of higher compression ratio and higher primary-zone velocities has produced a gradual deterioration in the environmental conditions of the ignition unit and igniter plug. At the same time, there has been an increasing demand for improvements in the performance, life, and reliability of ignition equipment. Thus, the problem of ignition, particularly in aircraft engines, is one of continuing importance and merits discussion in some detail.

5.12 Assessment of Ignition Performance

The ignition performance of an aircraft engine is usually expressed in terms of the range of flight conditions over which combustion can be re-established after a flameout at altitude. To determine the relighting capabilities of an engine, it is customary to carry out a series of combustor rig tests, in which the inlet parameters are varied to reproduce a range of flight conditions. The test procedure is very similar to that employed in deriving stability limits. For constant combustor inlet pressure, temperature, and air mass flow, ignition is attempted at various values of fuel/air ratio. Successful ignition is indicated by continued burning after the igniting source has been switched off. A maximum time, which is normally 10 seconds but could be as low as 3 seconds, is allowed for each ignition attempt. The procedure is repeated for a range of mass flows until a complete ignition loop can be drawn. A typical loop for an aircraft combustion chamber is shown in Figure 5.29. Usually,

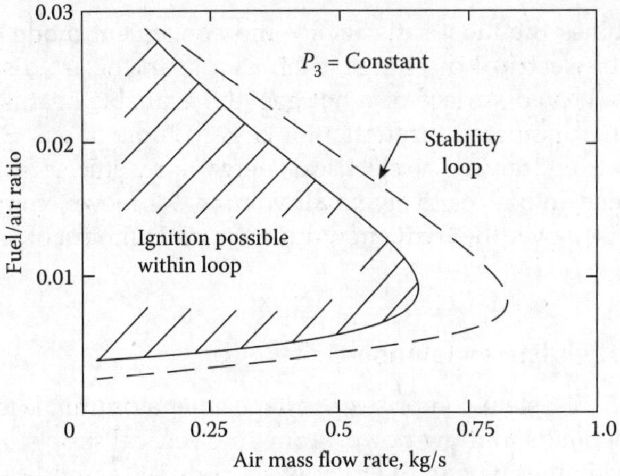

FIGURE 5.29
Typical combustor ignition loop.

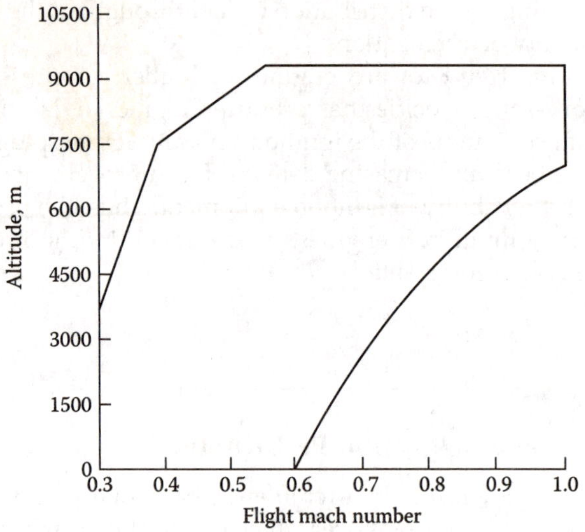

FIGURE 5.30
Altitude relight limits of a conventional annular combustor.

between four and six ignition loops, obtained at different levels of pressure, are sufficient to determine the complete relighting characteristics of the engine. The altitude relighting limits of a conventional annular combustor are shown in Figure 5.30.

5.13 Spark Ignition

For gas turbines the most satisfactory and convenient mode of ignition is some form of electrical discharge, such as a spark or arc discharge. With ignition by a heated surface or a hot gas, the available heat is used wastefully by being dissipated throughout a large volume of gas. Sparks or discharges, however, convert the electrical energy fairly efficiently into heat that is concentrated into a relatively small volume. Moreover, complete control can be exercised over the frequency, duration, and amount of energy in each discharge.

5.13.1 The High-Energy Ignition Unit

A basic ignition system comprises a voltage generator unit, lead, and igniter plug. Its function is to draw power from an electrical supply and to release energy to the igniter plug in the form of short-duration pulses.

The circuit diagram of a standard 12 J ignition unit is shown in a simplified form in Figure 5.31. An induction coil, operated by an electromechanical

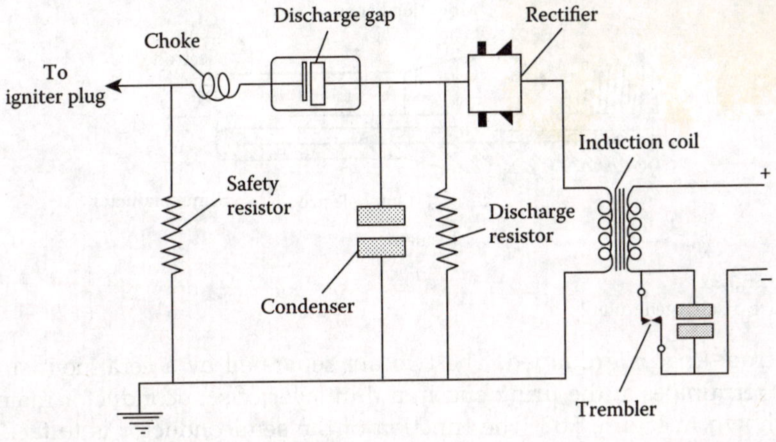

FIGURE 5.31
Standard high-energy ignition unit.

vibrator from the normally available 24 volt DC supply, charges a reservoir condenser through a high-voltage rectifier until the condenser voltage equals the breakdown voltage of the sealed discharge gap, which is usually around 2 kV. The condenser then discharges through the sealed barrier gap, a choke, and the surface discharge plug, which are all connected in a series. The purpose of the choke is to control the spark duration, whereas the safety resistor is fitted to ensure the dissipation of stored energy in the condenser, should it be left in a charged condition when the unit is not in use.

On more modern systems, the vibrating-contact voltage generator is replaced by a transistorized charging unit, thereby extending life that would otherwise be limited by contact wear [42].

Many types of ignition units are now produced to suit individual engine and aircraft requirements. Both single and twin channel units are available, having stored energy levels between 1 and 12 J. The normal spark rate of a typical ignition system is between 60 and 100 sparks per minute. In some designs, the spark is made unidirectional in order to increase the amount of energy released in the spark. Fuel-cooled units are used when ambient air temperatures are exceptionally high.

Units for small gas turbine engine applications have ratings of around 2 J, with a typical sparking rate of 250 per minute. For larger engines, the stored energy is normally between 4 and 12 J, the rate of sparking around one per second, and the energy released at the igniter tip is between 2 and 4 J, depending on the plug design.

5.13.2 The Surface Discharge Igniter

The high-energy system is most effective when used in combination with a *surface discharge* igniter plug. This consists of a central electrode and an outer

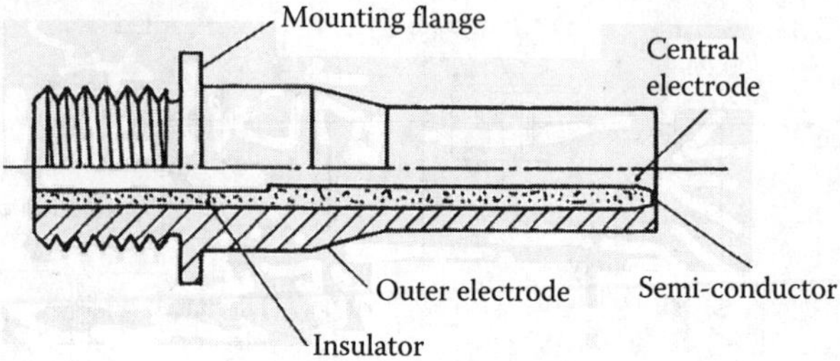

FIGURE 5.32
Surface discharge igniter.

electrode that is grounded. The two are separated by a ceramic insulator that terminates at the firing end in a thin layer of semiconductor material, as shown in Figure 5.32. The function of the semiconductor is to facilitate ionization of the spark gap and allow sparks to be produced from energy sources at relatively low voltage. An important characteristic of the semiconductor material is that its electrical resistance falls with an increase in temperature. This means that when a condenser voltage is applied, and current starts to flow through the semiconductor, it is soon concentrated into a fine filament that rapidly becomes incandescent, thus providing an ionization path across the electrodes. Once ionization has taken place, the main discharge then occurs in the form of an intense flame-like arc.

The surface discharge plug was developed at the Royal Aircraft Establishment, Farnborough, in the late 1940s, and by the early 1950s had become accepted as a standard piece of equipment on nearly all aircraft engines. Its performance is so markedly superior to all other modes of ignition that, in one form or another, it is very widely used.

Due to losses and leakages in the system, the energy reaching the plug tip is only about a quarter of the energy released by the condenser discharge. Moreover, Watson [43] has shown that only a small fraction of this, about one-third, directly heats the combustible mixture. Some of the remainder is lost by conduction to the plug face, but the largest portion is dissipated in the form of radiation and sound waves. Odgers and Coban [44] have examined the effects of gas pressure and other relevant variables on the amount of energy released at the plug. It was found that wetting the plug face with fuel can almost double the energy release, but excessive amounts of fuel reduce the spark energy and also quench the flame kernel. Optimum conditions for ignition are obtained with only a very thin layer of fuel on the plug face.

5.13.2.1 Igniter Performance

The performance of a surface discharge plug is usually expressed as the ratio of the energy in the spark to the stored energy in the condenser. As discussed above, this can only give an approximate guide because only a small proportion of the spark energy contributes directly to ignition, and this proportion varies between one design of plug and another. Nevertheless, for a constant storage capacitor, the energy release in the spark provides a useful and convenient yardstick of its "ignitability."

5.13.2.2 Igniter Design

Surface discharge plugs may be classified into two main types, according to whether the semiconductor is flush with the electrodes (*flush fire*) or recessed (*sunken fire*), as shown in Figure 5.33. Tests have demonstrated that spark energy is increased by an increase in gap width or a reduction in depth of recess. Thus, maximum ignitability is obtained from flush plugs whose gap width is large.

A flush-gap igniter is more difficult to manufacture than a recessed-gap type, is more expensive, and has less mechanical strength. On the other hand, it is more efficient, i.e., a larger proportion of the stored energy is available for ignition, and its operating life is longer. Another advantage over the recessed-gap igniter is susceptibility of the latter to "fuel wetting." This deficiency is most serious for multispool, bypass aero engines that are characterized by a long delay between starting the engine and achieving suitable lighting conditions in the primary zone. As a result, the igniter may be sparking under fairly wet conditions for a relatively long time. The effect of this fuel wetting on the recessed plug is to confine the discharge to the bottom of the recess,

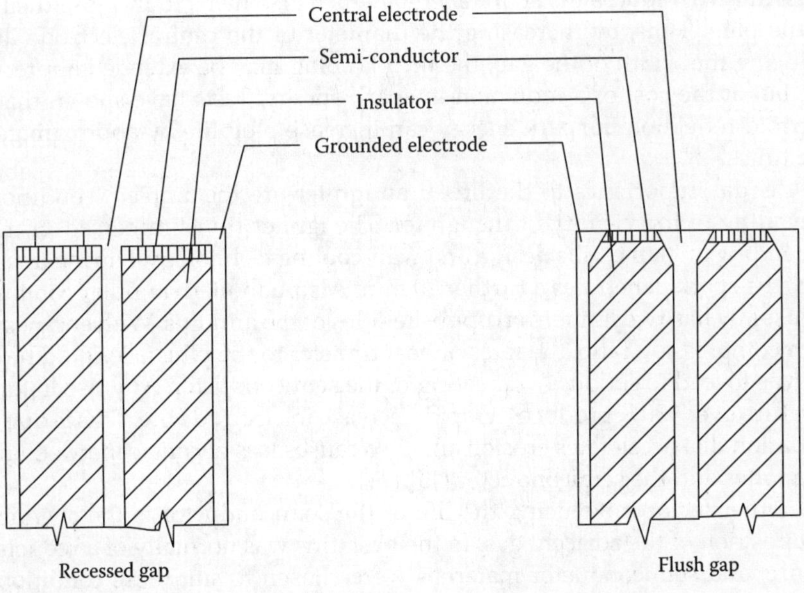

FIGURE 5.33
Two basic types of surface discharge igniters.

so that most of the spark energy is dissipated in heating the electrodes and causing serious erosion. Thus, under fuel wetting conditions, a recessed-gap igniter has poor ignitability and a fairly short life.

5.13.2.3 Igniter Life

West [45] has pointed out that as the service life of aircraft engines is continuously increasing, to allow maintenance costs and unscheduled replacements to be cut to a minimum, the requirement of operators is for an igniter that will last the full engine life. If this is impossible, it should at least have a life that will coincide with some convenient check period. To be worthwhile, any gain in plug life should be sufficient to enable the plug to fire satisfactorily until the next check period.

Experience has shown that the life of a surface discharge igniter is governed partly by the energy release in the spark and also by the environmental operating conditions. Each spark causes a small pit in the electrodes, which are gradually eroded away until eventually contact is broken between the electrodes and the semiconductor. Clearly, the larger the energy in the spark, the higher the rate of erosion.

Plug life can be extended by increasing the diameter of the central electrode, thereby increasing the volume of metal available for erosion. However, as it is very undesirable to increase the aerodynamic blockage of an igniter, its outer diameter must remain fixed, and any increase in the size of the central electrode can only be obtained at the expense of a reduction in gap width. This, in turn, reduces the spark energy, which also helps to increase the life of the plug. Thus, by increasing the diameter of the central electrode and reducing the width of the gap, the life of a plug may be extended appreciably, but at the cost of a reduction in spark energy. Tests have shown that a fourfold reduction in spark energy can increase plug life by approximately five times.

Of equal importance to the life of an igniter are the ambient conditions prevailing in the vicinity of the igniter. The igniter tip is immersed ±1 mm depending on flame tube design and wall-cooling technology. During operation, the spark penetrates a further 20 mm. Also, a modern annular combustor usually has two igniters on opposite sides of the annulus. With engines of high compression ratio, the rates of heat transfer to the plug face could be so high as to lead to serious overheating of the semiconductor. Any rise in temperature over 900 K produces a rapid decrease in electrode life. This is due to oxidation that accelerates erosion and also causes loss of contact between the electrodes and the semiconductor [45].

Another factor influencing the life of the semiconductor is the chemical composition of the adjacent gas. In the past, this was normally of a reducing nature, and semiconductor materials were chosen to suit these conditions. Nowadays, owing to the continuing trend toward leaner primary zones, the ambient gases tend to be more oxidizing.

The problems of overheating can be alleviated by directing a film of cooling air over the plug face. However, this is not a completely satisfactory solution because the "chilling" effect impairs ignition performance at extreme altitudes. Retractable plugs are sometimes used on large industrial engines, but they would pose formidable mechanical problems for aircraft engines. A better solution could emerge from the development of new semiconductor materials capable of operating for long periods in a high-temperature, oxidizing atmosphere. Further extensions in plug life could also be gained from reductions in spark-energy requirements and by the use of plugs that incorporate, as a basic design feature, electrodes of large volume.

5.14 Other Forms of Ignition

Although the combination of high-energy unit and surface discharge igniter is most widely used for gas turbine ignition, other means of ignition are available for certain special applications.

5.14.1 Torch Igniter

The torch igniter incorporates a spark plug and an auxiliary fuel jet in a common housing. The juxtaposition of these two components is such that ignition of the fuel spray by the spark produces a "torch" of burning droplets that ignites the main fuel spray. The performance of a torch igniter is fairly insensitive to its location. Usually, it is fitted into the annulus formed between the liner and the air casing near the upstream end of the chamber, but at least one annular combustor has been produced with a torch igniter mounted on the dome of the liner within the snout.

The main problem with torch igniters is that of fuel cracking and gumming when the atomizer is inoperative. This causes blockage of the discharge orifice, which is deliberately made very small (typically around 0.23 mm in diameter) to produce a well-atomized spray. The problem can be alleviated by fitting solenoid valves to turn off the fuel after lightup, and by the provision of clean purging air, but these items introduce additional weight and complexity.

Torch igniters are mandatory for vaporizer combustors. It is customary to provide at least two units. Typically, they are located in the two lower quadrants of the combustor in the four and eight o'clock positions. On some engines, supplementary fuel nozzles (without spark igniters) are fitted at the ten, twelve, and two o'clock positions to expedite the spread of flame throughout the primary combustion zone. After the vaporizing tubes are warmed and the flame is fully established, the pilot fuel and the spark igniter are turned off.

The factors governing the rate of flame spreading in annular vaporizer combustors have been discussed by Opdyke [46]. Inlet air temperature is an important consideration. In fact, the time required to reach 90% complete spreading increases severalfold when the temperature decreases from normal ambient to 220 K. Fuel volatility and fuel/air ratio also have a significant effect on the rate of flame spreading, presumably due to their influence on the rate of fuel evaporation. It is also of interest to note that the spacing of the vaporizer exits is an important variable; the wider the spacing, the slower the propagation rate.

5.14.2 Glow Plug

The function of a glow plug is to provide rapid reignition of the flame should extinction occur as a result of the sudden ingestion of water or ice, or through temporary fuel starvation. The dimensions of the plug are chosen to suit the size of the liner, but a typical plug would be in the form of a hollow cylinder, 25 mm in length and 15 mm in external diameter.

The glow plug is mounted on the liner wall and protrudes into the primary zone where it is immersed in flame gases at high temperature. This is the ideal location for relighting the fuel–air mixture in the event of a sudden flameout. An ideal plug material would combine high values of specific heat, density, and thermal conductivity with strong resistance to oxidation and thermal shock. Tests on platinum glow plugs were highly satisfactory, but this material is too expensive [47]. Nitride-bonded silicon carbide proved to be an effective substitute, and service lives exceeding 4000 hours have been realized in military aircraft. In tests carried out on Rolls Royce Proteus and Viper II engines, successful relights were achieved for periods of up to 12 seconds after engine shut down.

The main drawback to glow plugs is the obvious one—the risk of a plug, or part of a plug, becoming detached and damaging the turbine blades. This is why they have found so few applications. However, they appear to merit consideration for helicopter engines for which they seem almost ideally suited.

5.14.3 Hot-Surface Ignition

Although the ignition of a fuel spray by a hot surface is technically feasible and, in fact, has been demonstrated many times, it is not normally regarded as a practical proposition in gas turbines due to the very high rates of heat transfer that would be needed to evaporate the fuel and raise the vapor–air temperature to the point of ignition in the very short time that the mixture is in contact with the hot surface. This, of course, is why spark ignition is so successful, because it provides an almost instantaneous (<150 μs) transfer of heat to the fresh mixture. However, one practical form of hot surface igniter was developed by Saintsbury [48] for the PT6 engine. It featured electrical heating and is reported to have functioned satisfactorily for a multiplicity of fuels, including heavy diesel oil.

5.14.4 Plasma Jet

This method of ignition has been studied by Weinberg et al. [49–52]. The plasma jet used in their early experiments differs from a normal igniter in that the electrical discharge occurs in a small cavity that is supplied with a suitable plasma medium via a small capillary. The very high pressures and temperatures generated by the discharge cause the plasma to be ejected as a supersonic jet through an orifice located at the downstream end of the cavity. By varying the feed to the cavity, the energy input, and the size of the discharge orifice, it is possible to vary the temperature of the plasma jet and its velocity (and hence its penetration distance) [50]. The gases tested include nitrogen and hydrogen and their effectiveness as an ignition source is attributed to their high content of radicals. Kingdon and Weinberg [51] also studied minimum ignition energies for plasmas of various compositions, generated by focusing laser beams on minute targets of different materials located in the test section.

In a later study, Warris and Weinberg [52] employed a new concept, which is not only smaller and more efficient than previous designs, but also generates plasma jets of a rather different structure. Instead of a continuous high temperature jet, the new design produces a "lukewarm" gas that contains "hot pockets" of highly dissociated active species. The results obtained with this new device demonstrate significant improvements in both ignition and stability performance when used in fast-flowing combustible mixtures. This increased range of performance is achieved at the cost of a very small increase in electrical power.

Warris and Weinberg also investigated the ignition performance of pulsed plasma jets. The main advantage of using pulsed jets in place of continuous ones is that the electrical equipment tends to be lighter because large currents are required only transiently. Furthermore, it is readily available commercially because it is in widespread use for the high-energy, surface discharge igniter, as described earlier in this chapter.

The results obtained with pulsed plasma jets generally confirm those obtained with continuous plasma jets in showing that ignition and stable combustion can be achieved at fuel/air ratios much lower than the normal lean blowout limit. What is not clear from these studies is the degree of fuel conversion achieved, i.e., the level of combustion efficiency, when the combustor is operating at these exceptionally low fuel/air ratios.

5.14.5 Laser Ignition

For more than 30 years, it has been known that the focused output of a sufficiently powerful Q-switched laser beam can cause the electrical breakdown of gases. This phenomenon has obvious potential for the ignition of combustible mixtures and appears to have a number of advantages over conventional methods of ignition. It allows the ignition site to be accurately positioned at some point within the primary combustion zone where conditions for ignition are most favorable, and also avoids the various heat losses that are incurred when the igniter is located at the liner wall. These losses are especially important because the incipient flame kernel is in close proximity to the electrodes during its early and most vulnerable moments when it is also subject to the chilling effects of film-cooling air flowing over the face of the plug.

As with conventional spark ignition, the duration of the energy pulse is important. If the heating period is too long, heat losses to the surrounding gas will be excessive and the incipient flame kernel will be extinguished. On the other hand, a heating period that is too short will create a powerful blast wave that carries energy away from the ignition site.

The research carried out so far on laser-induced spark (LIS) ignition has been largely confined to pressure bombs and reciprocating engines [53,54]. An important conclusion from these studies is that LIS ignition can ignite fuel–air mixtures that are much leaner than can be ignited by conventional methods. This clearly has important implications for gas turbines, which are increasingly being called on to burn leaner mixtures to meet legislation aimed at reducing nitrogen oxide (NO_x) emissions (see Chapter 9).

With liquid fuels, the presence of fuel droplets raises important questions as to how these droplets will behave when irradiated by intense laser beams. Any process that leads to the destruction of the droplets at the focus of the laser beam and the formation of fuel vapor is likely to increase the probability of ignition. However, the presence of droplets in the converging beam, in a region where the laser intensity is insufficient to cause breakdown, will reduce the amount of energy reaching the focal point.

Greenhalgh and Gallagher [55] have reviewed the current status of LIS ignition and the problems that must be resolved before it can be applied to gas turbine engines. At the present time, much interest is being shown in this novel form of ignition, but it is clear that more research and development is needed before it can be regarded as a serious contender to the existing high-energy system.

5.14.6 Chemical Ignition

There are a number of chemicals that ignite spontaneously on contact with air and produce a high rate of energy release. Tests have shown that very small quantities of these so-called *pyrophoric* fuels (about 2 cc) are very effective when injected through a hypodermic tube into the primary zone [56].

Pyrophoric fuels include trimethyl aluminum, triethyl aluminum, and aluminum borohydride. To ease storage and injection problems, in many of the tests reported, the fuel was diluted with kerosine or mineral oil. Although these tests show that pyrophoric fuels are potentially very powerful sources of ignition, practical means of storage and injection must be found before they can be used in civil aircraft. They are obviously very dangerous in the event of a crash, and this would appear to restrict their use to military aircraft.

5.14.7 Gas Addition

The effect of gas addition on ignition limits has been studied by Xiong et al. [57] using a combustor supplied with kerosine fuel. In some experiments, the gaseous fuel (composition unspecified) was injected into the inlet air upstream of the combustor; in others, it was injected through the spark plug itself. The results show that ignition limits can be extended to lower levels of pressure by the local injection of gaseous fuel into the spark kernel. This finding is fully consistent with the observations of Rao and Lefebvre (Reference [41] in Chapter 2) to the effect that "the main obstacle to ignition is lack of evaporated fuel within the spark kernel."

5.14.8 Oxygen Injection

The beneficial effect of oxygen addition to almost all aspects of combustion performance is well known. In the context of ignition, this is clearly demonstrated in Figure 2.10, which shows how increasing the oxygen content of a propane–air mixture from the normal value of 21–50% reduced the ignition energy requirement by a factor of 40. This beneficial effect is found to also apply to the ignition of fuel sprays in gas turbine combustors, where ignition at low pressures is greatly facilitated by the injection of oxygen into the primary zone [58]. Tests carried out at Lucas (now Aero and Industrial Technology Ltd., Burnley, UK) on a Proteus chamber at a pressure of 14 kPa showed that an oxygen flow equivalent to 0.5% of the normal chamber airflow allowed an increase of the order of 200% in the limiting air mass flow for ignition [59]. Similar results were obtained on a tuboannular chamber, where an oxygen flow of 0.4% gave an increase of 250% in the limiting air mass flow for ignition. In both series of tests, the oxygen was injected through the atomizer air shroud. More recent work by Chin [60] on an engine combustor operating at simulated high-altitude conditions showed that flame blowout and relight performance were improved significantly by a small amount of oxygen addition.

As oxygen is normally carried on aircraft for other purposes, it would appear to be an attractive means of raising altitude relighting limits in circumstances where more conventional methods have proved inadequate.

5.15 Factors Influencing Ignition Performance

The main factors affecting ignition performance fall into the categories of ignition system, flow variables, and fuel parameters. All these factors have been investigated experimentally in a number of researches on both idealized and actual combustion chambers.

5.15.1 Ignition System

The most important characteristics of an ignition system are the energy, duration, and frequency of the spark discharge, which are dependent on the design of the ignition unit, the size of the storage condenser, and the design of the igniter.

5.15.1.1 Spark Energy

Generally, it is found that, of the total energy released during the condenser discharge, the proportion that appears in the spark increases with increase in pressure, gap width, and velocity. The beneficial effect of pressure and gap width stems from the fact that an increase in either of these, increases the number of molecules in the arc-conducting path. This raises the electrical resistance of the gap, which then demands a higher breakdown voltage, resulting in higher spark energy.

An increase in air velocity tends to increase the energy in the spark. This is due partly to "stretching" of the spark, which increases its electrical resistance and hence also the energy release, and also to a reduction in heat loss to the electrodes as the arc is displaced downstream of the plug face by the airflow.

5.15.1.2 Spark Duration

As discussed above, of the total stored energy in the condenser, only a small fraction is effective in heating the combustible mixture. Of the remainder, some is accounted for by losses in the ignition unit, e.g., condenser dielectric losses, and the rest is wasted in electromagnetic and acoustic radiation from the spark. When the spark discharge is rapid, these energy losses are very high; on the other hand, if the spark duration is too long, the energy is dispersed over a large volume of the flowing mixture, and gas temperatures are too low to cause ignition.

Swett [61], investigating ignition energies in flowing propane–air mixtures with a condenser discharge spark, found an optimum spark duration of around 100 μs. A similar conclusion was reached by Watson [43], who recommends a spark duration of not less than 100 μs, based on his experiments on stagnant mixtures using a conventional aircraft ignition unit.

Ballal and Lefebvre [62] conducted a number of experiments on gaseous fuels, where the flow conditions were kept constant and the minimum amount of energy needed to effect ignition was measured for several values of spark duration. The optimum spark duration was then defined as the value corresponding to the lowest measured value of ignition energy.

From an analysis of the heat losses suffered by a spark during the time of its discharge, it was concluded that after the initial shock wave, the main source of loss is by forced convection to the flowing stream. Based on these findings, expressions were derived for calculating the optimum spark duration for any specified mixture and flow conditions. Over the range of conditions investigated, the optimum spark duration was found to lie between 30 and 90 μs, the highest value being obtained with stoichiometric mixtures. It was also found that optimum spark duration was unaffected by turbulence, but decreased with an increase in velocity.

Lefebvre et al. [63–66] used the apparatus shown schematically in Figure 5.34 to study the influence of spark duration and other relevant parameters on the minimum ignition energy required to achieve ignition, E_{min}. Basically, it comprises the means for supplying air at normal atmospheric temperature to a test section of 75 mm^2 cross section and 350 mm long. The test section is fitted with glass side walls to allow the onset of the spark and subsequent growth of the flame kernel to be visually observed and photographed. At its upstream end are two tungsten electrodes of 3 mm diameter, mounted in insulating bushes. The upper electrode has a micrometer adjustment to provide fine control of the electrode gap width. The electrodes are connected to an ignition unit that provides electrical sparks whose energy and duration can be independently varied and measured. Downstream of the test section is an exhaust duct that is fitted with drain points to facilitate the removal of liquid fuel precipitated on the duct walls.

The experimental programs covered wide ranges of flow conditions and fuel spray properties. Fuel injection was accomplished using simplex pressure-swirl atomizers located at the center of the convergent entry duct. Five atomizers of different flow numbers were used to provide a wide range of drop sizes in the ignition zone. Mean drop sizes were measured using a diffractive light-scattering technique.

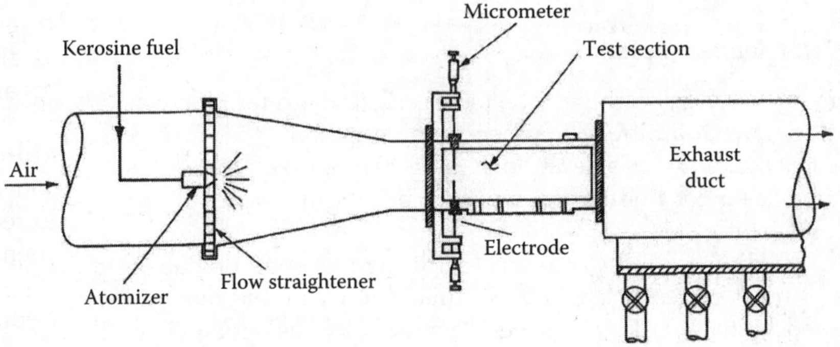

FIGURE 5.34
Schematic diagram of ignition test facility. (From Subba Rao, H.N. and Lefebvre, A.H., *Combustion Science and Technology*, 8(1), 95–100, 1973. Talyor & Francis Group. http://informaworld.com. With permission.)

Using this apparatus, Rao and Lefebvre [64] carried out a series of measurements of E_{min} for kerosine sprays injected into a flowing airstream. Their results confirmed those obtained with gaseous fuels in showing that optimum spark duration decreases with an increase in velocity. They also found that the optimum spark duration increases with an increase in the mean drop size of the spray.

5.15.1.3 Sparking Rate

During startup on the ground, the engine airflow and fuel flow both increase with time, but at different rates, resulting in wide variations in mixture strength adjacent to the igniter plug. Ignition can only occur if the spark discharge coincides with a local mixture strength that is well inside the limits of flammability. Under these conditions, an increase in sparking rate is likely to be far more effective for ignition than an increase in spark energy. An alternative method of eliminating lightup problems on the ground is by delaying or advancing the admission of fuel into the combustor, thereby changing the mixture strength conditions in the ignition zone.

Foster [67] examined the influence of spark frequency on altitude ignition performance in a J-33 combustion chamber, and demonstrated a slight improvement in relighting capability as the sparking rate was increased from 2 to 150 per second. Presumably, the effect of the sparking rate can only be really significant if it is sufficiently high for sparks to be generated in gas that has already been heated as a result of previous sparks.

Although a high sparking rate is advantageous, particularly under cranklighting conditions, for any given size of ignition unit it can only be obtained by a reduction in spark energy. Experience has shown that a spark frequency of between one and two per second, depending on the application, generally entails the minimum expenditure of power and gives the most compact form of ignition unit.

5.15.1.4 Igniter Location

In many early gas turbines, the location of the igniter plug was determined in a somewhat arbitrary manner, with accessibility for fitting and replacement as the main consideration. It is now recognized that the position of the igniter has a controlling influence on both ignition performance and plug life.

In deciding the best position for the igniter, one obvious stipulation is that it must be restricted to the primary zone, so that the hot kernel of gas created by the spark is returned upstream by the action of the flow reversal. This implies a mechanism for ignition whereby the burned pocket of gas is retained within the reversal, being continuously rotated and, at the same time, propagating outward until the entire primary zone is filled with flame.

Tests have shown that an excellent location for the igniter is close to the centerline of the liner, adjacent to the fuel nozzle [68]. Unfortunately, this is a very inconvenient position from the viewpoints of accessibility and interference with the airflow pattern. Moreover, the plug face is likely to become fouled by carbon deposits and damaged through overheating. The usual location for the plug is on the cylindrical portion of the liner, near the outer edge of the spray. However, it is very important that the igniter is not subjected to excessive fuel wetting, either by direct impingement from the spray or as a result of fuel flowing along the liner walls.

The igniter tip should protrude far enough through the liner to clear, or almost clear, the layer of cooling air flowing along the inside wall. Some tip-cooling air is needed to protect the tip face from overheating. On no account should the tip temperature be allowed to exceed 900 K. In general, increasing the depth of immersion of the plug into the hot gas stream improves its ignitability and reduces its life.

With most industrial engines, the ignition problem does not loom very large because the penalties of failure to achieve lightup are much less severe than in the case of aircraft. A common practice is to fit igniters that can be withdrawn when not in use. This approach has much to commend it because it allows the plug to be sited in the most advantageous position for ignition, and yet avoids all the problems of aerodynamic interference and plug life. If the engine is burning distillate fuel, a torch igniter is often used and is supplied with fuel from the main tank. With heavy fuel oils, however, a separate source of gaseous or distillate fuel must be employed.

5.15.2 Flow Variables

The main flow variables of concern are pressure, temperature, velocity, and turbulence.

5.15.2.1 Air Pressure

All the existing experimental data, obtained from stagnant mixtures, idealized flowing systems, and practical combustors, highlight the fact that an increase in pressure reduces minimum ignition energy. Typical of the results obtained for flowing gaseous mixtures are those of Ballal and Lefebvre [69] for propane, as illustrated in Figure 2.10.

For heterogeneous fuel–air mixtures, the effect of pressure on E_{min} may be appreciably less, depending on the extent to which fuel evaporation rates are limiting to the onset of ignition. Where the ignition process is fully controlled by chemical reaction rates, $E_{min} \propto P^{-2}$. If evaporation rates are controlling, then $E_{min} \propto P^{-0.5}$. Thus, the pressure exponent is always between −0.5 and −2.0, becoming closer to −2.0 with a reduction in pressure and/or fuel/air ratio.

Most of the published data on practical combustors were obtained by Foster et al. [67,70,71]. Although the results are few and show appreciable scatter, they all follow the same trend as noted above in regard to the adverse effect of a reduction in pressure on ignition performance.

5.15.2.2 Air Temperature

All the available evidence shows that a reduction in air temperature is detrimental to ignition. This is only to be expected because more energy is required to raise the fuel–air mixture to its reaction temperature, and also because at low air temperatures evaporation rates are slower and a larger proportion of the spark energy is absorbed in evaporating fuel drops.

The effect of inlet air temperature on minimum ignition energy is shown in Figure 5.35, which contains data obtained at a pressure of 20 kPa, a velocity of 15 m/s, and a constant SMD of 60 μm [66]. This figure shows that the ignition energy requirement decreases rapidly with an increase in air temperature and with an increase in fuel/air ratio toward the stoichiometric value.

As combustion chambers are normally required to operate at a constant value of $U/T^{0.5}$, any reduction in air temperature is accompanied by a

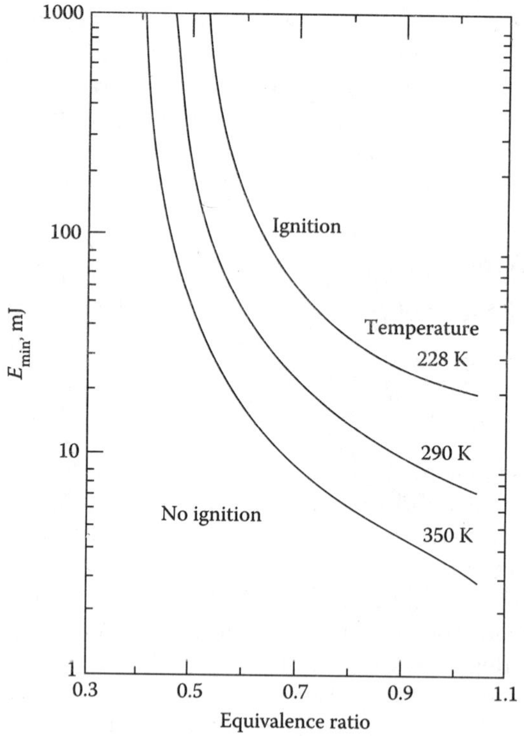

FIGURE 5.35
Effect of air temperature on minimum ignition energy; P=20 kPa, U=15 m/s, SMD=60 μm. (Reprinted from Ballal, D.R. and Lefebvre, A.H., *Combustion and Flame*, 31(2), 115–126, 1978. Copyright by The Combustion Institute.)

reduction in velocity. Thus, except at low fuel flows where evaporation rates predominate, the adverse effect of a reduction in air temperature is partly offset by the corresponding reduction in air velocity.

5.15.2.3 Air Velocity

One beneficial effect of an increase in velocity, as mentioned earlier, stems from stretching of the spark in a downstream direction, which increases the energy released during the spark discharge and reduces the loss of heat and active species from the spark to the electrodes. However, offsetting these advantages is the convective heat loss suffered by the spark kernel during the initial phase of its development when it is still "anchored" to the electrodes. This loss of heat, which increases almost linearly with velocity, impairs ignition performance unless it is compensated for by an increase in spark energy.

Figure 5.36 shows the variation of E_{min} with the SMD of the spray for stoichiometric kerosine–air mixtures at four levels of velocity. Each line represents the ignition limit for that particular velocity, with ignition being

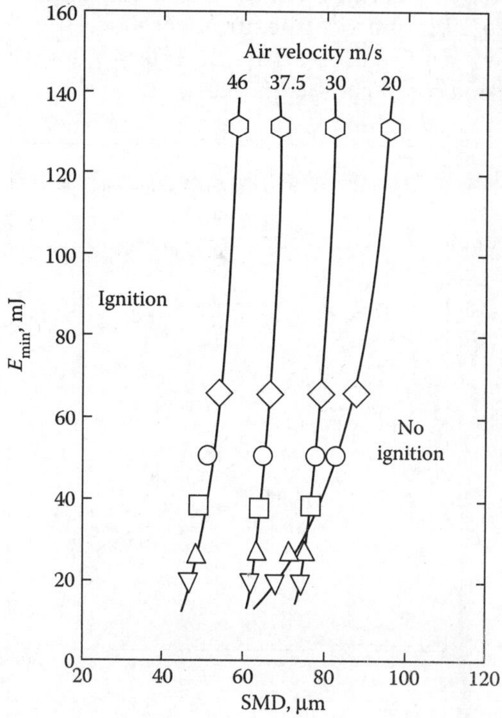

FIGURE 5.36
Variation of minimum ignition energy with mean drop size. (From Subba Rao, H.N. and Lefebvre, A.H., *Combustion Science and Technology*, 8(1), 95–100, 1973. Talyor & Francis Group. http://informaworld.com. With permission.)

possible only in the region to the left of the line. This figure illustrates the extent to which the adverse effect of an increase in velocity can be offset by an improvement in atomization.

The detrimental effect of an increase in velocity is also illustrated in Figure 5.37. Except at low velocities, where heat losses to the spark electrodes are significant, to maintain the same ignition performance, an increase in velocity must be accompanied by an increase in spark energy and/or a reduction in mean drop size.

Another adverse effect of an increase in velocity is that it gives the spark kernel less time in which to propagate throughout the primary zone before being swept downstream.

5.15.2.4 Turbulence

When the spark kernel separates itself from the igniter and enters the recirculation zone, being airborne it is no longer subject to the influence of velocity, per se. However, it still suffers loss of heat to its surroundings by turbulent diffusion to an extent that is governed by the fluctuating velocity component, u'. The adverse effect of turbulence on ignition is illustrated for gaseous mixtures in Figure 2.9. In practice, the turbulence level in the primary zone is determined by the pressure drop across the liner wall, which does not vary appreciably from one combustor to another.

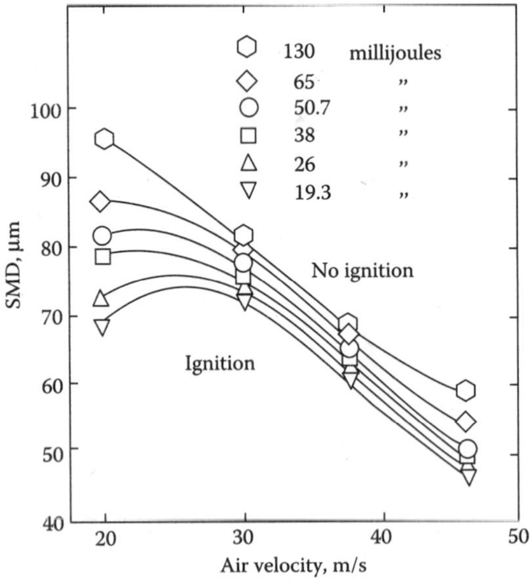

FIGURE 5.37
Influence of air velocity on ignition limits. (From Subba Rao, H.N. and Lefebvre, A.H., *Combustion Science and Technology*, 8(1), 95–100, 1973. Talyor & Francis Group. http://informaworld.com. With permission.)

5.15.3 Fuel Parameters

For the aero gas turbine to have some operational flexibility, it must be able to accept a fairly broad range of kerosine-type fuels. In a like manner, for the industrial gas turbine to retain its competitive edge over other forms of prime mover, it must uphold its reputation as being "omnivorous" of fuels. Thus, the effect of fuel properties on ignition performance is one of special interest and importance.

5.15.3.1 Fuel Type

In practical combustion systems, ignition performance is affected by fuel properties mainly through their influence on the concentration of fuel vapor in the vicinity of the spark plug and throughout the primary zone during the lightup sequence. The evaporation rates are governed by two main factors:

1. The fuel volatility, as indicated by Reid vapor pressure, the ASTM 10% evaporated temperature, the transfer number *B*, or the evaporation constant, λ.
2. The total surface area of the fuel spray, which is directly related to spray SMD, and hence to fuel viscosity

If the aviation turbine fuel specifications were broadened to include a larger fraction of the total crude petroleum, the most significant changes would be increased aromatic content and a higher final boiling point. These changes would lower the volatility of the fuel and, at the same time, increase its viscosity, thereby impairing atomization quality and reducing the surface area of the spray. Both of these effects would combine to lower the rate of fuel evaporation, thereby aggravating the lightup problem. These considerations are not a cause for concern at the present time, but they could become very significant if the inability of world petroleum production to keep pace with basic demand compels the use of synthetic and biojet fuels derived from coal, biomass, shale, and tar sands.

The detrimental effect of a reduction in fuel volatility on ignition energy requirement is illustrated in Figure 5.38. All the data in this figure were obtained for a constant SMD of 100 μm to exclude the atomizer characteristics (which normally tend to dominate ignition performance) and to demonstrate the effect of volatility acting alone.

5.15.3.2 Fuel/Air Ratio

As might be expected from simple considerations of all aspects of the ignition process, optimum conditions for ignition are when the primary-zone mixture strength is roughly stoichiometric, i.e., when flame speed and flame temperature are highest. Under lightup conditions, however, only vaporized

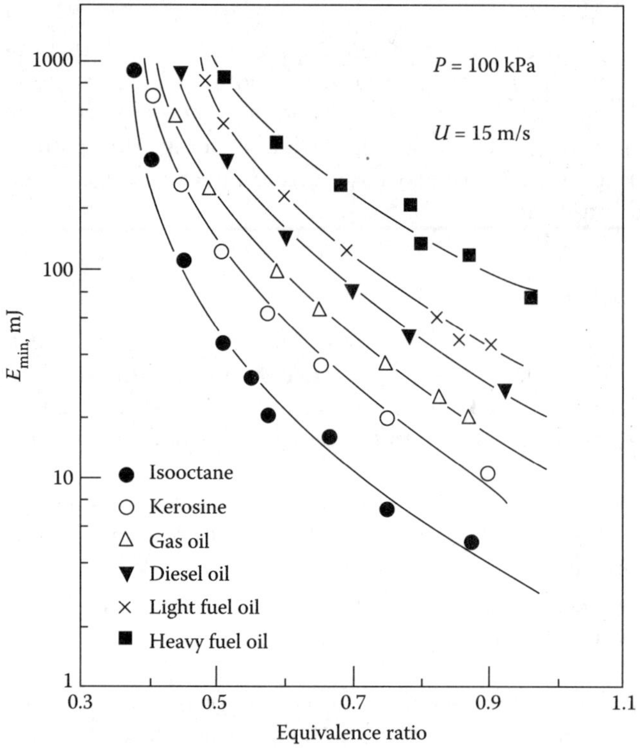

FIGURE 5.38
Influence of fuel type on minimum ignition energy. (Reprinted from Ballal, D.R. and Lefebvre, A.H., *Combustion and Flame*, 35(2), 155–168, 1979. Copyright by The Combustion Institute.)

fuel can participate in the ignition process and the fraction of the total fuel that is vaporized in the time available depends mainly on its volatility and on the quality of atomization. Thus, the average fuel/air ratio in the primary zone has no real significance, and ignition performance is governed by the effective fuel/air ratio that denotes the mass ratio of fuel *vapor* to air. Conditions for ignition are ideal when the effective fuel/air ratio in the primary zone is close to stoichiometric.

5.15.3.3 Spray Characteristics

It is well known that great improvements in engine starting characteristics can be obtained by changes to the fuel injector that reduces the mean drop size of the fuel droplets in the spray. Normal fuels are not sufficiently volatile to produce vapor in the amounts required for ignition and combustion, unless their surface area is vastly increased by atomizing the fuel into a large number of small drops. The smaller the drop size, the faster the rate of evaporation. The influence of drop size on the ignition process has been studied in some detail by Lefebvre et al. [63–66] and has been shown to be of paramount importance. This is apparent from Figure 5.36, which clearly illustrates the substantial increase in minimum ignition energy required by even a modest increase in mean drop size.

For ignition to be successful, the spatial distribution of fuel droplets is also important. Unfortunately, in practical combustion chambers, current methods of fuel injection tend to produce significant variations in mixture strength throughout the combustion zone. In particular, when pressure atomizers are operating at low combustion pressures, the mixture strength in the vicinity of the igniter plug tends to be appreciably greater than in the primary zone as a whole. This means that even if the spark kernel is fortunate enough to survive fuel quenching at the liner wall, it could still fail to achieve lightup due to lack of fuel near the centerline of the combustor. This problem is less serious with airblast atomizers because they tend to produce a more uniform radial fuel distribution. However, for all types of atomizer, ignition performance depends not only on the average fuel/air ratio in the primary zone, but also on the effective fuel/air ratio in the ignition zone.

5.15.3.4 Fuel Temperature

Ignition performance is adversely affected by a reduction in fuel temperature. This result may be explained qualitatively in terms of the effect of fuel temperature on evaporation rates. In general, evaporation rates increase with fuel temperature, partly because of the higher volatility, but also because of finer atomization due to the reduction in viscosity.

5.16 The Ignition Process

A significant advance in the understanding of gas turbine ignition came with the realization that instead of the simple single-step mechanism as hitherto supposed, the ignition process actually occurs in two or more distinct phases [72]. Phase 1 is the formation of a kernel of flame of sufficient size and temperature to be capable of propagation. Phase 2 is the subsequent propagation of flame from this kernel to all parts of the primary zone. Phase 3, which applies only to tubular and cannular designs of chamber, is the spread of flame from a lighted liner to an adjacent unlighted liner. The three-phase nature of the ignition process is illustrated in Figure 5.39.

A failure of any of the above steps will result in a failure to light up the combustor. Recognition of the multiphase nature of the ignition process helps to shed light on various apparent anomalies. For example, it explains why, in one instance, an increase in spark energy can improve ignition performance, while in another it has no effect. The explanation is simply that in

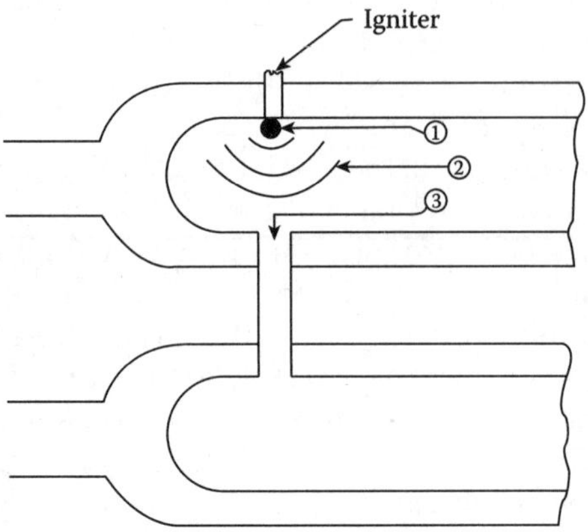

FIGURE 5.39
Three-phase nature of the ignition process.

the first case the bottleneck on performance is in phase 1, while in the second it is in phase 2 or phase 3. Similarly, a failure to achieve in flight the altitude relighting performance predicted from rig tests on a single segment of a multitube combustor, can usually be attributed to a breakdown of phase 3. Thus, recognition of the multiphase nature of the ignition process can usefully assist combustion chamber development.

5.16.1 Factors Influencing Phase 1

Survival of the kernel of hot gas created by the spark depends entirely on whether or not the rate of heat release by combustion within the kernel exceeds the rate of heat loss to the surroundings by radiation and turbulent diffusion.

The rate of heat release is governed mainly by the effective fuel/air ratio adjacent to the plug, which should be close to stoichiometric, and by the size and temperature of the kernel, which are, in turn, determined by the energy and duration of the spark. The rate of heat loss from the kernel is largely dictated by the local conditions of velocity and turbulence and by the quantity of excess fuel present in the ignition zone. This phase of the ignition process is also strongly affected by the design of the igniter plug—*flush fire* or *sunken fire*, by its location, and by the extent to which the plug tip protrudes through the liner wall.

5.16.2 Factors Influencing Phase 2

The location of the igniter is also important in phase 2 because it determines whether the hot kernel is entrained into the primary-zone reversal or is swept away downstream. This phase is also governed by all the factors that control flame stability. Thus, an increase in pressure and/or temperature, or a reduction in primary-zone velocity, or any change in fuel/air ratio toward the stoichiometric value, all of which are beneficial to stability, will also improve phase 2.

5.16.3 Factors Influencing Phase 3

The location of the interconnector tubes is of prime importance. Ideally, the tube entrance should coincide with the region of highest gas temperature in the liner, whereas the tube exit should be sited so as to ensure that the issuing hot gas flows directly into the recirculation zone of the adjacent liner. Care should be taken to minimize the flow of film-cooling air over the ends of the tube because this could interfere with the flow of hot gas and, more important, seriously reduce its temperature.

Phase 3 is aided by the use of interconnectors in which the flow area is made large in order to facilitate the passage of flame, and whose length is kept short to minimize heat loss by external convection to the annulus air. Basic data on quenching effects of relevance to interconnector performance is contained in Reference [73].

5.17 Methods of Improving Ignition Performance

If the ignition performance of a combustion chamber is unsatisfactory, the first step is to find out in which phase the bottleneck is arising. This information can be obtained quite readily by examining the position of the ignition loop in relation to the stability limits.

Since the flow properties that control stability also exercise a similar influence on ignition behavior, it might be expected that ignition and stability limits should coincide. Stability limits, however, relate essentially to burning conditions and high metal temperatures, whereas ignition is inevitably associated with cold liner walls and comparatively high heat losses. For this reason, the two limits can never be the same, but the object of ignition development is to ensure that they are separated only by the effects of heat loss.

If the ignition loop lies well inside the stability loop, this indicates that the limitation on ignition performance is arising in phase 1. This may be checked by changing the spark energy, which should produce a corresponding change in the ignition loop. If the ignition and stability loops lie in close proximity, the bottleneck on performance is almost certainly in phase 2. These points are illustrated in Figure 5.40. Failure in phase 3 is indicated when the maximum relighting altitude is significantly less than the value predicted from rig tests carried out on a single liner.

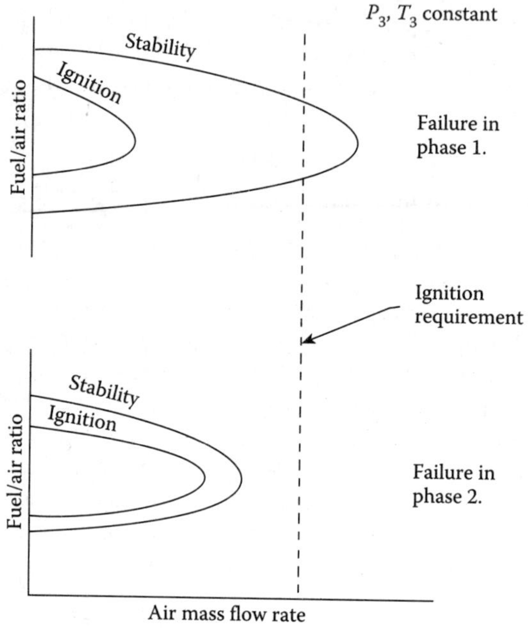

FIGURE 5.40
Curves illustrating the two main types of ignition failure.

5.17.1 Correlation of Experimental Data

From an analysis of lean lightup data acquired from a large number of aircraft combustion chambers [12–17], the following equation for lean lightup fuel/air ratio, q_{LLO}, was derived [11].

$$q_{LLO} = \left[\frac{B}{V_c}\right]\left[\frac{\dot{m}_A}{P_3^{1.5}\exp(T_3/300)}\right]\left[\frac{D_r^2}{\lambda_r H_r}\right], \tag{5.29}$$

where B is a constant whose value depends on the geometry and mixing characteristics of the combustion zone and also on the amount of air employed in primary combustion.

For each combustor, a value of B was chosen for insertion into Equation 5.29 that gave the best fit to the experimental data. These values are listed in Table 5.1. The level of agreement between predicted and measured values of lean lightup fuel/air ratio, as illustrated for three aircraft combustors in Figures 5.41 through 5.43, is considered satisfactory, especially in view of the well-known lack of consistency that is usually associated with experimental data on spark ignition.

It is of interest to note that Equation 5.29 is virtually identical to Equation 5.27 except for a higher pressure dependence; $P_3^{1.5}$ vs. $P_3^{1.3}$.

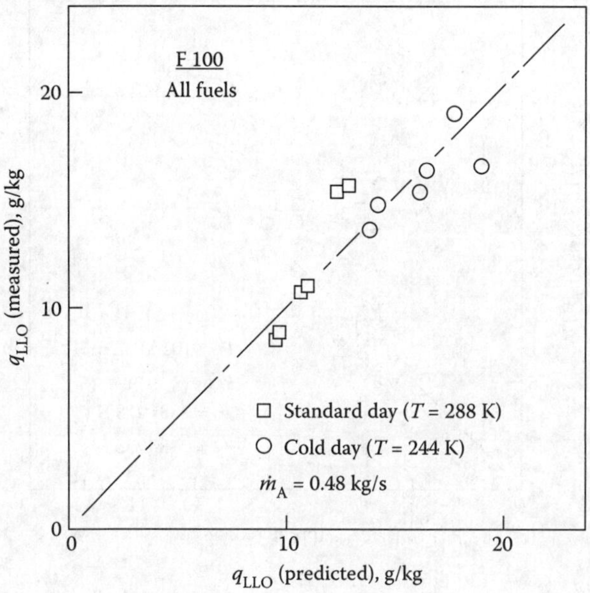

FIGURE 5.41
Comparison of measured and predicted values of q_{LLO} for an F 100 combustor. (From Lefebvre, A.H., *Journal of Engineering for Gas Turbines and Power*, 107, 24–37, 1985. With permission.)

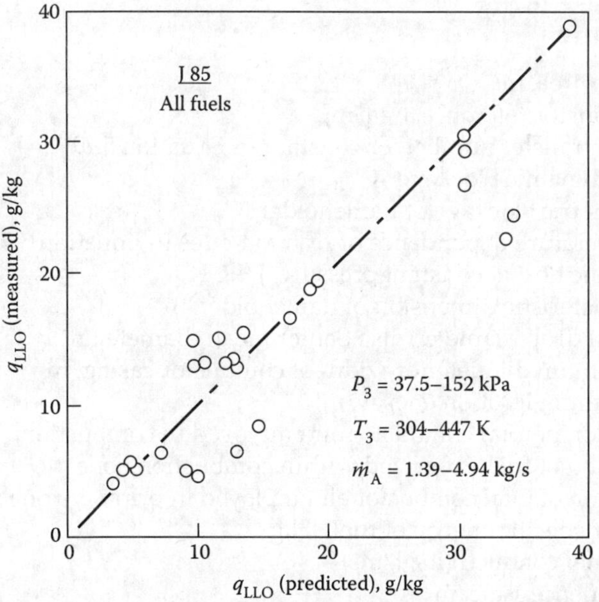

FIGURE 5.42
Comparison of measured and predicted values of q_{LLO} for a J 85 combustor. (From Lefebvre, A.H., *Journal of Engineering for Gas Turbines and Power*, 107, 24–37, 1985. With permission.)

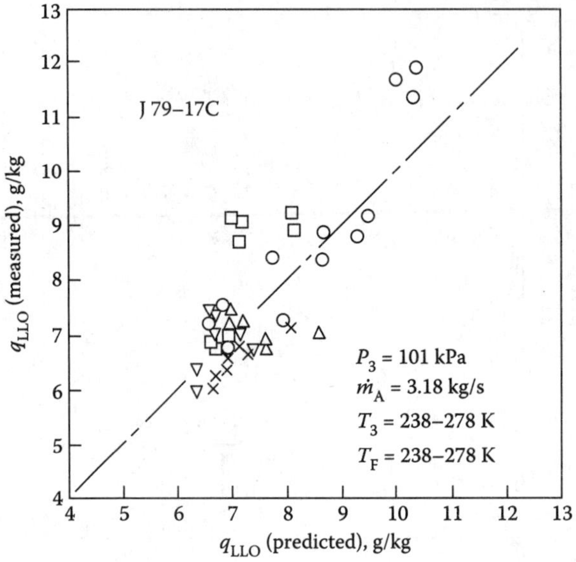

FIGURE 5.43
Comparison of measured and predicted values of q_{LLO} for a J 79 combustor. (From Lefebvre, A.H., *Journal of Engineering for Gas Turbines and Power*, 107, 24–37, 1985. With permission.)

Nomenclature

A_f	flame area, m^2
A_{ref}	combustor reference area, m^2
B	mass transfer number, or constant in Equation 5.29
B_a	aerodynamic blockage
B_g	geometric blockage of flameholder
b	temperature dependence of reaction rates in Equation 5.5
C_p	specific heat at constant pressure, J/kg K
D_c	characteristic dimension of flameholder, m
D_o	initial drop diameter, also Sauter mean diameter, m
D_{ref}	maximum diameter or width of combustor casing, m
E_{min}	minimum ignition energy, mJ
f_c	fraction of total combustor air employed in combustion
f_f	fraction of fuel vaporized within combustion zone
f_{pz}	fraction of total combustor air employed in primary-zone combustion
H	lower specific energy of fuel, J/kg
k	thermal conductivity, J/msK
l	turbulence scale, m
$\dot{m}$	mass flow rate, kg/s
n	number of drops, or reaction order

P	pressure, Pa (kPa in Equations 5.19, 5.20, 5.27, and 5.29)
q	fuel/air ratio by mass
q_{ref}	reference dynamic pressure, Pa
R	gas constant (286.9 m^2/s^2 K)
Re	drop Reynolds number ($u' D_o/\upsilon_g$)
SMD	Sauter mean diameter, m or μm
S_T	turbulent flame speed, m/s
T	temperature, K
U	velocity, m/s
U_j	jet velocity, m/s
U_{ref}	combustor reference velocity, m/s
u'	RMS component of fluctuating velocity, m/s
V	volume, m^3
w	gutter width, m
ΔP_L	liner pressure differential, Pa
ΔT	temperature rise due to combustion, K
ϕ	equivalence ratio
λ	evaporation constant, m^2/s (mm^2/s in Equation 5.20)
ν	kinematic viscosity, m^2/s
μ	dynamic viscosity, kg/ms
η_c	combustion efficiency
η_θ	combustion efficiency (reaction-controlled)
η_e	combustion efficiency (evaporation-controlled)
η_m	combustion efficiency (mixing-controlled)
ρ	density

Subscripts

A	air
F	fuel
c	combustion zone value
g	gas
ad	adiabatic
eff	effective value
LBO	lean blowout value
LLO	lean lightup value
o	initial value
oυ	overall value
p	preheater value
pz	primary-zone value
ref	reference value
st	stoichiometric value
WE	weak extinction value
3	combustor inlet value
∞	ambient value

References

1. Lefebvre, A. H., "Theoretical Aspects of Gas Turbine Combustion Performance," CoA Note Aero No. 163, Cranfield University, UK, 1966.
2. Greenhough, V. W., and Lefebvre, A. H., "Some Applications of Combustion Theory to Gas Turbine Development," *Sixth Symposium (International) on Combustion*, pp. 858–69, Reinhold, New York, 1957.
3. Lefebvre, A. H., and Halls, G. A., "Some Experiences in Combustion Scaling," AGARD Advanced Aero Engine Testing," *AGARD-ograph 37*, pp. 177–204, Pergamon, New York, 1959.
4. Odgers, J., Kretschmer, D., and Pearce, G. R., "The Combustion of Droplets Within Gas Turbine Combustors. Some Recent Observations on Combustion Efficiency," ASME Paper 92-GT-135, American Society of Mechanical Engineers, New York, NY, 1992.
5. Childs, J. H., "Preliminary Correlation of Efficiency of Aircraft Gas-Turbine Combustors for Different Operating Conditions," NACA RM E50F15, NACA, Washington DC, September 1950.
6. Avery, W. H., and Hart, R. W., "Combustor Performance with Instantaneous Mixing," *Journal of Industrial and Engineering Chemistry*, Vol. 45, No. 8, pp. 1634–37, 1953.
7. Bragg, S. L., "Application of Reaction Rate Theory to Combustion Chamber Analysis," ARC 16170, Aeronautical Research Council, UK, July 1953.
8. Longwell, J. P., and Weiss, M. A., "High Temperature Reaction Rates in Hydrocarbon Combustion," *Journal of Industrial and Engineering Chemistry*, Vol. 47, pp. 1634–43, 1955.
9. Chin, J. S., and Lefebvre, A. H., "Effective Values of Evaporation Constant for Hydrocarbon Fuel Drops," *Proceedings of 20th Automotive Technology Development Contractors Meeting*, P-120, pp. 325–31, Society of Automotive Engineers, Warrendale, PA, 1982.
10. Moses, C. A., "Studies of Fuel Volatility Effects on Turbine Combustor Performance," Joint Spring Meeting of Western and Central States Sections of the Combustion Institute, San Antonio, TX, 1975.
11. Lefebvre, A. H., "Fuel Effects on Gas Turbine Combustion—Ignition, Stability, and Combustion Efficiency," *Journal of Engineering for Gas Turbines and Power*, Vol. 107, pp. 24–37, 1985.
12. Gleason, C. C., Oiler, T. L., Shayeson, M. W., and Bahr, D. W., "Evaluation of Fuel Character Effects on J79 Engine Combustion System," AFAPL-TR-79-2015, Air Force Aero Propulsion Laboratories, Wright-Patterson Air Force Base, Dayton, OH, 1979.
13. Gleason, C. C., Oiler, T. L., Shayeson, M. W., and Bahr, D. W., "Evaluation of Fuel Character Effects on F101 Engine Combustion System," AFAPL-TR-79-2018, Air Force Aero Propulsion Laboratories, Wright-Patterson Air Force Base, Dayton, OH, 1979.
14. Vogel, R. E., Troth, D. L., and Verdouw, A. J., "Fuel Character Effects on Current, High-Pressure Ratio, Can-Type Combustion Systems," AFAPL-TR-79-2072, Air Force Aero Propulsion Laboratories, Wright-Patterson Air Force Base, Dayton, OH, 1980.

15. Gleason, C. C., Oiler, T. L., Shayeson, M. W., and Kenworthy, M. J., "Evaluation of Fuel Character Effects on J79 Smokeless Combustor," AFWAL-TR-80-2092, Air Force Aero Propulsion Laboratories, Wright-Patterson Air Force Base, Dayton, OH, 1980.
16. Oiler, T. L., Gleason, C. C., Kenworthy, M. J., Cohen, J. D., and Bahr, D. W., "Fuel Mainburner/Turbine Effects," AFWAL-TR-2100, Air Force Aero Propulsion Laboratories, Wright-Patterson Air Force Base, Dayton, OH, 1982.
17. Russell, P. L., "Fuel Mainburner/Turbine Effects," AFWAL-TR-2081, Air Force Aero Propulsion Laboratories, Wright-Patterson Air Force Base, Dayton, OH, 1982.
18. Lefebvre, A. H., and Halls, G. A., "Simulation of Low Combustion Pressures by Water Injection," *Seventh Symposium (International) on Combustion*, pp. 654–58, The Combustion Institute, Pittsburgh, PA, 1958.
19. Taylor, J. S., "Large-Scale Bluff Body Flame Stabilization," MSc thesis, Purdue University, 1980.
20. Rao, K. V. L., and Lefebvre, A. H., "Flame Blowoff Studies Using Large-Scale Flameholders," *Journal of Engineering for Power*, Vol. 104, pp. 853–57, 1982.
21. Rizk, N. K., and Lefebvre, A. H., "Influence of Laminar Flame Speed on the Blowoff Velocity of Bluff-Body Stabilized Flames," *AIAA Journal*, Vol. 22, No. 10, pp. 1444–7, 1984.
22. Rizk, N. K., and Lefebvre, A. H., "The Relationship Between Flame Stability and Drag of Bluff-Body Flameholders," *Journal of Propulsion and Power*, Vol. 2, No. 4, pp. 361–65, 1986.
23. Stwalley, R. M., and Lefebvre, A. H., "Flame Stabilization Using Flameholders of Irregular Shape," *Journal of Propulsion and Power*, Vol. 4, No. 1, pp. 4–13, 1988.
24. Baxter, M. R., and Lefebvre, A. H., "Weak Extinction Limits of Large Scale Flameholders," *Journal of Engineering for Gas Turbines and Power*, Vol. 114, No. 4, pp. 776–82, 1992.
25. Norster, E. R., "Subsonic Flow Flameholder Studies Using a Low Pressure Simulation Technique," in I. E. Smith ed., *Combustion in Advanced Gas Turbine Systems, Cranfield International Symposium Series*, Vol. X, pp. 79–93, Pergamon, London, 1968.
26. Sturgess, G. J., Heneghan, S. P., Vangsness, M. D., Ballal, D. R., and Lesmerises, A. L., "Lean Blowout in a Research Combustor at Simulated Low Pressures," ASME Paper 91-GT-359, American Society of Mechanical Engineers, New York, NY, 1991.
27. Longwell, J. P., Frost, E. E., and Weiss, M. A., "Flame Stability in Bluff Body Recirculation Zones," *Journal of Industrial and Engineering Chemistry*, Vol. 45, pp. 1629–33, 1953.
28. Zukowski, E. E., and Marble, F. E., "The Role of Wake Transition in the Process of Flame Stabilization on Bluff Bodies," *AGARD Combustion Researches and Reviews*, pp. 167–80, Butterworth, London, 1955.
29. Barrère, M., and Mestre, A., "Stabilisation des Flammes par des Obstacles," *Selected Combustion Problems: Fundamentals and Aeronautical Applications*, pp. 426–46, Butterworth, London, 1954.
30. De Zubay, E. A., "Characteristics of Disk-Controlled Flames," *Aero Digest*, Vol. 61, No. 1, pp. 54–56, 102–4, 1950.

31. Spalding, D. B., "Theoretical Aspects of Flame Stabilization," *Aircraft Engineering*, Vol. 25, pp. 264–76, 1953.
32. Ballal, D. R., and Lefebvre, A. H., "Weak Extinction Limits of Turbulent Flowing Mixtures," *Journal of Engineering for Power*, Vol. 101, No. 3, pp. 343–48, 1979.
33. Ballal, D. R., and Lefebvre, A. H., "Weak Extinction Limits of Turbulent Heterogeneous Fuel/Air Mixtures," *Journal of Engineering for Power*, Vol. 102, No. 2, pp. 416–21, 1980.
34. Ballal, D. R., and Lefebvre, A. H., "Some Fundamental Aspects of Flame Stabilization," *Fifth International Symposium on Air Breathing Engines*, pp. 48.1–48.8, AIAA, Washington, DC, 1981.
35. Plee, S. L., and Mellor, A. M., "Characteristic Time Correlation for Lean Blowoff of Bluff-Body-Stabilized Flames," *Combustion and Flame*, Vol. 35, pp. 61–80, 1979.
36. Longwell, J. P., "Flame Stabilization of Bluff Bodies and Turbulent Flames in Ducts," *Fourth Symposium (International) on Combustion*, pp. 90–97, Williams and Wilkins, Baltimore, MD, 1953.
37. Herbert, M. V., "Aerodynamic Influences on Flame Stability," in J. Ducarme, M. Gerstein, and A. H. Lefebvre, eds., *Progress in Combustion Science and Technology*, Vol. 1, pp. 61–109, Pergamon, New York, 1960.
38. Lefebvre, A. H., "A Method of Predicting the Aerodynamic Blockage of Bluff Bodies in a Ducted Airstream," CoA Report Aero. 188, College of Aeronautics, Cranfield University, UK, 1965.
39. Williams, G. C., Hottel, H. C., and Scurlock, A. C., "Flame Stabilization and Propagation in High-Velocity Gas Streams," *Third Symposium (International) on Combustion*, pp. 21–40, Williams and Wilkins, Baltimore, MD, 1949.
40. Ganji, A. T., and Sawyer, R. F., "Turbulence, Combustion, Pollutant, and Stability Characterization of a Premixed Step Combustor," NASA Contractor Report 3230, 1980.
41. Ibrahim, A. R. A. F., Benson, N. C., and Lefebvre, A. H., "Factors Affecting Fresh Mixture Entrainment in Bluff-Body Stabilized Flames," *Combustion and Flame*, Vol. 10, pp. 231–39, 1966.
42. Wharton, E., and Carr, E., "Starting Gas Turbine Engines," *Journal of the Institute of Mechanical Engineers*, Vol. 63, pp. 10–19, 1972.
43. Watson, E. A., Ignition Research Work Carried Out by the Lucas Organization with Special Reference to High Altitude Problems, Report L5988, Lucas Aerospace, Ltd., Hempstead, UK, 1954.
44. Odgers, J., and Coban, A., "The Energy Release to Static Gas from a 12 Joule High Energy Ignition System," ASME Paper 77-GT-18, American Society of Mechanical Engineers, New York, NY, 1977.
45. West, H. E., "Development of High Energy Igniters for Gas Turbines," SAE Preprint 660346, 1966.
46. Opdyke, G., "Development of an Annular Reverse-Flow Combustor," SAE Preprint 444E, 1962.
47. Anon., KLG Ignition Equipment, Igniters and Glow Plugs, Smiths Aviation Division Maintenance Manual 74-20-102/01, 1961.
48. Saintsbury, J. A., "A Glow Plug Ignition System for the Gas Turbine," SAE Paper 670937, Society of Automotive Engineers, Warrendale, PA, 1967.
49. Weinberg, F. J., Hom, K., Oppenheim, A. K., and Teichman, K., "Ignition by Plasma Jet," *Nature*, Vol. 272, pp. 341–3, 1978.

50. Orrin, J. E., Vince, I. M., and Weinberg, F. J., "A Study of Plasma Jet Ignition Mechanisms," *Eighteenth Symposium (International) on Combustion*, pp. 1755–65, Combustion Institute, Pittsburgh, PA, 1981.
51. Kingdon, R. G., and Weinberg, F. J., "The Effect of Plasma Constitution on Laser Ignition Energies," *Sixteenth Symposium (International) on Combustion*, pp. 747–56, Combustion Institute, Pittsburgh, PA, 1976.
52. Warris, A. M., and Weinberg, F. J., "Ignition and Flame Stabilization by Plasma Jets in Fast Gas Streams," *Twentieth Symposium (International) on Combustion*, pp. 1825–31, Combustion Institute, Pittsburgh, PA, 1984.
53. Hickling, R., and Smith, W. R., "Combustion Bomb Tests of Laser Ignition," SAE Paper 740144, Society of Automotive Engineers, Warrendale, PA, 1974.
54. Dale, J. D., Smy, P. R., and Clements, R. M., "Laser Ignited Internal Combustion Engine—An Experimental Study," SAE Paper 780329, Society of Automotive Engineers, Warrendale, PA, 1978.
55. Greenhalgh, D., and Gallagher, D., "Laser Ignition: Development and Application to Gas Turbine Combustors—A Literature Review," unpublished work, Cranfield University, UK, 1997.
56. Rudey, R. A., and Grobman, J. S., *Adaptation of Combustion Principles to Aircraft Propulsion, Basic Considerations in the Combustion of Hydrocarbon Fuels with Air*, Vol. 1, NACA RM E54J07, National Advisory Committee for Aeronautics, Washington DC, 1955.
57. Xiong, T. Y., Xuang, Z. X., and Wang, Y. Z., "Studies of Fuel Spray Ignition in a Gas Turbine Combustor," unpublished report, Institute of Thermophysics, Chinese Academy of Science, Beijing, 1979.
58. Pavia, R. E., and Rosenthal, J., "The Extension and Extinction Limits of Derwent V and Avon Combustors by Alternative Methods of Oxygen Addition," ARL/ME-Note-204, Aeronautical Research Laboratory, Melbourne, Australia, 1955.
59. Barlow, J., unpublished report, Lucas Aerospace, Ltd., Hempstead, UK, 1958.
60. Chin, J. S., "The Analysis of the Effect of Oxygen Addition on Minimum Ignition Energy," AIAA Paper 82-1160, 1982.
61. Swett, C. C., "Effect of Gas Stream Parameters on the Energy and Power Dissipated in a Spark and on Ignition," *Third Symposium on Combustion Flame and Explosion Phenomena*, pp. 353–61, Williams and Wilkins, Baltimore, MD, 1949.
62. Ballal, D. R., and Lefebvre, A. H., "Spark Ignition of Turbulent Flowing Gases," AIAA 77-185, Presented at AIAA Fifteenth Aerospace Sciences Meeting, Los Angeles, CA, 1977.
63. Subba Rao, H. N., and Lefebvre, A. H., "Ignition of Kerosine Fuel Sprays in a Flowing Air Stream," *Combustion Science and Technology*, Vol. 8(1), pp. 95–100, Talyor & Francis Group. http://informaworld.com, 1973.
64. Rao, K. V. L., and Lefebvre, A. H., "Minimum Ignition Energies in Flowing Kerosine-Air Mixtures," *Combustion and Flame*, Vol. 27, No. 1, pp. 1–20, 1976.
65. Ballal, D. R., and Lefebvre, A. H., "Ignition and Flame Quenching of Flowing Heterogeneous Fuel-Air Mixtures," *Combustion and Flame*, Vol. 35, No. 2, pp. 155–68, 1979.
66. Ballal, D. R., and Lefebvre, A. H., "Ignition of Liquid Fuel Sprays at Subatmospheric Pressures," *Combustion and Flame*, Vol. 31, No. 2, pp. 115–26, 1978.

67. Foster, H. H., "Effect of Spark Repetition Rate on the Ignition Limits of a Single Tubular Combustor," NACA RM E51J18, 1951.
68. Armstrong, J. C., and Wilsted, H. D., "Investigation of Several Techniques for Improving Altitude-Starting Limits of Turbojet Engines," NACA RM E52103, National Advisory Committee for Aeronautics, Washington DC, 1952.
69. Ballal, D. R., and Lefebvre, A. H., "Ignition and Flame Quenching in Flowing Gaseous Mixtures," *Proceedings of the Royal Society London Series* A, Vol. 357, No. 1689, pp. 163–81, 1977.
70. Foster, H. H., and Straight, D. M., "Effect of Fuel Volatility Characteristics on Ignition Energy Requirements in a Turbojet Combustor," NACA RM E52J21, 1953.
71. Foster, H. H., "Ignition Energy Requirements in a Single Tubular Combustor," NACA RM E51A24, National Advisory Committee for Aeronautics, Washington DC, 1951.
72. Lefebvre, A. H., and Halls, G. A., "Mechanism of Ignition in Gas Turbire Combustion Chambers," unpublished Rolls-Royce Report No. 32, 1957.
73. Odgers, J., White, I., and Kretschmer, D., "The Experimental Behavior of Premixed Flames in Tubes—The Effect of Diluent Gases," ASME Paper 79-GT-168, American Society of Mechanical Engineers, New York, NY, 1979.

6

Fuel Injection

6.1 Basic Processes in Atomization

6.1.1 Introduction

The atomization process is essentially one in which bulk fuel is converted into small drops. It represents a disruption of the consolidating influence of surface tension by the action of internal and external forces. In the absence of such disruptive forces, surface tension tends to pull the liquid into the form of a sphere, which has the minimum surface energy. Liquid viscosity has an adverse effect on atomization because it opposes any change in system geometry. On the other hand, aerodynamic forces acting on the liquid surface promote the disruption process by applying an external distorting force to the bulk liquid. Breakup occurs when the magnitude of the disruptive force just exceeds the consolidating surface tension force.

The atomization process is generally regarded as comprising two separate processes—*primary* atomization, in which the fuel stream is broken up into shreds and ligaments, and *secondary* atomization, in which the large drops and globules produced in primary atomization are further disintegrated into smaller droplets. These processes together determine the detailed characteristics of the fuel spray in regard to droplet velocities and drop-size distributions. In practice, they are markedly affected by the internal geometry of the atomizer, the properties of the gaseous medium into which the fuel stream is discharged, and the physical properties of the fuel itself.

In the following, we shall discuss the various mechanisms whereby a jet or sheet of fuel issuing from an atomizer is broken down into drops. A distinction is made between the two basic mechanisms of atomization—*classical* and *prompt*. Almost all the theoretical and experimental studies carried out in the past have been devoted to the various classical mechanisms of jet and sheet breakup, but it is now becoming increasingly recognized that for many practical atomizers, the prompt mechanism is the primary mode of atomization over much of their normal operating range.

The process of jet disintegration is of great importance for the design of plain-orifice pressure nozzles and plain-jet airblast atomizers, whereas the mechanism of sheet breakup has direct relevance to the performance of pressure-swirl and prefilming airblast atomizers. In view of their importance to both jet and sheet disintegration, the various mechanisms of drop breakup are considered first.

6.1.2 Breakup of Drops

The atomization process usually involves several interacting mechanisms, including the splitting up of the larger drops during the final stages of disintegration. A parameter of prime importance is the Weber number, We, which is the ratio of the disruptive aerodynamic force, represented by $0.5\rho_A U_R^2$, to the consolidating surface tension force, σ/D. The higher the Weber number, the larger the deforming external pressure forces, compared with the restoring surface tension forces.

The critical condition for drop breakup is achieved when the aerodynamic drag is just equal to the surface tension force, i.e.,

$$C_D(\pi/4)D^2 0.5\rho_A U_R^2 = \pi D\sigma, \tag{6.1}$$

where C_D is the drag coefficient of the drop.

Rearranging these terms provides the dimensionless group

$$(\rho_A U_R^2 D/\sigma)_{crit} = 8/C_D, \tag{6.2}$$

where subscript "crit" denotes that a critical condition has been reached. As the first term in Equation 6.2 is the Weber number, the equation may be written as

$$We_{crit} = 8/C_D. \tag{6.3}$$

For low-viscosity fuels, experimental data on C_D indicate an average value for $8/C_D$ of around 12, i.e.,

$$We_{crit} = 12. \tag{6.4}$$

For a relative velocity, U_R, the maximum stable drop size is obtained from Equation 6.2 as

$$D_{max} = 12\sigma/\rho_A U_R^2. \tag{6.5}$$

This equation has some practical significance because in many calculations on spray evaporation times it is only necessary to know the diameter of the largest drop in the spray.

To account for the influence of liquid viscosity on drop breakup, Hinze [1] used a dimensionless group known as Ohnesorge number, Oh, which is defined as

$$\mathrm{Oh} = (\mathrm{We})^{0.5}/\mathrm{Re}, \tag{6.6}$$

where $\mathrm{We} = (\rho_L U_R^2 D/\sigma)$ and $\mathrm{Re} = (\rho_L U_R D/\mu_L)$.

Substituting for We and Re in Equation 6.6 gives

$$\mathrm{Oh} = \mu_L/(\rho_L \sigma D)^{0.5}. \tag{6.7}$$

According to Brodkey [2], the effect of viscosity on the critical Weber number can be expressed as

$$\mathrm{We}_{crit} = \mathrm{We}_{crit} + 14\mathrm{Oh}^{1.6}, \tag{6.8}$$

where We_{crit} is the critical Weber number for zero viscosity.

6.1.2.1 Drop Breakup in Turbulent Flow Fields

The foregoing discussion is based on the assumption of a high relative velocity between the fuel drop and the surrounding air. However, in many practical situations, a high relative velocity may not exist due to the drops becoming airborne shortly after leaving the nozzle. It is then more logical to assume that the dynamic pressure forces of the turbulent motion of the airstream determine the size of the largest drop. The critical Weber number thus becomes

$$\mathrm{We}_{crit} = \rho_A \bar{u}^2 D_{max}/\sigma, \tag{6.9}$$

where $\bar{u}$ is the RMS value of the velocity fluctuations.

According to Sevik and Park [3], $\mathrm{We}_{crit} = 1.04$. Hence

$$D_{max} = 1.04\sigma/\rho_A \bar{u}^2. \tag{6.10}$$

6.2 Classical Mechanism of Jet and Sheet Breakup

This mechanism relies on the creation of small disturbances, either within or on the surface of a fuel jet or sheet, which promote the formation of waves that eventually lead to disintegration into ligaments and then drops. To understand the role of viscosity in classical atomization, it is important to recognize that the primary and secondary atomization processes both require a finite amount of time for the instabilities, which are an essential prerequisite for breakup to develop. If, for any reason, this time is lengthened, the various breakup processes will take place further downstream from the nozzle under conditions that are less conducive to atomization. In particular, the relative velocity between air and fuel will be lower, resulting in larger drops. An increase in fuel viscosity is one parameter that extends the breakup time. Thus, when atomization proceeds by the classical mechanism, an increase in fuel viscosity is always accompanied by an increase in mean drop size.

6.2.1 Breakup of Fuel Jets

Rayleigh [4] was among the first to theoretically study the breakup of liquid jets. He considered the simple situation of a laminar jet issuing from a circular orifice and postulated a mechanism for breakup, which is illustrated in Figure 6.1. According to Rayleigh, the growth of small disturbances leads to breakup when the fastest growing disturbance attains a wavelength, λ_{opt}, of 4.51d, where d is the initial jet diameter. After breakup, the cylinder of length 4.51d becomes a spherical drop, so that

$$4.51d \times (\pi/4)d^2 = (\pi/6)D^3, \tag{6.11}$$

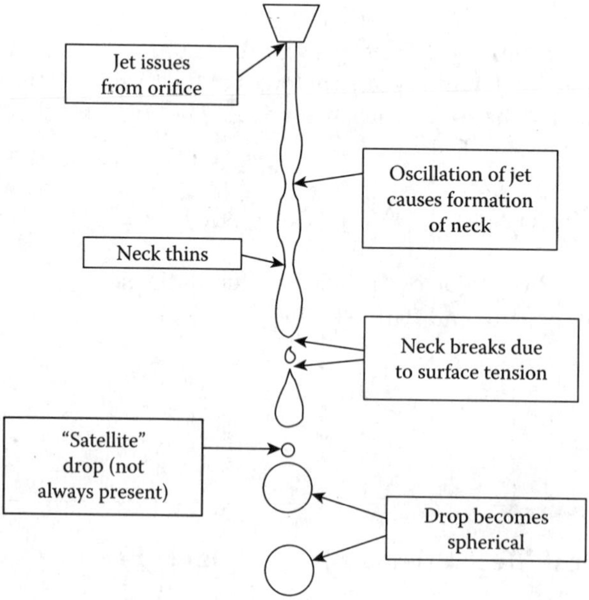

FIGURE 6.1
Breakup of a plain circular jet.

hence *D*, the drop diameter, is obtained as

$$D = 1.89d. \tag{6.12}$$

Weber [5] later extended Rayleigh's work to include the effect of viscosity on the disintegration of jets into drops. We have

$$\lambda_{opt} = 4.44d(1+3\mathrm{Oh})^{0.5}. \tag{6.13}$$

Note that the value for the constant in this expression of 4.44 corresponds closely to Rayleigh's value of 4.51. For low values of Oh, it leads to a drop diameter, *D*, of 1.88*d* as opposed to Rayleigh's 1.89*d*. The effect of an increase in viscosity is to raise Oh, thereby increasing the optimum wavelength for jet breakup.

Weber also examined the effect of increasing jet velocity on drop size. He found that raising the jet velocity from 0 to 15 m/s reduced λ_{opt} from 4.44*d* to 2.8*d*, which produced a reduction in *D* from 1.88*d* to 1.61*d*. Thus, the effect of increasing the relative velocity between the fuel jet and the surrounding air is to reduce the optimum wavelength for jet breakup, which results in a smaller drop size.

At even higher velocities, the atomization process is enhanced by the effect of relative motion between the surface of the jet and the surrounding air [6]. This aerodynamic interaction causes irregularities in the previously smooth surface, which become amplified and eventually detach themselves from the jet surface, as illustrated in the detailed photograph of Figure 6.2, which was obtained for a water jet by Taylor and Hoyt [7]. The ligaments formed in this manner subsequently disintegrate into drops, and as the jet velocity increases, the diameter of the ligaments decreases. When they collapse, smaller droplets are formed, in accordance with Rayleigh's theory.

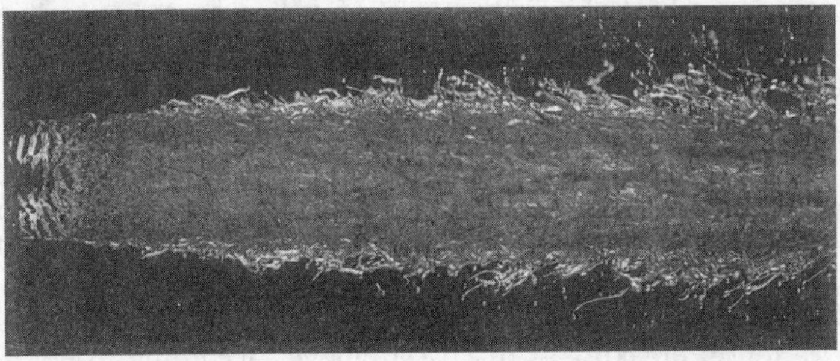

FIGURE 6.2
High-speed photograph of water jet showing surface wave instabilities and drop formation. (From Taylor, J.J. and Hoyt, J.W., *Journal of Experimental Fluids*, 1, 113–20, 1983. With permission from Springer Science+Business Media.)

Thus, the various modes of atomization may be classified into four main groups according to the magnitude of the relative velocity between the jet and the surrounding air:

1. At low velocities, the growth of axisymmetric oscillations on the jet surface causes the jet to disintegrate into drops of fairly uniform size. This is the Rayleigh mechanism of breakup. Drop diameters are roughly twice the initial jet diameter.
2. With an increase in jet velocity, the basic mechanism of breakup remains the same, but the interaction between the jet and the surrounding air reduces the optimum wavelength for jet breakup, which results in a smaller drop size. Drop diameters are about the same as the jet diameter.
3. With a further increase in jet velocity, droplets are produced by the unstable growth of small waves on the jet surface caused by interaction between the jet and the surrounding air. These waves become detached from the jet surface to form ligaments that disintegrate into drops. Drop diameters are much smaller than the initial jet diameter.
4. At very high jet velocities, atomization occurs rapidly and is complete within a short distance from the nozzle. Mean drop diameters are usually less than 80 μm.

Modes 1 to 3 follow the classical mechanisms of atomization, and drop sizes are very dependent on fuel viscosity, ambient air density, and initial jet diameter. Mode 4 corresponds to prompt atomization, and drop sizes are strongly dependent on surface tension, but are fairly insensitive to variations in viscosity, ambient air density, and initial jet diameter.

6.2.2 Breakup of Fuel Sheets

Most atomizers discharge fuel in the form of a conical sheet. In pressure-swirl and prefilming airblast atomizers, conical sheets are generated when fuel issues from an orifice with a tangential velocity component resulting from its passage through a number of tangential or helical slots. With pressure-swirl atomizers, the relative velocity required for atomization is achieved by injecting the conical sheet of fuel at high velocity into slow-moving air or gas, whereas in airblast atomizers, one or more high-velocity airstreams (usually swirling) impinge on a slow-moving, conical sheet of fuel.

The basic mechanisms of sheet integration are broadly the same as those responsible for jet breakup, as discussed above. According to Fraser et al. [8], if the relative velocity between the fuel sheet and the surrounding air is fairly low, a wave motion is generated on the sheet, which causes rings of

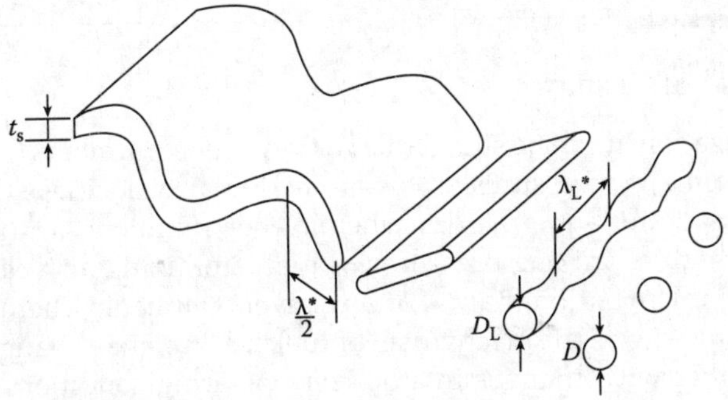

FIGURE 6.3
Successive stages in the idealized breakup of a fuel sheet. (From Fraser, R.P., Eisenklam, P., Dombrowski, N., and Hasson, D., *AIChE Journal*, 8(5), 672–80, 1962. Reproduced with permission from John Wiley & Sons, Inc.)

fuel to break away from its leading edge. The volume of fuel contained in the rings can be estimated as the volume of a ribbon cut out of the sheet with a thickness equal to that of the sheet at the breakup distance and a width equal to one-half wavelength of the oscillation ($\lambda_{opt}/2$). These cylindrical ligaments then disintegrate into drops of uniform size according to the Rayleigh mechanism, as illustrated in Figure 6.3.

With a continuous increase in relative velocity, sheet breakup occurs closer to the nozzle. Also, the optimum wavelength for breakup becomes smaller and ligament and drop diameters are reduced accordingly. Finally, at very high relative velocities, atomization starts at the nozzle exit and the mode of sheet disintegration changes from classical to prompt.

6.3 Prompt Atomization

In situations where breakup takes place very rapidly, for example, when high-velocity air jets impinge on a fuel jet or sheet at an appreciable angle, or when fuel is discharged at very high velocity into stagnant or slow-moving air, the atomization process may be described as "prompt" [9]. Under these conditions, the jet or sheet has no time to develop a wavy structure, but is immediately torn into fragments by its vigorous interaction with the surrounding air. An essential feature of this mode of atomization is that the rapid and violent disruption of the fuel ensures that the ensuing drop sizes are largely independent of the initial fuel dimension (jet diameter or sheet thickness). Furthermore, drop sizes must also be independent of viscosity, which cannot slow down a process that, by definition, occurs instantaneously.

6.4 Classical or Prompt?

As a generalization, it can be stated that for low Weber numbers (corresponding to low atomizing pressures or low atomizing air velocities), the classical mechanism is dominant, whereas when Weber numbers are high (corresponding to high atomizing pressures or high atomizing air velocities), the prompt mechanism is dominant. For any given atomizer, whether it is airblast or pressure-swirl, the mechanism of fuel breakup may change from one mode to another with changes in atomizer operating conditions. Consider, for example, airblast atomizers of the plain-jet or prefilming types. At low atomizing air velocities, these devices function mainly as pressure atomizers. With an increase in air velocity, they start to perform as airblast atomizers, with atomization taking place via the classical mechanism sequence of wave formation: wave breakup → ligament formation → ligament breakup → drops. As air velocities rise even further, the prompt mechanism starts to intervene in the atomization process. Finally, a velocity is reached above the point at which the prompt mechanism is dominant.

This change in the mechanism of breakup with change in atomizer operating conditions would be of academic interest only were it not for the marked differences exhibited by these two modes of atomization in regard to the influence on mean drop size of fuel viscosity, initial jet or sheet dimensions, and atomizing air pressure [9,10]. Increasing the angle at which the atomizing air impinges on the fuel stream promotes the changeover from classical to prompt atomization. As mentioned above, a key parameter in determining which of these two different atomization modes will dominate at any given operating condition is the Weber number. For both pressure and airblast atomizers, prompt atomization is promoted by an increase in the Weber number.

6.5 Drop-Size Distributions

Due to the random and chaotic nature of the atomization process, the threads and ligaments formed by the various mechanisms of jet and sheet disintegration vary widely in diameter, and their subsequent breakup yields a correspondingly wide range of drop sizes. Most practical atomizers produce droplets in the size range from a few microns up to several hundred microns. Thus, in addition to mean drop size, another parameter of importance in the definition of a spray is the distribution of drop sizes it contains.

6.5.1 Graphical Representation of Drop-Size Distributions

A simple method of illustrating the distribution of drop sizes in a spray is to plot a histogram where each ordinate represents the number of droplets whose dimensions fall between the limits $D - \Delta D/2$ and $D + \Delta D/2$. A typical histogram of this type is shown in Figure 6.4, in which $\Delta D = 17$ μm. As ΔD is made smaller, the histogram assumes the form of a frequency distribution curve, as shown in Figure 6.5, provided that it is based on a sufficiently large sample. Figure 6.6 illustrates the use of this type of curve to

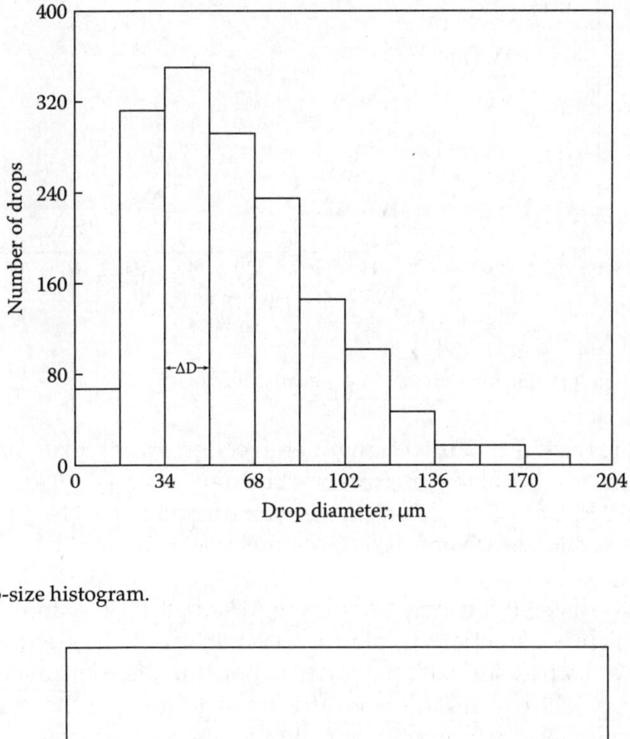

FIGURE 6.4
Typical drop-size histogram.

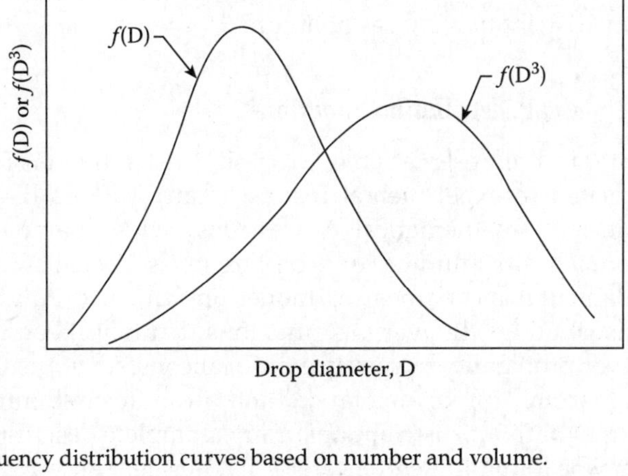

FIGURE 6.5
Drop-size frequency distribution curves based on number and volume.

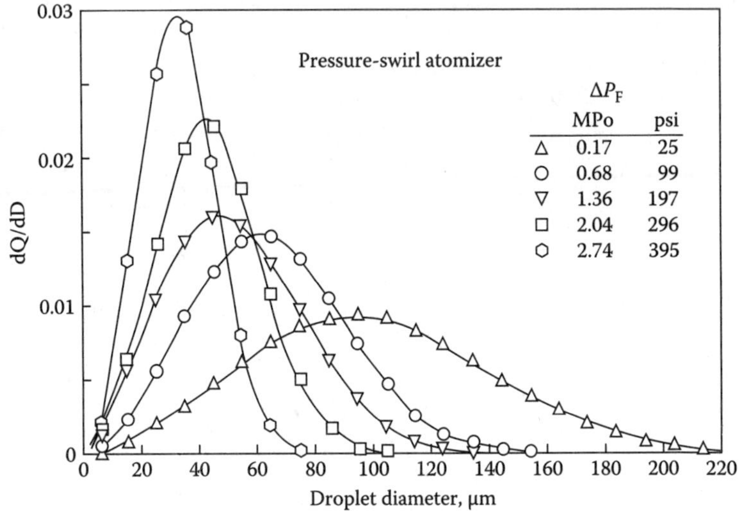

FIGURE 6.6
Influence of fuel-injection pressure on drop-size distributions.

demonstrate the effect of increasing fuel-injection pressure ΔP_F on drop-size distributions for a simplex pressure-swirl atomizer. It is well known that an increase in ΔP_F leads to a reduction in mean drop size, and Figure 6.6 shows that this reduction is accomplished mainly by eliminating the largest drops in the spray.

Drop-size distribution may also be represented by a cumulative distribution, which is essentially a plot of the integral of the frequency curve. Cumulative distribution curves plotted on arithmetic coordinates have the general shape shown in Figure 6.7; the ordinate may be the percentage of drops by number, surface area, or volume whose diameter is less than a given drop size. Figure 6.8 shows cumulative distributions corresponding to the frequency distribution curves of Figure 6.6.

6.5.2 Mathematical Distribution Functions

Since the graphical representation of drop-size distribution is laborious and not easily related to experimental results, many workers have attempted to replace it with mathematical expressions whose parameters can be obtained from a limited number of drop-size measurements. In the absence of any fundamental mechanism or model on which to build a theory of drop-size distributions, the various functions that have been proposed are based on either probability or purely empirical considerations. They include normal, log-normal, Nukiyama and Tanasawa, Rosin–Rammler, upper-limit, and log-hyperbolic distributions. Fairly complete descriptions of these functions may be found in References [11–15]. It is generally accepted that no

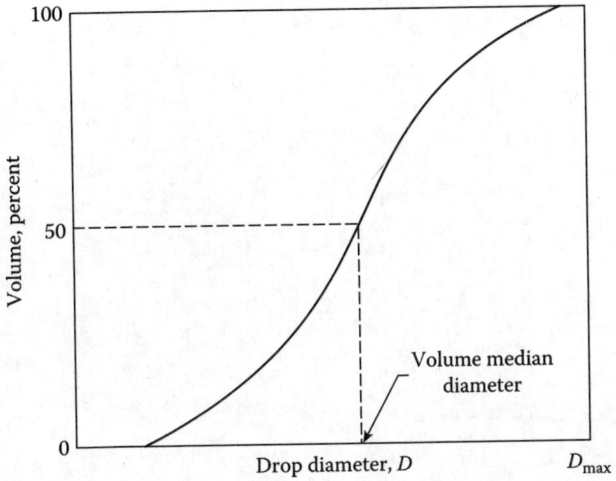

FIGURE 6.7
Typical shape of cumulative drop-size distribution curve.

single parameter can represent all drop-size data. In practice, it may be necessary to test several distribution functions to find the best fit to any given set of experimental data.

6.5.3 Rosin–Rammler

The most widely used expression for drop-size distribution is one that was originally developed for powders by Rosin and Rammler [16]. It may be expressed in the form

$$1-Q=\exp-(D/X)^q, \tag{6.14}$$

where Q is the fraction of the total volume contained in drops of diameter less than D, and X and q are constants that are determined experimentally. Thus, the Rosin–Rammler relationship describes the drop-size distribution in terms of the two parameters, X and q. The exponent q provides a measure of the spread of drop sizes. The higher the value of q, the more uniform the spray. If q is infinite, all the drops in the spray are the same size. For most practical sprays, the value of q lies between 1.8 and 3.0.

Although it assumes an infinite range of drop sizes, the Rosin–Rammler expression has the virtue of simplicity. Moreover, it permits data to be extrapolated into the range of very fine droplets, where measurements are most difficult and least accurate.

A typical Rosin–Rammler plot is shown in Figure 6.9. The value of q is obtained as the slope of the line, whereas X is given by the value of D for which $1-Q=\exp-1$. Solution of this equation yields the result that $Q=0.632$;

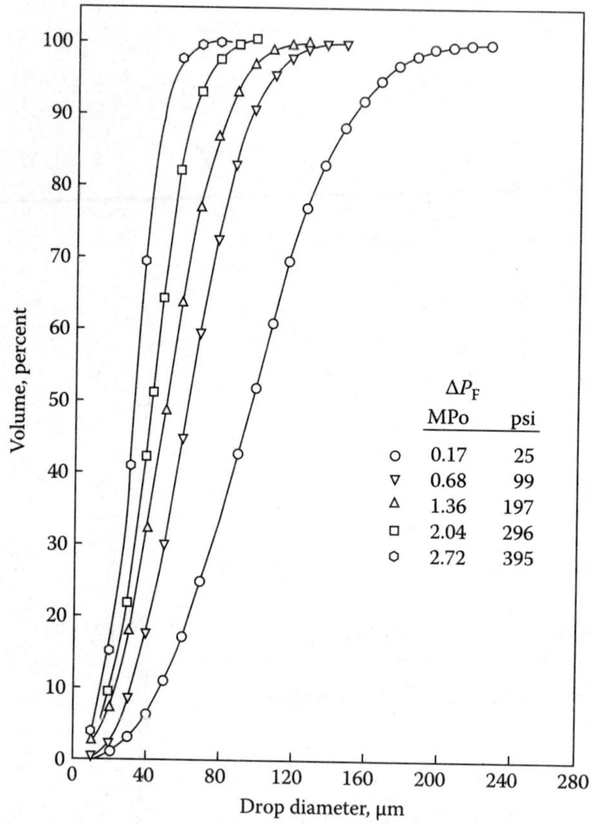

FIGURE 6.8
Drop-size distributions from a pressure-swirl atomizer in cumulative form.

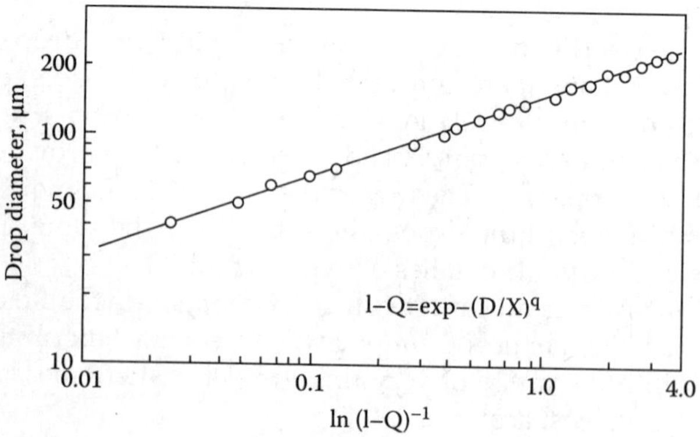

FIGURE 6.9
Typical Rosin–Rammler plot.

that is, X is the drop diameter such that 63.2% of the total liquid volume is in drops of smaller diameter.

6.5.4 Modified Rosin–Rammler

From analysis of a considerable body of drop-size data obtained with pressure-swirl nozzles, Rizk and Lefebvre [17] found that although the Rosin–Rammler expression provides an adequate data fit over most of the drop-size range, there is occasionally a significant deviation from the experimental data for the larger drop sizes. By rewriting the Rosin–Rammler equation in the form

$$1 - Q = \exp - \left(\ln D / \ln X\right)^q, \tag{6.15}$$

a much better fit to the drop-size data is usually obtained, as illustrated in Figure 6.10 from Rizk [18].

The Rosin–Rammler parameter is now widely used in both normal and modified forms. In a recent study by Han et al. [19], it was found that changing to this parameter substantially improved the prediction of sprays from a pressure-swirl atomizer, whereas Rizk and his colleagues have used the modified form successfully in a number of investigations [20–22]. However, as noted above, it would be unwise to assume that the Rosin–Rammler parameter gives the best representation in all cases.

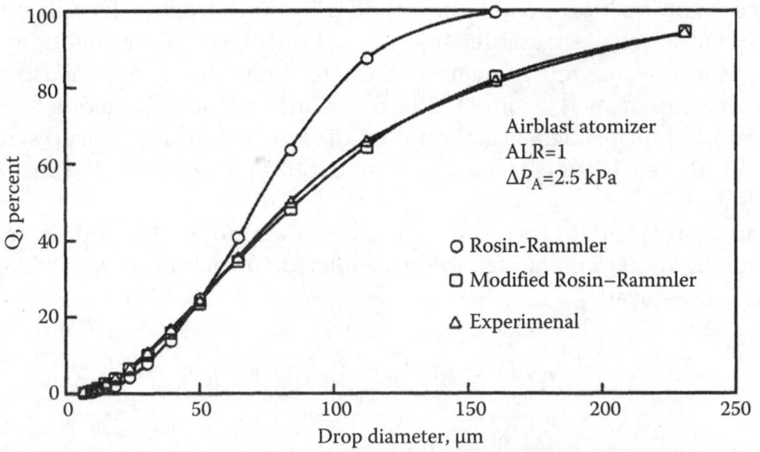

FIGURE 6.10
Comparison of Rosin–Rammler and modified Rosin–Rammler distributions. (From Rizk, N.K., Allison Gas Turbine Engines Report Nos. AR 0300-90 and AR 0300-91, 1984. With permission.)

6.5.5 Mean Diameters

In many calculations of mass transfer and spray evaporation it is convenient to work only with mean or average diameters instead of the complete drop-size distribution. The concept of mean diameter has been generalized and its notation standardized by Mugele and Evans [11]. The most important mean diameter for combustion applications is the Sauter mean diameter, which is usually abbreviated to SMD or D_{32}. This is the diameter of a drop within the spray whose ratio of volume to surface area is the same as that of the whole spray.

6.5.6 Representative Diameters

There are many possible choices of representative diameter, each of which could play a role in defining the drop-size distribution. The various possibilities include the following:

$D_{0.1}$ = drop diameter such that 10% of the total liquid volume is in drops of smaller diameter.

$D_{0.5}$ = drop diameter such that 50% of the total liquid volume is in drops of smaller diameter. This is generally known as the volume (or mass) median diameter (VMD or MMD).

$D_{0.632}$ = drop diameter such that 63.2% of the total liquid volume is in drops of smaller diameter. This is X in Equation 6.14.

$D_{0.9}$ = drop diameter such that 90% of the total liquid volume is in drops of smaller diameter.

For most combustion purposes, the distribution of drop sizes in a spray may be concisely represented as a function of two parameters, one of which is a mean or representative diameter, the other is a measure of the range of drop sizes. In some instances, it may be advantageous to introduce another term, such as a parameter to express minimum drop size, but basically there must be at least two parameters to describe the drop-size distribution.

If the drop-size data correspond to a Rosin–Rammler distribution, all the representative diameters are uniquely related to each other via q [23]. For example, we have

$$\text{MMD} = \text{SMD}(0.693)^{1/q}\,\tau(1-1/q), \tag{6.16}$$

where τ denotes the gamma function.

$$D_{0.1} = \text{MMD}(0.152)^{1/q}, \tag{6.17}$$

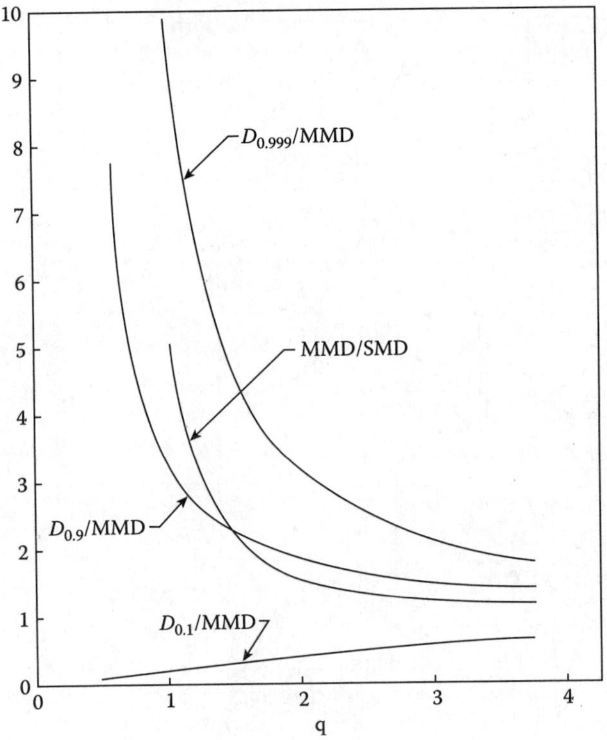

FIGURE 6.11
Relationship between Rosin–Rammler distribution parameter q and various spray properties. (From Chin, J.S., and Lefebvre, A.H., *International Journal of Turbo and Jet Engines*, 3(4), 293–300, 1986. With permission.)

$$D_{0.9} = \text{MMD}(3.32)^{1/q}, \tag{6.18}$$

$$D_{0.999} = \text{MMD}(9.968)^{1/q}. \tag{6.19}$$

These relationships are shown plotted in Figure 6.11. They serve to illustrate that when the Rosin–Rammler expression is used, the ratio of any two representative diameters is always a unique function of q.

It is important to bear in mind that no single parameter can completely define a drop-size distribution. For example, two sprays are not necessarily similar just because they have the same SMD or the same mass median diameter (MMD). This point is demonstrated in Figure 6.12, which contains two drop-size frequency distribution curves for the same SMD of 50 μm. The Rosin–Rammler distribution parameter is 2 in one case and 3 in the other. The figure shows that when q is 3, the spray contains no drops larger in diameter than 130 μm, whereas when q is 2, a significant proportion of the total spray volume is contained in drops larger than 130 μm.

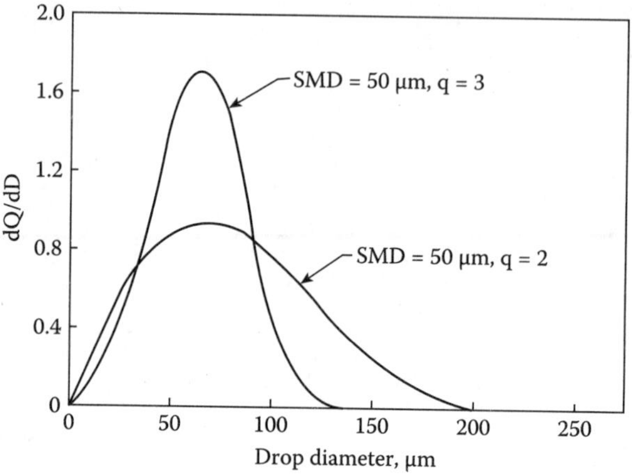

FIGURE 6.12
Influence of Rosin–Rammler parameter *q* on drop-size distributions for a constant SMD.

6.5.7 Prediction of Drop-Size Distributions

Initial work in this area by Sellens and Brustowski [24] considered the breakup of a thin sheet into drops. They obtained an expression for drop-size distributions in sprays by applying the maximum entropy formalism to the conservation laws for mass, momentum energy, and kinetic energy. It was assumed that the directed kinetic energy and the surface energy are separately conserved during the disintegration of the liquid sheet. The derived drop-size distribution has an adjustable parameter, which includes the initial liquid sheet thickness. Unfortunately, this is not an easily measurable quantity. In a later publication, Ahmadi and Sellens [15] simplified the previous approach to obtain only the drop-size distribution in a spray, rather than the joint drop-size and velocity distribution. It was found that measured drop-size distributions in a water spray agreed well with the proposed model. These workers also provided a useful review of similar theoretical studies on the prediction of drop-size distributions carried out before 1993, including those of Li and Tankin [25], Bhatia et al. [14], and Xu et al. [26]. In a more recent publication, Cousin et al. [27] combined maximum entropy formalism with the classical linear theory of sheet breakup to predict drop-size distributions in the sprays produced by a number of pressure-swirl atomizers. The results obtained showed excellent agreement between experimental and calculated drop-size distributions.

These theoretical approaches to the problem of probability distribution of droplet sizes in sprays, as outlined above, could play an increasingly important role in the analysis and representation of the drop-size distribution data that are now being generated in increasing amounts as a result of the more widespread use of phase-Doppler anemometry in spray interrogation.

6.6 Atomizer Requirements

Liquid fuel has to be delivered to the combustor, where it is thoroughly mixed with air before combustion. For liquid fuels, there are two distinct methods of doing this: vaporizers and fuel spray nozzles. Also, industrial engines have an additional requirement of multifuel (gaseous and liquid fuels) capability. These dual fuel nozzles are evolved from liquid spray nozzles, gas-only fuel injectors operating at lower pressures, and a hybrid liquid–gas fuel arrangement.

For large diameter annular combustors, a flow distribution valve is often required to compensate for the gravity head across the fuel manifold at low fuel pressures, to ensure that all fuel injectors discharge an equal quantity of fuel, especially at ignition conditions. This ensures that all sectors of the annular combustor operate in the same way, giving repeatability in the temperature distribution seen by the high pressure turbine stage. An ideal atomizer would possess all the following characteristics:

1. Ability to provide good atomization over a wide range of fuel flow rates
2. Rapid response to changes in fuel flow rate
3. Freedom from flow instabilities
4. Low power requirements
5. Capability for scaling, to provide design flexibility
6. Low cost, light weight, ease of maintenance, and ease of removal for servicing
7. Low susceptibility to damage during manufacture and installation
8. Low susceptibility to blockage by contaminants in the fuel and to carbon buildup on the nozzle face
9. Low susceptibility to gum formation by heat soakage
10. Uniform radial and circumferential fuel distribution

6.7 Pressure Atomizers

As their name suggests, pressure atomizers rely on the conversion of pressure into kinetic energy to achieve a high relative velocity between the fuel and the surrounding air or gas. Many of the atomizers in general use are of this type. They include plain-orifice and simplex nozzles, as well as the dual-orifice injector. These various types of pressure atomizers are discussed below.

6.7.1 Plain Orifice

The atomization of a low-viscosity fuel is most easily accomplished by passing it through a small circular hole, as illustrated in Figure 6.13a. If the velocity is low, the liquid emerges as a thin distorted pencil, but if the liquid pressure exceeds the ambient gas pressure by about 150 kPa, a high-velocity fuel jet is formed, which rapidly disintegrates into a well-atomized spray. Disintegration of the jet is promoted by an increase in fuel-injection pressure, which increases both the level of turbulence in the fuel jet and the aerodynamic forces exerted by the surrounding medium.

Perhaps the best known application of plain-orifice atomizers is to afterburners (reheat systems), where the fuel-injection system normally consists of one or more circular manifolds supported by struts inside the jet pipe. Fuel is supplied to the manifolds by feed pipes in the support struts and is sprayed into the flame zone from holes drilled in the manifolds. Sometimes "stub pipes" are used instead of manifolds, and many fuel injector arrays consist of stub pipes mounted radially on circular manifolds. In all cases, the objective is to provide a uniform distribution of well-atomized fuel throughout the portion of the gas stream that flows into the combustion zone.

6.7.2 Simplex

The narrow spray cone angles exhibited by plain-orifice atomizers are disadvantageous for most practical applications. Much wider cone angles are achieved in the pressure-swirl atomizer, in which a swirling motion is imparted to the fuel so that, under the action of centrifugal force, it spreads out in the form of a conical sheet as soon as it leaves the orifice.

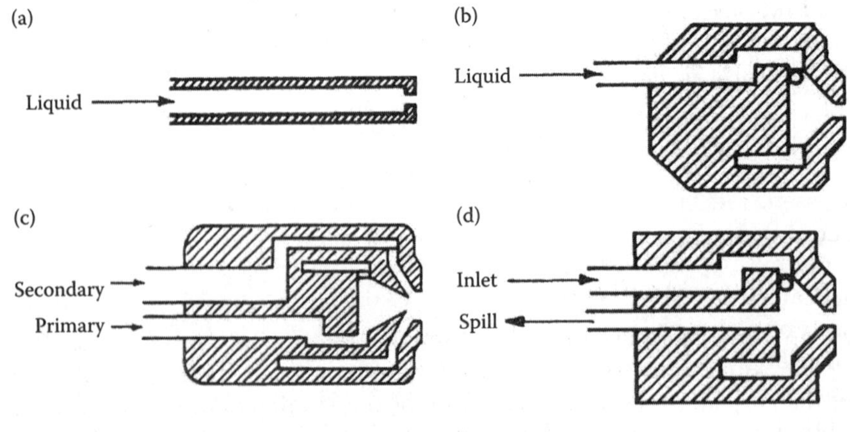

FIGURE 6.13
Schematic drawings of pressure-swirl atomizers: (a) plain orifice; (b) simplex; (c) dual orifice; (d) spill return.

The simplest form of pressure-swirl atomizer is the *simplex* atomizer, as illustrated in Figure 6.13b. Fuel is fed into a swirl chamber through tangential ports that give it a high angular velocity, thereby creating an air-cored vortex. The outlet from the swirl chamber is the final orifice, and the rotating fuel flows through this orifice under both axial and radial forces to emerge from the atomizer in the form of a hollow conical sheet.

The development of the spray passes through several stages as the fuel-injection pressure is increased from zero.

1. Fuel dribbles from the orifice.
2. Fuel leaves as a thin distorted pencil.
3. A cone forms at the orifice, but is contracted by surface tension forces into a closed bubble. This is known as the "onion" stage.
4. The bubble opens into a hollow "tulip" shape, terminating in a ragged edge, where the fuel disintegrates into fairly large drops.
5. The curved surface straightens to form a conical sheet. As the sheet expands, its thickness diminishes, and it soon becomes unstable and disintegrates into ligaments and then drops in the form of a well-defined hollow-cone spray.

A major drawback of the simplex atomizer is that its flow rate varies as the square root of the injection pressure differential, ΔP_F. Thus, doubling the flow rate demands a fourfold increase in injection pressure. For low-viscosity fuels, the lowest injection pressure at which atomization can be achieved is about 0.1 MPa (15 psi). This means that an increase in flow rate to some 20 times the minimum value would require an injection pressure of 40 MPa, which is beyond the capability of most pumps. This basic drawback of the simplex nozzle has led to the development of various "wide-range" atomizers, the most notable example being the dual-orifice atomizer, in which ratios of maximum to minimum flow rate in excess of 20 can readily be achieved with injection pressures not exceeding 7 MPa (1000 psi).

6.7.3 Dual Orifice

The essential features of a dual-orifice atomizer are shown in Figure 6.13c. In order to deal effectively with problems of mechanical integrity, differential thermal expansion, heat shielding, and carbon deposition on the nozzle face, practical atomizers tend to be more complex, as illustrated in Figure 6.14. This type of nozzle has been widely used on many types of aircraft and industrial gas turbines.

Essentially, a dual-orifice atomizer comprises two simplex nozzles that are fitted concentrically, one inside the other. The primary (or pilot) nozzle is mounted on the inside, and the juxtaposition of primary and secondary (or main) is such that the primary spray does not interfere with either the

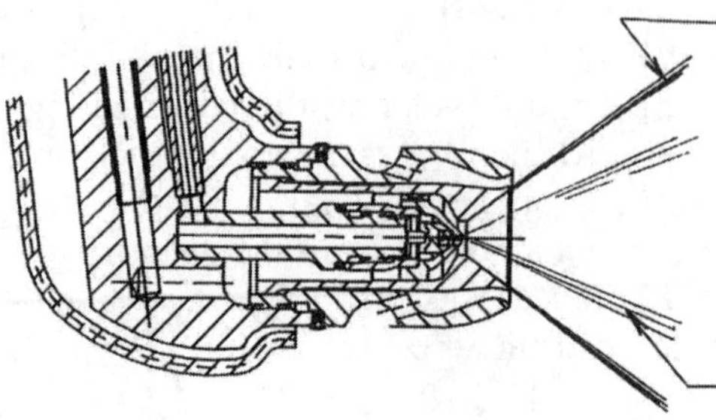

FIGURE 6.14
Dual-orifice atomizer. (Courtesy of Parker Hannifin Corporation.)

secondary orifice or the secondary spray within the orifice. When the fuel delivery is low, it all flows through the primary nozzle, and atomization quality tends to be high because a fairly high fuel pressure is needed to force the fuel through the small ports in the primary swirl chamber. As the fuel supply is increased, a fuel pressure is eventually reached at which a valve opens and admits fuel to the secondary nozzle. At this point, atomization quality is poor because the secondary fuel pressure is low. With further increases in fuel flow, the secondary fuel pressure increases, and atomization quality starts to improve. However, there is an inevitable range of fuel flows, starting from the point at which the valve opens, over which drop sizes are relatively large. To alleviate this problem, it is customary to arrange for the primary spray cone angle to be slightly wider than the secondary spray cone angle, so that the two sprays coalesce and share their energy within a short distance from the atomizer.

6.7.4 Spill Return

This is basically a simplex atomizer, except that the rear wall of the swirl chamber, instead of being solid, contains a passage through which the fuel that is surplus to combustion requirements enters the spill line and is returned to the fuel tank, as shown in Figure 6.13d. The main attraction of this system is that the fuel-injection pressure is always high, even at the lowest fuel flow rate, thus atomization quality is always excellent. Other attractive features include an absence of moving parts and, because the flow passages are designed to handle large flows all the time, freedom from "plugging" by contaminants in the fuel. The principal drawbacks of the spill-return atomizer are high fuel-pump power requirements and a wide variation in spray cone angle with change in fuel flow rate.

Another disadvantage of the spill system is that problems of metering the flow rate are more complicated than with other types of atomizer, and a larger-capacity pump is needed to handle the large recirculating flows. For these reasons, interest in the spill-return atomizer for gas turbines has gradually declined, and for the past several decades its main application has been to large industrial furnaces. However, if the aromatic content of gas turbine fuels continues to rise, it could pose serious problems of blockage by gum formation of the small passages of conventional pressure atomizers. The spill-return atomizer, having no small passages, is virtually free from this defect. Furthermore, by the judicious application of swirling air flowing around the nozzle, it is possible to maintain a fairly constant spray cone angle regardless of changes in fuel flow rate.

6.8 Rotary Atomizers

By far, the best known rotary atomizer is the "slinger" system, which was developed by the Turbomeca company in France. It is used in conjunction with a radial-annular combustion chamber, as illustrated in Figure 6.15. Fuel is supplied at low pressure along the hollow main shaft and is discharged radially outward through holes drilled in the shaft. These injection holes vary in number from 9 to 18 and in diameter from 2.0 to 3.2 mm. The holes may be drilled in the same plane as a single row, but some installations feature a double row of holes. The holes never run full; they have a capacity that is many times greater than the required flow rate. They are made large to obviate blockage. However, it is important that the holes be accurately machined and finished because experience has shown that uniformity of flow between one injection hole and another depends very much on their dimensional accuracy and surface finish. Clearly, if one injection hole supplies more fuel than the others, it will produce a rotating "hot spot" in the exhaust gases, with disastrous consequences for the particular turbine blade

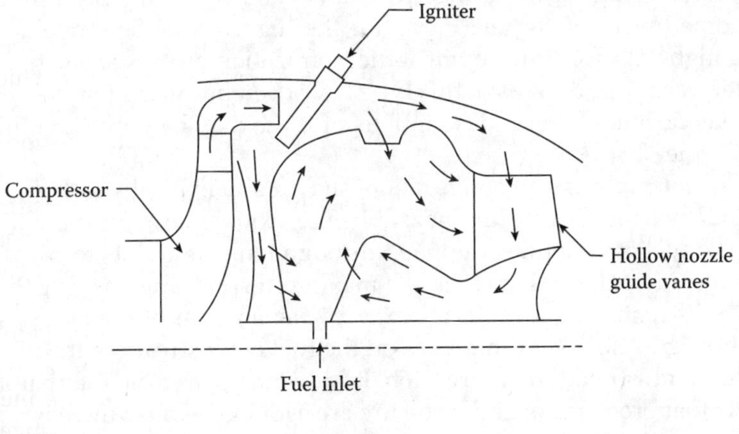

FIGURE 6.15
Turbomeca slinger system.

on which the hot spot happens to impinge. Flow uniformity is also critically dependent on the flow path provided for the fuel inside the shaft, especially in the region near the holes. Where there are two rows of holes, it is very important to achieve the correct flow division between the two rows. The internal geometry of the shaft is important in this regard.

The main advantages of the slinger system are its cheapness and simplicity. Only a low-pressure fuel pump is needed, and the quality of atomization is always satisfactory, even at speeds as low as 10% of the rated maximum. The influence of fuel viscosity is small, so the system has a potential multifuel capability.

The main problems with the system appear to be those of igniter-plug location, poor high-altitude relighting performance and, because of the long flow path, slow response to changes in fuel flow. Wall cooling could also pose a major problem if the system were applied to engines of high pressure ratio.

The system seems ideally suited for small engines of low compression ratio, and this has been its main application to date. As the success of the system depends on high rotational speeds, usually greater than 350 rps, it is clearly less suitable for large engines where shaft speeds are much lower. In the United States, slinger-type systems have been used successfully on several engines produced by the Williams Research Corporation.

6.9 Air-Assist Atomizers

As discussed earlier, a basic drawback of the simplex nozzle is that if the swirl ports are sized to pass the maximum fuel flow at the maximum fuel-injection pressure, then the fuel pressure differential will be too low to give good atomization at the lowest fuel flow condition. This problem can be overcome by sizing the fuel ports for the highest fuel flow rate and then using high-velocity air to augment the atomization process at low fuel flows. A wide variety of designs of this type have been produced for use in industrial gas turbines. Useful descriptions of these may be found in Mullinger and Chigier [28].

In the *internal-mixing* configuration, shown schematically in Figure 6.16a, air and fuel mix within the nozzle before exiting through the outlet orifice. The fuel is sometimes supplied through tangential slots to encourage a conical spray pattern. However, the maximum spray angle is usually about 60°. As its name suggests, in the *external-mixing* form of the air-assist nozzle, the high-velocity air impinges on the fuel downstream of the discharge orifice, as illustrated in Figure 6.16b. Its advantage over the internal-mixing type is that problems of back pressure are avoided because there is no internal communication between air and fuel. However, it is less efficient than the internal-mixing concept, and higher airflow rates are needed to achieve the

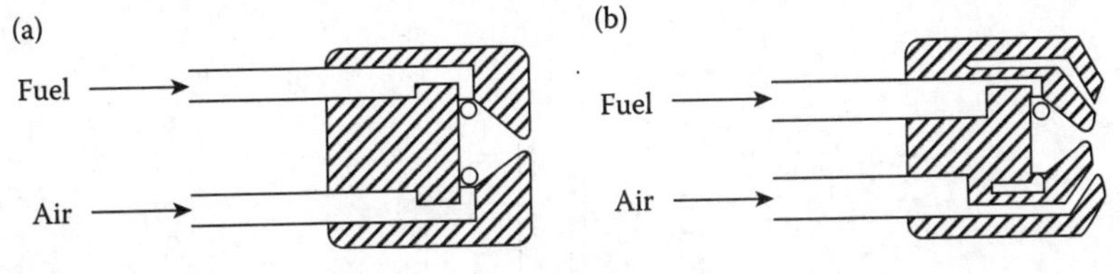

FIGURE 6.16
Schematic drawings of air-assist atomizers: (a) internal mixing; (b) external mixing.

same degree of atomization. Both types of nozzles can effectively atomize high-viscosity liquids.

6.10 Airblast Atomizers

In principle, the airblast atomizer functions in exactly the same manner as the air-assist atomizer because both employ the kinetic energy of a flowing airstream to shatter the fuel jet or sheet into ligaments and then drops. The main difference between the two systems lies in the quantity of air employed and its atomizing velocity. With the air-assist nozzle, where the air is supplied from a compressor or a high-pressure cylinder, it is important to keep the airflow rate down to a minimum. However, as there is no special restriction on air pressure, the atomizing air velocity can be made very high. Thus, air-assist atomizers are characterized by their use of a relatively small quantity of very high-velocity air. However, because the air velocity through an airblast atomizer is limited to a maximum value (usually around 120 m/s), corresponding to the pressure differential across the combustor liner, a larger amount of air is required to achieve good atomization. However, this air is not wasted because after atomizing the fuel, it conveys the drops into the combustion zone, where it meets and mixes with the additional air employed in combustion.

Airblast atomizers have many advantages over pressure atomizers, especially in their application to combustion systems operating at high pressures. They require lower fuel-pump pressures and produce a finer spray. Moreover, because the airblast atomization process ensures thorough mixing of air and fuel, the ensuing combustion process is characterized by very low soot formation and a blue flame of low luminosity, resulting in relatively low flame radiation and a minimum of exhaust smoke. The merits of the airblast atomizer have led to its installation in a wide range of aircraft, marine, and industrial gas turbines.

6.10.1 Plain-Jet Airblast

This is perhaps the simplest form of airblast atomizer, as illustrated in Figure 6.17. It features a round jet of fuel that is injected along the axis of

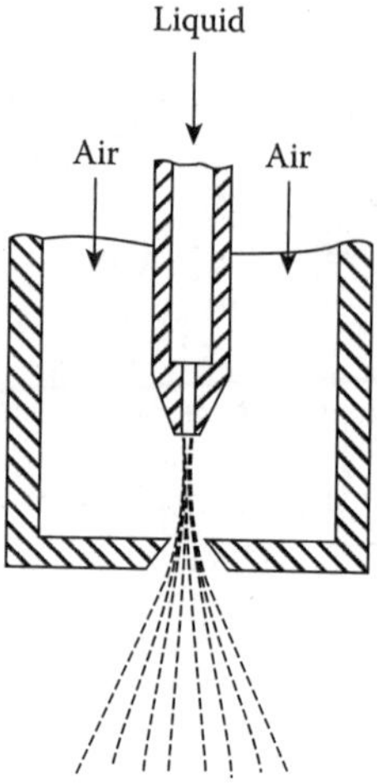

FIGURE 6.17
Plain-jet airblast atomizer.

a generally co-flowing round jet of air. Although this type of atomizer has relatively few applications in gas turbine combustion, it has some practical significance because much of our present knowledge on the effects of air and fuel properties on the mean drop sizes produced in airblast atomization was obtained with this type of atomizer, including the pioneering study of Nukiyama and Tanasawa [29].

6.10.2 Prefilming Airblast

Most of the airblast atomizers now in service are of the prefilming type, in which the fuel is first spread out into a thin continuous sheet and then subjected to the atomizing action of high-velocity air. One example of a prefilming airblast atomizer designed for gas turbines is shown in Figure 6.18. In this design, the atomizing air flows through two concentric air passages that generate two separate swirling airflows at the nozzle exit. The fuel flows through a number of equispaced tangential ports onto a prefilming surface where it spreads into a thin, circumferentially uniform sheet before being discharged at the atomizing "lip" or "edge" into the interface between the two swirling airstreams. The amount of air employed in atomization is constrained by the need to restrict atomizer size, partly in order to reduce weight, but also to avoid weakening the combustor casing by large insertion holes. Thus, modern prefilming airblast atomizers normally operate with a maximum air/fuel ratio (AFR) of around 3.

An important design choice is whether the two swirling airstreams should be co-rotating or counter-rotating. The advantage of co-rotation is that the two airstreams support each other in helping to create a strong primary-zone

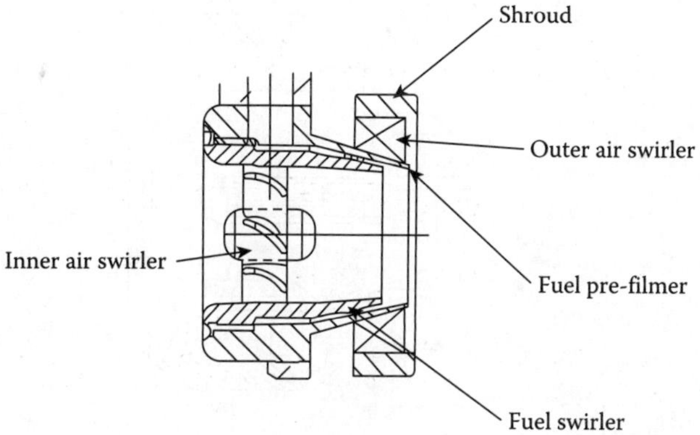

FIGURE 6.18
Basic components of prefilming airblast atomizer. (Courtesy of Parker Hannifin Corporation.)

flow recirculation. The advantage of counter-rotation is that it promotes a shearing action between the fuel and the atomizing air, which is beneficial to both atomization and fuel–air mixing. However, because the two swirl components are in opposite directions, the resulting swirl strength may be so small that the atomizing air does little to promote primary-zone flow recirculation. This drawback can be alleviated by making one airstream, usually the outer, much stronger than the other.

Chin et al. [30] carried out an experimental study using a prefilming injector that had the capability of reversing the direction of rotation of each of the two air swirlers used in the design. The results demonstrated that a combination of co-rotating inner airstream and counter-rotating outer airstream with respect to the rotational direction of the liquid film, yields the lowest SMD, as compared with other swirler configurations. The worst atomization was achieved when both airstreams were swirling in opposite directions to that of the liquid film.

6.10.3 Piloted Airblast

This device is also known as a "hybrid" injector because it consists of a prefilming airblast atomizer with a simplex pressure-swirl nozzle located on its centerline, as shown in Figure 6.19. The design objective is to overcome the airblast atomizer's inherent drawbacks of poor lean blowout performance (see Chapter 5) and poor atomization during engine startup when atomizing air velocities are low. At low fuel flows, all fuel is supplied through the pilot nozzle, and a well-atomized spray is obtained, giving efficient combustion at startup and idling. On aero engines, it also ensures good high-altitude relight performance. At higher power settings, fuel is supplied to both the airblast and pilot nozzles. The relative amounts are

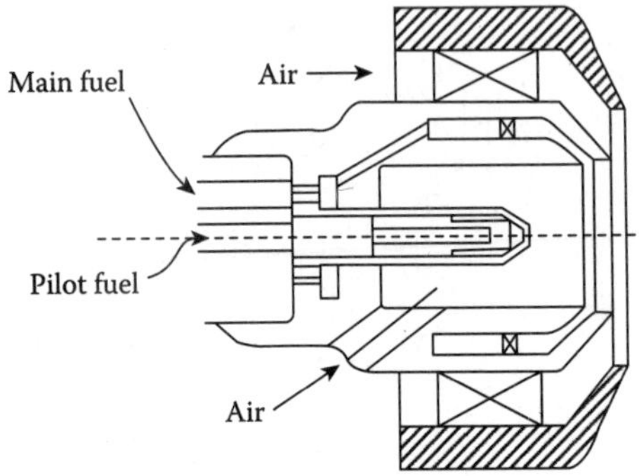

FIGURE 6.19
Piloted airblast atomizer. (From Rizk, N.K., Chin, J.S., and Razdan, M.K., AIAA Paper 96-2628, 1996. With permission.)

such that at the highest fuel flow conditions most of the fuel is supplied to the airblast atomizer. By this means, the performance requirements of good atomization at low fuel flows and low exhaust smoke at high fuel flows are both realized.

Chin et al. have carried out a number of experimental and modeling studies on the performance of hybrid atomizers (see, e.g., References [31] and [32]). These studies have focused on the interaction between the two separate sprays and on the influence of various atomizer design features on drop-size distributions in the combined spray. The results obtained provide detailed information for the modeling of combustors featuring hybrid atomizers and also on the methods available to the designer for optimizing atomization performance at various key combustor operating conditions.

6.10.4 Airblast Simplex

In its simplest form, the airblast simplex (ABS) atomizer comprises a simplex pressure-swirl nozzle surrounded by a co-flowing stream of swirling air. Essentially, it is the same as an external-mixing air-assist atomizer; the only difference is that atomizing air is supplied continuously and not just as and when required. It also has much in common with the hybrid airblast atomizer, except that in the latter the pressure-swirl nozzle supplies only a small fraction of the fuel at high-power conditions, whereas with the ABS concept the pressure-swirl nozzle supplies all the fuel at all conditions. According to Benjamin et al. [33], ABS atomizers offer the following advantages for aero-engine applications:

1. They are easier and cheaper to manufacture than prefilming airblast atomizers.
2. The heat shielding required to inhibit fuel coking is simpler to design and implement for the fuel passages of ABS atomizers than for the

small gaps between the inner and outer swirlers of prefilming airblast atomizers.

3. A simplex nozzle has a higher altitude relight capability than a prefilming airblast atomizer for a given fuel pressure drop.

The main barrier to the practical implementation of ABS nozzles has been that simplex atomizer sprays are known to "collapse" at elevated ambient pressures. This drawback would appear to rule them out for application to modern high-performance engines. However, Benjamin et al. have shown that spray collapse is not significant if the mass ratio of atomizing air to fuel is maintained above about 3.

Suyari and Lefebvre [34] investigated the atomizing performance of an ABS nozzle of the type shown in Figure 6.20. Measurements of SMD were carried out using water, gasoline, kerosine, and diesel oil. Some of the results obtained for kerosine are shown in Figure 6.21. From these and other data, they drew the following conclusions:

1. The key factors governing atomization quality are the dynamic pressure of the atomizing air and the relative velocity between the fuel and the surrounding air.
2. For any given value of air velocity, a continual increase in fuel flow rate from an initial value of zero produces an increase in SMD up to a maximum value, beyond which further increases in fuel flow rate causes SMD to decline.
3. The fuel flow rate at which the SMD attains its maximum value increases with an increase in atomizing air velocity.

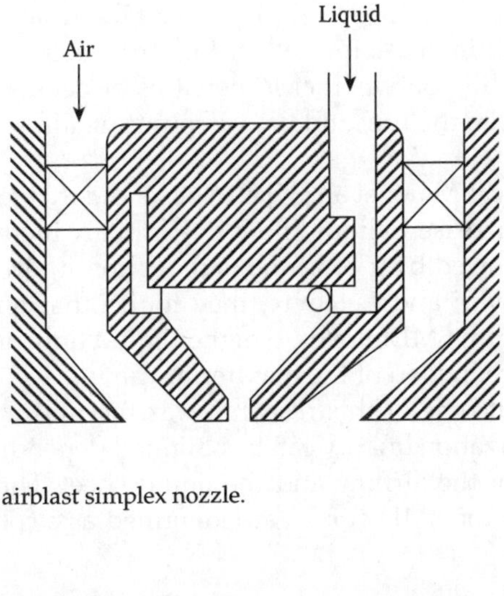

FIGURE 6.20
Schematic drawing of airblast simplex nozzle.

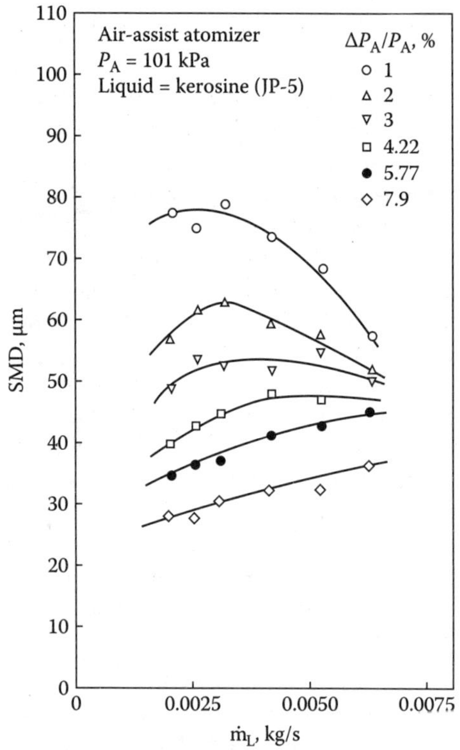

FIGURE 6.21
Influence of liquid flow rate and atomizing air velocity on mean drop size. (From Suyari, M. and Lefebvre, A.H., Paper presented at Central States Combustion Institute Spring Meeting, NASA-Lewis Research Center, Cleveland, OH, 1986. With permission.)

4. Whereas an increase in air velocity is usually beneficial to atomization quality, an increase in fuel velocity may help or hinder atomization, depending on whether it increases or decreases the relative velocity between the fuel and the surrounding air.

In a more recent study, Maier et al. [35] also observed that for any given value of air velocity, an increase in liquid flow rate initially increases the SMD up to a maximum, followed by a continuous reduction in droplet size. Also in agreement with Suyari and Lefebvre, they found that with increasing air velocity, the maximum SMD moves to higher liquid flow rates, accompanied by a simultaneous reduction of the maximum value.

A most useful outcome of this research was the finding that substantial differences in atomization quality can be obtained depending on the relative swirl orientations of the airflow and the liquid sheet. For counter-rotating swirls, the peaks in the SMD curves are identified as a collapse of the tulip shape of the liquid sheet into the onion shape. For co-rotating swirls, the tendency to collapse is much smaller [35].

6.11 Effervescent Atomizers

All the twin-fluid atomizers described above, in which air is used either to augment atomization or as the primary driving force for atomization, have one important feature in common: the bulk liquid to be atomized is first transformed into a jet or sheet before being exposed to high-velocity air. An alternative approach is to introduce the air directly into the bulk liquid at some point upstream of the nozzle discharge orifice. This air is injected at low velocity and forms bubbles that produce a two-phase bubbly flow at the discharge orifice. When the air bubbles emerge from the nozzle, they expand so rapidly that the surrounding liquid is shattered into droplets.

The advantages offered by effervescent atomization in gas turbine applications include the following:

1. Atomization is very good even at very low injection pressures and low airflow rates. When operating at a typical AFR of 0.03, mean drop sizes are comparable to those obtained with airblast atomizers at an AFR of 3.0.
2. The system has large holes and passages so that problems of "plugging" are greatly reduced. This could be an important advantage for combustion systems that burn residual fuels, slurry fuels, or any other type of fuel where atomization is impeded by the necessity of using large hole and passage sizes to avoid plugging of the nozzle.
3. The aeration of the spray created by the presence of the air bubbles could prove beneficial in alleviating soot formation and exhaust smoke.
4. The basic simplicity of the device lends itself to good reliability, easy maintenance, and low cost.

One drawback to effervescent atomization is that the resulting spray is characterized by a wide distribution of drop sizes, which typically correspond to a Rosin–Rammler distribution parameter q of about 2. A more serious drawback, however, is the need for a separate supply of atomizing air, which must be provided at essentially the same pressure as that of the fuel. Although this air requirement is small, about 1% by mass of what is required by a prefilming airblast atomizer, it necessitates a separate compressor. This drawback would appear to rule it out for aircraft applications, but it should not be a serious impediment to its installation in automotive, marine, and industrial gas turbines.

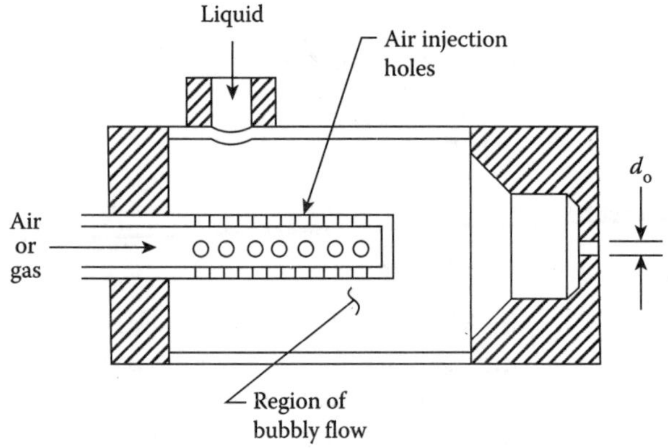

FIGURE 6.22
Plain-orifice effervescent atomizer.

Most of the research carried out on effervescent atomization [36–43] has used atomizers of the plain-orifice type shown in Figure 6.22. A drawback to this simple concept is that the spray cone angle is fairly small, typically around 20° [43]. Most gas turbine combustors require injectors that distribute the fuel in the form of a conical spray of approximately 90° included angle. Whitlow et al. [41] have studied several different types of effervescent atomizers designed to produce wide-angle sprays. One design was essentially the same as the atomizer shown in Figure 6.22, except that the single-hole orifice was replaced with four equispaced holes drilled at an angle of 40° from the axis of the mixing tube. Tests carried out on this four-hole design showed that the total liquid flow was uniformly distributed between the four holes to within a few percent. Using four holes instead of one had no deleterious effect on atomization quality, as Figure 6.23 clearly shows.

Whitlow et al. also found that a wide-angled spray could be produced by replacing the normal circular discharge orifice with a suitably angled annular passage. With this arrangement, the two-phase mixture is ejected from the atomizer in the form of a hollow-cone spray, with most of the droplets concentrated around the outer periphery. One of the advantages of this simple configuration is that the annular discharge passage can be designed to produce a hollow-cone spray having virtually any desired spray angle. The performance of this type of conical-sheet effervescent atomizer was investigated over wide ranges of pressure, air/liquid ratio, and annular gap width. A satisfactory and stable spray was observed at all operating conditions.

More detailed information on the design aspects of various types of effervescent atomizers, including single-hole, multi-hole, conical-spray, and annular spray, may be found in References [42] and [44].

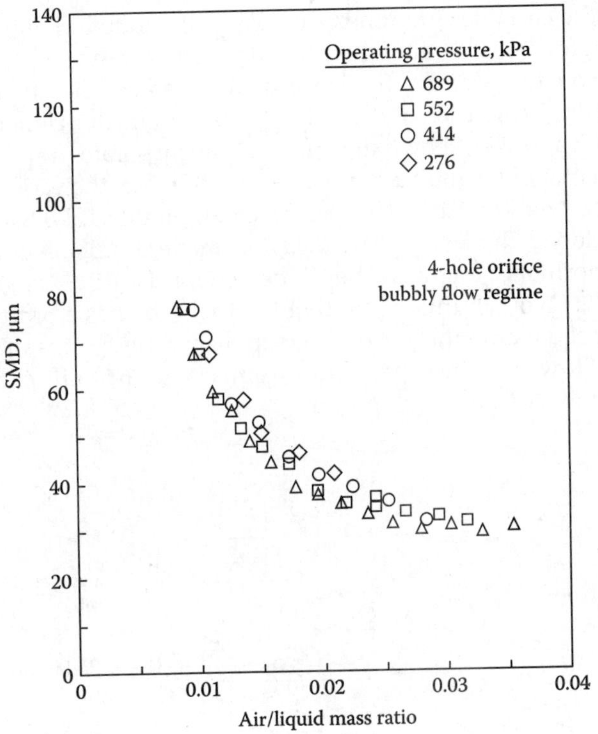

FIGURE 6.23
Influence of injection pressure and air/liquid ratio on mean drop size. (From Whitlow, J.D., Lefebvre, A.H., and Rollbuhler, J.R., *Fuels and Combustion Technology for Advanced Aircraft Engines,* AGARD Conference Proceedings, 536, 38, 1993.)

6.12 Vaporizers

Apart from the various atomization methods already discussed, an alternative method of preparing a liquid fuel for combustion is by heating it above the boiling point of its heaviest hydrocarbon ingredient, so that it is entirely converted to vapor before combustion. This method is, of course, applicable only to such high-grade fuels as can be completely vaporized, leaving no solid residue (see Chapter 1).

An alternative and much simpler method of vaporization is to inject the fuel, along with some air, into tubes that are immersed in the flame. The injected fuel–air mixture is heated by the tube walls and, under ideal conditions, emerges as a mixture of vaporized fuel and air. The remainder of the combustion air is admitted through apertures in the liner wall and reacts with the fuel–air mixture issuing from the tubes.

Some of the early designs, one of which is illustrated in Figure 6.24, were generally known as "walking stick" or "candy cane" vaporizers. They were used on the Mamba, Sapphire, and Viper engines in the UK, and on the Curtiss-Wright J65 and Westinghouse J46 in the United States. The Lycoming T vaporizer, shown in Figure 6.25, incorporates a splitter that runs down the center of the inlet leg and, in effect, converts the vaporizer into two back-to-back "walking sticks" sharing a common inlet. SNECMA and Rolls Royce developed this concept further and made it suitable for application to high-performance engines. The RR design for the RB199 engine is shown in Figure 6.26. Note in this figure that the fuel tube has a bifurcated end to ensure that equal amounts of fuel are supplied to both arms of the T vaporizer. The AFR within the tubes varies from around 6 at idling conditions to between 2 and 3 at maximum power.

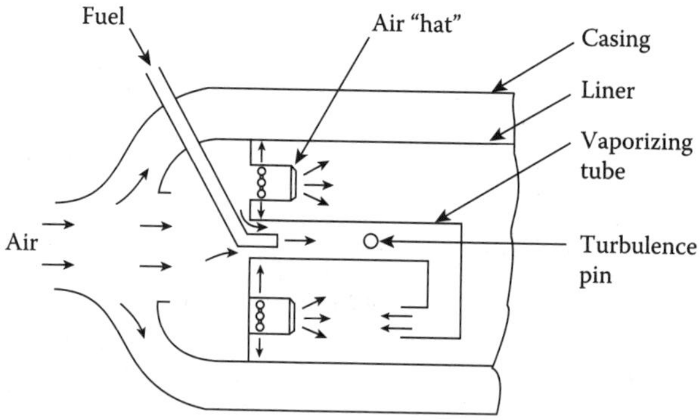

FIGURE 6.24
"Walking stick" vaporizing system.

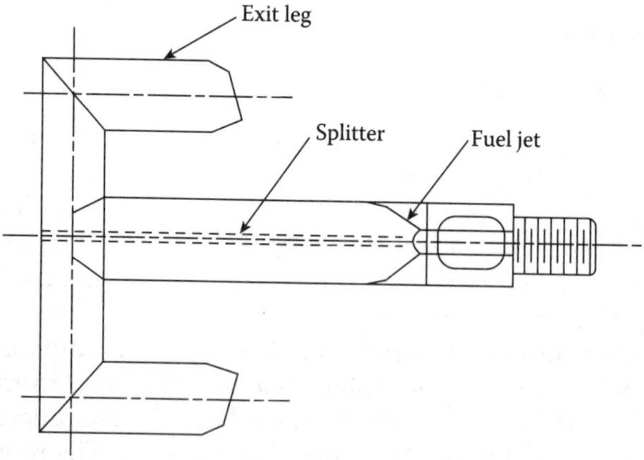

FIGURE 6.25
Lycoming T vaporizer. (Reproduced by permission of Lycoming Engines, a div. of Avco Corporation.)

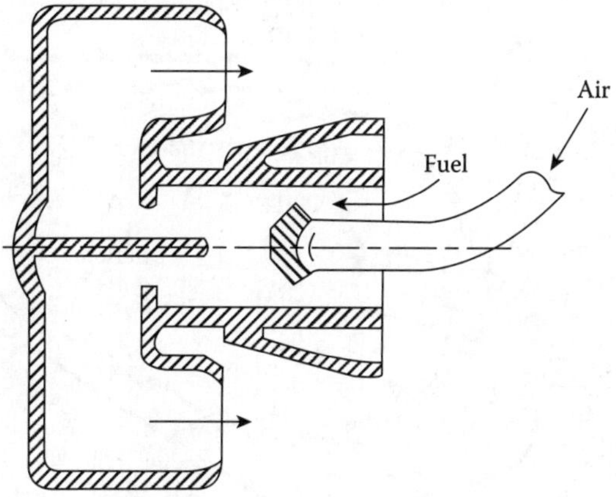

FIGURE 6.26
RB 199 vaporizer. (Courtesy of Rolls Royce plc.)

Vaporizing systems have useful advantages in terms of low cost, modest fuel-pump pressure requirements, and fairly low soot formation. Their drawbacks include risk of thermal damage to the vaporizing elements and sensitivity to variation in fuel type. Moreover, during the starting cycle, the tubes are too cold to effect vaporization and a torch igniter is needed to initiate combustion. Usually, this takes the form of a plain-orifice, pressure-jet atomizer adjacent to an igniter plug. A further drawback is that during rapid engine acceleration the sudden addition of more fuel can overcool the tubes, thereby lowering evaporation rates and reducing combustion efficiency.

It is now widely recognized that the term vaporizer is largely a misnomer because at high-power conditions the heat transferred to the tubes is insufficient to vaporize more than a small fraction of the fuel. Thus, only at the lowest fuel flows can the system be regarded as a true vaporizer. Where vaporizers are used on modern engines, their main function appears to be that of providing a satisfactory distribution of fuel throughout the primary combustion zone.

Vaporizing systems are now in service on a number of Rolls Royce engines, including the Pegasus, Olympus, and RB199. Useful descriptions of these can be found in papers by Parnell and Williams [45], Low [46], Sotheran [47], and Jasuja and Low [48]. A typical modern vaporizing combustor is shown in Figure 6.27.

In regard to future applications, it is important to bear in mind that vaporizer elements survive only because they are fuel-cooled. This can be a problem during rapid engine decelerations when the fuel is cut off quickly and the only available coolant is the combustor inlet air, which, temporarily at least, is still at a high temperature. As engine pressure ratios continue to rise, and combustor inlet air temperatures along with them, the cooling effectiveness of this air will diminish correspondingly. This clearly has important implications for the mechanical integrity of any future vaporizer design.

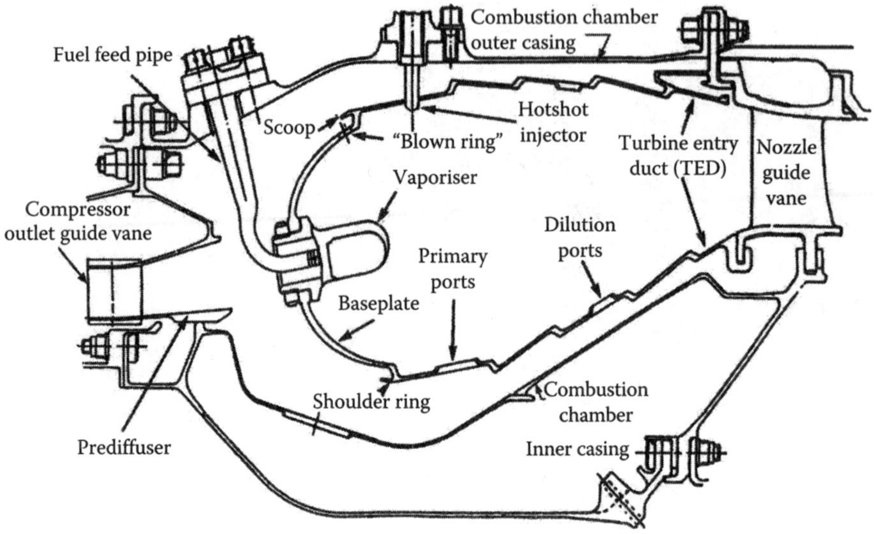

FIGURE 6.27
Modern vaporizing combustor. (Courtesy of Rolls Royce plc.)

6.13 Fuel Nozzle Coking

This problem is by no means new, but it is becoming especially serious for advanced turbojet engines because of the growing use of fuel as a heat sink for cooling the airframe, avionics, and engine lubricating oil. It is further exacerbated by the fact that the fuel feed arm is immersed in the compressor efflux air. This high-pressure, high-velocity airflow causes convective heating, which further raises the temperature of the fuel before it flows into the fuel injector. The combined effect of all these various inputs is that by the time the fuel is sprayed into the combustion zone, its temperature is appreciably higher than when it left the fuel tank.

From a combustion viewpoint, this elevation in fuel temperature is not altogether undesirable because it reduces fuel viscosity, thereby promoting finer atomization. Unfortunately, high fuel temperatures stimulate oxidation reactions, which lead to the formation of gums and other insoluble materials (including carbon) that tend to deposit on the walls of the passages and metering orifices within the nozzle. The rate of deposition is governed mainly by fuel temperature, but is also enhanced by an increase in wall temperature [49,50]. These deposits can distort the fuel spray and create appreciable nonuniformities in spray patternation [51,52].

The problems created by the deposition of carbonaceous materials, generally referred to as "coke," within the fuel nozzle are of special importance for pressure-swirl nozzles because they contain small internal passages that are especially prone to plugging and blockage. Coke agglomerates, formed either upstream of the nozzle tip or within the nozzle itself, can break off and be carried into the metering passages. Airblast atomizers are inherently less susceptible than pressure-swirl atomizers to the problems of fuel coking because they employ much larger fuel passages in the nozzle tip. However, the inability of airblast atomizers to always meet the requirements of cold day starting has prevented pure airblast systems from completely displacing pressure-swirl atomizers from engine designs.

The effects of partially or totally blocked fuel metering passages on the fuel-air distributions produced by an airblast atomizer have been examined by McCaldon et al. [52]. They found that as more and more fuel metering holes are obstructed with increasing operating time, more fuel is forced through the remaining nozzles. Consequently, engine damage may be caused by those injectors that, if tested individually, still flow within tolerances.

In recent years, the U.S. Naval Air Propulsion Center has sponsored an Innovative High-Temperature Fuel Nozzle Program with the objective of designing and evaluating fuel nozzles capable of operating satisfactorily despite extreme fuel and air inlet temperatures. As part of this program, Stickles et al. [53] evaluated 27 different nozzle designs, all of which were based on the production GE F404 fuel nozzle. Heat transfer analysis highlighted the following design rules for reducing wetted-wall temperatures:

1. Reduce fuel flow passage area to increase fuel velocity
2. Add air gaps
3. Substitute ceramics for metal parts
4. Avoid bends and steps in the fuel flow path

Sample tube coking test results showed the importance of surface finish on the fuel coking rate. Reducing the surface roughness from 3.1 to 0.25 μm reduced the deposition rate by 26%. In summary, Stickles et al. found that reduced passage flow area, reduced surface roughness, additional insulating air gaps, and replacement of metallic tip components with ceramics, minimized the wetted-wall temperature, thereby reducing the rate of deposition.

Thermal modeling studies carried out by Myers et al. [54] as part of the same U.S. Navy program showed that the two major sources of heat absorption into the fuel nozzle are the air swirler vanes and any surface exposed to the flame. At an altitude cruise condition, for example, the predicted heat flux entering the nozzle face from flame radiation is more than 20 times that absorbed by conduction and convection through the burner feed arm. The frontal area exposed to the flame is thus a key element in nozzle thermal loading.

Myers et al. concluded that substantial reductions in wetted-wall temperatures can be realized at extreme fuel and air inlet temperatures by using simple air gaps as thermal barriers. Detailed thermal analysis and simple thermal barriers, rather than exotic cooling schemes, can produce dramatic improvements in thermal protection.

The problem of fuel coking and its strong adverse effects on spray uniformity and pollutant emissions is one of growing concern because of the anticipated gradual deterioration in fuel quality and the continuing trend toward higher temperature engines.

6.14 Gas Injection

Provided that their energy density is reasonably high (say not less than 6 MJ/m^3), gaseous fuels present no special problems, at least from a combustion standpoint. They are usually characterized by clean combustion, with low rates of formation of soot and nitric oxides. The main problem is that of achieving the optimal level of mixing in the combustion zone. A mixing rate that is too high produces narrow stability limits, but a mixing rate that is too low may make the system prone to combustion-induced pressure oscillations. On engines designed to operate on both gaseous and liquid fuels, it is important that the gas flow pattern be matched to that of the liquid fuel; otherwise, some variation in the temperature distribution of the outlet gases could occur during the changeover from one fuel to the other. During this period, careful control over the liquid and gas flow rates is required to avoid flame blowout or overtemperature.

Many different methods have been used to inject gas into conventional combustion chambers, including plain orifices, slots, swirlers, and venturi nozzles. Good descriptions of these methods can be found in Winterfeld et al. [55]. The various methods of gas injection employed in modern low-emissions combustors are described in Chapter 9.

6.15 Equations for Mean Drop Size

For any given atomizer type, mean drop sizes are largely dependent on atomizer size, design features, and operating conditions. Atomization quality is also highly dependent on the physical properties of the fuels employed and on the properties of the surrounding gaseous medium.

The three fuel properties of relevance to atomization are density, surface tension, and viscosity. In practice, the significance of density for atomization performance is diminished by the fact that most gas turbine fuels exhibit only minor differences in this property. Surface tension is important in atomization because it resists the formation of new surface area, which is fundamental to the atomization process. Whenever atomization occurs under conditions where surface tension is important, the Weber number is a useful dimensionless parameter for correlating drop-size data. From a practical standpoint, viscosity is the most important fuel property. An increase in viscosity lowers the Reynolds number of the flow inside the atomizer, thickens the fuel sheet produced at the atomizer exit, opposes the development of instabilities in the fuel jet or sheet, and generally delays the onset of atomization. This delay causes atomization to occur further downstream from the nozzle where conditions are less conducive to the production of small drops. Another important practical consideration is that although the variations normally encountered in surface tension are only about 15%, the corresponding variations in viscosity are more than an order of magnitude.

The most important air property influencing atomization is density. With air-assist and airblast atomizers, if breakup occurs by the classical mechanism, an increase in air density generally improves atomization by raising the Weber number. With pressure-swirl atomizers, the effect of an increase in ambient air density on atomization is more complex. The concomitant increase in the Weber number is again beneficial, but this effect is opposed by a decrease in spray cone angle, which reduces the interaction between the developing sheet and the surrounding air, and increases the initial sheet thickness (see Figure 6.28). Furthermore, an increase in air density reduces the breakup length, so that breakup occurs closer to the nozzle where the fuel sheet is thicker. As SMD is proportional to the square root of the sheet thickness at breakup, the increase in sheet thickness produced by these two separate effects acting together must result in larger drops. Thus, an increase in air density can either raise or lower the SMD depending on whether the beneficial effect of increasing the Weber number outweighs the adverse effect of the increase in sheet thickness. Usually, it is found that if the ambient air density is increased continuously from its normal atmospheric level, the cone angle gradually falls until a value of density is reached beyond which there is no further reduction in the cone angle [56,57]. Moreover, as the breakup length declines with increase in air density, a condition is eventually reached where breakup occurs directly at the nozzle exit, or even within the nozzle itself. Beyond this point, sheet thickness has little or no effect on mean drop size. The net effect of all these separate influences is that drop sizes generally increase with ambient air density up to a maximum value (which

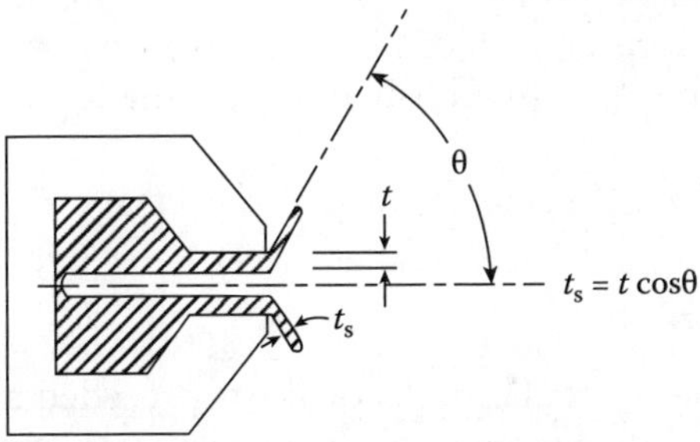

FIGURE 6.28
Relationship between sheet thickness and spray cone angle.

roughly corresponds to the condition at which the breakup length becomes zero) and then slowly decline with further increases in air density [57].

Unfortunately, the physical processes involved in atomization are not sufficiently well understood for mean diameters to be expressed in terms of equations derived from first principles. In consequence, the majority of investigations into the drop-size distributions produced in atomization have been empirical in nature and have resulted in empirical equations for mean drop size. The most authentic of these equations are those in which mean drop size is expressed in terms of dimensionless groups, such as Reynolds number, Weber number, or Ohnesorge number.

Most of the mean drop-size equations published before the 1970s should be regarded as suspect because of deficiencies in the methods available for drop-size measurements. Even equations based on accurate experimental data should only be used within the ranges of air properties, liquid properties, and atomizer operating conditions employed in their derivation. Extrapolation to other conditions is fraught with risk because changes in any of these variables could produce a change in the mode of atomization, which could have a significant effect on the manner and extent to which variations in the relevant flow parameters affect the drop-size distributions in the spray.

The following equations for SMD have been selected from the large number available in the literature. More detailed information on drop-size equations for all types of atomizers may be found in Lefebvre [58].

6.16 SMD Equations for Pressure Atomizers

6.16.1 Plain Orifice

With this device, a simple circular orifice is used to inject a round jet of high-velocity liquid into the surrounding air or gas. Finest atomization is achieved with small orifices but, in practice, the difficulty of keeping liquids free from foreign particles usually limits the minimum orifice size to around 0.3 mm.

Due to the formidable problems involved in making drop-size measurements in the dense sprays produced by plain-orifice nozzles, few equations for mean drop size have been published. According to Elkotb [59]

$$\mathrm{SMD} = 3.08\upsilon_L^{0.385}\left(\sigma\rho_L\right)^{0.737}\rho_A^{0.06}\Delta P_L^{-0.54} \tag{6.20}$$

6.16.2 Pressure Swirl

In this type of nozzle, a circular outlet orifice is preceded by a swirl chamber that causes the liquid to emerge from the nozzle as an annular sheet, which spreads radially outward to form a hollow conical spray. Despite its apparent simplicity, the various physical phenomena involved in pressure-swirl atomization are highly complex. For most of the past half century, mean drop sizes have been correlated using empirical equations of the form

$$\mathrm{SMD} = \mathrm{constant}\,\sigma^{a}\mu_{L}^{b}\dot{m}_{L}^{c}\Delta P_{L}^{-d}. \tag{6.21}$$

For example, Radcliffe's equation [60] is

$$\mathrm{SMD} = 7.3\sigma^{0.6}\mu_{L}^{0.2}\rho_{L}^{-0.2}\dot{m}_{L}^{0.25}\Delta_{L}^{-0.4}, \tag{6.22}$$

whereas subsequent work by Jasuja [61] yielded the expression

$$\mathrm{SMD} = 4.4\sigma^{0.6}\mu_{L}^{0.16}\rho_{L}^{-0.16}\dot{m}_{L}^{0.22}\Delta P_{L}^{-0.43}. \tag{6.23}$$

It is noteworthy that in the experiments of Radcliffe and Jasuja, the variation in surface tension was quite small and was accompanied by wide variations in viscosity. Thus, the surface tension exponent of 0.6 has no special significance in Equations 6.22 and 6.23.

Another example of this type of equation, which has an advantage over most others in that it is dimensionally correct, is the following [58]:

$$\mathrm{SMD} = 2.25\sigma^{0.25}\mu_{L}^{0.25}\dot{m}_{L}^{0.25}\Delta \mathrm{P}_{L}^{-0.5}\rho_{A}^{-0.25}. \tag{6.24}$$

It is now generally accepted that for both pressure and airblast nozzles, the relative velocity between the liquid and the surrounding air has a profound effect on atomization. It generates the protuberances on the liquid surface that are a prerequisite to atomization and also furnishes the energy needed to convert these protuberances into ligaments and then drops. However, another important factor in atomization, as discussed above, is the contribution made to sheet or jet disintegration by the instabilities created within the liquid itself, which are very dependent on liquid velocity. In airblast atomization, where high-velocity air impacts on a slow-moving liquid, the only factor promoting atomization is the relative velocity between the air and the liquid. This is equally important in pressure atomization but, by achieving this relative velocity through liquid motion instead of air motion, an important advantage is gained in that the liquid now makes an additional and independent contribution to its own disintegration, an effect that is either absent or negligibly small in airblast and air-assist atomization.

These arguments highlight the special importance of velocity in pressure atomization. The velocity at which the liquid is discharged from the nozzle has two separate effects on atomization. One important effect, which is dependent on the *absolute* velocity U_L, is in generating the turbulence and instabilities within the liquid stream that contribute to the first stage of the atomization process. The other effect, which depends on the *relative* velocity U_R, is in promoting the atomization mechanisms that occur on the liquid surface and in the adjacent ambient air.

Based on these considerations, Lefebvre [62] adopted an alternative approach to the derivation of an equation for mean drop size. For the purpose of analysis, the atomization process was treated in two separate stages. The first stage represents the generation of surface instabilities due to the combined effects of internal hydrodynamic and external aerodynamic forces. The second stage is the conversion of surface protuberances into ligaments and drops. This subdivision allows the formulation of an equation for mean drop size as

$$\mathrm{SMD} = 4.52\left(\sigma\mu_L^2/\rho_A \Delta P_L^2\right)^{0.25}(t\cos\theta)^{0.25} + 0.39\left(\sigma\rho_L/\rho_A \Delta P_L\right)^{0.25}(t\cos\theta)^{0.75}, \quad (6.25)$$

where t is the film thickness within the final discharge orifice (see Figure 6.28) and θ is the half-angle of the spray.

This equation takes into account all the factors that are known to affect the drop sizes produced in pressure-swirl atomization, including the cone angle of the spray. An increase in the cone angle improves atomization by reducing the thickness of the liquid sheet after it is discharged from the nozzle, as illustrated in Figure 6.28.

The values of the constants 4.52 and 0.39 in Equation 6.25 were obtained from a detailed experimental study carried out by Wang and Lefebvre [63], in which measurements of SMD were made using six simplex nozzles of different sizes and spray cone angles. Several different liquids were employed to provide a range of viscosity from 10^{-6} to 18×10^{-6} kg/ms (1–18 cS) and a range of surface tension from 0.027 to 0.073 kg/s^2 (27–73 dyn/cm). Figure 6.29

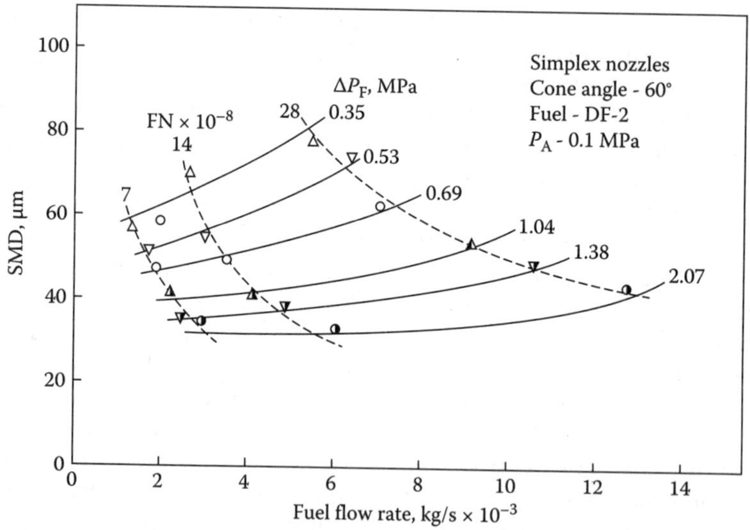

FIGURE 6.29
Graphs illustrating relationship between SMD and nozzle operating variables for a spray cone angle of 60°. (From Wang, X.F. and Lefebvre, A.H., *Journal of Propulsion and Power*, 3(1), 11–18, 1987. With permission.)

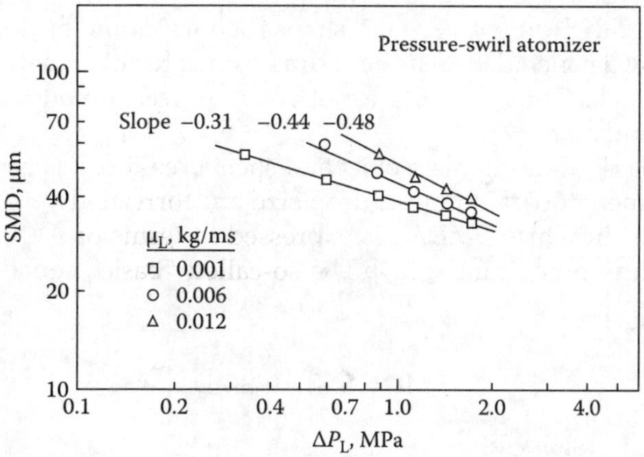

FIGURE 6.30
Influence of liquid viscosity on relationship between SMD and injection pressure. (From Wang, X.F., and Lefebvre, A.H., *Journal of Propulsion and Power*, 3(1), 11–18, 1987. With permission.)

is typical of the results obtained from this investigation. It shows the effect of variations in fuel-injection pressure, fuel flow rate, and nozzle flow number, on SMD for light diesel oil.

Inspection of Equation 6.25 reveals some interesting features that are discussed in detail in Reference [58]. For example, it suggests that liquids of high viscosity should exhibit a higher dependence of SMD on injection pressure differential ΔP_L than liquids of low viscosity, and this is borne out by the results presented in Figure 6.30.

6.17 SMD Equations for Twin-Fluid Atomizers

The first major study of twin-fluid atomization was conducted a half-century ago by Nukiyama and Tanasawa [29] on a plain-jet airblast atomizer. Drop sizes were measured by collecting samples of the spray on oil-coated glass slides. The experimental data were correlated by the following empirical equation for SMD:

$$\text{SMD} = 0.585\left(\sigma/\rho_L U_R^2\right)^{0.5} + 53\left(\mu_L^2/\sigma\rho_L\right)^{0.225}\left(Q_L/Q_A\right)^{1.5}. \tag{6.26}$$

This equation is not dimensionally correct, but could be made so by introducing some atomizer characteristic dimension, L_c, raised to the power 0.5. An obvious choice for this dimension is either the diameter of the liquid discharge orifice or the diameter of the air nozzle at exit. However, from tests carried out with different sizes and shapes of nozzles and orifices, Nukiyama and Tanasawa concluded that these dimensions have virtually no effect on mean drop size. Thus, the absence of any atomizer dimension is a notable feature of Equation 6.26.

For the classical mechanism of jet and sheet breakup, it is generally found that experimental data on mean drop size are correlated very satisfactorily by equations in which SMD/L_c is expressed in terms of ALR, Weber number, and Ohnesorge number [64]. The so-called "basic" equation is usually expressed as

$$\mathrm{SMD} = \mathrm{SMD}_1 + \mathrm{SMD}_2. \tag{6.27}$$

Analysis of the factors governing SMD_1 and SMD_2 leads to

$$\mathrm{SMD}/L_c\left(1+\mathrm{ALR}^{-1}\right) = A\mathrm{We}^{-0.5} + B\mathrm{Oh}^{0.5}, \tag{6.28}$$

or

$$\mathrm{SMD}/L_c = \left[1+(\mathrm{ALR})^{-1}\right]\left\{A\left(\sigma/\rho_A U_A^2 D_p\right)^{0.5} + B\left(\mu_L^2/\sigma\rho_L D_P\right)^{0.5}\right\}, \tag{6.29}$$

where A and B are constants whose values depend on atomizer design. For plain-jet atomizers, L_c is the initial liquid jet diameter, d_0. For prefilming atomizers, L_c is the initial thickness of the liquid sheet.

With practical atomizers, various design features and internal flow effects tend to modify the basic equation for SMD to forms as shown in Equations 6.30 and 6.31. Thus, for example, Rizk and Lefebvre [64] used their measured values of SMD to derive the following dimensionally correct equation for the mean drop sizes produced by a plain-jet airblast atomizer.

$$\begin{aligned}\mathrm{SMD} = {} & 0.48 d_0\left[\sigma/\rho_A U_R^2 d_0\right]^{0.4}\left[1+(\mathrm{ALR})^{-1}\right]^{0.4} \\ & +0.15 d_0\left[\mu_L^2/\sigma\rho_L d_0\right]^{0.5}\left[1+(\mathrm{ALR})^{-1}\right].\end{aligned} \tag{6.30}$$

This equation was shown to provide an excellent data correlation, especially for low-viscosity fuels. Lorenzetto and Lefebvre [65] and Jasuja [66] also derived very similar expressions for plain-jet airblast atomizers, thereby confirming the general validity of this form of predictive equation.

For prefilming airblast atomizers, El-Shanawany and Lefebvre [67] found that mean drop sizes could be correlated satisfactorily by the following dimensionally correct equation:

$$\text{SMD}/D_\text{h} = \left[1+(\text{ALR})^{-1}\right]\left[0.33\left(\sigma/\rho_\text{A}U_\text{R}^2D_\text{p}\right)^{0.6}\left(\rho_\text{A}/\rho_\text{L}\right)^{0.1} + 0.068\left(\mu_\text{L}^2/\sigma\rho_\text{L}D_\text{p}\right)^{0.5}\right], \quad (6.31)$$

where D_h is the hydraulic diameter of the air exit duct and D_p is the prefilmer diameter. Equations 6.30 and 6.31 show that SMD always increases with an increase in liquid viscosity, although the effect may be small for liquids of low viscosity due to the relatively small magnitude of the SMD2 term in these equations.

Usually it is found that an increase in surface tension serves to increase the mean drop size (by reducing the Weber number), but this is because most liquid hydrocarbon fuels tend to have relatively low viscosities and the SMD is dominated by the first term on the right hand side of Equations 6.30 and 6.31. However, these equations also predict that an increase in liquid viscosity causes the influence of surface tension on SMD to decline until a critical value of viscosity is eventually attained, above which any further increase in surface tension actually serves to reduce the mean drop size. The physical explanation for these seemingly contradictory effects is that surface tension forces assist viscosity in damping oscillations for the short-wavelength disturbances associated with liquids of low viscosity, but enhance oscillation growth for the long-wavelength disturbances associated with liquids of high viscosity.

Equations 6.30 and 6.31 also show that a continuous increase in relative velocity, U_R, causes SMD_1 to decline, so that SMD becomes more sensitive to changes in liquid viscosity (via SMD_2). This result conflicts with the findings of Buckner and Sojka [68], Sattelmayer and Wittig [69] and Beck et al. [10], all of whom observed only a small effect of liquid viscosity on SMD at high atomizing air velocities. The reason for this apparent contradiction is that Equations 6.30 and 6.31 are based implicitly on the notion that droplets are produced by the classical mechanisms of jet and sheet breakup and, in fact, the experimental data used to derive these equations were obtained using atomizers in which the air and liquid were essentially co-flowing (see Figures 6.31 and 6.32). As discussed above, these conditions are highly conducive to the classical mechanisms of jet and sheet breakup.

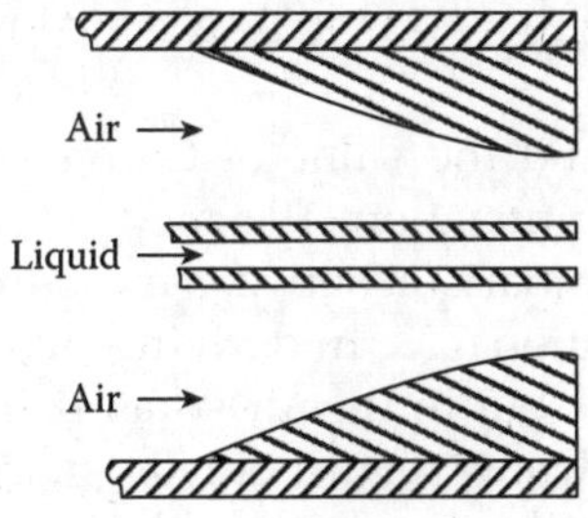

FIGURE 6.31
Schematic of co-flowing, plain-jet airblast atomizer.

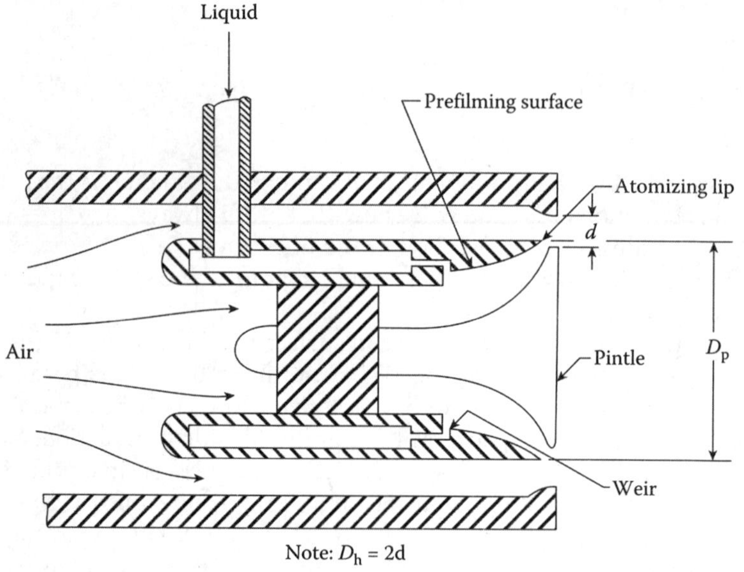

FIGURE 6.32
Schematic of co-flowing, prefilming airblast atomizer.

6.18 SMD Equations for Prompt Atomization

For sheet disintegration we have [9]

$$\text{SMD}/t = 3\{1 + 0.00175\text{We}/(1 + \text{ALR}^{-1})\}^{-1}, \tag{6.32}$$

where $\text{We} = \rho_L U_A^2 t/\sigma$.

For jet breakup by the prompt mechanism, the corresponding expression is

$$\text{SMD}/d_0 = 1.5\{1 + C\text{We}/(1 + \text{ALR}^{-1})\}^{-1}, \tag{6.33}$$

where $\text{We} = \rho_L U_A^2 d_0/\sigma$, and the value of C depends on the various design features that govern the utilization efficiency of the atomizing air.

Figure 6.33 shows a comparison between measured values of SMD and predicted values from Equation 6.33. In deriving this plot, Goris [70] employed a value for C of 0.000144, but in the air-spray paint nozzle he used, a large proportion of the total airflow served only as shaping air and made no contribution to the atomization process. If this is taken into account, a more accurate value for C would be around 0.00084. However, as the efficiency of

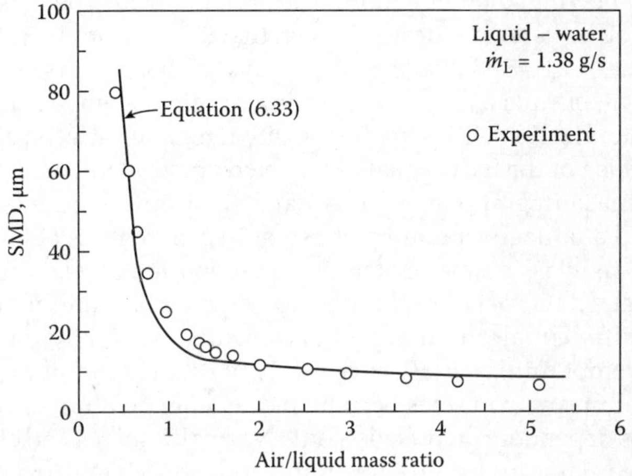

FIGURE 6.33
Comparison of measured values of SMD and predicted values from Equation 6.33. (From Goris, N.H., MSME Thesis, Purdue University, 1990.)

air utilization can vary widely from one atomizer design to another, for any given atomizer design the value of C should be determined experimentally.

According to Equations 6.29 through 6.31, for liquids of low viscosity, the mean drop size in the spray should diminish with the increase in ambient air pressure, according to the relationship SMD $\alpha P_A^{-0.5}$. Pressure exponents close to −0.5 have, in fact, been obtained by a number of workers [64,66,67,71,72], using atomizer designs of the type shown in Figures 6.31 and 6.32, in which the air and liquid are co-flowing. However, tests carried out by Zheng et al. [73,74] on a more practical form of airblast atomizer (see Figure 6.18) in which the fuel film is injected into the highly turbulent region created at the interface between two counter-rotating swirling airflows, showed that SMD was virtually independent of P_A over the entire test range from 1 to 12 bar. This result conforms to the predictions of Equation (6.32) for the prompt mechanism of sheet breakup.

6.18.1 Comments on SMD Equations

The atomization literature is replete with equations for correlating and predicting the mean drop sizes produced by various types of pressure and twin-fluid atomizers. All these equations should be used with caution, with proper judgment being exercised in regard to the accuracy of the experimental data and the ranges of fuel and air properties and atomizer operating conditions covered in the experiments.

It should be noted that no single equation can satisfactorily predict the drop sizes produced by any given type of atomizer over its entire range of operation. If a twin-fluid atomizer is designed to produce fuel and airstreams that are essentially co-flowing, then Equations 6.30 and 6.31 would be most appropriate. If, on the other hand, the design is such that it favors prompt atomization, then Equations 6.32 or 6.33 would be more suitable. For most twin-fluid atomizers, it is inevitable that a range of operating conditions will exist over which the mode of atomization will be in the transition regime between classical and prompt and none of the SMD equations quoted above would be satisfactory.

The same reasoning applies with equal force to pressure atomizers. For pressure-swirl atomizers operating at pressure differentials below around 1 MPa (145 psi), the classical mode of breakup predominates, and drop sizes are markedly affected by variations in fuel viscosity. With a continuous increase in pressure differential, the mode of atomization gradually changes from classical to prompt until, at a ΔP_F of around 3 MPa, the prompt mechanism is dominant and mean drop sizes become more dependent on surface tension and much less dependent on fuel viscosity. Normally, there is no clear demarcation between classical and prompt atomization; the change from one mode to the other taking place slowly as the pressure differential is either gradually increased from a low value or gradually reduced from a high value. The situation is analogous to airblast atomizers in that prompt atomization is promoted by increases in Δ/V and reductions in liquid viscosity, corresponding to increases in the Weber number and the Reynolds number, respectively.

6.19 Internal Flow Characteristics

In twin-fluid atomizers of the airblast and air-assist types, atomization and spray dispersion tend to be dominated by air momentum forces, with hydrodynamic processes playing only a secondary role. With pressure-swirl nozzles, however, the internal flow characteristics are of primary importance, because they govern the thickness and uniformity of the annular fuel film formed in the final discharge orifice, as well as the relative magnitude of the axial and tangential components of velocity of this film. It is, therefore, of great practical interest to examine the inter-relationships that exist between internal flow characteristics, nozzle design variables, and important spray features such as cone angle and mean drop size.

6.20 Flow Number

The effective flow area of a pressure atomizer is usually described in terms of a flow number, which is expressed as the ratio of the nozzle throughput to the square root of the fuel-injection pressure differential. Two definitions of flow number are in general use: a British version, based on the volume flow rate, and an American version, based on the mass flow rate. They are

$$FN_{UK} = \frac{\text{Flow rate, UK gals./hr}}{(\text{Injection pressure differential, psi})^{0.5}}, \tag{6.34}$$

and

$$FN_{USA} = \frac{\text{Flow rate, 1b/hr}}{(\text{Injection pressure differential, psi})^{0.5}}. \tag{6.35}$$

Note that 1 UK gallon = 1.2 US gallons.

Equations 6.34 and 6.35 have the advantage of being expressed in units that are in general use. Unfortunately, they are basically unsound. For example, they do not allow a fixed and constant value of flow number to be assigned to any given nozzle. Thus, although it is customary to stamp or engrave a value of flow number on the body of a simplex atomizer, this value is correct only when the nozzle is flowing a standard calibrating fluid of density 765 kg/m^3. In the past, this has posed no problems with aircraft gas turbines because 765 kg/m^3 roughly corresponds to the density of aviation kerosine. However, for fuels of other densities, these two definitions of flow number could lead to appreciable errors when used to calculate mass flow rates or injection pressures.

The basic deficiency in Equations 6.34 and 6.35 is the omission of fuel density. Inclusion of this property would not only allow these equations to be rewritten in a dimensionally correct form, but would also enable the flow number to be defined in a much more positive and useful manner than at present, namely, as the effective flow area of the nozzle. Thus, the flow number of any given nozzle would have a fixed and constant value for all liquids.

By including density, the flow number in square meters is obtained as

$$FN_{SI} = \frac{\text{Flow rate, kg/s}}{(\text{Pressure differential, Pa})^{0.5}(\text{Liquid density, kg/m}^3)^{0.5}}. \tag{6.36}$$

The standard UK and US flow numbers may be calculated from Equation 6.36 using the formulae:

$$FN_{UK} = 0.66 \times 10^8 \times \rho_L^{-0.5} \times FN, \tag{6.37}$$

$$FN_{US} = 0.66 \times 10^6 \times \rho_L^{-0.5} \times FN. \tag{6.38}$$

By combining Equations 6.36, 6.40, and 6.45, the flow number of a pressure-swirl atomizer is obtained in terms of atomizer dimensions as

$$FN = 0.389 d_0^{1.25} A_p^{0.5} D_s^{-0.25}. \tag{6.39}$$

6.21 Discharge Coefficient

The discharge coefficient of a pressure atomizer is governed partly by the pressure losses incurred in the nozzle flow passages and also by the extent to which the fuel flowing through the final discharge orifice makes full use of the available flow area. Discharge coefficient is related to nozzle flow rate by the equations

$$m_F = C_D A_0 (2\rho_F \Delta P_F)^{0.5}, \tag{6.40}$$

$$m_F = 1.11 C_D d_0^2 (\rho_F \Delta P_F)^{0.5}. \tag{6.41}$$

6.21.1 Plain-Orifice Atomizers

Measurements of discharge coefficient carried out on various orifice configurations over wide ranges of operating conditions indicate that the most important parameters are Reynolds number, length/diameter ratio, injection pressure differential, ambient gas pressure, inlet chamfer (or radius), and cavitation.

For noncavitating flow, it is found that discharge coefficients generally increase with an increase in Reynolds number, until a maximum value is attained at a Reynolds number of around 7000. Beyond this point, the value of C_D remains sensibly constant at its maximum value, regardless of Reynolds number. Maximum values of C_D are shown plotted against l_o/d_o in Figure 6.34. The experimental data on which this figure is based were drawn from Reference [75], but actual data points have been omitted for clarity. Figure 6.34 shows $C_{D(max)}$ rising steeply from about 0.61 to a maximum value of about 0.81 as l_o/d_o increases from 0 to 2. Further increase in l_o/d_o causes $C_{D(max)}$ to slowly decline in a nearly linear fashion to about 0.74 at $l_o/d_o = 10$. For the range of l_o/d_o between 2 and 10, Lichtarowicz et al. [75] proposed the following expression, which is claimed to fit the experimental data to within about 1%.

$$C_{D(max)} = 0.827 - 0.0085(l_o/d_o). \tag{6.42}$$

In flow regions of low static pressure, gas or vapor may be released from the fuel to form bubbles that can have a pronounced effect on discharge

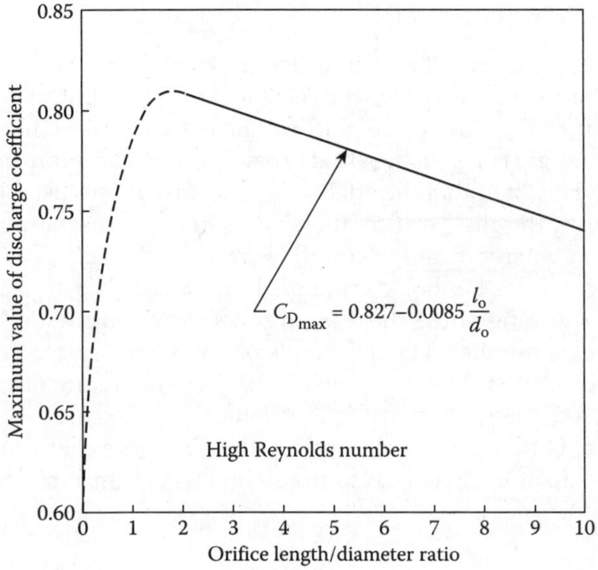

FIGURE 6.34
Variation of $C_{D(max)}$ with orifice l_o/d_o ratio.

coefficient. Bergwerk [76] was the first to carry out a systematic study of cavitation in plain-orifice atomizers. Several others have since investigated the influence of cavitation on discharge coefficient. The main findings of these studies have been reviewed by Ohrn et al. [77]. They show that when cavitation is present, C_D is governed primarily by the vapor pressure of the fuel and the pressure drop across the nozzle.

It is perhaps worthy of mention that the influence of cavitation on injector performance is not confined solely to its effect on discharge coefficient. For example, Ruiz and Chigier [78] have asserted that cavitation is more important than turbulence in promoting the initial disturbances necessary for jet atomization, whereas Reitz and Bracco [79] claim that cavitation, although not a necessary component for atomization, has a marked influence when present.

The main conclusion from the experiments of Ohrn et al. on nominally sharp-edged inlets is that the most important factor influencing the discharge coefficient is the inlet edge condition. Examination of many scanning electron microscope (SEM) photographs revealed that even minor deviations from a sharp-edged inlet, such as roughness or a slight local radius, could produce a significant increase in discharge coefficient. These workers also observed that increasing the orifice inlet radius raises the discharge coefficient, as noted in previous studies [76,80,81], and also causes C_D to increase slightly with an increase in Reynolds number up to around 30,000.

6.21.2 Pressure-Swirl Atomizers

The discharge coefficient of a swirl atomizer is inevitably low, owing to the presence of the air core, which effectively blocks off the central portion of the orifice. Radcliffe [82] studied the performance of a family of injectors based on common design rules, using fluids that covered wide ranges of density and viscosity. He noted that the effect of an increase in viscosity is to thicken the fluid film in the final orifice, thereby raising the discharge coefficient. This effect can be significant at low flow rates with nozzles of small flow number. However, for Reynolds numbers larger than 3000, that is, over most of the normal working range, the discharge coefficient is practically independent of Reynolds number. Thus, for fuels of low viscosity, the convention is to disregard conditions at low Reynolds number and assume that any given atomizer has a constant discharge coefficient.

According to Giffen and Muraszew [83], the discharge coefficient of a pressure-swirl atomizer is related to atomizer dimensions and the area of the air core by the equations

$$C_D = 1.17[(1-X)^3/(1+X)]^{0.5}, \tag{6.43}$$

where

$$X = \frac{\text{Air core area}}{\text{Discharge orifice area}} = \frac{A_a}{A_o} = \frac{(d_o - 2t)^2}{d_o^2}, \tag{6.44}$$

where t is the fuel film thickness in the discharge orifice.

Several other equations for discharge coefficients have been derived [58]. The following relationship, which is based on the analysis of a large amount of experimental data by Rizk and Lefebvre [84], is illustrated in Figure 6.35.

$$C_D = 0.36\left(\frac{A_p}{D_s d_o}\right)^{0.5}\left(\frac{D_s}{d_o}\right)^{0.25}. \tag{6.45}$$

6.21.3 Film Thickness

In pressure-swirl atomizers, the fuel emerges from the nozzle as a thin conical sheet that rapidly attenuates as it spreads radially outward, finally disintegrating into ligaments and then drops. In prefilming airblast atomizers, the fuel is also spread out into a thin continuous sheet before being exposed to high-velocity air. It is of interest, therefore, to examine the factors that govern the thickness of this fuel film.

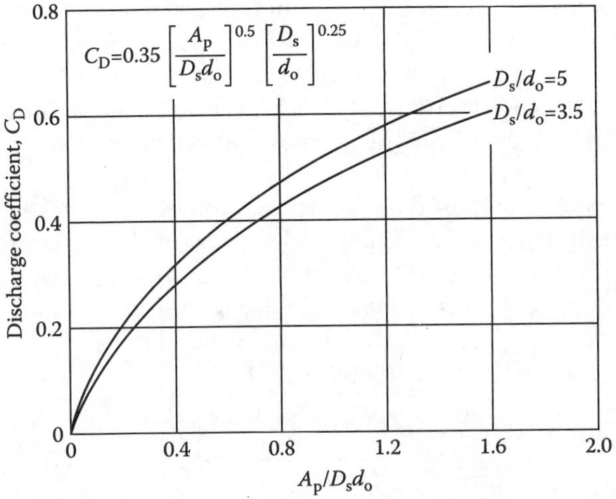

FIGURE 6.35
Relationship between discharge coefficient and atomizer dimensions.

For both pressure-swirl and airblast types of atomizers, it has long been recognized that the thickness of the annular fuel film produced at the nozzle exit has a strong influence on the mean drop size of the spray. In pressure-swirl atomizers, the thickness of this film is directly related to the area of the air core, as indicated in Equation 6.44. Giffen and Muraszew's analysis of the flow conditions within a simplex nozzle, assuming a nonviscous fluid, led to the following relationship between atomizer dimensions and the size of the air core:

$$\left[\frac{A_p}{D_s d_o}\right]^2 = \frac{\pi^2}{32}\frac{(1-X)^3}{X^2}. \tag{6.46}$$

After calculating X from Equation 6.46, the corresponding value of film thickness t is then obtained from Equation 6.44.

A similar relationship between atomizer dimensions and the size of the air core was derived by Suyari and Lefebvre [85].

$$0.09\left[\frac{A_p}{D_s d_o}\right]^2\left[\frac{D_s}{d_o}\right]^{0.5} = \frac{(1-X)^3}{1+X}. \tag{6.47}$$

Simmons and Harding [86] derived the following simple equation for fuel film thickness in terms of nozzle flow number and spray cone angle.

$$t = \frac{0.00805\sqrt{\rho_L}\,\mathrm{FN}}{d_o \cos\theta}. \tag{6.48}$$

Suyari and Lefebvre tested the validity of these and other equations for film thickness by comparing predicted values with their measured values using water as the working fluid. They found that Equations 6.47 and 6.48 provide a good fit to the experimental data. However, these equations do not take into account the effects of fuel properties and nozzle pressure drop on film thickness.

Rizk and Lefebvre [84] used a theoretical approach to investigate the internal flow characteristics of pressure-swirl atomizers. A general expression for film thickness was derived in terms of atomizer dimensions, liquid properties, and liquid injection pressure as

$$t^2 = \frac{1560\dot{m}_L \mu_L}{\rho_L d_o \Delta P_L} \frac{1+X}{(1-X)^2}. \tag{6.49}$$

Rizk and Lefebvre [84] used this equation to calculate film thicknesses for different nozzle dimensions and operating conditions. Some of their results are shown in Figure 6.36 as plots of film thickness against injection pressure differential. Also shown in this figure are the measured values of Kutty et al. [87,88]. Theory and experiment both indicate that a higher pressure drop produces a thinner film. Thus, the improvement in atomization quality

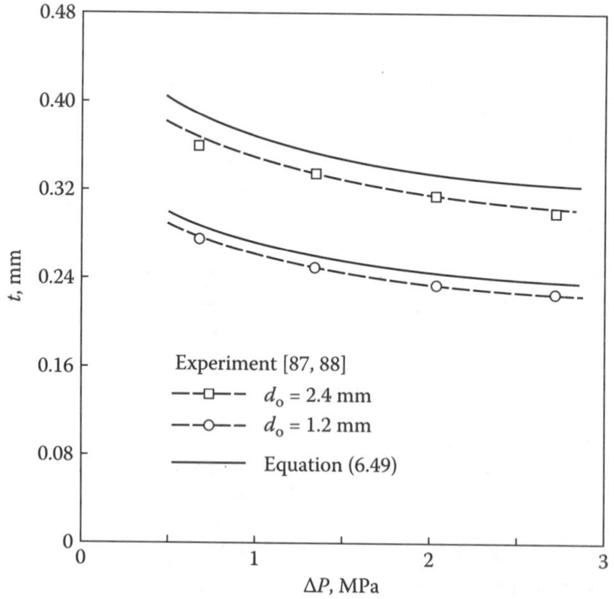

FIGURE 6.36
Variation of film thickness with injection pressure for different orifice diameters. (From Rizk, N.K., and Lefebvre, A.H., *Journal of Propulsion and Power*, 1(3), 93–199, 1985. With permission.)

that always accompanies an increase in nozzle pressure drop is due in some measure to the concomitant decrease in film thickness.

A drawback to Equation 6.49 is that because X is dependent on t (see Equation 6.44), some trial-and-error procedures are involved in its solution. However, if $t/d_o \ll 1$, it can be written more succinctly, while still retaining its essential features, as

$$t = 2.7\left[\frac{d_o \mathrm{FN} \mu_L}{(\Delta P_L \rho_L)^{0.5}}\right]^{0.25}. \tag{6.50}$$

6.22 Spray Cone Angle

An important aspect of atomizer design, in addition to achieving the desired drop-size distribution, is to ensure that the droplets formed in atomization are discharged from the nozzle in the form of a symmetrical uniform spray. In general, an increase in spray cone angle increases the exposure of the droplets to the surrounding air or gas, leading to improved atomization and to higher rates of heat and mass transfer.

6.22.1 Plain-Orifice Atomizers

With plain-orifice atomizers, the cone angle is narrow and the drops are fairly evenly dispersed throughout the entire spray volume. The angle of the spray is normally defined as the angle formed by two straight lines drawn from the discharge orifice to the outer periphery of the spray at a distance 60 d_o downstream of the nozzle. Several formulae have been derived to express the spray angle in terms of nozzle dimensions and the relevant air and liquid properties. The simplest expression for the spray angle is given by the jet mixing theory of Abramovich [89] as

$$\tan\theta = 0.13\left(1 + \frac{\rho_A}{\rho_L}\right). \tag{6.51}$$

According to Reitz and Bracco [90], the spray angle can be determined by combining the radial velocity of the fastest growing of the unstable surface waves with the axial injection velocity. This hypothesis results in the following expression for the spray angle:

$$\tan\theta = \frac{2\pi}{\sqrt{3A}}\left(\frac{\rho_A}{\rho_L}\right)^{0.5}. \tag{6.52}$$

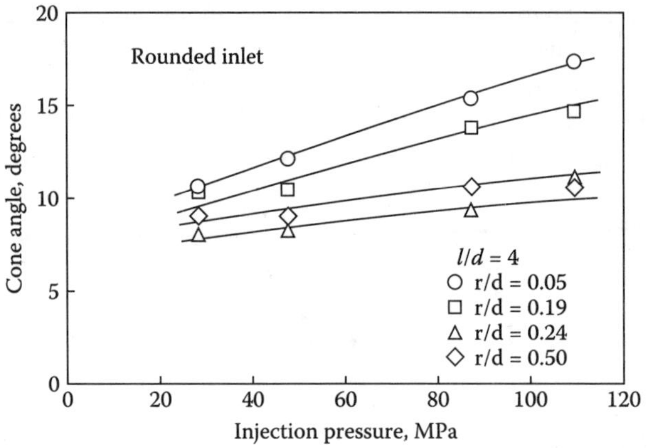

FIGURE 6.37
Spray cone angles for rounded inlets; $l_o/d_o = 4$.

Yokota and Matsuoka [91] and Hiroyasu and Arai [92] have derived correlations for their experimental data on spray angles obtained at high ambient air pressures. These and other equations for the spray cone angles of plain-orifice atomizers are presented and discussed in References [58] and [92].

Ohrn et al. [93] used 40 different plain-orifice atomizers to examine the effects of nozzle geometry and flow conditions on spray cone angle. Some of their results, showing the effects of nozzle pressure differential and orifice length/diameter ratio on the cone angle for round-edged inlets, are presented in Figure 6.37. The main conclusion from this study is that the cone angle increases with injection pressure for sharp-edged and slightly radiused inlet nozzles, but is largely independent of injection pressure for highly radiused inlets.

6.22.2 Pressure-Swirl Atomizers

In most combustion applications, the spray is in the form of a hollow cone of wide angle, with most of the drops concentrated at the periphery. A major difficulty in the definition and measurement of the cone angle is that the spray cone has curved boundaries, owing to the effects of air interaction with the spray. To overcome this problem, the cone angle is often given as the angle formed by two straight lines drawn from the discharge orifice to cut the spray contours at some specified distance from the atomizer face.

6.22.2.1 Theoretical Aspects

During the last half-century, several expressions for the spray cone angle have been derived, usually with the assumption of inviscid flow. Taylor's [94] inviscid theory showed that the spray cone angle is determined solely by the swirl chamber geometry and is a unique function of the ratio of the inlet ports area to the product of swirl chamber diameter and orifice diameter, $A_p/D_s d_o$, as shown

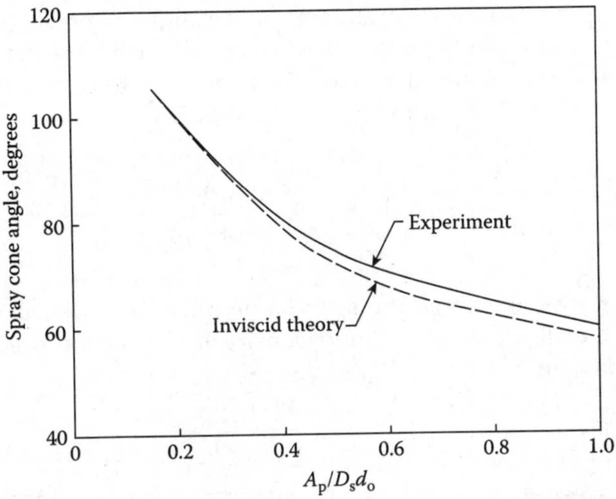

FIGURE 6.38
Relationship between spray cone angle and atomizer geometry.

in Figure 6.38. The solid curve in this figure corresponds to experimental data obtained by several workers (see Reference [58]). Giffen and Muraszew's [83] analysis also assumed a nonviscous fluid, which allowed the spray cone angle to be expressed as a function of nozzle dimensions only. It led to the following expression for the mean value of the spray cone half angle θ.

$$\sin\theta = \frac{(\pi/2)(1-X)^{1.5}}{K(1+\sqrt{X})(1+X)^{0.5}}, \tag{6.53}$$

where $K = A_p/D_s d_o$ and $X = A_a/A_0$.

Rizk and Lefebvre's [84] inviscid flow analysis also led to a unique relationship between the cone angle and X of the form

$$\cos^2\theta = \frac{1-X}{1+X}, \tag{6.54}$$

where X depends solely on the atomizer dimensions, as indicated in Equation 6.44.

In the above equations, θ is the cone half-angle, as measured close to the nozzle. As the spray in this region has a small but definite thickness, the cone angle formed by the outer boundary of the spray is defined as 2θ, whereas $2\theta_m$ represents the mean cone angle in this near-nozzle region. By assuming a constant axial velocity across the liquid film, the maximum cone angle is related to the mean cone angle by the expression [84]

$$\tan\theta_m = 0.5\tan\theta(1+\sqrt{X}). \tag{6.55}$$

The equations quoted above for the spray cone angles of pressure-swirl atomizers are valid only for liquids of low viscosity, such as water or kerosine. Rizk and Lefebvre [95] used a theoretical approach to derive the following dimensionally correct equation for viscous liquids:

$$2\theta = 6\left(\frac{D_s d_o}{A_p}\right)^{0.15}\left(\frac{\Delta P_L d_o^2 P_L}{\mu_L^2}\right)^{0.11}. \tag{6.56}$$

According to this equation, the spray cone angle is widened by increases in discharge orifice diameter, liquid density, and injection pressure, and is diminished by an increase in liquid viscosity.

6.23 Radial Fuel Distribution

The symmetry of the spray patterns produced in atomization is of considerable importance because the fuel must be distributed uniformly throughout the combustion zone to achieve high combustion efficiency, low pollutant emissions, and a uniform distribution of temperature in the combustor efflux gases. Although the visible spray cone angle gives some indication of spray symmetry and the total dispersion of a spray, it provides little or no information on how the fuel mass flux is distributed radially and circumferentially within the spray volume.

The term adopted by the gas turbine industry for the purpose of defining spray distribution is "patternation" and the instruments used to measure fuel flux distributions in sprays are commonly referred to as "patternators." A typical radial patternator consists of a number of small collection tubes oriented equidistant radially from the origin of the spray, as shown schematically in Figure 6.39. The sampling tubes are allowed to fill with fuel until one of the tubes is nearly full. At that point, the fuel supply is turned off and the volume of fuel in each tube is measured by visually locating the meniscus between lines scribed into the clear plastic of the patternator. Radial distribution curves are made by plotting fuel volume as the ordinate and the corresponding angular location of the sampling tubes as the abscissa, as illustrated in Figure 6.39. A typical plot is shown in Figure 6.40; it illustrates how the spray cone angle of a pressure-swirl atomizer contracts with an increase in ambient air pressure.

To more succinctly describe the effect of changes in operating parameters on fuel distribution, a radial distribution curve may be reduced to a single numerical value, called the *effective* or *equivalent* spray angle [56,96]. The effective spray angle, 2θ, is the sum of two angles, $2\theta = \theta_L + \theta_R$, where θ_L (or θ_R) is the value of θ that corresponds to the position of the center

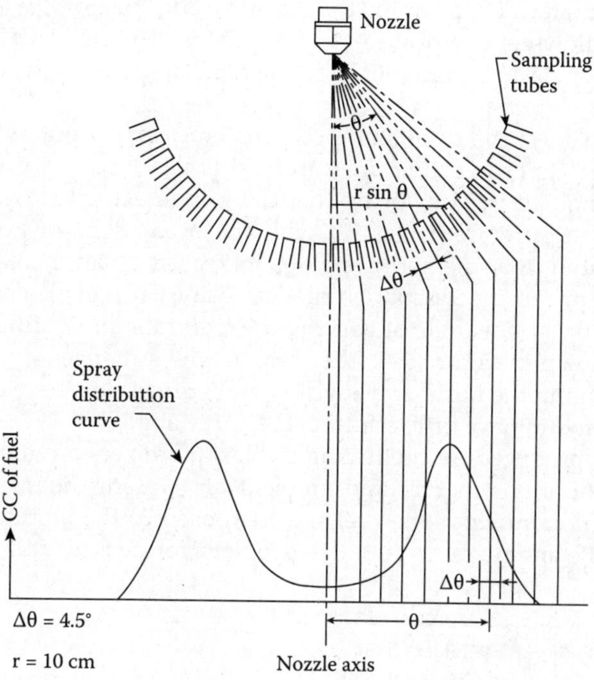

FIGURE 6.39
Measurement of radial fuel distribution.

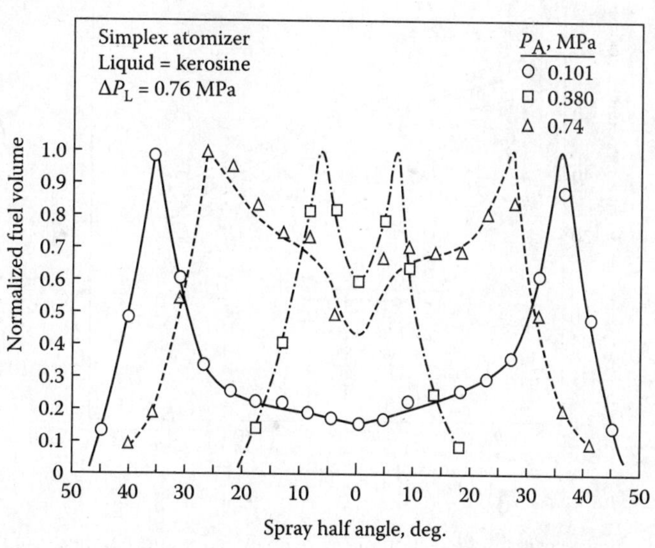

FIGURE 6.40
Influence of ambient air pressure on radial fuel distribution. (From Ortman, J., and Lefebvre, A.H., *Journal of Propulsion and Power*, 1(1), 11–15, 1985. With permission.)

of mass of a material system for the left (or right) lobe of the distribution curve. For hollow-cone, pressure-swirl atomizers, the effective spray angle tends to be from 5° to 15° smaller than the normal spray angle, as observed visually.

Figure 6.41 shows the results of measurements carried out by Chen et al. [97] on the effects of injection pressure and liquid viscosity on equivalent spray cone angle. It is noteworthy that the trends exhibited by the curves drawn in this figure are consistent with the predictions of Equation 6.56 in demonstrating that the spray angle is increased by an increase in injection pressure and/or a reduction in fuel viscosity. Chen et al. also examined the effect on the spray angle of varying the l_o/d_o ratio of the final discharge orifice. As shown in Figure 6.42, the spray angle is reduced by an increase in l_o/d_o. Changing the number of swirl-chamber feed slots between one and three was found to have little effect on the spray angle.

The sprays produced by prefilming airblast atomizers are much less susceptible to variations in fuel and ambient air pressure than sprays from pressure-swirl atomizers. The fuel sheet exposed to the atomizing air has relatively little momentum and the droplets formed in atomization are

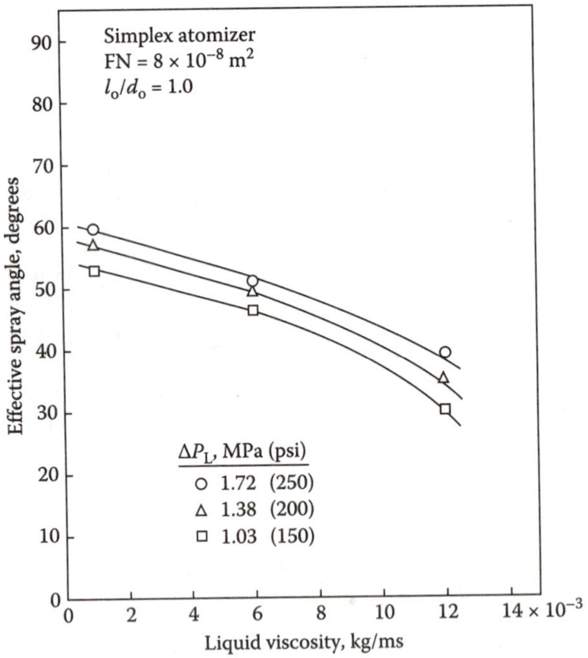

FIGURE 6.41
Influence of liquid viscosity and injection pressure on effective spray angle. (From Chen, S.K., Lefebvre, A.H., and Rollbuhler, J., *Journal of Engineering for Gas Turbines and Power*, 114, 97–103, 1992. With permission.)

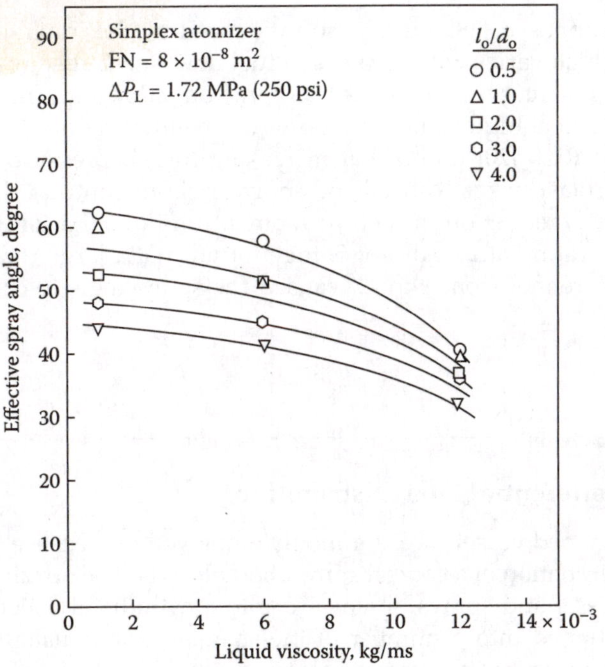

FIGURE 6.42
Influence of discharge orifice length/diameter ratio on effective spray angle. (From Chen, S.K., Lefebvre, A.H., and Rollbuhler, J., *Journal of Engineering for Gas Turbines and Power*, 114, 97–103, 1992. With permission.)

largely dependent on the kinetic energy of the atomizing air to transport them away from the nozzle. This means that droplet trajectories are governed mainly by the air movements created by air swirlers and other aerodynamic devices that form an integral part of the nozzle configuration. Thus, the spray structure of airblast atomizers is not overly sensitive to the physical properties of the fuel and the surrounding gaseous medium.

Recent work by Zheng et al. [73,74] on a counter-rotating, prefilming airblast atomizer flowing kerosine fuel has shown that the main factor governing the spray cone angle is the fuel/air momentum ratio. Increases in this parameter result in wider cone angles, which indicate a displacement of the fuel flux profile toward the outer boundaries of the spray. For the engine fuel nozzle used in this investigation, it was found that increasing the air pressure from 1 to 12 bar at a constant AFR caused the initial fuel spray angle to widen from 85° to 105°. No general conclusions should be drawn from these results because they may relate only to the type of nozzle tested. What they do demonstrate is that the radial distribution of fuel droplets throughout the spray volume is not solely dictated by the atomizing airflow pattern, as hitherto supposed [58], but is also influenced by the initial angle of the fuel sheet, which varies with changes in fuel flow rate and engine operating conditions. This intervention of the fuel spray on radial patternation is especially significant at high values of fuel/air momentum ratio.

Custer and Rizk [20] found that increasing the air pressure differential across an airblast nozzle caused the spray angle to contract slightly (presumably due to reduction in fuel/air momentum ratio). For one prefilming atomizer, operating at an AFR of 5, they found that a large increase in air pressure differential from 1 to 4% caused the spray angle to contract from 100° to 80°.

6.24 Circumferential Fuel Distribution

Circular-sectioned vessels are commonly employed to measure the circumferential patternation of a conical spray about its axis. The nozzle is centered above the vessel and sprays downward into a cylindrical collection vessel that is partitioned into a number of pie-shaped sectors, usually 12 or 16. Each sector drains into a separate sampling tube. The duration of each test is determined as the time required for one of the sampling tubes to become nearly full. After the level of fuel in each tube is measured and recorded, the values are averaged to get a mean height. The levels of the tubes are normalized against the mean, and the standard deviation of the normalized values is calculated. The normalized standard deviation is indicative of the circumferential irregularity of the nozzle spray. More sophisticated spray patternators have been devised that are capable of high-resolution measurements of the mass flux distributions produced by gas turbine fuel injectors [98,99]. These methods are not necessarily more accurate than the simple patternators described above, but they do allow a large amount of data to be collected in a relatively short time.

6.24.1 Pressure-Swirl Atomizers

Very little information is available in the literature on the circumferential patternation of sprays produced by pressure-swirl atomizers. Chen et al. [100] used several different pressure-swirl nozzles to examine the effects of variations in liquid properties, operating conditions, and atomizer design features, on spray patternation. Figure 6.43 shows the effects of liquid viscosity and nozzle injection pressure differential on spray uniformity. In this figure, the circumferential maldistribution is expressed in terms of a standard deviation, σ. If the circumferential distribution of liquid within the spray were completely uniform, the value of σ would be zero. Figure 6.43 shows that σ declines, i.e., patternation improves, with increases in injection pressure

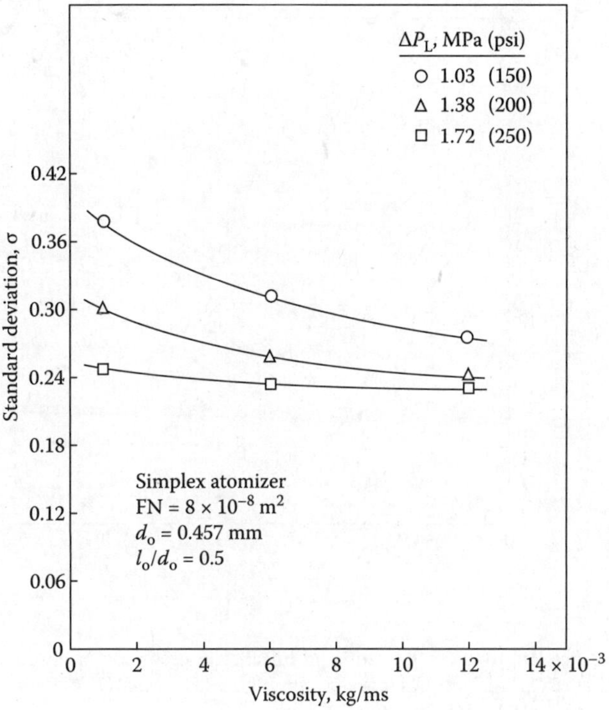

FIGURE 6.43
Influence of liquid viscosity and injection pressure on circumferential patternation. (From Chen, S.K., Lefebvre, A.H., and Rollbuhler, J.R., *Journal of Engineering for Gas Turbines and Power*, 115, 447–52, 1993. With permission.)

and viscosity, although the influence of the latter declines with the increase in injection pressure. The generally beneficial effect of viscosity in promoting spray uniformity is attributed to its influence on film thickness in the discharge orifice (see Equation 6.50). By thickening this film, an increase in viscosity makes the flow less susceptible to surface imperfections in the final discharge orifice. Further evidence showing that patternation is improved by an increase in injection pressure is contained in Figure 6.44. Presumably, this improvement is due to higher injection pressures promoting more turbulence and better mixing in the swirl chamber.

Of special interest in Figure 6.44 are the results obtained for different values of length/diameter ratio of the final discharge orifice. This figure shows that the circumferential distribution is most uniform for a value of l_o/d_o of 2. An optimum value of around 2 was found to apply at all operating conditions and to all liquids tested, some of which varied in viscosity by a factor of 12. This result is of considerable practical interest because for many years the trend in pressure-swirl atomizer design has been toward lower values of l_o/d_o in order to reduce internal losses, thereby improving atomization, as illustrated in Figure 6.45 [101]. Most current designs have values of l_o/d_o

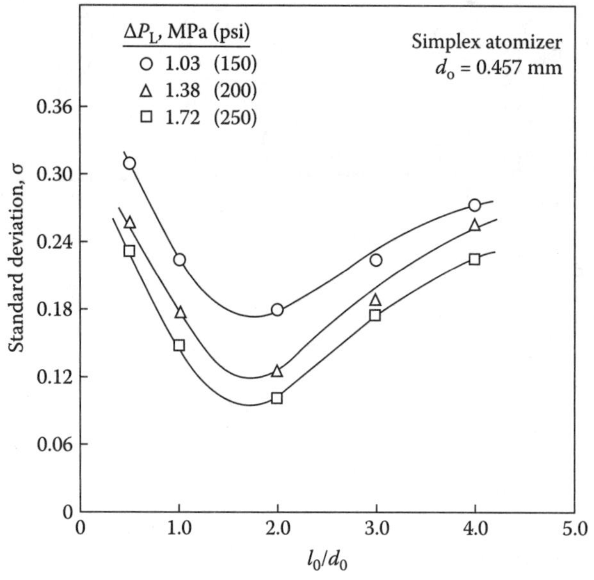

FIGURE 6.44
Influence of orifice length/diameter ratio and injection pressure on circumferential patternation. (From Chen, S.K., Lefebvre, A.H., and Rollbuhler, J.R., *Journal of Engineering for Gas Turbines Power*, 115, 447–52, 1993. With permission.)

of 0.5 or less. Inspection of Figures 6.44 and 6.45 confirms the importance of l_o/d_o to the performance of pressure-swirl atomizers. Reducing l_o/d_o below 2 improves atomization quality, but worsens the circumferential patternation.

Another factor that is known to influence the circumferential uniformity of the spray patterns produced by pressure-swirl atomizers is the degree of eccentricity between the swirl chamber and the final discharge orifice [102]. Manufacturing quality is also important, and spray patternation may be impaired by poor surface finish, orifice imperfections, plugged or contaminated flow passages, eccentric alignment of key nozzle components, and other manufacturing defects. Chen et al. [100] found that reducing the number of feed slots from three to two had only a slight adverse effect, but further reduction down to one slot produced a marked deterioration in circumferential patternation.

6.24.2 Airblast Atomizers

In order to determine the relative importance of various geometrical features on the velocity profiles produced downstream of an airblast atomizer, Rosfjord and Eckerle [103] measured velocity and turbulence levels downstream of eight variations of the same basic nozzle design. These variations included misaligned swirlers, changes to the number of vanes in a swirler,

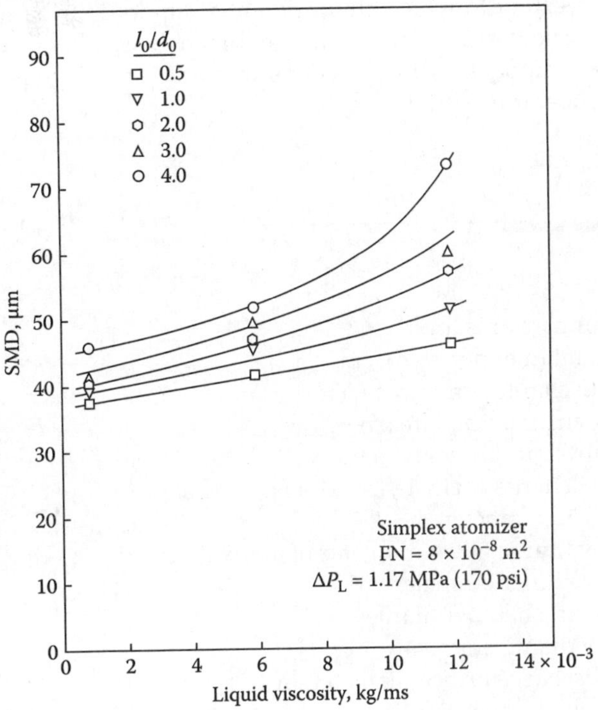

FIGURE 6.45
Influence of discharge orifice length/diameter ratio on mean drop size. (From Chen, S.K., Rollbuhler, J., and Lefebvre, A.H., *Atomization and Sprays*, 1(1), 1–22, 1991.)

and contouring of the trailing edge of swirl vanes. The results showed that significant variations in the airflow profile at the nozzle exit mix out rapidly to produce a uniform profile within three exit diameters downstream of the nozzle. It was also found that the swirler passages dominate in establishing the velocity and turbulence fields downstream of the nozzle. This means that upstream disturbances are not easily transmitted through the nozzle. These findings led Rosfjord and Eckerle to conclude that because airflow profiles are very axisymmetric, whereas fuel spray patterns are much less so, the observed nozzle patternation quality must be mainly dependent on the degree of uniformity of the fuel distribution at the nozzle exit. This aspect was investigated in a separate study by Rosfjord and Russel [104] using a nozzle that delivered swirling airflows on either side of an annular fuel sheet. Their results showed that small variations in the fuel annulus gap (< 0.05 mm) can severely compromise the circumferential uniformity of the ensuing spray. It was also noted that small imperfections in the prefilming surface could be detrimental to patternation. Point impressions of 0.2 mm (0.008 in) depth noticeably degraded the fuel profile.

The observations of Rosfjord et al., along with those of other workers (e.g., Wang et al. [105,106]), have highlighted the need for more information on the manner and extent to which circumferential liquid distributions are influenced by nozzle design features, nozzle dimensions, eccentric alignment of key nozzle components, imperfections in surface finish, liquid properties, and nozzle operating conditions.

Nomenclature

A_a	air core area, m^2
A_P	total inlet ports area, m^2
A_o	discharge orifice area, m^2
A_s	swirl chamber area, m^2
ALR	air/liquid ratio by mass
C_D	discharge coefficient or drag coefficient
D	drop diameter, m
D_h	hydraulic mean diameter of air exit duct, m
D_P	prefilmer diameter, m
D_s	swirl chamber diameter, m
d	jet diameter, m
d_o	discharge orifice diameter, m
FN	flow number, m^2
K	atomizer constant (= A_p/d_oD_s)
L_c	characteristic dimension of airblast atomizer, m
$\dot{m}$	flow rate, kg/s
MMD	mass median diameter, m
Oh	Ohnesorge number
P	total pressure, Pa
ΔP	pressure differential across nozzle, Pa
Q	volumetric flow rate, m^3/s
q	Rosin–Rammler drop-size distribution parameter
Re	Reynolds number
SMD	Sauter mean diameter, m
t	film thickness in final orifice, m
t_s	sheet thickness at nozzle exit, m
U	velocity, m/s
VMD	volume median diameter, m
We	Weber number
X	A_a/A_o
λ	wavelength
β	dynamic viscosity, kg/ms
θ	maximum spray cone half-angle, degrees
θ_m	mean spray cone half-angle, degrees
υ	kinematic viscosity, m^2/s

ρ	density, kg/m^3
σ	standard deviation, or surface tension, kg/s^2

Subscripts

A	air
F	fuel
L	liquid
R	air relative to liquid

References

1. Hinze, J. O., "Fundamentals of the Hydrodynamic Mechanism of Splitting in Dispersion Processes," *AIChE Journal*, Vol. 1, No. 3, pp. 289–95, 1995.
2. Brodkey, R. O., *The Phenomena of Fluid Motions*, Addison-Wesley, Reading, MA, 1967.
3. Sevik, M., and Park, S. H., "The Splitting of Drops and Bubbles by Turbulent Fluid Flow," *Journal of Fluids Engineering*, Vol. 95, pp. 53–60, 1973.
4. Rayleigh, L., "On the Instability of Jets," *Proceedings of London Mathematical Society*, Vol. 10, pp. 4–13, 1878.
5. Weber, C., "Disintegration of Liquid Jets," *Zeitschrift für Angewandte Mathematik und Mechanik*, Vol. 11, No. 2, pp. 136–59, 1931.
6. Castleman, R. A., "The Mechanism of the Atomization Accompanying Solid Injection," NACA Report 440, 1932.
7. Taylor, J. J., and Hoyt, J. W., "Water Jet Photography—Techniques and Methods," *Journal of Experimental Fluids*, Vol. 1, pp. 113–20, 1983.
8. Fraser, R. P., Eisenklam, P., Dombrowski, N., and Hasson, D., "Drop Formation from Rapidly Moving Sheets," *AIChE Journal*, Vol. 8, No. 5, pp. 672–80, 1962.
9. Lefebvre, A. H., "Energy Considerations in Twin-Fluid Atomization," *Journal of Engineering for Gas Turbines and Power*, Vol. 114, No. 1, pp. 89–96, 1992.
10. Beck, J. E., Lefebvre, A. H., and Koblish, T. R., "Airblast Atomization at Conditions of Low Air Velocity," *Journal of Propulsion and Power*, Vol. 7, No. 2, pp. 207–12, 1991.
11. Mugele, R., and Evans, H. D., "Droplet Size Distributions in Sprays," *Journal of Industrial and Engineering Chemistry*, Vol. 43, No. 6, pp. 1317–24, 1951.
12. Marshall, W. R., Jr., (ed.) "Mathematical Representation of Drop-Size Distributions of Sprays," *Atomization and Spray Drying*, Chapter VI, *Chemical Engineering Progress Monograph Series*, Vol. 50, No. 2, American Institute of Chemical Engineers, Pittsburgh, PA, pp. 215–48, 1954.
13. Miesse, C. C., and Putnam, A. A., "Mathematical Expressions for Drop-Size Distributions," *Injection and Combustion of Liquid Fuels*, Section II, WADC Technical Report 56-344, Battelle Memorial Institute, Colombus, OH, March 1957.
14. Bhatia, J. C., Domnick, J., Durst, F., and Tropea, C., "Phase-Doppler Anemometry and the Log-Hyperbolic Distribution Applied to Liquid Sprays," *Particle System Characterization*, Vol. 5, pp. 153–64, 1988.

15. Ahmadi, M., and Sellens, R. W., "A Simplified Maximum-Entropy-Based Drop Size Distribution," *Atomization and Sprays*, Vol. 3, No. 3, pp. 291–310, 1993.
16. Rosin, P., and Rammler, E., "The Laws Governing the Fineness of Powdered Coal," *Journal of the Institute of Fuel*, Vol. 7, No. 31, pp. 29–36, 1933.
17. Rizk, N. K., and Lefebvre, A. H., "Drop-size Distribution Characteristics of Spill-Return Atomizers," *Journal of Propulsion and Power*, Vol. 1, No. 1, pp. 16–22, 1985.
18. Rizk, N. K., "Spray Characteristics of the LHX Nozzle," Allison Gas Turbine Engines Report Nos. AR 0300-90 and AR 0300-91, 1984.
19. Han, Z., Parrish, S., Farrell, P. V., and Reitz, R. D., "Modeling Atomization Processes of Pressure-Swirl Hollow-Cone Fuel Sprays," *Atomization and Sprays*, Vol. 7, No. 6, pp. 663–84, 1997.
20. Custer, J. R., and Rizk, N. K., "Effect of Design Concept and Liquid Properties on Fuel Injector Performance," *Journal of Propulsion and Power*, Vol. 4, No. 4, pp. 378–84, 1988.
21. Rizk, N. K., and Mongia, H. C., "Calculation Approach Validation for Airblast Atomizers," *Journal of Engineering for Gas Turbines and Power*, Vol. 114, pp. 386–94, 1992.
22. Rizk, N. K., and Mongia, H. C., "Performance of Hybrid Airblast Atomizers under Low Power Conditions," AIAA Paper 92-0463, 1992.
23. Chin, J. S., and Lefebvre, A. H., "Some Comments on the Characterization of Drop-Size Distributions in Sprays," *International Journal of Turbo and Jet Engines*, Vol. 3, No. 4, pp. 293–300, 1986.
24. Sellens, R. W., and Brustowski, T. A., "A Prediction of the Drop Size Distribution in a Spray from First Principles," *Atomization and Spray Technology*, Vol. 1, pp. 85–102, 1985.
25. Li, X., and Tankin, R. S., "Derivation of Droplet Size Distribution in Sprays by using Information Theory," *Combustion Science and Technology*, Vol. 60, pp. 345–57, 1988.
26. Xu, T-H., Durst, F., and Tropea, C., "The Three-Parameter Log-Hyperbolic Distribution and Its Application to Particle Sizing," *Atomization and Sprays*, Vol. 3, No. 1, pp. 109–24, 1993.
27. Cousin, J., Yoon, S. J., and Dumouchel, C., "Coupling of Classical Linear Theory and Maximum Entropy Formalism for Prediction of Drop Size Distribution in Sprays: Application to Pressure-Swirl Atomizers," *Atomization and Sprays*, Vol. 6, No. 5, pp. 601–22, 1996.
28. Mullinger, P. J., and Chigier, N. A., "The Design and Performance of Internal Mixing Multi-Jet Twin-Fluid Atomizers," *Journal of the Institute of Fuel*, Vol. 47, pp. 251–61, 1974.
29. Nukiyama, S., and Tanasawa, Y., "Experiments on the Atomization of Liquids in an Airstream," *Transactions of the Society of Mechanical Engineers, Japan*, Vol. 5, pp. 68–75, 1939.
30. Chin, J. S., Rizk, N. K., and Razdan, M. K., "Effect of Inner and Outer Air Flow Characteristics on High Liquid Pressure Prefilming Airblast Atomization," XIII ISABE Conference, Paper 11, 1997.
31. Chin, J. S., Rizk, N. K., and Razdan, M. K., "Experimental Investigation of Hybrid Airblast Atomizer," ASME Paper 96-GT-464, 1996.
32. Rizk, N. K., Chin, J. S., and Razdan, M. K., "Influence of Design Configuration on Hybrid Atomizer Performance," AIAA Paper 96-2628, 1996.

33. Benjamin, M. A., McDonell, V. G., and Samuelsen, G. S., "Effect of Fuel/Air Ratio on Air Blast Simplex Nozzle Performance," ASME Paper 97-GT-150, 1997.
34. Suyari, M., and Lefebvre, A. H., "Drop-Size Measurements in Air-Assist Swirl Atomizer Sprays," Paper presented at Central States Combustion Institute Spring Meeting, NASA-Lewis Research Center, Cleveland, OH, 1986.
35. Maier, G., Willmann, M., and Wittig, S., "Performance of Air-Assisted Pressure Swirl Atomizer," Paper presented at 13th Annual Conference on Liquid Atomization and Spray Systems, Florence, Italy, July 1997.
36. Lefebvre, A. H., Wang, X. F, and Martin, C. A., "Spray Characteristics of Aerated-Liquid Pressure Atomizers," *Journal of Propulsion and Power*, Vol. 4, No. 4, pp. 293–8, 1988.
37. Lefebvre, A. H., "A Novel Method of Atomization With Potential Gas Turbine Applications," *Indian Defence Science Journal*, Vol. 38, No. 4, pp. 353–62, 1988.
38. Roesler, T. C., and Lefebvre, A. H., "Studies on Aerated-Liquid Atomization," *International Journal of Turbo and Jet-Engines*, Vol. 6, Nos. 3 and 4, pp. 221–9, 1989.
39. Chen, S. K., Lefebvre, A. H., and Rollbuhler, J. R., "Influence of Ambient Air Pressure on Effervescent Atomization," *Journal of Propulsion and Power*, Vol. 9, No. 1, pp. 10–15, 1993.
40. Whitlow, J. D., and Lefebvre, A. H., "Effervescent Atomizer Operation and Spray Characteristics," *Atomization and Sprays*, Vol. 3, No. 2, pp. 137–56, 1993.
41. Whitlow, J. D., Lefebvre, A. H., and Rollbuhler, J. R., "Experimental Studies on Effervescent Atomizers with Wide Spray Angles," *Fuels and Combustion Technology for Advanced Aircraft Engines*, AGARD Conference Proceedings, Vol. 536, Paper No. 38, 1993.
42. Chin, J. S., and Lefebvre, A. H., "A Design Procedure for Effervescent Atomizers," *Journal of Engineering for Gas Turbines and Power*, Vol. 117, No. 2, pp. 266–71, 1995.
43. Chen, S. K., and Lefebvre, A. H., "Spray Cone Angles of Effervescent Atomizers," Paper presented at ILASS-Americas 6th Annual Conference on Liquid Atomization and Spray Systems, Worcester, MA, 1993.
44. Li, J., Lefebvre, A. H., and Rollbuhler, J. R., "Effervescent Atomizers for Small Gas Turbines," ASME Paper 94-GT-495, 1994.
45. Parnell, E. C., and Williams, M. R., "A Survey of Annular Vaporizing Combustion Chambers," in E. R. Norster, ed., *Combustion and Heat Transfer in Gas Turbine Systems*, Cranfield International Symposium Series, Vol. 11, pp. 91–104, Pergamon, London, 1971.
46. Low, H. C., "Recent Research on the Efflux of the Rolls-Royce Vaporizer Fuel Injector," AGARD Conference Proceedings, Vol. 353, Paper 11, 1984.
47. Sotheran, A., "The Rolls Royce Annular Vaporizer Combustor," *Journal of Engineering for Gas Turbines and Power*, Vol. 88, pp. 106–114, 1984.
48. Jasuja, A. K., and Low, H. C., "Spray Performance of a Vaporizing Fuel Injector," AGARD Conference Proceedings, Vol. 422, Paper No. 9, 1987.
49. Chin, J. S., Lefebvre, A. H., and Sun, F. T-Y, "Temperature Effects on Fuel Thermal Stability," *Journal of Engineering for Gas Turbines and Power*, Vol. 114, pp. 353–8, 1992.
50. Chin, J. S., and Lefebvre, A. H., "Influence of Flow Conditions on Deposits From Heated Hydrocarbon Fuels," *Journal of Engineering for Gas Turbines and Power*, Vol. 115, pp. 433–38, 1993.

51. Nickolaus, D., and Lefebvre, A. H., "Fuel Thermal Stability Effects on Spray Characteristics," *Journal of Propulsion and Power*, Vol. 3, No. 6, pp. 502–7, 1987.
52. McCaldon, K., Prociw, L. A., and Sampath, P., "Design Aspects for Small Aircraft Gas Turbine Fuel Injectors," *Fuels and Combustion Technology for Advanced Aircraft Engines*, AGARD Conference Proceedings, Vol. 536, Paper No. 21, 1993.
53. Stickles, R. W., Dodds, W. J., Koblish, T. R., Sager, J., and Clouser, S., "Innovative High-Temperature Aircraft Engine Fuel Nozzle Design," *Journal of Engineering for Gas Turbines and Power*, Vol. 115, pp. 439–46, 1993.
54. Myers, G. D., Armstrong, J. P., White, C. D., Clouser, S., and Harvey, R. J., "Development of an Innovative High-Temperature Gas Turbine Fuel Nozzle," *Journal of Engineering for Gas Turbines and Power*, Vol. 114, pp. 401–8, 1992.
55. Winterfeld, G., Eickhoff, H. E., and Depooter, K., "Fuel Injectors," in A. M. Mellor, ed., *Design of Modern Gas Turbine Combustors*, pp. 323–9, Academic Press, San Diego, CA, 1990.
56. Ortman, J., and Lefebvre, A. H., "Fuel Distributions from Pressure-Swirl Atomizers," *Journal of Propulsion and Power*, Vol. 1, No. 1, pp. 11–15, 1985.
57. Wang, X. F., and Lefebvre, A. H., "Influence of Ambient Air Pressure on Pressure-Swirl Atomization," *Atomization and Spray Technology*, Vol. 3, pp. 209–26, 1987.
58. Lefebvre, A. H., *Atomization and Sprays*, Hemisphere, New York, 1989.
59. Elkotb, M. M., "Fuel Atomization for Spray Modeling," *Progress in Energy and Combustion Science*, Vol. 8, pp. 61–91, 1982.
60. Radcliffe, A., "Fuel Injection," in W. R. Hawthorne and W. T. Olson, eds., *High Speed Aerodynamics and Jet Propulsion*, XID, Princeton University Press, Princeton, NJ, 1960.
61. Jasuja, A. K., "Atomization of Crude and Residual Fuel Oils," *Journal of Engineering for Power*, Vol. 101, No. 2, pp. 250–58, 1979.
62. Lefebvre, A. H., "The Prediction of Sauter Mean Diameter for Simplex Pressure-Swirl Atomizers," *Atomization and Spray Technology*, Vol. 3, pp. 37–51, 1987.
63. Wang, X. F., and Lefebvre, A. H., "Mean Drop Sizes from Pressure-Swirl Nozzles," *Journal of Propulsion and Power*, Vol. 3, No. 1, pp. 11–18, 1987.
64. Rizk, N. K., and Lefebvre, A. H., "Spray Characteristics of Plain-Jet Airblast Atomizers," *Journal of Engineering for Gas Turbines and Power*, Vol. 106, pp. 639–44, 1984.
65. Lorenzetto, G. E., and Lefebvre, A. H., "Measurements of Drop Size on a Plain Jet Airblast Atomizer," *AIAA Journal*, Vol. 15, pp. 1006–10, 1977.
66. Jasuja, A. K., "Plain-Jet Airblast Atomization of Alternative Liquid Petroleum Fuels under High Ambient Air Pressure Conditions," ASME Paper 82-GT-32, 1982.
67. El-Shanawany, M. S. M. R., and Lefebvre, A. H., "Airblast Atomization: The Effect of Linear Scale on Mean Drop Size," *Journal of Energy*, Vol. 4, pp. 184–9, 1980.
68. Buckner, H. N., and Sokja, P. E., "Effervescent Atomization of High-Viscosity Fluids: Part I. Newtonian Liquids," *Atomization and Sprays*, Vol. 1, pp. 239–52, 1991.
69. Sattelmayer, T., and Wittig, S., "Internal Flow Effects in Prefilming Airblast Atomizers; Mechanisms of Atomization and Droplet Spectra," *Journal of Engineering for Gas Turbines and Power*, Vol. 108, pp. 465–72, 1986.
70. Goris, N. H., "Operational Characteristics and Energy Considerations in Pneumatic Atomizers," MSME thesis, Purdue University, 1990.

71. Fraser, R. P., Dombrowski, N., and Routley, J. H., "The Production of Uniform Liquid Sheets from Spinning Cups; The Filming of Liquids by Spinning Cups; The Atomization of a Liquid Sheet by an Impinging Air Stream," *Journal of Chemical Engineering Science*, Vol. 18, pp. 315–21, 323–37, 339–53, 1963.
72. Bryan, R., Godbole, P. S., and Norster, E. R., "Characteristics of Airblast Atomizers," in E. R. Norster, ed., *Combustion and Heat Transfer in Gas Turbine Systems*, Cranfield International Symposium Series, Vol. 11, pp. 343–59, Pergamon, London, 1971.
73. Zheng, Q. P., Jasuja, A. K., and Lefebvre, A. H., "Influence of Air and Fuel Flows on Gas Turbine Sprays at High Pressures," *Twenty-Sixth Symposium (International) on Combustion*, pp. 2757–62, The Combustion Institute, Pittsburgh, PA, 1996.
74. Zheng, Q. P., Jasuja, A. K., and Lefebvre, A. H., "Structure of Airblast Sprays under High Ambient Pressure Conditions," *Journal of Engineering for Gas Turbines and Power*, Vol. 119, No. 3, pp. 512–18, 1997.
75. Lichtarowicz, A., Duggins, R. K., and Markland, E., "Discharge Coefficients for Incompressible Non-Cavitating Flow Through Long Orifices," *Journal of Mechanical Engineering Science*, Vol. 7, No. 2, pp. 210–19, 1965.
76. Bergwerk, W., "Flow Pattern in Diesel Spray Holes," *Proceedings of the Institution of Mechanical Engineers*, Vol. 173, No. 25, pp. 655–60, 1959.
77. Ohrn, T. R., Senser, D. W., and Lefebvre, A. H., "Geometrical Effects On Discharge Coefficients for Plain-Orifice Atomizers," *Atomization and Sprays*, Vol. 1, No. 2, pp. 137–54, 1991.
78. Ruiz, F., and Chigier, N., "The Mechanics of High-Speed Cavitation," *Proceedings of the Third International Conference on Liquid Atomization and Spray Systems*, pp. V1B/3/1–15, London, 1985.
79. Reitz, R. D., and Bracco, F. V, "Mechanism of Atomization of a Liquid Jet," *Physics of Fluids*, Vol. 25, No. 10, pp. 1730–42, 1982.
80. Zucrow, M. J., "Discharge Characteristics of Submerged Jets," Bull. No. 31, Engineering Experimental Station, Purdue University, West Lafayette, IN, 1928.
81. Spikes, R. H., and Pennington, G. A., "Discharge Coefficient of Small Submerged Orifices," *Proceedings of the Institution of Mechanical Engineers*, Vol. 173, No. 25, pp. 661–5, 1959.
82. Radcliffe, A., "The Performance of a Type of Swirl Atomizer," *Proceedings of the Institution of Mechanical Engineers*, Vol. 169, pp. 93–106, 1955.
83. Giffen, E., and Muraszew, A., *Atomization of Liquid Fuels*, Chapman and Hall, London, 1953.
84. Rizk, N. K., and Lefebvre, A. H., "Internal Flow Characteristics of Simplex Swirl Atomizers," *Journal of Propulsion and Power*, Vol. 1, No. 3, pp. 93–199, 1985.
85. Suyari, M., and Lefebvre, A. H., "Film Thickness Measurements in a Simplex Swirl Atomizer," *Journal of Propulsion and Power*, Vol. 2, pp. 528–33, 1986.
86. Simmons, H. C., and Harding, C. F, "Some Effects on Using Water as a Test Fluid in Fuel Nozzle Spray Analysis," ASME Paper 80-GT-90, American Society of Mechanical Engineers, New York, NY, 1980.
87. Sankaran Kutty, P., Narasimhan, M. V., and Narayanaswamy, K., "Design and Prediction of Discharge Rate, Cone Angle, and Air Core Diameter of Swirl Chamber Atomizers," *Proceedings of the 1st International Conference on Liquid Atomization and Spray Systems*, Tokyo, pp. 93–100, 1978.

88. Narasimhan, M. V., Sankaran Kutty, P., and Narayanaswamy, K., "Prediction of the Air Core Diameter in Swirl Chamber Atomizers," unpublished report, 1978.
89. Abramovich, G. N., *Theory of Turbulent Jets,* MIT Press, Cambridge, MA, 1963.
90. Reitz, R. D., and Bracco, F. V, "On the Dependence of Spray Angle and Other Spray Parameters on Nozzle Design and Operating Conditions," SAE Paper 790494, 1979.
91. Yokota, K., and Matsuoka, S., "An Experimental Study of Fuel Spray in a Diesel Engine," *Transactions of the Japanese Society of Mechanical Engineers*, Vol. 43, No. 373, pp. 3455–64, 1973.
92. Hiroyasu, H., and Arai, M., "Fuel Spray Penetration and Spray Angle in Diesel Engines," *Transactions of the Japanese Society of Mechanical Engineers*, Vol. 21, pp. 5–11, 1980.
93. Ohrn, T. R., Senser, D. W., and Lefebvre, A. H., "Geometrical Effects On Spray Cone Angle for Plain-Orifice Atomizers," *Atomization and Sprays*, Vol. 1, No. 3, pp. 253–68, 1991.
94. Taylor, G. I., "The Mechanics of Swirl Atomizers," *Seventh International Congress of Applied Mechanics*, Vol. 2, Pt. 1, pp. 280–5, 1948.
95. Rizk, N. K., and Lefebvre, A. H., "Prediction of Velocity Coefficient and Spray Cone Angle for Simplex Swirl Atomizers," *Proceedings of the Third International Conference on Liquid Atomization and Spray Systems,* London, 11 lC/2/1–16, 1985.
96. De Corso, S. M., and Kemeny, G. A., "Effect of Ambient and Fuel Pressure on Nozzle Spray Angle," *Transactions of the American Society of Mechanical Engineers*, Vol. 79, No. 3, pp. 607–15, 1957.
97. Chen, S. K., Lefebvre, A. H., and Rollbuhler, J., "Factors Influencing the Effective Spray Cone Angle of Pressure-Swirl Atomizers," *Journal of Engineering for Gas Turbines and Power*, Vol. 114, pp. 97–103, 1992.
98. McVey, J. B., Russell, S., and Kennedy, J. B., "High-Resolution Patternator for the Characterization of Fuel Sprays," *Journal of Propulsion and Power*, Vol. 3, No. 3, pp. 202–9, 1987.
99. McVey, J. B., Kennedy, J. B., and Russell, S., "Application of Advanced Diagnostics to Airblast Injector Flows," *Journal of Engineering for Gas Turbines and Power*, Vol. Ill, No. 1, pp. 53–62, 1989.
100. Chen, S. K., Lefebvre, A. H., and Rollbuhler, J., "Factors Influencing the Circumferential Liquid Distribution from Pressure-Swirl Atomizers," *Journal of Engineering for Gas Turbines and Power*, Vol. 115, pp. 447–52, 1993.
101. Chen, S. K., Rollbuhler, J., and Lefebvre, A. H., "Influence of Liquid Viscosity on Pressure-Swirl Atomizer Performance," *Atomization and Sprays*, Vol. 1, No. 1, pp. 1–22, 1991.
102. Tate, R. W., "Spray Patternation," *Journal of Industrial and Chemical Engineering,* Vol. 52, pp. 49–52, 1960.
103. Rosfjord, T. J., and Eckerle, W. A., "Aerating Fuel Nozzle Design Influences on Airflow Features," *Journal of Propulsion and Power*, Vol. 7, No. 6, pp. 849–56, 1991.
104. Rosfjord, T. J., and Russel, S., "Nozzle Design and Manufacturing Influences on Fuel Spray Circumferential Uniformity," *Journal of Propulsion and Power*, Vol. 5, No. 2, pp. 144–50, 1989.

105. Wang, H. Y., McDonell, V. G., and Samuelsen, G. S., "The Two-Phase Flow Downstream of a Production Engine Combustor Swirl Cup," *Twenty-Fourth Symposium (International) on Combustion*, pp. 1457–63, The Combustion Institute, Pittsburgh, PA, 1992.
106. Wang, H. Y., McDonell, V. G., and Samuelsen, G. S., "Influence of Hardware Design on the Flow Field Structures and the Patterns of Droplet Dispersion: Part I—Mean Quantities," *Journal of Engineering for Gas Turbines and Power*, Vol. 117, pp. 282–9, 1995.

7

Combustion Noise

7.1 Introduction

Combustion in a flowing turbulent fuel–air mixture is always accompanied by noise. In a gas turbine, the total noise resulting from normal combustion is often referred to as core noise. It comprises two components, (1) *direct* combustion noise that is generated solely by the combustion process itself, and (2) *indirect* combustion noise that is produced by the flow of hot combustion products through the turbine and exhaust nozzle [1]. In this chapter, we shall concentrate on direct combustion noise and also on the more insidious forms of noise that arise when combustion instabilities become coupled to acoustic modes in a combustion chamber.

Every combustor has natural frequencies that may be excited by the combustion process to produce oscillations, and the gas turbine combustor is no exception. The oscillations may be longitudinal, radial, or circumferential, or a combination of these modes. In the longitudinal, or so-called "organ pipe" mode, the resonant pressure waves are along the length of the combustor, whereas in the circumferential or "sloshing" mode, the waves are tangential to the main flow direction.

The topic of combustion instability has figured fairly prominently in the combustion literature, but usually in the context of rocket engines and jet-engine afterburners. Comparatively little has been published on acoustic oscillations in main combustors. Usually, the problem only arises during full-scale combustor testing, and sometimes instabilities do not show up until the combustor is fitted to an engine. Typically, the problem has been to eliminate or control an instability in a combustor at a comparatively late stage in its development, when relatively few changes can be made except at large expense and loss of time.

In recent years, the problem of sustained oscillating combustion, which gives rise to noise and engine vibrations, has become of increasing concern. The main reason for this is the continuing trend toward higher degrees of fuel–air premixing prior to combustion. The motivation is to further reduce pollutant emissions, in particular oxides of nitrogen, which can only be done by improving the mixedness of the fuel–air mixture entering the combustion zone. Unfortunately, developments in premixed combustion are generally accompanied by an increase in the occurrence of oscillating combustion. This has led to a resurgence of interest in all aspects of combustion noise, including its cause, the mechanisms involved in noise generation and suppression, methods of noise control, both passive and active, and computational models to provide guidance for developing control strategies. These topics form the basis of the material presented in this chapter.

7.2 Direct Combustion Noise

Direct combustion noise arises when a volume of gas expands at constant pressure as it is heated by combustion. The resulting expansion of the surrounding gas produces a sound wave that propagates outside the boundary of the flame. The pressure in the sound wave, and hence the sound intensity, depends on the rate of volume generation by the source. Such sources are described as acoustic monopoles. Since the size distribution of the eddies in a turbulent flow is governed by the statistical distribution of the turbulent mixing lengths, it follows that the sound generation in a highly turbulent flame is equivalent to a statistical distribution of monopole sources throughout the combustion zone [2,3].

Two parameters are of prime importance in the description of direct combustion noise. One is the *radiated sound power* and the other is the *thermoacoustic efficiency*, which is defined as the ratio of the radiated sound power to the heat released in combustion. Due to the destructive interference that occurs within the source region, the actual acoustic power emitted in gas turbine combustion tends to be only a minute fraction of the total thermal power [2,3].

The radiated sound power covers a broad spectrum of frequencies from around 100 to 2000 Hz, reaching a blunt peak between 300 and 500 Hz. This general shape of the sound power/frequency curve appears to be largely independent of combustor size, engine power, and flame temperature, although these factors strongly affect the radiated sound power level [1].

The most important single factor governing combustion noise is engine power. This is because both the mass flow rate through the combustor and the temperature level in the combustor increase with engine power. If the thermoacoustic efficiency remained constant, the radiated sound power would be proportional to the power developed by the engine. In practice, sound levels are higher than this simple relationship would predict because the thermoacoustic efficiency also increases with engine power [1].

7.2.1 Theory

In an early study by Bragg [2], the turbulent flame zone is assumed to comprise a region of uncorrelated flamelets that produce monopole-type sound on burning. His model shows that the radiated sound power should vary with the fuel reactivity and is proportional to the square of the mixture flow velocity. It also predicts a peak frequency in the region of 500 Hz and a thermoacoustic efficiency of around 10^{-6} for a typical hydrocarbon fuel.

The results obtained by Thomas and Williams [3] show that the normal burning velocity of a fuel has a substantial influence on both sound power and thermoacoustic efficiency. For example, it was found that increasing the burning velocity from 50 to 100 cm/s raised the thermoacoustic efficiency from 10^{-6} to 10^{-5}. These workers suggest that for practical flames, the thermoacoustic efficiency will be lower than these values and will probably lie in the range from 10^{-8} to 10^{-7}. This result would appear to have important implications for combustion noise in gas turbines.

As described by Ballal and Lefebvre [4,5], under conditions of high turbulence the combustion zone comprises a fairly thick region of burned gases interspersed with multitudinous small eddies of unburned mixture. Within each eddy, the burning velocity is greatly enhanced by the flow of heat and active species into the unburned mixture from the enveloping flame front. Sometimes, the acceleration of chemical reactions and flame speeds within an individual eddy may proceed to such an extent that combustion occurs almost instantaneously throughout its volume. It would, therefore, be a considerable oversimplification to regard the combustion process in a highly turbulent primary zone, where turbulent intensities can exceed 30%, as simply a collection of assorted flamelets in which the rate of expansion is determined primarily by the normal burning velocity of the unburned mixture. Thus, in gas turbine combustion, the thermoacoustic efficiency may be appreciably higher than the values indicated by Thomas and Williams. However, normal burning velocity may still provide a useful yardstick for the relative noise levels emitted by the turbulent combustion of different fuels.

In their review of theoretical work relating to direct combustion noise, Mahan and Karchmer [1] note that several parameters are common to nearly all theoretical developments: the total mass flow rate through the burner, the burner length and cross-sectional area, the fuel/air ratio, and some measure of the fuel reactivity. Although some theories present their findings in terms of burner pressure drop and/or temperature rise, these terms are directly relatable to the above parameters.

Among the most highly developed combustion noise theories is that of Strahle [6]. In its most practical form, this theory predicts that the sound power radiated by a can-type combustor is given by

$$S \alpha P_3 U_{\text{ref}}^2 T_3^{(-2\text{ to }-3)} q^2 N^{(0\text{ to }-1)} A_e^{1.5} L^{-1}, \tag{7.1}$$

where S is the sound power in watts, P_3 and T_3 are the combustor inlet pressure and temperature, respectively, U_{ref} is the combustor reference velocity, q is the fuel/air ratio, N is the number of fuel injectors, A_e is the combustor outlet area, and L is the combustor length. It is of interest to compare this theoretical prediction of sound power with the following equation in which the various exponents denote the experimental values obtained by Strahle and Muthukrishnan [7].

$$S = CP_3^{1.9} U_{ref}^{3.4} T_3^{-2.5} q^{1.3} N^{-0.78} A_e^{1.5} L^{-1}. \tag{7.2}$$

The constant C has the value of 0.047 if the various quantities in Equation 7.2 are expressed in S.I. units.

Comparison of Equations 7.1 and 7.2 shows reasonably good agreement between theory and experiment, except for P_3 and U_{ref}. The higher exponents for these two quantities in Equation 7.2 are attributed in Reference [7] to the intervention of jet noise in the experiments. Since jet noise varies as velocity to the eighth power, it is generally found that although theory tends to predict a velocity exponent of around 2, the experimentally determined values are usually close to 3 [8–11].

7.2.2 Core Noise Prediction Methods

The General Electric (GE) Company has used combustion noise theory, supplemented by engine data, to derive an equation that allows the overall sound power level (direct plus indirect combustion noise) to be calculated from a knowledge of just a few combustor and turbine operating parameters [12]. This equation was found to predict the core noise levels of a number of turboshaft, turbojet, and turbofan engines to an accuracy of within 5 dB. Of special interest in the GE equation is that it identifies combustion pressure and combustor temperature rise, along with mass flow rate, as the key factors governing direct combustion noise.

The Pratt and Whitney Company has made extensive use of combustion noise theory to develop a prediction method that yields both the overall sound power level and the peak frequency [13]. Since it employs more parameters than the GE expression, including several operating and geometrical variables, it has a potentially broader range of application. The method has been applied successfully to several engine types, including the JT8, JT8D, and JT9D. The data presented in Reference [13], which cover wide ranges of operating conditions for these engines, show close agreement between the predicted and measured sound levels, with a standard deviation of 1.7 dB.

A full description of the GE and P&W noise prediction methods is beyond the scope of this chapter. For further information, reference should be made to the original publications [12,13] and to the review article on core noise by Mahan and Karchmer [1].

7.3 Combustion Instabilities

Combustion instabilities are usually the result of interactions between the combustion process and the acoustic fields within the combustor. Some combustion instabilities are caused by fluctuations in the air supply to the combustor, others by aerodynamic disturbances created within the combustor itself, and the remainder are due either to variations in the fuel supply to the nozzles or to maldistributions of fuel in the combustion zone, which give rise to a cycle of extinction and reignition in localized regions of the flame.

If, for any reason, the combustion heat release process is periodic, the resulting acoustic pressure waves emanating from the combustion zone are periodic with the same frequency. The presence of a combustor liner causes these pressure waves to be returned to the combustion zone with a time delay that depends on the chamber shape and size and the average speed of sound in the combustor. Energy is added to the pressure waves at any frequency for which the instantaneous peak in acoustic pressure in the combustion zone coincides with the instantaneous peak in heat release. This coupling between the combustion process and the acoustic field causes energy to be added to the system with each cycle, which results in oscillations that grow in amplitude until dissipative viscous losses arrest further growth. This situation conforms to the well-known Rayleigh criterion for combustion oscillations [14]. In a similar manner, energy is removed from a pressure wave at any frequency for which the wave is 180° out of phase with the periodic heat release in the flame zone [1].

7.3.1 Descriptions of Acoustic Oscillations

Various terms are used to describe the sound emitted from a combustor when combustion instabilities are present. There are no general guidelines laid down; for example, in some engine companies, the term "rumble" is used to describe all audible acoustic oscillations, regardless of their frequency, whereas in others, the description "rumble" or "growl" is reserved for the noise emitted in the low frequency range from 50 to 180 Hz, which usually occurs at sub-idle conditions. For the higher frequencies associated with engine speeds around and above idle, the term "howl" or "humming" is generally considered to be more appropriate.

7.3.1.1 Growl

The characteristics of growl vary from engine to engine and between engine startups, but its onset may occur soon after ignition is accomplished and it may persist at engine speeds up to idle. Growl is considered to be undesirable because it lengthens engine startup time and reduces compressor stall margins. According to Seto [15], some compressors are quite tolerant to growl, whereas others appear to have growl-related stall problems.

Certain engine operating parameters affect growl. An increase in combustor inlet air temperature decreases the speed range and intensity of growl, whereas an increase in combustion pressure has the opposite effect of promoting growl [15].

Methods of alleviating growl include improvements in primary-zone flow patterns and structure, and modifications to the fuel-injection system to raise the fuel delivery pressure. As growl is normally most prevalent when the primary-zone fuel/air ratio is near the weak extinction limit, any change to the acceleration schedule that raises the fuel/air ratio in the primary zone, or any change in fuel spray characteristics that lowers the lean blowout limit, such as a reduction in spray cone angle, will tend to suppress growl.

7.3.1.2 Howl

The phenomenon of howl or humming is closely related to growl, but it occurs at higher engine speeds. Its frequency is usually in the range from 200 to 500 Hz. As with growl, its intensity is dependent on ambient air temperature and falls off rapidly as engine inlet air temperatures rise above normal atmospheric values (around 288 K). It is sensitive to fuel type and diminishes in severity with an increase in fuel volatility.

With growl, engine compressor instabilities play an important role and may even be the trigger for growl, but with howl, the compressor is much less aerodynamically involved. The primary cause of howl appears to be from fuel pressure perturbations. Isolating the feedback mechanism tends to eliminate howl.

7.3.2 Characteristic Times

The heat release in a gas turbine combustor does not take place immediately. The fuel–air mixture travels at least part way around the recirculation zone in the dome region of the combustor before releasing the major portion of its heat [16]. The time that elapses between the injection of fuel and the region of maximum heat release represents the characteristic combustion time, which is obtained as the sum of the characteristic times for fuel evaporation, mixing of fuel vapor with air and combustion products to reach a critical reaction temperature, and chemical reaction.

$$t_{\text{combustion}} = t_{\text{mixing}} + t_{\text{evaporation}} + t_{\text{reaction}}. \tag{7.3}$$

The system becomes prone to instabilities when the overall combustion time becomes equal to a characteristic acoustic time of the combustor.

Very detailed studies would be required to quantify time scales in Equation 7.3 for specific gas turbine applications but, even without this information,

Equation 7.3 does allow some qualitative assessments to be made of the impact on combustion oscillations of changes in various relevant fuel and combustion parameters.

7.3.3 Influence of Fuel Type

The arguments presented above suggest that a change in fuel type or composition may either increase or reduce noise amplitudes, depending on whether the resulting change in chemical reaction time moves the overall combustion time closer or further away from the characteristic acoustic time. Tests carried out by Janus et al. [17] on a subscale combustor burning varying proportions of natural gas, propane, and hydrogen showed that instability regimes are markedly affected by changes in fuel composition. Keller et al. [18] also observed changes in both oscillation frequency and amplitude with change in gaseous fuel composition. As mixing times are largely unaffected by variations in fuel chemistry, these findings are clearly due to the change in chemical reaction time resulting from the change in fuel composition.

With liquid fuels, the influence of evaporation time must also be considered in addition to reaction time. The amount of available information on this effect is not large, but Vandsburger et al. [19] observed that the oscillating behavior of step-stabilized flames was appreciably different for various combinations of spray vs. gaseous combustion, whereas Mehta et al. [20] have demonstrated the influence of fuel volatility on oscillating combustion in aero engines. If the difference in evaporation time between one fuel and another is significant (say, more than half the acoustic period), then the overall combustion time of the two fuels will also be quite different. In consequence, with one fuel the coupling between the combustion and acoustic fields may be weak or nonexistent, resulting in relatively noise-free combustion, whereas the other fuel may produce strong coupling and large acoustic amplitudes.

7.3.4 Influence of Combustor Operating Conditions

Janus et al. [17] have conducted a number of experiments to determine the influence of inlet air temperature on combustion stability characteristics. Their tests were conducted on a subscale combustor burning a variety of gaseous fuels. The results of this study are presented in the form of stability maps. Examination of these maps indicates that the combustor is highly unstable over a wide operating range at 273 K inlet temperature, but relatively stable over the same operating range at 394 K inlet temperature. This result can readily be explained by reference to Equation 7.3. An increase in inlet air temperature decreases the chemical reaction portion of the total combustion time. The new combustion time represents a condition at which acoustic loss exceeds acoustic gain. It is of interest to note on the stability maps presented by Janus et al. [17] that the instability region has not disappeared. It has simply moved to a new location on the map where one or more components of the total combustion time have increased enough to offset the decrease in reaction time caused by the increase in inlet air temperature.

These considerations suggest that an increase in combustion pressure, which also enhances chemical reaction rates, should again serve to alleviate acoustic oscillations. However, this effect tends to be more than offset by the corresponding increase in combustion energy that sustains the oscillations (note that for a constant fuel/air ratio, the combustor heat release rate is directly proportional to pressure), so the net result is that combustion noise usually increases with an increase in pressure [15].

From the above discussion, it would appear that any change that reduces the overall combustion time, such as an increase in fuel volatility that reduces the evaporation time, or an increase in inlet air temperature that reduces the reaction time, will tend to alleviate instabilities, and more often than not this is the case. However, this should not be regarded as a general result, and many exceptions have been observed in laboratory tests and on engine hardware. Much depends on the frequency of the oscillation. For the exceptionally high frequencies associated with small combustors, an increase in inlet air temperature tends to promote instabilities. As the key factor is the ratio of the overall combustion time to the relevant acoustic time, it is hard to generalize the effect of inlet air temperature or, in fact, any other operational parameter on combustion oscillations. Any change in the characteristic combustion time could possibly move the combustor from a stable to an unstable region, and vice versa [21].

7.3.5 Influence of Ambient Conditions

The laboratory tests carried out by Janus et al. [17] generally confirmed the beneficial effect of an increase in ambient air temperature in suppressing noise and also showed that increased ambient humidity decreases combustor pressure oscillations. This latter effect was attributed to the additional heat capacity of the water molecules, which lowers the peak flame temperature, thereby reducing the reaction rate. However, the main conclusion to be drawn from their investigation is that gas turbine manufacturers and users need to be aware of the effects of ambient conditions (and fuel composition) on combustion oscillations. According to these workers, significant variations in combustor stability performance could occur due to changes in geographic climate and seasonal weather conditions.

7.3.6 Aerodynamic Instabilities

Experimental and numerical studies on combustor aerodynamics have revealed that the shear layers created by counter-rotating swirling airflows and the breakup regions of air jets are characterized by the presence of coherent large-scale structures. These structures play an important role in promoting the high mixing rates that are an essential prerequisite for high volumetric heat release rates. However, they can also give rise to unsteady heat release that could, in turn, induce combustion instability. An important factor in this instability mechanism is the time delay between the formation of a coherent vortex structure and the instant that energy is released due to combustion in the vortex. This delay may provide the proper phase relationship between the oscillating pressure field and unsteady heat release to drive the instability [22].

Another, and more prevalent cause of pressure oscillations in gas turbine combustors, is when the inherent aerodynamic stability of the combustor is too low. The primary-zone airflow pattern is of major importance to both flame and aerodynamic stability. Many different types of airflow patterns are employed in gas turbine combustors, but one feature common to all is the creation at the upstream end of the liner of a toroidal flow reversal that entrains and recirculates a portion of the hot combustion products to mix with the incoming air and fuel. These vortices are continually replenished by air flowing through the dome swirler and holes pierced in the liner walls. Additional air is supplied through flare-cooling slots and from air employed in airblast atomization. A satisfactory airflow pattern can only be achieved by good matching of these various modes of air admission. In particular, it is important to ensure that the large-scale flow reversals induced by the swirler air and the primary air jets merge and blend in such a manner that each one complements the other to produce an aerodynamically strong and stable recirculation zone. Any significant mismatch between these two main sources of primary air can lead to the creation of localized regions within the primary zone in which the flow is sluggish and disorganized. Rates of combustion and heat release in such regions are highly susceptible to perturbations in flow and pressure.

Even when a strong and stable recirculation zone has been achieved, imperfections in fuel spray patternation and/or airflow distribution can lead to the formation of local regions in which the fuel/air ratio is appreciably lower than the average value. Changes in engine power setting may cause flame extinction to occur in these regions, but the combustor continues to function because elsewhere in the primary zone the fuel/air ratio lies inside the normal stability limits. However, cessation of combustion in part of the primary zone may cause the recirculating airflow to redistribute itself in such a way that mixture strengths in these local regions now fall within the normal burning range and reignition occurs. Thus, the conditions for local flame extinctions are restored, and a cycle of flame extinction and reignition in local pockets of mixture is established. If the pressure pulses generated by this sequence of events become coupled with the combustor's acoustic field, the oscillations are strengthened and sustained.

An example that illustrates the importance of a stable primary airflow pattern to the attainment of noise-free combustion has been provided by Scalzo et al. [23]. The field conversion of two Westinghouse 104 MW W50 1D5 gas turbines to burn medium Btu synthetic fuel gas resulted in excessive 100 Hz airborne sound and unacceptable engine vibration when burning natural gas with steam injection. The combustors fitted in these engines contain no air swirlers and the basic recirculation flow pattern is established using primary air scoops. In the original design, the fuel and steam injection holes were drilled at angles that promoted good mixing with the primary air, but did not directly oppose the recirculating flow. As a result of the modification to accommodate the syngas fuel, the angles of natural gas and steam injection were both reduced, in one case to directly oppose the recirculating air pattern, and in the other, to provide more opposition to the recirculating airflow. This led to a disruption of the primary recirculation flow pattern and a dramatic increase in combustion noise.

A number of tests were carried out that demonstrated the importance of the recirculating primary scoop flow and fuel gas momentum vectors to combustion noise. Increasing the included angle of the natural gas injection holes by 40°, and blanking off the central gas injection hole, caused the fuel and steam momentum vectors to, once again, be "in sympathy" with the primary air recirculation pattern. This decreased the noise level from 115 dB to 97 dB, an intensity reduction of 64–1.

7.3.7 Fuel-Injector Instabilities

It is well established that fluctuations in fuel flow rate are a major cause of combustion oscillations in many combustion systems, notably rocket engines and gas turbines. Depending on the noise frequency, it may be called "chugging" or "rumble," but in all cases it refers to an interaction between the acoustics of the combustor and the fuel-injection system.

Suppose one of the combustor's acoustic modes produces pressure pulsations at the fuel nozzle(s). If the fuel supply pressure is low, these pressure pulsations will create oscillations in the fuel flow rate, and these, in turn, will produce oscillations in the heat-release rate. If these heat-release oscillations are properly located and in phase with the acoustic mode, they will add energy to the mode and sustain the oscillating condition [16].

In addition to their influence on fuel flow rates, fluctuations in fuel pressure can alter various spray characteristics, such as mean drop size, drop-size distribution, and spray cone angle in ways that also affect the heat-release rate.

The problems of oscillating combustion in gas turbines began to emerge and gain significance in the 1960s and 1970s, during the period when most engines were fitted with dual-orifice fuel nozzles. With these injectors, there is always a range of fuel flows, starting from the point at which the pressurizing valve opens, over which the secondary fuel delivery pressure is low and the system is highly susceptible to the onset of combustion oscillations.

Methods of alleviating this source of combustion noise include changing the primary nozzle flow number and/or changing the pressurizing valve opening pressure [24]. Note that these changes can also affect both the combustor pattern factor (via the secondary head effect) and the ability of the fuel nozzle to provide good atomization over the entire operating range.

Oscillating combustion is by no means confined to dual-orifice pressure nozzles. It can arise with any type of fuel injector at engine operating conditions where the fuel delivery pressure is so low that fuel flow rates and spray characteristics are affected by fluctuations in combustion pressure. One of the main advantages of airblast atomizers is their ability to provide good atomization at low fuel-injection pressures, which makes them especially prone to this mechanism of noise generation. Fitting a pilot nozzle (see Chapter 6) should eliminate the problem at low engine speeds, but it could reappear at higher speeds when the fuel flow rate is again just above the pressurizing valve opening point.

The discussion so far has focused on liquid fuel injectors, but pulsations in gaseous fuel delivery pressures and flow rates are equally effective in promoting combustion oscillations. One method of eliminating fuel injector-induced noise in multinozzle combustors, which is equally efficacious for both liquid and gaseous fuels, is by fuel staging, or sector burning, as discussed in Chapter 9. With this technique, fuel is supplied only to selected combinations of nozzles. This enriches the localized combustion zones, thereby moving their operating point away from the lean blowout limit and, at the same time, raises the fuel-injection pressure, which reduces the sensitivity of the fuel supply to acoustic oscillations. The main drawback to sector burning is that it creates a circumferentially nonuniform exit temperature distribution, with consequent loss of turbine efficiency and lengthening of engine startup time. In fact, although there are many tools at the combustion engineer's disposal for eliminating or alleviating this category of noise, they all need careful consideration for their impact on other important aspects of combustor and engine performance.

As discussed below, nonuniformities in fuel delivery not only trigger combustion oscillations, but may also be used to eradicate them. Thus, the need to understand and predict the dynamic behavior of fuel nozzles is becoming increasingly important and has provided the motivation for a number of experimental studies on the influence of fuel flow perturbations on spray characteristics (see, e.g., Ibrahim et al. [25]).

7.3.8 Compressor-Induced Oscillations

The possibility that combustion noise might be due to pressure fluctuations in the air supply from the compressor should not be overlooked. At or near the design point, the compressor efflux is smooth and continuous, apart from turbulence fluctuations that generally have a beneficial effect by inhibiting flow separations within the combustor diffuser (see Chapter 3). However, at certain "off-design" or transient conditions, such as, for example, during engine acceleration, the compressor may operate close to its stall line. In consequence, the air supply to the combustor may contain pressure fluctuations that are amplified in the combustion process to produce noise. Compressor-induced oscillations of this type can be a trigger for low-frequency noise (e.g., growl).

7.3.9 LPM Combustor Noise

As pointed out by Richards et al. [26,27], lean premixed (LPM) combustors are especially susceptible to instabilities because acoustic losses are smaller due to the absence of liner holes and wall-cooling slots, and very little acoustic energy is absorbed by the hard liner walls. Moreover, its basic requirement to operate at mixture strengths near the lean blowout limit means that small perturbations in fuel/air ratio tend to produce disproportionately large variations in the rate of heat release (see also Keller [28]). However, this does not mean that instabilities in LPM combustors are confined solely to operate at very weak mixture strengths. Janus et al. [17] observed that combustion oscillations can occur over the entire burning range and not just near the lean blowout limit, as in conventional combustors. Presumably, this is because when the latter are operating close to lean blowout, maldistributions in fuel–air mixing can give rise to a cycle of extinction and reignition in localized regions of the flame, and hence to oscillating combustion, as discussed above. However, once the primary-zone fuel/air ratio has risen to a level at which the mixture strength is within the normal burning limits in all regions of the combustion zone, regardless of maldistributions in fuel–air mixing, the main driving force for oscillating combustion is no longer present. By contrast, LPM combustors are characterized by complete homogeneity in the fuel–air mixture entering the combustion zone. In this situation, there is no reason why the mechanism for instability near the weak extinction limit should not also be operative at higher fuel/air ratios where, although the response of the heat release rate to fluctuations in fuel/air ratio is lower, more energy is available to sustain the instability.

7.3.10 Test Rig Simulations

Practical solutions to instability problems are seldom clearly defined and are often fraught with doubts concerning the specific mechanism(s) driving a given oscillation. One complicating factor is that the acoustic modes and inlet conditions may be different between the test rig and the final engine design. Furthermore, for any given engine or test rig, a change in ambient conditions, or a change in fuel of the same type but from a different source, may serve as a trigger for oscillating combustion.

Many important aspects of combustion performance, such as combustion efficiency, extinction limits, and pollutant emissions, can be studied effectively on simple test rigs comprising, for example, 60° or 90° sectors of annular combustors that contain only a small fraction of the total number of fuel nozzles. Much useful experimental data can be acquired from such test rigs, with the advantage of considerable savings in the cost of fuel and air supplies.

Unfortunately, this approach has little to offer when dealing with combustion instabilities. Rig tests on combustor sectors, or even full-scale combustors, cannot fully reproduce the acoustic properties of the engine combustor, which are influenced by the inlet air conditions and by the geometry of the hot sections downstream. That is why remedial combustor and/or fuel nozzle modifications must often be performed at great cost and inconvenience at a late stage of engine development.

From their work on the development of LPM combustors, Richards et al. [29] have suggested that some aspects of engine oscillations can be studied in single-nozzle test rigs provided that steps are taken to replicate the engine flame geometry and to minimize acoustic losses in the test device having a natural frequency corresponding to the oscillating frequency observed (or expected) on the engine. Their measurements on a single-nozzle test device showed similar oscillations to those on the engine at comparable operating conditions. However, these workers recognize that such test devices cannot reproduce the arbitrary oscillations that occur on the engine, nor can they simulate oscillations that are controlled by transverse acoustic modes. Nevertheless, they contend that the simplicity of single-nozzle testing makes the approach very attractive for a preliminary assessment of how changes to individual fuel nozzles will affect combustor stability. An interesting by-product of this research is the finding that two key parameters in the simulation of oscillating combustion are reference velocity and inlet air temperature.

7.4 Control of Combustion Instabilities

Sustained combustion oscillations are the result of resonant interaction between two or more physical processes. A driving process generates the perturbations of the flow, whereas a feedback process couples this perturbation to the driving mechanism and produces the resonant interaction that may lead to oscillatory combustion [30]. This coupling is shown schematically in Figure 7.1. Acoustic wave propagation is usually responsible for the feedback mechanism that relates the downstream flow to the upstream region where the perturbations are initiated. Methods for alleviating or eliminating combustion instabilities generally fall into the two main categories of "passive" and "active" control, which are described below.

7.4.1 Passive Control

Passive control techniques have been widely used in industrial burners for many years [31]. The passive methods employed for stabilizing combustion oscillations in liquid rocket engines have been described in publications

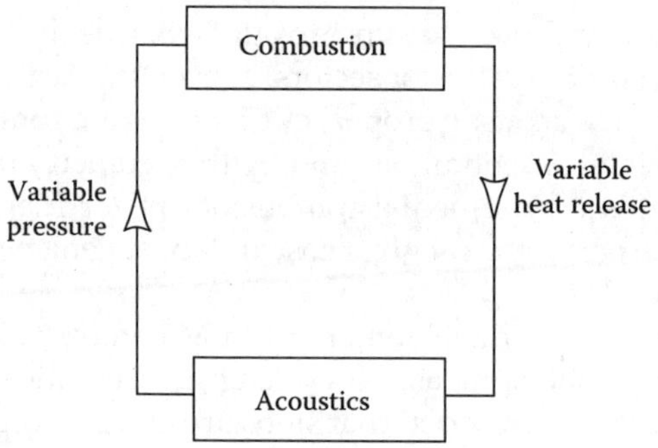

FIGURE 7.1
Representation of coupling between combustion and acoustic processes.

by Harrje and Reardon [32] and, more recently, by Yang and Anderson [33]. Passive methods for controlling combustion instabilities in ramjet dump combustors have been studied by Gutmark et al. [34]. Their application typically involves modifications to the fuel injector or combustor hardware to eliminate the source of the variation in heat release or to increase the acoustic damping in the system, thereby reducing the amplitude of any pressure oscillations. They include baffles, resonators, and acoustic liners of the type that have proved highly successful in suppressing screech in afterburner systems. These devices are much less effective at the low frequencies encountered in main combustors, where the emphasis changes to the fuel delivery system, as discussed above, in particular to those aspects that govern fuel pressure and fuel distribution pattern. Inevitably, passive control techniques tend to be applied on a costly trial-and-error basis.

In recent years, increasing interest has been shown in the control and suppression of combustion instabilities by actively and continuously perturbing in real time the processes responsible for coupling the heat release with the acoustic pressure. The growing importance of these active control techniques merits their discussion in some detail.

7.4.2 Active Control

The active control measures employed to dampen or eliminate acoustic oscillations in combustion systems appear to vary widely, both in regard to the theoretical basis for the control system and the actual hardware employed. They are described in two important review papers by Candel [30] and McManus et al. [22]. A recent paper by Richards et al. [35] also outlines some of the strategies and devices used in the area of active control.

Active control methods were derived from a number of studies on rocket motor instabilities carried out in the early 1950s. One outcome of this work was the notion of introducing perturbations into the combustor using an actuator to decouple the physical processes responsible for the oscillations.

However, it is only within the past decade that the practical demonstration of this concept has been realized [30].

Active control systems fall under the two main headings of "open-loop" and "closed-loop."

7.4.2.1 Open-Loop Systems

A key feature of an open-loop system is that the control action is independent of the combustor's response to the control input. Essentially, they provide a fixed stimulus to the combustor in order to disrupt the coupling of the combustion instability mechanism. Usually, an oscillatory signal with a fixed amplitude and frequency is applied to the combustor, using some form of control actuator. The objective is to introduce a perturbation in some physical variable that has a strong influence on the combustion process, such as the acoustic pressure field or the inlet flow velocity profile.

The types of actuators commonly used include acoustic drivers, flow valves with rotating or oscillating elements to produce periodic flows, and shaking devices to produce mechanical oscillations. The relative merits of these different types of control actuators for various applications have been discussed by McManus et al. [22]. According to these workers, the main advantage of open-loop controllers is that they tend not to suffer from control system instabilities. However, their effectiveness is very dependent on their calibration, which can be a difficult task. For this reason, the more versatile closed-loop system is generally preferred.

7.4.2.2 Closed-Loop Systems

The most distinguishing feature of closed-loop controllers is their use of feedback to control combustion oscillations. The basic idea of feedback in its application to continuous-flow combustion systems is that, by monitoring both the output and input of a combustor, appropriate adjustment of the input can be made that will eliminate any instabilities in the output. The feedback signal is produced by a sensor, usually a pressure transducer or flame detector, which monitors some time-varying property of the combustion process, and is then passed on to the controller. The output signal from the controller is applied to the combustor via an appropriate form of actuator, as described above, in order to eliminate the acoustic oscillations.

7.4.3 Examples of Active Control

McManus et al. [36] used an open-loop configuration to control combustion instabilities in a dump combustor. The control system consisted of an actuator that applied a cross-stream velocity perturbation to the inlet boundary layer of the combustor test section. The actuator was driven by amplified sine-wave signals in order to create a periodic oscillation in the boundary layer.

Application of active control with a 160 Hz sine-wave signal produced a 30% reduction in the RMS pressure fluctuation level, due mainly to a decrease in amplitude of the acoustic mode associated with the instability.

Choudhury et al. [37] used a similar dump combustor in their active control experiments. The actuator employed in this investigation was a row of pulsed gas jets located just upstream of the rearward-facing step. Periodic forcing at a fixed frequency and amplitude was achieved by inserting a rotating valve into the gas supply line. The disruptive effect of the pulsed jets on the recirculation zone created by the step produced a significant decrease in the combustion oscillations.

Bloxsidge et al. [38] used a 250 kW model jet-engine afterburner test rig to investigate the possibility of closed-loop control of instabilities by varying the inlet flow area to the combustor. The apparatus consisted of a long duct with a centerbody flameholder located in the downstream portion of the duct. A premixed ethylene fuel–air mixture entered the test section through a nozzle whose flow area was determined by the axial position of a translating plug located at the center of the nozzle. A mechanical shaker connected to the plug allowed the inlet flow area to be modulated, thereby serving as a control system actuator. A pressure transducer located downstream of the nozzle provided the feedback signal required to control the instability. Without control, the combustion process was characterized by high-amplitude oscillations in the frequency range from 80 to 300 Hz. These oscillations were attributed to the coupling between unsteady heat release and longitudinal acoustic modes of the duct. Activation of the control system effectively suppressed the oscillations.

The same apparatus was used by Langhorne et al. [39] to investigate the possibility of eliminating combustion oscillations by adding fuel out of phase with the oscillation. For these experiments, the translating plug was removed to give a fixed-area inlet nozzle, and an additional fuel-injection manifold was fitted just upstream of the flameholder to provide a controlled unsteady injection of fuel in addition to the main ethylene–air mixture flowing through the inlet nozzle. It was found that a closed-loop control, using the pressure signal to modulate just 3% of the total fuel flow, reduced the noise level by 12 dB. Sivasegaram and Whitelaw [40] also used periodic fuel injection, in both open- and closed-loop configurations, to reduce the oscillating pressure amplitude in a laboratory test rig by values up to 15 dB, depending on the type of control used.

Schadow et al. [41] have reported on a number of control techniques, including fuel flow modulation. In a subsequent paper, Schadow et al. [42] compared fuel modulation with a markedly different approach, whereby a spark discharge ignites a portion of the fuel before it enters the main flame of a dump combustor. Finally, Richards et al. [35] have described the application of active control to a premixing fuel nozzle, using natural gas fuel. Cyclic injection of 14% control fuel was found to produce a 30% (10 dB) reduction in oscillating pressure amplitude at 300 Hz.

7.4.4 Influence of Control Signal Frequency

In their study of combustion instabilities in a dump combustor, McManus et al. [36] found that maximum reduction in pressure oscillations level was obtained when using a control signal frequency of 160 Hz, as noted above, but they also observed that pressure oscillations could be reduced when active control was applied over a range of frequencies from 100 to 1000 Hz. Several other workers have reported that effective control can be achieved at frequencies much lower than that of the acoustic oscillation. For example, Brouwer et al. [43] found that a slow modulation of atomizing air (less than 2 Hz) reduced combustion oscillation in a liquid-fuelled gas turbine combustor, whereas Richards et al. [35] were able to stabilize oscillating combustion using a low-frequency modulation of the natural gas fuel. Gemmen et al. [44] also observed similar behavior when modulating a pilot flame. Oscillation control was again achieved by changing the flame conditions at a frequency that was much lower than the acoustic oscillation. These results are interesting because they do not conform to the generally held view that fluctuations in heat release can only be smoothed out by repeated perturbations that occur at about the same frequency as the fluctuations.

7.5 Modeling of Combustion Instabilities

Many attempts have been made to develop analytical models for predicting combustion instabilities along with their amplitudes and frequencies, but these efforts have met with only partial success. The problems involved are formidable because audible engine noise is usually the result of complex instability mechanisms that cannot be modeled or described using standard analytic techniques until many simplifications have been made to render the problem tractable [22]. Clearly, this approach could result in an oversimplified view of the problem. At the present time, there is no universal model for predicting combustion instabilities, thus modeling of combustion instabilities is usually carried out on a case by case basis.

Much of the early work on the modeling of combustion oscillations was motivated by the instabilities encountered in liquid-fueled rocket engines. The so-called τ–n analysis of Crocco and Cheng [45], which relates heat release and pressure oscillations by a time lag, τ, and an interaction index, n, has proved useful in analyzing rocket engine data. It invokes the Rayleigh criterion that heat release and acoustic fluctuations should be in phase to drive oscillations and out of phase to dampen oscillations [14]. The basis of the τ–n model is that acoustic disturbances produce a change in heat release rate, but delayed by a time interval, τ. Given the correct value of τ, and assuming the gain is sufficiently large, the heat-release fluctuations will drive the pressure fluctuations. The magnitude of the gain is determined by the value of n.

Similar time lag models have been applied to a variety of industrial burners (see the review by Putnam [31]), and to ramjets [46,47]. More recent extensions of the τ–n model to include nonlinear acoustics in combustors have been described by Culick [48]. To a large extent, the success of the τ–n approach depends on predicting the time lag. The main uncertainties lie in the estimation of mixing rates and, in liquid-fuelled systems, evaporation rates also.

Advances in computational fluid dynamics (CFD) allow the time history of complex reacting flow fields to be computed without the need to specify a combustion time lag. For example, Menon et al. [49,50] have used CFD analysis to describe combustion instabilities in ramjets. The main drawback to this approach, in addition to the long computational time, is that the results obtained tend to be combustor-specific, so that few general conclusions can be drawn.

Janus and Richards [51] have developed a simple, nonlinear model in which the combustion process is represented by a well-stirred reactor with finite kinetics. The model was developed to provide explanations for specific experimental observations on LPM gas turbine combustors, and to provide guidance in developing active control strategies. Conservation equations for the combustor and fuel injector provide a set of ordinary differential equations that can be solved on a personal computer. Comparison with experimental data shows good agreement with the predictions of the model that, according to Janus and Richards, can be used to examine stability trends associated with changes in fuel/air ratio, mass flow rate, geometry, ambient conditions, and other relevant parameters.

Recently, Lieuwen and Yang [52] have provided an excellent overview of the operational experience, fundamental mechanisms, and modeling of combustion instabilities in gas turbine engines. They state that combustion instabilities routinely constrain the operating envelope and power output of gas turbines for power generation and damage combustor liners, transition pieces, and fuel nozzles. For aero engines, instabilities appear in a few cases, mostly in the development stage. Military engines, however, have experienced major problems with low-frequency instabilities in augmentors. This work describes research performed through 2005.

References

1. Mahan, J. R., and Karchmer, A., "Combustion and Core Noise," *Aero Acoustics of Flight Vehicles: Theory and Practice*, Vol. 1; *Noise Sources*, WRDC Technical Report 90-3052, 1991.

2. Bragg, S. L., "Combustion Noise," *Journal of the Institute of Fuel*, January, pp. 12–16, 1963.
3. Thomas, A., and Williams, G. T., "Flame Noise: Sound Emission From Spark-Ignited Bubbles of Combustible Gas," *Proceedings of the Royal Society, London, Series A*, Vol. 294, No. 1439, pp. 449–66, 1966.
4. Ballal, D. R., and Lefebvre, A. H., "Turbulence Effects on Enclosed Flames," *Acta Astronautica*, Vol. 1, pp. 471–83, 1974.
5. Ballal, D. R., and Lefebvre, A. H., "The Structure and Propagation of Turbulent Flames," *Proceedings of the Royal Society, London, Series A*, Vol. 344, pp. 217–34, 1975.
6. Strahle, W. C., "Some Results in Combustion Generated Noise," *Journal of Sound and Vibration*, Vol. 23, No. 1, pp. 113–25, 1972.
7. Strahle, W. C., and Muthukrishnan, M., "Correlation of Combustor Rig Sound Power Data and Theoretical Basis of Results," *AIAA Journal*, Vol. 18, No. 3, pp. 269–74, 1980.
8. Shivashankara, B. N., and Crouch, R. W., "Noise Characteristics of a Can-Type Combustor," *Journal of Aircraft*, Vol. 14, No. 8, pp. 751–6, 1977.
9. Kazin, S. B. and Emmerling, J. J., "Low Frequency Core Engine Noise," ASME Paper 74-WA/Aero-2, 1974.
10. Ho, P. Y., and Tedrick, R. N., "Combustion Noise Prediction Techniques for Small Gas Turbine Engines," in M. J. Crocker, ed., *INTER-NOISE 72 Proceedings*, Inst. of Noise Control Engineering, Poughkeepsie, New York, pp. 507–12, 1972.
11. Strahle, W. C., and Shivashankara, B. N., "Combustion Generated Noise in Gas Turbine Combustors," NASA CR-134843, 1974.
12. Ho, P. Y., and Doyle, V. L., "Combustion Noise Prediction Update," AIAA Paper 79-0588, 1979.
13. Mathews, D. C., and Rekos, N. F., Jr., "Prediction and Measurement of Direct Combustion Noise in Turbopropulsion Systems," *Journal of Aircraft*, Vol. 14, No. 9, pp. 850–9, 1977.
14. Rayleigh, Lord, *The Theory of Sound*, Vol. 2., Dover, New York, p. 226, 1945.
15. Seto, S. P., private communication, 1989.
16. Kenworthy, M. J., Bahr, D. W., Mungur, P., Burrus, D. L., Mehta, J. M., and Cifone, A. J., "Dynamic Instability Characteristics of Aircraft Turbine Engine Combustors," AGARD CP-450, ISBN 92-835-0503-4, 1989.
17. Janus, M. C., Richards, G. A., Yip, M. J., and Robey, E. H., "Effects of Ambient Conditions and Fuel Composition On Combustion Stability," ASME Paper 97-GT-266, 1997.
18. Keller, J. O., Bramlette, T. T., Westbrook, C. K., and Dec, J. E., "Pulse Combustion: The Importance of Characteristic Times," *Combustion and Flame*, Vol. 75, pp. 33–44, 1989.
19. Vandsburger, U., McManus, K., and Bowman, C., "Effects of Fuel Spray Vaporization on the Stability Characteristics of a Dump Combustor," AIAA Paper 89-2436. Presented at the 25th Joint Propulsion Conference, July 10–12, Monterey, CA, 1989.
20. Mehta, J., Mungur, P., Dodds, W., Bahr, D., and Clouser, S., "Fuel Effects on Gas Turbine Combustor Dynamics," AIAA Paper 90-1957. Presented at the 26th Joint Propulsion Conference, Orlando, FL, 1990.

21. Janus, M. C., and Richards, G. A., "Results of a Model for Premixed Combustion Oscillations," Presented at the American Flame Research Committee Meeting, September 30–October 1, Baltimore, MD, 1996.
22. McManus, K. R., Poinsott, T., and Candel, S. M., "A Review of Active Control of Combustion Instabilities," *Progress in Energy and Combustion Science*, Vol. 19, pp. 1–29, 1993.
23. Scalzo, A. J., Sharkey, W. T., and Emmerling, W. C., "Solution of Combustor Noise in a Coal Gasification Cogeneration Application," *Turbomachinery International*, September/October, pp. 22–7, 1989.
24. Hudson, R. H., "Pressure Oscillations in a Combustor with Dual-Orifice Fuel Injection," ASME Paper 67-GT-23, 1967.
25. Ibrahim, M., Darling, D., Sanders, T., and Zaller, M., "Dynamic Response of Fuel Nozzles for Liquid-Fueled Gas Turbine Combustors," ASME Paper 96-GT-54, 1996.
26. Richards, G. A., "Gas Turbine Combustion Instability." Proceedings of the Spring Technical Meeting of the Central States Section of the Combustion Institute, May 5–7, St. Louis, MO, 1996.
27. Richards, G. A., and Janus, M. C., "Characterization of Oscillations During Premix Gas Turbine Combustion," ASME Paper 97-GT-244, 1997.
28. Keller, J. J., "Thermoacoustic Oscillations in Combustion Chambers of Gas Turbines," *AIAA Journal*, Vol. 33, No. 12, pp. 2280–7, 1995.
29. Richards, G. A., Gemmen, R. S., and Yip, M. J., "A Test Device for Premixed Gas Turbine Combustion Oscillations," DOE/METC-96/1027 (DE96004367), 1996.
30. Candel, S. M., "Combustion Instabilities Coupled by Pressure Waves and Their Active Control," *Twenty-Fourth (International) Symposium on Combustion*, The Combustion Institute, Pittsburgh, PA, pp. 1277–96, 1992.
31. Putnam, A. A., *Combustion Driven Oscillations in Industry*, American Elsevier, New York, 1971.
32. Harrje, D. T., and Reardon, F. H., "Liquid Propellant Rocket Combustion Instability," NASA SP-194, 1972.
33. Yang, V., and Anderson, W., eds., *Liquid Rocket Combustion Instability*, AIAA, Cambridge, MA, 1995.
34. Gutmark, E., Wilson, K. J., Schadow, K. C., Stalnaker, R. A., and Smith, R. A., "Combustion Characteristics and Passive Control of an Annular Dump Combustor," AIAA Paper 93-1772, 1993.
35. Richards, G. A., Yip, M. J., Robey, E., Cowell, L., and Rawlins, D., "Combustion Oscillation Control by Cyclic Fuel Injection," *Journal of Engineering for Gas Turbines and Power*, Vol. 119, pp. 340–3, 1997.
36. McManus, K. R., Vandsburger, U., and Bowman, C. T., "Combustor Performance Enhancement Through Direct Shear Layer Excitation," *Combustion and Flame*, Vol. 82, pp. 75–92, 1990.
37. Choudhury, P. R., Gerstein, M., and Mojaradi, R., "A Novel Feedback Concept for Combustion Instability in Ramjets," 22nd JANNAF Combustion Meeting, 1985.
38. Bloxsidge, G. J., Dowling, A. P., Hooper, N., and Langhorne, P. J., "Active Control of Reheat Buzz," *AIAA Journal*, Vol. 26, pp. 783–90, 1988.
39. Langhorne, P. J., Dowling, A. P., and Hooper, N., "Practical Active Control System for Combustion Oscillations," *Journal of Propulsion and Power*, Vol. 6, No. 3, pp. 324–33, 1990.

40. Sivasegaram, S., and Whitelaw, J. H., "Active Control of Combustors with Several Frequency Modes," *American Society of Mechanical Engineers DSC*, Vol. 38, pp. 69–74, 1992.
41. Schadow, K. C., Hendricks, E. W., and Hansen, R. J., "Recent Progress in the Implementation of Active Combustion Control," *Proceedings of the 18th ICAS Congress*, Beijing, China, September 20–25, Vol. 1, Paper No. ICAS-92-2.5.3, pp. 942–52, 1992.
42. Schadow, K. C., Wilson, K. J., Gutmark, E., Yu, K., and Smith, R. A., "Periodic Chemical Energy Release for Active Control," *11th ISABE-International Symposium on Air Breathing Engines*, Vol. 1, AIAA, Washington, DC, pp. 479–85, 1993.
43. Brouwer, J., Ault, B. A., Bobrow, J. E., and Samuelsen, G. S., "Active Control for Gas Turbine Combustors," *The Twenty-Third (International) Symposium on Combustion*, The Combustion Institute, Pittsburgh, PA, pp. 1087–92, 1990.
44. Gemmen, R. S., Richards, G. A., Yip, M. J., and Norton, T. S., "Combustion Oscillation Chemical Control Showing Mechanistic Link to Recirculation Zone Purge Time," *Eastern States Section Meeting of the Combustion Institute*, Worcester, MA, 1995.
45. Crocco, L., and Cheng, S. I., "Theory of Combustion Instability in Liquid Propellant Rocket Motors," *AGARD Monograph*, No. 8, Butterworths, London, 1956.
46. Reardon, F. H., "Very Low-Frequency Oscillations in Liquid Fuel Ramjets," NATO AGARD Conference Proceedings No. 450, 1989.
47. Yu, K. H., Trouve, A., and Daily, J. W., "Low-Frequency Pressure Oscillations in a Model Ramjet Combustor," *Journal of Fluid Mechanics*, Vol. 232, pp. 47–72, 1991.
48. Culick, F. E. C., "Some Recent Results for Nonlinear Acoustics in Combustion Chambers," *AIAA Journal*, Vol. 32, No. 1, pp. 146–69, 1994.
49. Menon, S., and Jou, W. H., "Large-Eddy Simulations of Combustion Instabilities in an Axisymmetric Ramjet Combustor," *Combustion Science and Technology*, Vol. 84, pp. 51–79, 1991.
50. Menon, S., "Secondary Fuel Injection Control of Combustion Instability in a Ramjet," *Combustion Science and Technology*, Vol. 100, pp. 385–93, 1994.
51. Janus, M. C., and Richards, G. A., "A Model for Premixed Combustion Oscillations," Technical Note DOE/METC-96/1026 (DE96004366), 1996.
52. Lieuwen, T.C. and Yang, V., *Combustion Instabilities in Gas Turbine Engines*, Progress in Astronautics and Aeronautics, Vol. 210, AIAA, Reston, VA, 2005.

Bibliography

Throughout this chapter, frequent reference has been made to the paper on combustion and core noise by Mahan and Karchmer and to the articles on the active control of combustion instabilities by Candel and McManus et al. These publications are listed above as References [1], [30], and [22], respectively. They are strongly recommended as additional reading for anyone seeking a comprehensive and in-depth review of these important topics. For further information on current research activities relating to combustion oscillations in modern premix gas turbine combustors, refer to [52] which summarizes the most up to date research in this area through 2005.

8

Heat Transfer

8.1 Introduction

In a modern combustor, the temperature of the gases released by the combustion process may peak over 2100°C; this is much higher than the melting point of the combustor flame tube and turbine blades. Therefore, the designer must adequately cool all the metal surfaces exposed to the hot gases and improve structural integrity and durability. Furthermore, the amount of cooling air must be minimized to maximize the air available for emissions control.

The mechanical stresses experienced by the combustor liner are small in comparison with those to which many other engine components are subjected. However, it is called on to withstand high temperatures and steep temperature gradients that threaten its structural integrity. To ensure a satisfactory liner life, it is important to keep temperatures and temperature gradients down to an acceptable level. Just what these levels are has never been clearly defined, but for the nickel- or cobalt-based alloys in common use, such as Nimonic 75, Hastelloy X, and HS188, the maximum operating temperature should not exceed 1100 K. The mechanical strength of these materials declines rapidly at temperatures above this level. The practical implication of this limit is that some means must be provided to supplement the removal of heat from the liner walls, which normally occurs through radiation to the combustor casing and convection to the annulus air. The traditional method has been to provide a film of cool air along the inner surface of the liner.

Since the early 1960s, the need to develop more efficient methods of liner-wall cooling has become increasingly important. The reasons for this are fourfold.

1. Substantial reductions in engine fuel consumption continue to be achieved by the use of higher pressure ratios, higher turbine entry temperatures and, on aircraft engines, by higher bypass ratios. Unfortunately, increases in pressure ratio raise the amount of heat transferred to the liner walls by radiation. Moreover, the accompanying increase in combustor inlet temperature impairs the ability of the

annulus air to cool the walls by convection. Thus, as pressure ratios rise, the problem of wall cooling becomes more severe, to the extent that on many engines more than one-third of the total combustor airflow is used in film cooling the liner.

2. During the past 20 years, as the regulations governing pollutant emissions have become increasingly stringent, the requirement for lower levels of nitric oxides has resulted in more air being allocated to combustion. This trend has led to a continuing decline in the amount of air available for liner-wall cooling. Furthering this trend was the knowledge that reductions in film-cooling air are highly beneficial in lowering the emissions of carbon monoxide and unburned hydrocarbons (see Chapter 9).
3. Increases in turbine inlet temperature call for improvements in combustor pattern factor in order to maintain the integrity of the hot sections downstream. Since film-cooling air flows along the liner wall, it makes no contribution to mixing. Thus, reductions in wall-cooling requirements make a direct contribution to pattern factor improvement by releasing more air for dilution zone mixing.
4. As noted by Dodds and Bahr [1], as combustor operating temperatures have increased, component durability expectations have also risen. Life expectations for combustor liners of early engines were only a few hundred hours between repairs. Today, the demand from customers is for many thousands of operating hours before combustor maintenance is required.

For all these reasons, it is important to improve the effectiveness of the various cooling devices that are now in widespread use, and to develop new cooling schemes that are even more economical in their use of cooling air. Also, heat transfer modeling of metal temperatures is necessary to determine the displacement, thermal stresses, and component life. To calculate metal temperatures, it is necessary to input material property data, engine performance data, air system data, and experimentally validated heat transfer coefficients.

In this chapter, the heat-transfer processes that govern liner-wall temperatures are discussed in some detail, along with the methods employed to combat the heat flux to the wall, thereby achieving an acceptable liner life.

8.2 Heat-Transfer Processes

For the purpose of analysis, a liner may be regarded as a container of hot flowing gases surrounded by a casing, with air flowing between the container

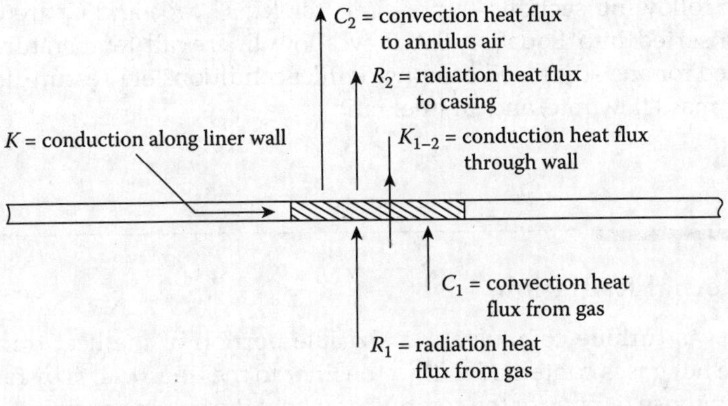

FIGURE 8.1
Basic heat-transfer processes.

and the casing. Broadly, the liner is heated by radiation and convection from the hot gases inside it; it is cooled by radiation to the outer casing and by convection to the annulus air. The relative magnitudes of the radiation and convection components depend on the geometry and operating conditions of the system. Under equilibrium conditions, the liner temperature is such that the internal and external heat fluxes at any point are just equal. Loss of heat by conduction along the liner wall is comparatively small and, usually, may be neglected.

The heat-transfer model shown in Figure 8.1 includes only the axial variation of properties. All properties are assumed to be constant around the circumference at any axial location.

Under steady-state conditions, the rate of heat transfer into a wall element must be balanced by the rate of heat transfer out. Therefore, for an element with inside surface area ΔA_{w_1},

$$(R_1 + C_1 + K)\Delta A_{w_1} = (R_2 + C_2)\Delta A_{w_2} = K_{1-2}\Delta A_{w_1}. \tag{8.1}$$

K, the heat conduction along the liner wall, is always negligibly small compared to the radiation and convection terms. Also, the liner wall is usually so thin that $\Delta A_{w_1} \cong \Delta A_{w_2}$. Thus, Equation 8.1 can be simplified to

$$R_1 + C_1 = R_2 + C_2 = K_{1-2}, \tag{8.2}$$

where K_{1-2} is the conduction heat transfer through a *solid* liner wall due to a temperature gradient within the wall, i.e.,

$$K_{1-2} = \frac{k_w}{t_w}\left(T_{w_1} - T_{w_2}\right). \tag{8.3}$$

In the following sections, expressions for R_1, C_1, R_2, and C_2 are derived. When inserted into Equation 8.2, they allow liner-wall temperatures to be calculated for any stipulated combustor inlet conditions of pressure, temperature, air mass flow rate, and air/fuel ratio.

8.3 Internal Radiation

In most gas turbine combustors, a sizeable portion of the heat transferred from the hot gases contained within the liner to the liner wall is by radiation. In fact, in those regions of the combustor where the cooling air injected at the walls forms an effective barrier between the hot gas and the wall, radiation represents the only mechanism by which heat can be transferred from the gas to the wall. For the combustion gases generated by gas turbine fuels, the total emitted radiation has two components: (1) the "nonluminous" radiation that emanates from certain heteropolar gases, notably carbon dioxide and water vapor, and (2) the "luminous" radiation that depends on the number and size of the solid particles (mainly soot) in the flame. Both modes of radiant heat transfer enter into calculations of the liner-wall temperature and the amount of air to be employed in liner-wall cooling.

8.3.1 Radiation from Nonluminous Gases

The rate of heat transfer by nonluminous radiation from a gas to its enclosure can be calculated from knowledge of the size and shape of the gas volume and its mean or "bulk" conditions of pressure, temperature, and chemical composition. Let us consider the radiation heat exchange between a gas at temperature T_g and the surface of a black-body container at temperature T_{w_1}. While the black surface emits and absorbs heat at all wavelengths, the gas emits only a few narrow bands of wavelengths and absorbs only those wavelengths included in its emission bands. The net radiant heat transfer is given by

$$R_1 = \sigma\left(\varepsilon_g T_g^4 - \alpha_g T_{w_1}^4\right), \tag{8.4}$$

where σ is the Stefan–Boltzmann constant = 5.67×10^{-8} W/(m^2 K^4), ε_g is the gas emissivity at temperature T_g, and α_g is the gas absorptivity at temperature T_{w_1}.

Both ε_g and α_g are functions of gas composition. However, ε_g relates to the *emission* of radiation from the gas to the wall and depends on T_g, but α_g applies to the *absorption* by the gas of radiation from the wall, and hence depends on T_{w_1}.

In practice, the surface exposed to the flame is not black, but has an effective absorptivity that is less than unity. For most practical purposes, this effect may be accounted for by introducing the factor $0.5(1 + \varepsilon_w)$ to obtain

$$R_1 = 0.5\sigma(1+\varepsilon_w)(\varepsilon_g T_g^4 - \alpha_g T_{w_1}^4), \tag{8.5}$$

in which ε_w is dependent on the material, temperature, and degree of oxidation of the wall. Approximate mean values of ε_w at typical liner-wall temperatures for Nimonic and stainless steel are 0.7 and 0.8, respectively.

Investigation over a wide range of values has shown [2] that to a sufficiently close approximation

$$\frac{\alpha_g}{\varepsilon_g} = \left(\frac{T_g}{T_{w_1}}\right)^{1.5}. \tag{8.6}$$

Hence, Equation 8.5 may be rewritten as

$$R_1 = 0.5\sigma(1+\varepsilon_w)\varepsilon_g T_g^{1.5}(T_g^{2.5} - T_{w_1}^{2.5}). \tag{8.7}$$

In practice, the composition and temperature of the hot gases are far from homogeneous. Thus, the variables in the above expression must be represented by mean or "effective" values. Improvements in accuracy could be achieved by the method of zoning, as described by Hottel [3], but this would demand a more exact knowledge of the distribution of fuel and temperature in the combustion zone than is available for most current chamber designs.

The bulk or mean gas temperature, T_g, is obtained as the sum of the chamber entry temperature, T_3, and the temperature rise due to combustion ΔT_{comb}:

$$T_g = T_3 + \Delta T_{comb}. \tag{8.8}$$

ΔT_{comb} may be obtained from standard temperature-rise curves. When these curves are used, the appropriate value for the fuel/air ratio is the product of the local fuel/air ratio and the local level of combustion efficiency. Most heat-transfer calculations are carried out at high pressures, for which it is reasonable to assume a combustion efficiency of 100%.

Values of ε_g for nonluminous flames may be obtained from the following approximate formula due to Reeves [4].

$$\varepsilon_g = 2 - \exp\left[-290P(ql_b)^{0.5} T_g^{-1.5}\right], \tag{8.9}$$

where P is the gas pressure (kPa), T_g is the gas temperature (K), l_b is the beam length (m), and q is the fuel/air ratio by mass.

The beam length, l_b, is determined by the size and shape of the gas volume. For most practical purposes, it is given to sufficient accuracy [5] by the expression

$$l_b = 3.4\,(\text{volume})/(\text{surface area}). \tag{8.10}$$

For tubular systems, the above expression yields values of beam length, l_b ranging from 0.6 D_L to 0.9 D_L, depending on the length/diameter ratio of the liner. For annular combustors, l_b, is 1.0 D_L for the inner liner and 1.2 D_L for the outer liner.

8.3.2 Radiation from Luminous Gases

When a hydrocarbon fuel is burned in a combustion chamber, soot particles are formed; these particles have an important effect on the nature of the radiation from the flame. At atmospheric pressure, the soot particles are too small in number and size to radiate appreciable energy. However, some of the radiation from these hot, glowing particles falls in the visible spectrum and gives rise to the name "luminous flame." With increasing pressure, the luminous radiation increases in intensity, and the banded spectra from water vapor and carbon dioxide become less pronounced. At the high levels of pressure encountered in modern gas turbines, the soot particles can attain sufficient size and concentration to radiate as black-bodies in the infrared region, and the flame is then characterized by a predominance of luminous radiation. It is under these conditions that severe radiant heating and its attendant problem of liner durability are encountered.

According to Lefebvre and Herbert [2], the influence of luminosity on gas emissivity may be accounted for by including a luminosity factor L into Equation 8.9 to obtain

$$\varepsilon_g = 1 - \exp\left[-290PL(ql_b)^{0.5}\,T_g^{-1.5}\right]. \tag{8.11}$$

The original equation [2] for L is

$$L = 7.53(C/H - 5.5)^{0.84}, \tag{8.12}$$

which was later modified to

$$L = 3(C/H - 5.2)^{0.75}, \tag{8.13}$$

where C/H is the carbon to hydrogen ratio of the fuel by mass.

Kretschmer and Odgers [6] recommend the following expression:

$$L = 0.0691(C/H - 1.82)^{2.71}. \tag{8.14}$$

Lefebvre [7] has correlated modern engine combustor data using:

$$L = 336/H^2, \tag{8.15}$$

where H is the fuel hydrogen content (by mass) in percent.

For any given fuel type, the luminosity factor L may be calculated from one of these equations and inserted in Equation 8.11 to obtain the luminous flame emissivity ε_g. Substituting this value of ε_g into Equation 8.7 gives the radiation heat flux from the flame to the liner wall.

8.4 External Radiation

The radiation heat transfer, R_2, from the liner wall to the outer casing can be approximated by assuming gray surfaces with emissivities ε_w and ε_c and assuming that T_{w_2} and T_c are approximately uniform in the axial direction. The net radiation heat transfer from the liner is then given by

$$R_2 A_w = \frac{\sigma\left(T_{w_2}^4 - T_c^4\right)}{(1-\varepsilon_w)/\varepsilon_w A_w + 1/A_w F_{wc} + (1-\varepsilon_c)/\varepsilon_c A_c}, \tag{8.16}$$

where A_w is the surface area of the liner wall, A_c is the surface area of the casing, and E_{wc} is the geometric shape factor between liner and casing.

The amount of heat transferred from the liner to the casing is usually quite small compared with C_2, the external convective heat transfer [2]. Its significance increases with liner-wall temperature, and at low values it can often be neglected. It can be estimated only approximately because of a lack of accurate knowledge of wall emissivities. For this reason, it is sufficient to use the cooling-air temperature, T_3, in place of the unknown temperature of the outer casing. Also, for radiation across a long annular space, the geometric shape factor can be assumed to be equal to unity. The expression for the net radiation flux then reduces to:

$$R_2 = \sigma \frac{\varepsilon_w \varepsilon_c}{\varepsilon_c + \varepsilon_w (1-\varepsilon_c)(A_w/A_c)}\left(T_{w_2}^4 - T_3^4\right). \tag{8.17}$$

For a tubular chamber, A_w/A_c is equal to the ratio of liner diameter to casing diameter at the section considered. For tuboannular systems, in which the depth of the annulus varies from point to point around the liner, an average value of 0.8 should be used. For an annular chamber, the ratio A_w/A_c is slightly greater than unity for the inner liner, and slightly less than unity for the outer liner.

Accurate values for the emissivity of various materials may be obtained from Reference [8]. However, for most practical purposes, the following expressions, based on typical values of emissivity and diameter ratio, should suffice:

$$R_2 = 0.4\sigma\left(T_{w_2}^4 - T_3^4\right), \tag{8.18}$$

for an aluminum air casing and

$$R_2 = 0.6\sigma\left(T_{w_2}^4 - T_3^4\right), \tag{8.19}$$

for a steel air casing.

8.5 Internal Convection

Of the four heat-transfer processes that together determine the liner temperature, internal convection is the most difficult to estimate accurately. In the primary zone, the gases involved in heat transfer are at high temperature and are undergoing rapid physical and chemical change. Further difficulty is introduced by the existence within the primary zone of steep gradients of temperature, velocity, and composition. Uncertainties regarding the airflow pattern, the state of boundary-layer development, and the effective gas temperature make the choice of a realistic model almost arbitrary.

In the absence of more exact data, it is reasonable to assume that some form of the classical heat-transfer relation for straight pipes will hold for conditions inside a liner, provided the Reynolds-number index is consistent with established practice for conditions of extreme turbulence. This leads to an expression of the form

$$C_1 = 0.020\frac{k_g}{d_{h1}^{0.2}}\left(\frac{\dot{m}_g}{A_L\mu_g}\right)^{0.8}\left(T_g - T_{w_1}\right), \tag{8.20}$$

where d_{h1} is the hydraulic diameter of the liner:

$$d_{h1} = 4\frac{\text{Cross-sectional flow area}}{\text{Wetted perimeter}} = D_L.$$

Hence,

$$C_1 = 0.020 \frac{k_g}{D_L^{0.2}} \left(\frac{\dot{m}_g}{A_L \mu_g} \right)^{0.8} (T_g - T_{w_1}). \tag{8.21}$$

Several difficulties arise when Equation 8.21 is applied to the primary combustion zone. These have been discussed in detail elsewhere [2], so reference need only be made to the three factors that must be taken into account in calculations of wall temperature in the primary zone.

The primary zone contains, by design, a reversal of flow, so that only in a region adjacent to the wall does the direction of flow correspond to the assumed pipe analogy. A mean value $\dot{m}_g/A_L$ across the whole section may be obtained by summing the upstream and downstream flow components, irrespective of sign. A further complication is that, if a swirler is used, the local gas velocity at the wall is greater than the downstream component by the factor $(\cos\ \beta)^{-1}$, where β is the angle that the local velocity makes with the combustor axis.

Of equal importance is the question of which gas temperature is to be used in Equation 8.21. The bulk gas temperature, T_g, is appropriate for radiation, but the conventional primary-zone radial temperature profile is deliberately arranged to provide lower-than-average gas temperatures near the wall. To account for this, the value of the constant in Equation 8.21 is reduced from 0.020 to 0.017, so that for calculations in the primary zone, Equation 8.21 becomes

$$C_{1.pz} = 0.017 \frac{k_g}{D_L^{0.2}} \left(\frac{\dot{m}_{pz}}{A_L \mu_g} \right)^{0.8} (T_g - T_{w_1}). \tag{8.22}$$

8.6 External Convection

In estimating this component, Re is now based on the hydraulic mean diameter, D_{an}, of the annulus air space. This diameter is given by

$$D_{an} = 4 \frac{\text{Cross-sectional area of flow}}{\text{Wetted perimeter}},$$

which, for a tubular chamber, gives

$$D_{an} = D_{ref} - D_L,$$

and for an annular chamber,

$$D_{an} = 2 \times \text{local annulus height.}$$

Again, the contents of the annulus may be assumed sufficiently stirred for fully developed turbulent transfer to occur. Hence,

$$C_2 = 0.020 \frac{k_a}{D_{an}^{0.2}} \left(\frac{\dot{m}_{an}}{A_{an} \mu_a} \right)^{0.8} \left(T_{w_2} - T_3 \right). \tag{8.23}$$

The fluid properties are now evaluated at the annulus air temperature, T_3. In practice, the cooling-air temperature will increase during the passage downstream, but normally this increase amounts to no more than a few degrees and can reasonably be neglected.

8.7 Calculation of Uncooled Liner Temperature

Summarizing, equations for all four heat-transfer processes have now been derived, namely,

$$R_1 = 0.5\sigma \left(1 + \varepsilon_w\right) \varepsilon_g T_g^{1.5} \left(T_g^{2.5} - T_{w_1}^{2.5} \right), \tag{8.7}$$

$$C_1 = 0.020 \frac{k_g}{D_L^{0.2}} \left(\frac{\dot{m}_g}{A_L \mu_g} \right)^{0.8} \left(T_g - T_{w_1} \right), \tag{8.21}$$

or, in the primary zone,

$$C_{1.pz} = 0.017 \frac{k_g}{D_L^{0.2}} \left(\frac{\dot{m}_{pz}}{A_L \mu_g} \right)^{0.8} \left(T_g - T_{w_1} \right), \tag{8.22}$$

$$R_2 = Z\sigma \left(T_{w_2}^4 - T_3^4 \right), \tag{8.24}$$

where Z depends on the casing emissivity, and

$$C_2 = 0.020 \frac{k_a}{D_{an}^{0.2}} \left(\frac{\dot{m}_{an}}{A_{an} \mu_a} \right)^{0.8} \left(T_{w_2} - T_3 \right). \tag{8.23}$$

For equilibrium,

$$R_1 + C_1 = R_2 + C_2 = \frac{k_w}{t_w}\left(T_{w_1} - T_{w_2}\right) = K_{1-2}. \tag{8.25}$$

8.7.1 Method of Calculation

The uncooled liner-wall temperature may be calculated as follows:

1. Estimate the mean fuel/air ratio for the zone under consideration. If the maximum possible value of T_w in the combustion zone is required, the fuel/air ratio should be assumed to be stoichiometric.
2. Obtain R_1 as a function of T_{w_1} from Equation 8.7, in which the gas emissivity may be obtained from Equation 8.11.
3. Obtain R_2 as a function of T_{w_2} from Equation 8.24.
4. Calculate C_1 as a function of T_{w_1} from Equation 8.21. or Equation 8.22, using the values of k and μ for combustion products at temperature T_g.
5. Calculate C_2 as a function of T_{w_2} from Equation 8.23, using values of k and μ for air at temperature T_3.
6. Solve Equation 8.25 for T_{w_1} and T_{w_2}.

Example

We wish to estimate the liner-wall temperature that could be expected in the primary zone of a tubular combustor if no film cooling was used, given the following information:

$P_3 = 30$ atm $= 3040$ kPa
$T_3 = 880$ K
Casing diameter = 0.192 m
Liner outer diameter = 0.1344 m
Liner wall thickness = 0.0012 m
Liner inner diameter = 0.132 m
$\varepsilon_c = 0.4$ (aluminum casing)
$\varepsilon_w = 0.7$ (Nimonic 75 liner material)
$k_w = 26$ W/(m K)
$\dot{m}_{an} = 7.074$ kg/s
$\dot{m}_{pz} = 2.62$ kg/s
$q_{pz} = 0.0588$
$L = 1.7$ (kerosine fuel)

The primary-zone gas temperature, T_g, is determined using Equation 8.8. A primary-zone combustion efficiency of 85% is assumed. Thus, the effective value of q_{pz} is $0.85 \times 0.0588 = 0.050$. From temperature-rise curves for kerosine–air mixtures, for $T_3 = 880$ K, $P_3 = 3040$ kPa, and $q = 0.050$, we obtain $\Delta T_{comb} = 1455$ K.

A correction must now be made for the heat lost in evaporating the unburned fuel and raising its temperature to that of the surrounding hot gas. This is estimated at 55 K. Hence,

$$T_g = 880 + 1455 - 55 = 2280\text{K}.$$

The beam length, l_b, is obtained as

$$l_b = 0.6D_L = 0.0792\text{m}.$$

For the calculation of R_1, we have

$$\sigma = 5.67 \times 10^{-8}\ \text{W/(m}^2 \cdot \text{K}^4\text{)},$$

$$\varepsilon_w = 0.7,$$

$$T_g = 2280\,\text{K},$$

$$L = 1.7,$$

$$P = 3040\text{kPa},$$

$$q = 0.0588,$$

$$l_b = 0.0792\text{m}.$$

Substitution of these values into Equations 8.11 and 8.7 yields

$$\varepsilon_g = 0.61 \text{ and } R_1 = 794460 - 0.0032T_{w_1}^{2.5}\ \text{W/m}^2. \tag{8.26}$$

For the calculation of R_2, we use Equation 8.17

$$R_2 = \sigma \frac{\varepsilon_w \varepsilon_c}{\varepsilon_c + \varepsilon_w (1 - \varepsilon_c) D_L / D_{ref}} \left(T_{w_2}^4 - T_3^4\right),$$

where

$\sigma = 5.67 \times 10^{-8}\ \text{W/(m}^2 \cdot \text{K}^4\text{)}$,
$\varepsilon_w = 0.7$,
$\varepsilon_c = 0.4$,
$D_L / D_{ref} = 0.1344/0.192 = 0.7$,
$T_c = T_3 = 880\text{K}$.

The equation for R_2 thus becomes

$$R_2 = 2.29\left(\frac{T_{w_2}}{100}\right)^4 - 13{,}715\ \text{W/m}^2, \tag{8.27}$$

The calculation of C_1 proceeds from the Equation 8.22

$$C_{1.pz} = 0.017 \frac{k_g}{D_L^{0.2}} \left(\frac{\dot{m}_{pz}}{A_L \mu_g} \right)^{0.8} \left(T_g - T_{w_1} \right).$$

The gas properties at $T_g = 2280$ K and $P = 3040$ kPa are

$$k_g = 0.157\,\text{W/(m·K)},$$

$$\mu_g = 7.05 \times 10^{-5}\,\text{kg/(m·s)},$$

$$\dot{m}_{pz} = 2.62\,\text{kg/s},$$

$$D_L = 0.132\,\text{m},$$

$$A_L = \pi/4(0.132)^2 = 0.01368\,\text{m}^2.$$

Substitution of these values into Equation 8.22 yields

$$C_{1.pz} = 565\left(T_g - T_{w_1}\right) = 1{,}280{,}500 - 562T_{w_1}\ \text{W/m}^2. \tag{8.28}$$

For the calculation of C_2 via Equation 8,23

$$C_2 = 0.020 \frac{k_a}{D_{an}^{0.2}} \left(\frac{\dot{m}_{an}}{A_{an} \mu_a} \right)^{0.8} \left(T_{w_2} - T_3 \right),$$

we have

$$T_c = T_3 = 880\,\text{K},$$

$$k_a = 0.0553\,\text{W/(m · K)},$$

$$\mu_a = 3.89 \times 10^{-5}\,\text{kg/(m · s)},$$

$$\dot{m}_{an} = 7.074\,\text{kg/s},$$

$$A_{an} = (\pi/4)(0.192^2 - 0.1344^2) = 0.01476\,\text{m}^2,$$

$$A_{an} = 0.192 - 0.1344 = 0.0576\,\text{m}.$$

Hence,

$$C_2 = 291(T_{w_2} - 880) = 921T_{w_2} - 810{,}480\ \text{W/m}^2. \tag{8.29}$$

K_{1-2} is calculated from Equation 8.3:

$$K_{1-2} = \frac{26}{0.0012}(T_{w_1} - T_{w_2}) = 21{,}667(T_{w_1} - T_{w_2}).$$

Finally, substitution of the calculated expressions for R_1, C_1, R_2, C_2, and K_{1-2} into Equation 8.25 yields

$$T_{w_1} = 1640\text{K} \quad T_{w_2} = 1603\text{K}.$$

In the early work of Lefebvre and Herbert [2], which was confined to low combustion pressures and low heat-transfer rates, the temperature difference across the liner wall could be ignored and only negligible errors would result. However, as this example shows, for modern engines of high compression ratio, neglect of this temperature difference could give rise to significant discrepancies in heat-transfer calculations.

8.7.2 Significance of Calculated Uncooled Liner Temperatures

The liner-wall temperatures obtained above are fundamental to the geometric design and airflow distribution of the chamber. No account, however, is taken of supplementary film-cooling arrangements, which comprise an additional variable in any basic chamber design.

Although the calculation neglects such supplementary cooling, the result does not necessarily represent the maximum possible metal temperature. The calculation is based on the assumption of uniform annulus velocity, and, in regions where the velocity is appreciably below the mean value, the liner wall can attain exceptionally high temperatures. Severe hot spots often arise from the combined effect of a localized low annulus velocity on one side of the wall and a coincident breakdown in the cooling layer on the other.

Variations in flow conditions around the liner periphery produce corresponding variations in metal temperature. Nevertheless, calculated average values of liner-wall temperature, based on complete uniformity of flow conditions, serve a useful purpose in providing a comparative measure of the amount of supplementary cooling required. The qualitative effect of any change in inlet conditions on metal temperature can be readily predicted, and the operating conditions under which this temperature is a maximum can be estimated.

8.8 Film Cooling

Although many methods of supplementing the removal of heat from the liner involve a film of cooling air on the inner surface of the liner wall, the name *film cooling* is usually reserved for those schemes that employ a number of annular slots through which air is injected axially along the inner wall of the liner to provide a protective film of cooling air between the wall and the hot combustion gases. The cool film is gradually destroyed by turbulent mixing with the hot gas stream, so normal practice is to provide a succession of slots at about 40–80 mm intervals along the length of the liner. At the downstream end of the liner, the flow acceleration in the nozzle tends to suppress the hot stream turbulence, and the cooling film can persist for a much greater distance.

The main advantage of the method is that the cooling slots can be designed to withstand severe pressure and thermal stresses at high temperatures for periods up to several thousand hours. Moreover, the stiffness provided by the cooling slots results in a liner construction that is both light in weight and mechanically robust. A basic limitation of the method is that it does not allow a uniform wall temperature. The wall is coolest near each slot and increases in temperature in a downstream direction to the next slot. Thus, the method is inherently wasteful of cooling air.

The most widely used film-cooling devices are wigglestrips, stacked rings, splash-cooling rings, and machined rings.

8.8.1 Wigglestrips

In some combustors, the *static* pressure drop across the liner is too low to provide the desired amount of film-cooling air. In this situation, recourse must be made to devices that utilize the *total* pressure drop across the liner. The advantage of this approach is that it can always provide an adequate amount of cooling air, regardless of the static pressure drop across the liner. Its basic drawback is that variations in annulus velocity around the liner produce corresponding variations in the supply of cooling air.

Usually, the liner is made up of several sections, with an annular clearance between each section and the next. In one early concept, the sections overlapped each other and were joined together by "fluting," i.e., corrugating the larger diameter and spot-welding the flutes to the upstream section. This design failed to provide a satisfactory liner life and was soon replaced by a configuration that employs a corrugated spacer, known as a "wigglestrip," to connect the overlapping sections, as shown in Figure 8.2a. This method of construction provides a strong mechanical structure, but the poor aerodynamic quality of the cooling film encourages hot gas entrainment. Thermal paint tests typically indicate the presence of long hot streaks downstream of the cooling slots. A further drawback to the wigglestrip design is that

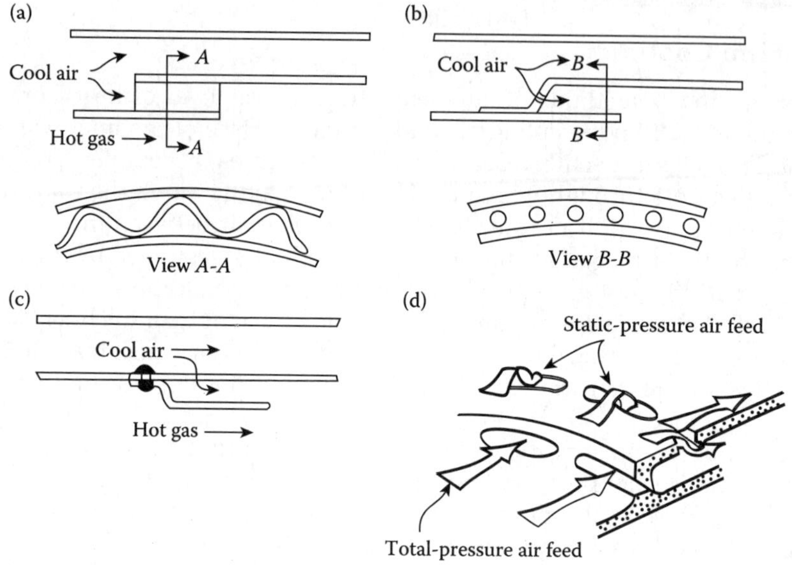

FIGURE 8.2
Film-cooling devices: (a) wigglestrip, (b) stacked ring, (c) splash-cooling ring, (d) machined ring.

wide variations in cooling-air quantity can occur between seemingly identical liners, owing to slight differences in wigglestrip material thickness. Even small variations in metal thickness, within normal manufacturing tolerances, can have a marked effect on coolant airflow rate. Nevertheless, by careful control of weld quality, and by flow-testing to check dimensional accuracy, film-cooling devices of the wigglestrip type have been used successfully on a large number of American and British aero engines at pressure ratios up to 18–1. Rolls Royce engines featuring wigglestrip cooling include the Avon, Spey, cannular Olympus, and Pegasus.

8.8.2 Stacked Ring

Another cooling device that uses total pressure feed is the "stacked ring," as shown in Figure 8.2b. Although it provides a less rigid form of liner construction than wigglestrips because the air-admission holes are drilled or punched, their dimensional accuracy is higher, resulting in smaller variations in cooling airflow rate. The total flow area of these holes is calculated to meter the required amount of cooling air. The aft end of the previous liner panel provides a plenum in which turbulence is dissipated and the individual jets coalesce to form a single annular sheet of air. At its downstream end, the gap width is dimensioned to give the required cooling-air velocity. Thus, a useful asset of this arrangement is that the cooling-air velocity can be fixed at the optimum value for maximum cooling effectiveness, regardless of the actual pressure drop across the liner.

8.8.3 Splash-Cooling Ring

This device uses only the static pressure drop across the liner wall as the driving force for the injection of film-cooling air (see Figure 8.2c). The cooling air is bled from the annulus through a row of small holes in the wall and is directed along the inside surface of the liner by means of an internal deflector "skirt" or "lip" that is attached to the wall by riveting or welding. The function of the skirt is again to provide space in which the separate air jets can merge to form a continuous sheet at the slot exit. A typical skirt length is about four times the slot depth, which is usually of the order of 1.5–3.0 mm.

8.8.4 Machined Ring

One concern with the stacked ring is the quality of the braze joint where the rings are connected [1]. Conduction of heat through this joint is essential to liner-wall cooling, and voids in the braze filler material can lead to local hot spots. This problem does not arise in the "machined ring" liner, which is machined either from a single piece of metal or from several rings welded together. Rows of holes are then drilled to allow annulus air to enter the cooling slot by either total-head feed, static pressure differential, or a combination of both, as illustrated in Figure 8.2d.

The machined ring offers advantages in terms of more accurate control of cooling-air quantity and a marked improvement in the mechanical strength of the liner, which is particularly important for large annular combustors. Machined rings have acquired many millions of hours of operational service on the Rolls Royce RB211 engine, and are specified for the RR Trent combustor, where they will be used in conjunction with augmented external convection and selective angled effusion cooling (AEC).

8.8.5 Rolled Ring

A drawback to both stacked and machined ring liners is that steep temperature gradients exist between the slot lip and the metal adjacent to the cooling air feed holes. The lip inevitably has a high temperature because the cooling air from the previous slot has lost its effectiveness, entrained hot gas, and is now heating the liner wall instead of cooling it. On the other hand, the metal near the cooling holes is immersed in air at combustor inlet temperature. The resulting thermal gradients produce high stresses that can lead to liner distortion and cracking [1].

The General Electric rolled-ring liner, shown in Figure 8.3a, is fabricated from a series of rings that are rolled into shape and welded together. In this design, the static-pressure fed air jets provide impingement cooling to the rolled ring before emerging from the slot as an effective cooling film. Similar design principles are employed in the Pratt and Whitney "double-pass" ring shown in Figure 8.3b.

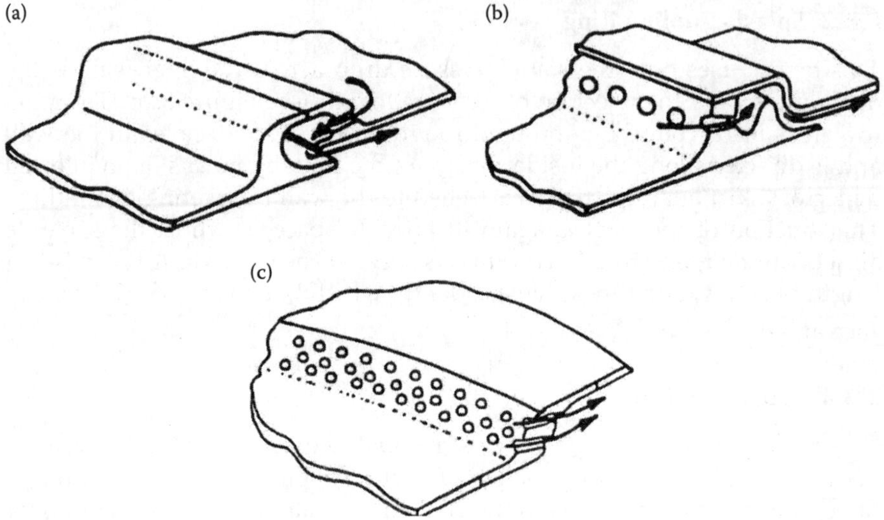

FIGURE 8.3
Film-cooling devices: (a) GE rolled ring, (b) P&W double-pass ring, (c) RR Z ring.

8.8.6 Z Ring

As discussed above, the function of the skirt or lip is to allow the individual cooling air jets to coalesce and form a continuous film. Clearly, any reduction in the initial diameter of these jets would allow a corresponding reduction in the length of lip required. The extreme case of zero lip length is the Z-ring slot, as illustrated in Figure 8.3c. This design was made possible by the increased availability of EDM and laser drilling (or trepanning) techniques. Using a large number of closely pitched, small-diameter holes ensures that the jets coalesce quickly to form a uniform film without needing the protection of a skirt.

In addition to its superior cooling performance in comparison to more conventional cooling slots, the Z-ring design has another obvious advantage in that it eliminates the life limitation due to skirt cracking. It also lends itself to manufacture using ring rolling techniques, which give good material utilization and produce contoured shells that need little machining.

One drawback to the Z ring is the high cost of drilling a large number of small holes. Improved manufacturing methods should alleviate this problem. There is also a need to control the land width between adjacent holes and other critical dimensions to ensure satisfactory mechanical integrity without loss of cooling performance.

Liners featuring Z-ring cooling have been fitted to a number of Rolls Royce military engines.

8.9 Correlation of Film-Cooling Data

Almost all the theoretical and experimental studies of film cooling carried out so far have been aimed at finding parameters to describe the temperature of an adiabatic wall at any point downstream of the coolant injection. The results of these investigations will now be examined, and later, it will be shown how the data may be applied to a liner that is nonadiabatic due to the heat fluxes produced by flame radiation and external cooling.

When a wall is film cooled by injecting a stream of air between the surface and the hot mainstream flow, three separate flow regions may be identified, as illustrated in Figure 8.4. According to Stollery and El-Ehwany [9], the first flow region comprises a *potential core*, in which the wall temperature remains close to the coolant-air temperature. This is followed by a zone where the velocity profile is similar to that of a *wall jet*. Further downstream, the flow conditions approximate those in a *turbulent boundary layer*. The relative lengths of the three regions are governed mainly by the velocity ratio U_a/U_g. For $U_a < U_g$, the second zone is nonexistent, and a turbulent boundary-layer model is appropriate for all regions downstream of the potential core.

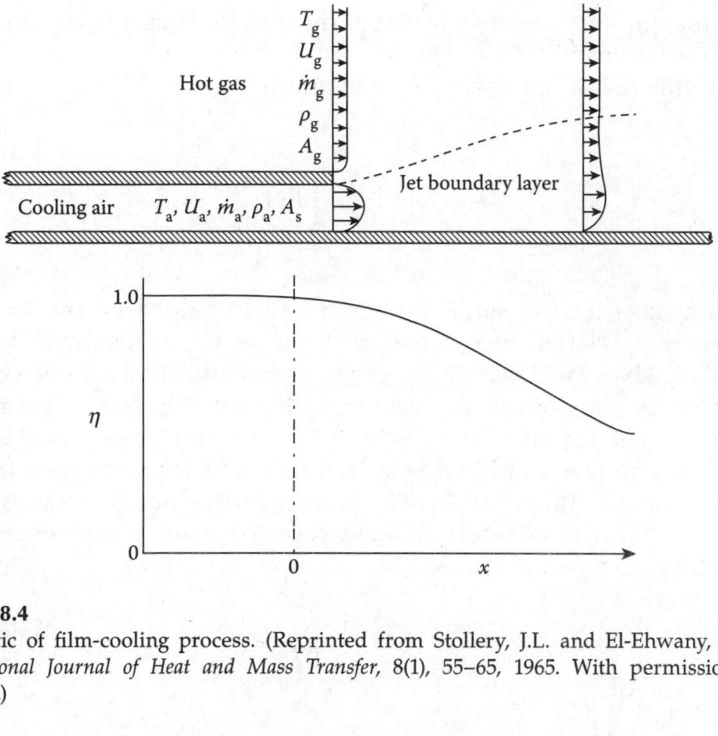

FIGURE 8.4
Schematic of film-cooling process. (Reprinted from Stollery, J.L. and El-Ehwany, A.A.M., *International Journal of Heat and Mass Transfer*, 8(1), 55–65, 1965. With permission from Elsevier.)

8.9.1 Theories Based on Turbulent Boundary-Layer Model

Stollery and El-Ehwany [9] used a turbulent boundary-layer model to derive the following expression for film-cooling effectiveness.

$$\eta = 3.09S^{-0.8}, \tag{8.30}$$

where η, the film-cooling effectiveness, is

$$\eta = \frac{T_g - T_{w.ad}}{T_g - T_a}, \tag{8.31}$$

and

$$S = \frac{x}{ms}\mathrm{Re}_s^{0.25}, \tag{8.32}$$

where s is the depth of the film-cooling slot, x is the distance downstream of the slot, m is $(\rho U)_a/(\rho U)_g$, and Re_s is the slot Reynolds number = $U_a\rho_a s/\mu_a$.

When Stollery and El-Ehwany compared their theoretical formula with the experimental data of other workers, they found that a better correlation could be obtained by increasing the value of the constant from 3.09 to 3.68 and slightly modifying the correlation group to give

$$\eta = 3.68\left(\frac{x}{ms}\right)^{-0.8}\left(\mathrm{Re}_s\frac{\mu_a}{\mu_g}\right)^{0.2}. \tag{8.33}$$

A drawback to this equation and to all others based on the Blasius skin-friction relationship is that they relate to an idealized turbulent boundary layer far downstream of the slot. Thus, they cannot satisfactorily describe the flow situation near to the slot, which is of prime interest for the gas turbine. This suggests that a better model for this region would be one that employed skin-friction coefficients obtained by direct measurement in this zone. From a study of skin-friction data, Ballal and Lefebvre [10] derived the following expression for effectiveness in the near-slot region:

$$\eta = 0.6\left(\frac{x}{ms}\right)^{-0.3}\left(\mathrm{Re}_s\frac{m\mu_a}{\mu_g}\right)^{0.15}. \tag{8.34}$$

This equation was found to predict, to ±5% accuracy, all the available experimental data within the following ranges of conditions:

Parameter	Range
m	0.5–1.3
ρ_a/ρ_g	0.8–2.5
s	0.19–0.64 cm
x/s	0–150

8.9.2 Theories Based on Wall-Jet Model

If the velocity of the cooling air is significantly higher than that of the main stream, the flow emerging from the slot behaves more like a jet than a boundary layer. This jet model applies, of course, only in regions close to the slot. Further downstream, the flow conditions revert to the boundary-layer type.

The wall-jet model of Ballal and Lefebvre [10] leads to the following expressions for effectiveness, for $1.3 < m < 4.0$:

For $x/ms < 8$:

$$\eta = 1.0.$$

For $8 < x/ms < 11$:

$$\eta = \left(0.6 + 0.05\frac{x}{ms}\right)^{-1}. \tag{8.35}$$

For $x/ms > 11$:

$$\eta = 0.7\left(\frac{x}{s}\right)^{-0.3}\left(\mathrm{Re}_s\frac{\mu_a}{\mu_g}\right)^{0.15} m^{-0.2}. \tag{8.36}$$

Values of effectiveness based on these expressions agree quite closely with published experimental data [10].

Equations 8.35 and 8.36 may be applied to the near-slot regions of all two-dimensional "clean" slots, provided that the thickness of the slot lip is small in relation to the slot height. However, for mechanical integrity, the lip is sometimes made quite thick; in that case, a wake region is created behind it, which tends to shorten the length of the potential core and extend the transition zone.

From an analysis of available experimental data concerning the influence of slot-lip thickness, t, on effectiveness, Ballal and Lefebvre [11] derived an empirical "correction factor" that, when applied to Equations 8.34 and 8.36, gives, for $0.5<m<1.3$,

$$\eta = 1.10m^{0.65}\left(\frac{\mu_a}{\mu_g}\right)^{0.15}\left(\frac{x}{s}\right)^{-0.2}\left(\frac{t}{s}\right)^{-0.2}, \tag{8.37}$$

and for $1.3<m<4.0$,

$$\eta = 1.28\left(\frac{\mu_a}{\mu_g}\right)^{0.15}\left(\frac{x}{s}\right)^{-0.2}\left(\frac{t}{s}\right)^{-0.2}. \tag{8.38}$$

It should be noted that Equation 8.38 is recommended for all values of x/ms because the additional accuracy given in Equation 8.35 for thin-lipped systems is lost when t/s exceeds 0.2, owing to the influence of lip thickness on effectiveness in this very-near-slot region [12]. Thus, the use of two separate equations for the wall-jet model is unnecessary in thick-lip situations.

A comparison of Equations 8.37 and 8.38 shows that, apart from a small difference in the value of the constant, the only significant difference is in regard to the influence of m. According to Equation 8.37, an increase in m should improve effectiveness, whereas Equation 8.38 implies that effectiveness is independent of the value of m. This apparent anomaly arises because of the contrasting effects produced by a change in m, which depend on whether the initial value of m is greater or less than unity. When $m<1$, an increase in m improves film-cooling effectiveness in two ways: (1) through a relative increase in the amount of coolant air, and (2) through a reduction in the rate of mixing between the coolant and mainstream gases as m approaches unity. However, when $m>1$, any further increase has two opposing influences on cooling effectiveness. The relative increase in coolant flow is again beneficial, but this effect is countered by a more rapid rate of mixing between the coolant and mainstream flows as m departs further from unity. This causes an increase in the thickness of the boundary layer, which consequently embraces a larger amount of mainstream gas. The net result is that effectiveness is sensibly independent of m.

In many early designs of gas turbine combustion chambers, unfortunately the cooling slots are not clean, but contain obstructions of various kinds. These reduce effectiveness by generating turbulence and enhancing mixing rates.

The effectiveness of such practical film-cooling slots has been investigated in detail by Sturgess [13]. His equation for the prediction of effectiveness may be written in a slightly simplified form for machined-ring devices as

$$\eta = 1.0 - 0.12S_N^{0.65}, \tag{8.39}$$

and for stacked-ring devices as

$$\eta = 1.0 - 0.094 S_N^{0.65}, \tag{8.40}$$

with

$$S_N = \frac{x - x_p}{ms}\left(\text{Re}\frac{\mu_a}{\mu_g}\right)^{-0.15}\frac{A_o}{A_{eff}},$$

where x_p is the potential core length, A_o is the slot outlet area, and A_{eff} is the slot overall effective area.

These equations were found by Sturgess to correlate measured values of effectiveness from a number of machined-ring and stacked-ring geometries to an accuracy of within 10%.

To summarize, for "dirty" cooling slots of the type still used in many gas turbine combustors, it is recommended that Equations 8.39 or 8.40, due to Sturgess, be used to predict effectiveness. For thick-lipped systems in which the slot geometry is sensibly "clean," it is suggested that Equations 8.37 and 8.38 be used for $m<1.3$ and $m>1.3$, respectively, over the range of x/s from 0 to 50.

8.9.3 Calculation of Film-Cooled Wall Temperature

In the calculation of film-cooled wall temperatures, the previously derived expressions for R_1, R_2, and C_2 remain the same, but the internal-convection component, C_1, is altered because the coolant flow changes both the velocity and temperature of the hot gas near the wall. Dealing with velocity first, we have [11] for $0.5<m<1.3$,

$$Nu = 0.069\left(\text{Re}_s\frac{x}{s}\right)^{0.7}, \tag{8.41}$$

leading to

$$C_1 = 0.069\frac{k_a}{x}\text{Re}_x^{0.7}(T_{w.ad} - T_{w_1}), \tag{8.42}$$

where

$$\text{Re}_x = U_a\rho_a\frac{x}{\mu_a}.$$

For $m>1.3$,

$$Nu = 0.10\mathrm{Re}_s^{0.8}\left(\frac{x}{s}\right)^{0.44}, \tag{8.43}$$

leading to

$$C_1 = 0.10\frac{k_a}{x}\mathrm{Re}_x^{0.8}\left(\frac{x}{s}\right)^{-0.36}(T_{w.ad} - T_{w_1}). \tag{8.44}$$

The gas temperature, $T_{w,ad}$, at the wall is obtained from the definition of η, given in Equation 8.31 as:

$$\eta = \frac{T_g - T_{w.ad}}{T_g - T_a},$$

where η is the effectiveness value calculated from Equations 8.37 or 8.38 for clean slots, and Equations 8.39 or 8.40 for dirty slots.

Example

For the combustor of the previous example, let us calculate the liner-wall temperature at a distance x downstream of the slot, where $x/s = 18$.

We first calculate the slot parameters. From the liner geometry, we know that

$$\frac{x}{s} = 18 \quad \frac{t}{s} = 0.4 \quad s = 0.00145\,\mathrm{m},$$

$$A_s = \pi D_L S = 5.95\times10^{-4}\,\mathrm{m}^2.$$

Thus

$$x = 18\times0.00145 = 0.0261\,\mathrm{m},$$

$$\dot{m}_s = \rho_a U_a A_s = 0.289\,\mathrm{kg/s}.$$

Hence

$$\rho_a U_a = 485.7\,\mathrm{kg/(m^2\cdot s)}.$$

For air at 880 K and a pressure of 3040 kPa,

$$\mu_a = 3.89\times10^{-5}\,\mathrm{kg/(m\cdot s)},$$

$$k_a = 0.0553\,\mathrm{W/(m\cdot K)}.$$

Therefore,

$$\text{Re}_s = \frac{\rho_a U_a s}{\mu_a} = 1.81\times10^4,$$

$$\text{Re}_x = \frac{\rho_a U_a x}{\mu_a} = 3.26\times10^5.$$

To determine the mainstream parameters, we have,

$$A_L = 0.0137\text{m}^2,$$

$$T_g = 2280\text{K},$$

$$\mu_g = 7.05\times10^{-5}\,\text{kg/(m}\cdot\text{s)},$$

$$q_{pz} = 0.05,$$

$$\dot{m}_g = \rho_g U_g A_L = 2.62\text{kg/s},$$

$$k_g = 0.157\,\text{W/(m}\cdot\text{K)}.$$

Hence,

$$\rho_g U_g = \frac{2.62}{0.137} = 191\ \text{kg/(m}^2\cdot\text{s)}.$$

To calculate the film-cooling effectiveness, we note that

$$m = \frac{\rho_a U_a}{\rho_g U_g} = \frac{485.7}{191} = 2.54.$$

Since m>1.3, it is appropriate to use Equation 8.38:

$$\eta = 1.28\left(\frac{\mu_a}{\mu_g}\right)^{0.15}\left(\frac{xt}{s^2}\right)^{-0.2} = 1.28\left(\frac{3.89}{7.05}\right)^{0.15}(18\times0.4)^{-0.2} = 0.789.$$

Now

$$\eta = \frac{T_g - T_{w,ad}}{T_g - T_a} = \frac{2280 - T_{w.ad}}{2280 - 880}.$$

Hence, $T_{w,ad}$ = 1176 K.

We begin the heat-transfer calculation by computing C_1. Since $m > 1.3$, we use Equation 8.44:

$$C_1 = 0.10\frac{k_a}{x}(\mathrm{Re}_x)^{0.8}\left(\frac{x}{s}\right)^{-0.36}(T_{w,ad} - T_{w1})$$

$$= 0.10\frac{0.157}{0.0261}(3.26\times10^5)^{0.8}18^{-0.36}(1176 - T_{w1})$$

$$= 1926(1176 - T_{w1})\,\mathrm{W/m^2}.$$

From the previous example,

$$R_1 = 794{,}460 - 0.0032T_{w1}^{2.5}\,\mathrm{W/m^2},$$

$$R_2 = 2.29\left(\frac{T_{w2}}{100}\right)^4 - 13{,}715\,\mathrm{W/m^2},$$

$$C_2 = 921T_{w2} - 810{,}400\,\mathrm{W/m^2},$$

$$K_{1-2} = 21{,}667(T_{w1} - T_{w2}).$$

Substitution of these expressions for C_1, R_1, C_2, R_2, and K_{1-2} into Equation 8.25 yields

$$T_{w1} = 1283\mathrm{K} \quad T_{w2} = 1265\mathrm{K}.$$

A comparison of the results for T_w with and without film cooling shows that, for the conditions chosen in this example, the use of film-cooling air produces a decrease in T_{w_1} of 357 K. It may also be of interest to note that, for the same conditions, an increase in luminosity factor from 1.7 to 4.0 would produce an increase in T_w of approximately 80 K, whereas a decrease in film-cooling effectiveness from 0.789 to 0.700 would increase T_w by approximately 66 K.

The results of numerous calculations of film-cooled wall temperature, carried out on various representative types of gas turbine combustors [11], show that changes in operating conditions have the same general effect on wall temperature as for uncooled liners [2]. In particular, it is found that liner-wall temperatures increase with

1. An increase in pressure
2. An increase in inlet temperature
3. A decrease in air mass flow
4. An increase in liner size
5. A decrease in fuel hydrogen content (see Figure 8.5)

Figure 8.5 also illustrates the beneficial effect of film-cooling air in reducing liner-wall temperatures for a General Electric F101 combustor. In this figure, the dashed lines represent the results of calculations of uncooled wall

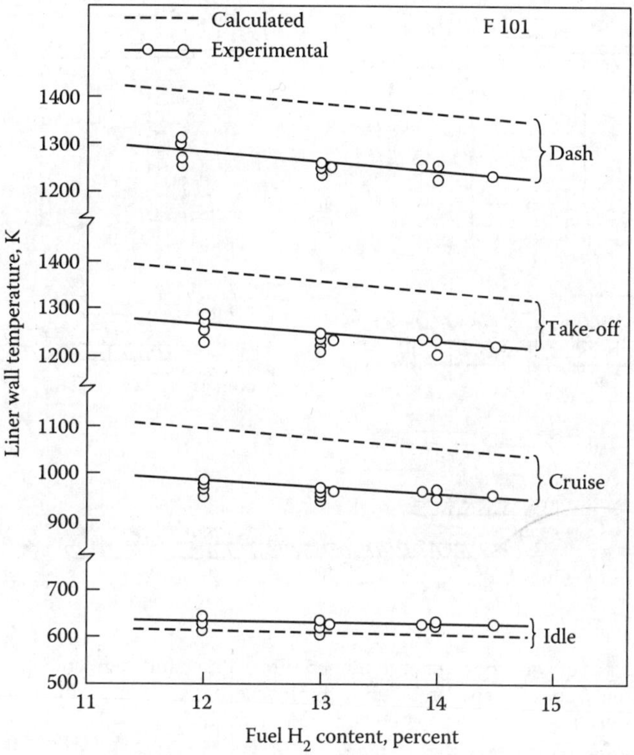

FIGURE 8.5
Comparison of measured and predicted values of wall temperature for a GE F101 combustor. (From Lefebvre, A.H., AFWAL-TR-84-2104, 1985.)

temperatures, whereas the full lines are drawn through points representing measured values of film-cooled wall temperature for different fuel types ranging from JP4 to DF2. The influence of fuel hydrogen content on flame emissivity was accounted for by combining Equations 8.11 and 8.15 to obtain

$$\varepsilon_g = 1 - \exp\left[-97440 P_3 (\%H_2)^{-2} (q l_b)^{0.5} T_g^{-1.5}\right]. \tag{8.45}$$

In Figure 8.5, the calculated values of T_w are generally higher than the corresponding measured values, as would be expected due to the neglect of internal film cooling. Only at low power conditions, where the errors incurred through neglect of internal wall cooling are partially balanced by the assumption of 100% combustion efficiency in the combustion zone, do the measured and calculated wall temperatures roughly coincide.

8.9.4 Film Cooling with Augmented Convection

Studies have shown that a substantial reduction in the film-cooling airflow requirement can be achieved by augmenting the convective heat transfer

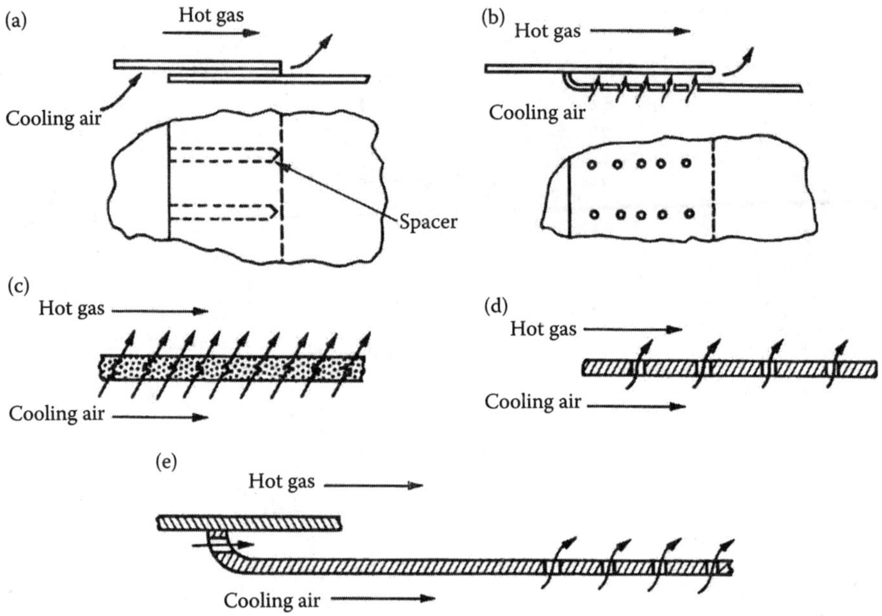

FIGURE 8.6
Alternative types of wall cooling: (a) combined film cooling and augmented external convection, (b) multijet impingement combined with film cooling, (c) transpiration cooling, (d) effusion cooling, (e) effusion cooling combined with film cooling.

on the coolant side of the liner [14,15]. Figure 8.6a illustrates the use of a double-walled construction to provide film cooling and augmented external convection.

Further augmentation of C_2 by roughening the inside surfaces of the double-walled slot (e.g., by chemical etching) has resulted in additional improvement of the total cooling effectiveness [16].

8.9.5 Impingement Cooling

Another method of augmenting the conventional film-cooling process is by means of an impingement film system, as illustrated in Figure 8.6b. This is similar to the convection system described above, except that the double-walled passage is blocked at its upstream end, and the outer wall in the double-walled region is perforated. The advantage of the method derives from its use of cooling air to serve a dual purpose. First, the air is shaped into multiple small jets that provide impingement cooling to one section of the liner wall, and then the jets merge to form an annular sheet that operates in a conventional film-cooling mode to cool a further section of the liner wall. Another advantage of impingement cooling is that the impingement jets can be positioned to provide extra cooling on liner hot spots [17].

Impingement cooling requires a double-wall construction with attendant penalties in terms of cost and weight. Another drawback stems from the fundamental difference in temperature between the two walls. This causes problems of differential expansion that can lead to buckling of the inner wall if the local hot spots become too severe. Moreover, the high heat-transfer coefficients that are normally associated with impingement cooling cannot be realized in full, because the film of air formed on the inner wall from the upstream air jets tends to reduce the efficacy of the impingement cooling from the downstream air jets.

8.9.6 Transpiration Cooling

An ideal wall-cooling system would be one in which the entire liner was maintained at the maximum temperature of the material because cooler regions would represent a wasteful use of cooling air. The method that comes closest to this ideal is known as transpiration cooling, whereby the liner wall is constructed from a porous material that provides a large internal area for heat transfer to the air passing through it, as illustrated in Figure 8.6c. Since the pores are uniformly dispersed over the surface of the wall, the tiny air jets emerging from each pore rapidly coalesce to form a protective layer of cool air over the entire inner surface of the liner. In this way, the convective heat transfer from the hot gas to the liner walls can be drastically reduced, resulting in substantial economies in film-cooling air.

Even if the protective film of air is completely successful in preventing the hot gases from making physical contact with the inner liner wall, the latter will still be exposed to intense radiation from the flame. The only means by which this heat can be removed is by transfer to the coolant air during its passage through the porous wall. This means that, in addition to acting as a porous medium, the wall must also have good heat-transfer properties and be of adequate thickness. A problem this poses is that, in order to form a stable boundary layer on the inner surface of the wall, the coolant flow should emerge with as low a velocity as possible, whereas for maximum heat transfer within the wall, a high velocity is required.

8.10 Practical Applications of Transpiration Cooling

Although transpiration cooling is potentially the most efficient method of liner cooling, its practical implementation has been hampered by the limitations of available porous materials. The porous materials developed to date have failed to demonstrate the required tolerance to oxidation, which has led to the small passages becoming blocked. These passages are also prone to blockage by airborne debris.

The search for more practical methods of increasing the internal heat transfer within the liner wall has led to the development of multilaminate sheets from which a "quasi-transpiration" cooled wall can be manufactured. The two best-known examples of this approach are Transply and Lamilloy, as described below.

8.10.1 Transply

Transply is the name given to a composite liner structure developed by the Rolls Royce Company [18]. It is produced by brazing together two or more laminates of a high-temperature alloy containing a multiplicity of interconnecting flow passages, as illustrated in Figure 8.7. These passages are designed for maximum heat transfer between the wall material and the air flowing through them. After cooling the wall structure, the air emerges in the form of numerous tiny jets flowing through evenly spaced holes on the hot gas side of the liner wall. The air jets are deflected downstream, gradually forming an insulating blanket between the hot gas and the wall. Thus, the liner is cooled partly by the air passing through the wall, and partly through the leaving air acting as a protective film.

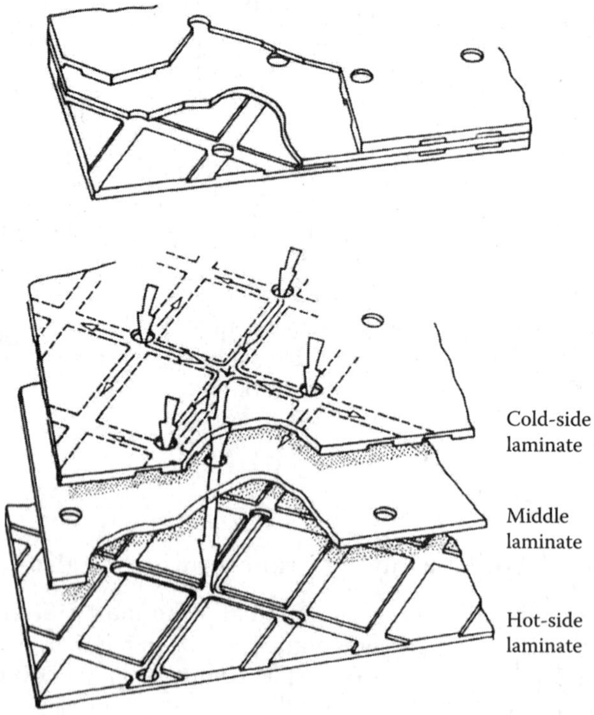

FIGURE 8.7
Constructional features of Rolls Royce Transply. (From Wassell, A.B., and Banghu, J.K., ASME Paper 80-GT-66, 1980. With permission.)

An important requirement is that the alloy selected for the sheets should have good brazing characteristics, to ensure the mechanical integrity of the liner structure, and high oxidation resistance at normal operating temperatures. Intergranular oxidation of the thin-walled sections would reduce the strength of the structure, whereas excessive oxide formation on the internal surfaces would lead to partial blockage of the cooling passages and hence to local overheating.

A Transply combustor was developed for the RR Spey Mk 512 engine and entered service in the BA 1-11 aircraft. It achieved a 70% reduction in cooling airflow requirement relative to the conventional film-cooled Spey combustor. This large saving in cooling air allowed the liner internal airflow distribution to be optimized for minimum pollutant emissions.

8.10.2 Lamilloy

In the United States, the Allison Engine Company has pursued a similar development of quasi-transpirate materials and has produced a multilaminate porous structure, known as Lamilloy, which is fabricated from two or more diffusion-bonded, photoetched metal sheets, as illustrated in Figure 8.8.

In general, Lamilloy has the same high cooling potential and the same practical problems and limitations as Transply. Both materials require further development to reduce manufacturing costs and achieve better process control. A more basic deficiency is their lack of mechanical strength in comparison with more conventional cooling methods such as, e.g., machined and rolled rings. This drawback is likely to militate against their application to large annular combustors.

At the time of their emergence and initial development in the 1970s, both Transply [18] and Lamilloy [19] were viewed as replacements for the various film-cooling configurations then in widespread use. Today, with the advent of new cooling concepts, along with major advances in manufacturing techniques, the competition is far more keen. It will, therefore, be of great interest to see what the future holds for these and other quasi-transpiration cooling schemes.

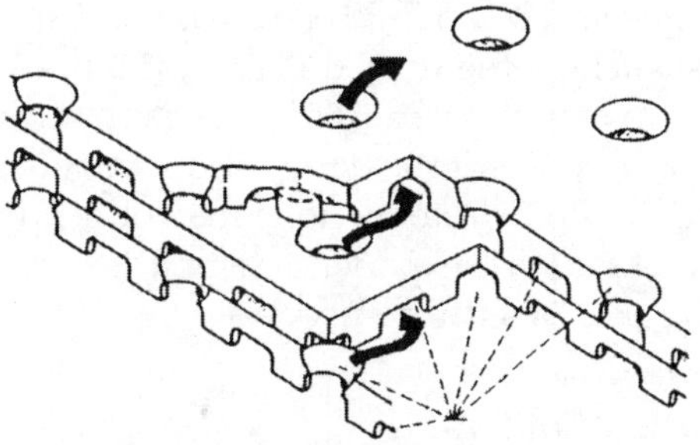

FIGURE 8.8
Lamilloy construction and airflow path. (From Nealy, D.A., *Gas Turbine Combustor Design Problems*, Hemisphere, Washington, DC, 151–85, 1980.)

8.10.3 Effusion Cooling

Another, and perhaps the simplest approach to a practical form of transpiration cooling, is a wall perforated by a large number of small holes, as shown in Figure 8.6d. Ideally, the holes should be large enough to remain free from blockage by impurities, but small enough to prevent excessive penetration of the air jets. Provided that the jet penetration is small, it is possible to produce, along the inner surface of the liner, a fairly uniform film of cooling air. If, however, the penetration is too high, the air jets rapidly mix with the hot gases and provide little cooling of the wall downstream.

Effusion cooling can be applied to all or any portion of the liner wall, but because it is somewhat lavish in its use of cooling air, it is best used for treating local hot spots in the liner wall. Another useful role is in improving the effectiveness of a conventional film-cooling slot. As the film of air from this slot moves downstream, its temperature gradually rises due to entrainment of the surrounding combustion gases. Eventually, it becomes so hot that it starts to heat the liner wall instead of cooling it. If effusion cooling is applied before this point is reached, the injection of cold air into the film enables it to maintain its cooling effectiveness for a further distance downstream. A typical arrangement is shown in Figure 8.6e.

Few published data are available on the performance of conventional effusion cooling systems. This is not a serious omission because recent developments in AEC have rendered them virtually obsolete.

8.11 Advanced Wall-Cooling Methods

The above discussion on wall-cooling methods has concentrated largely on concepts that fall into the general category of film cooling. This is appropriate because almost all of the combustors now in service employ film cooling in one form or another. However, few of the combustors being designed today for application to high-performance gas turbines use film cooling, except perhaps in local areas, such as the liner dome. At the present time, interest is focused mainly on two new and widely different approaches—AEC and tiled walls. These new concepts merit the designation "advanced," partly because of their potential for significant reductions in cooling-air requirements, but also because much more service experience with these devices must be gained before they can be regarded as orthodox designs.

8.11.1 Angled Effusion Cooling

With conventional effusion cooling, the holes are drilled normal to the liner wall. The advantages to be gained from drilling the holes at a more shallow angle are twofold, as described by Dodds and Ekstedt [20].

1. An increase in the internal surface area available for heat removal. This area is inversely proportional to the square of the hole diameter and the sine of the hole angle. Thus, for example, a hole drilled at 20° to the liner wall has almost three times the surface area of a hole drilled normal to the wall.
2. Jets emerging from the wall at a shallow angle have low penetration and are better able to form a film along the surface of the wall. The cooling effectiveness of this film also improves as the hole size and angle are decreased.

From this brief description of AEC, it is clear that its practical implementation is highly dependent on an ability to accurately, consistently, and economically manufacture large numbers of oblique holes of very small diameter. Advances in laser drilling have made this possible, and AEC is now regarded as a viable and economically acceptable cooling technique. At the present time, the lower limit on hole diameter is about 0.4 mm, whereas the lowest attainable hole angle is just below 20°.

AEC is perhaps the most promising contender among the various advanced combustor cooling techniques that are being actively developed for the new generation of industrial and aeronautical gas turbines. It is a technique that can be used locally, to supplement other forms of wall cooling, or it can be applied to the entire liner. It is used extensively on the GE 90 combustor, where it has reduced the normal cooling-air requirement by 30%.

The known drawbacks to AEC include an increase in liner weight of around 20%, which stems from the need for a thicker wall to achieve the required hole length and to provide buckling strength. Cost is also an important consideration, and this is allied to the need to drill the holes consistently and to specification in a production environment. Other concerns include the "repairability" of AEC liners and their durability. These issues can only be fully resolved by extensive service experience.

Future developments in AEC will tend to focus on the optimization of hole geometry, a topic that cannot be separated from hole manufacture. A diffuser-shaped expansion at the exit portion of the hole has been shown to improve cooling effectiveness [21], presumably because the lower exit velocity reduces the penetration of the air jet into the hot gas stream. As an extension of this approach, lateral expansion of the hole exit (fanning) improves the lateral spreading of the jet to give better coverage of the wall surface. However, a cost effective method of producing shaped holes has yet to be developed.

8.11.2 Tiles

The principle of using tiles is well established for large industrial engines where size and weight are of minor importance and it is thus practicable to line the combustor with refractory bricks to reduce the heat flux to the liner wall.

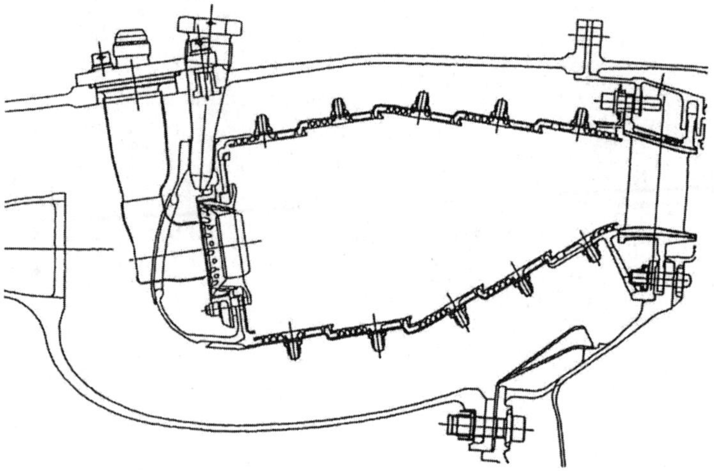

FIGURE 8.9
P&W V2500 tiled combustor.

Refractory bricks are clearly too heavy and cumbersome for application to aero and most industrial engines, but metallic tiles offer an attractive solution. The V2500 engine is now in service with a tiled combustor, as shown in Figure 8.9, and P&W is also using tiles on its radially staged combustor for the PW4000.

The method of construction is to mount a large number of tiles on a support shell. The shell is protected by the tiles, but the tiles themselves are exposed to the hot combustion gases. This effectively decouples the mechanical stresses, which are taken by the shell, from the thermal stresses, which are taken by the tiles. Methods for alleviating the thermal load on tiles involve various heat-transfer features, such as multiple pedestals (see below) on their rear surface to enhance the convective heat transfer. Metered air flows over the pedestals and is then ejected at the ends of the tiles to form a protective film over their front surface. Thermal barrier coatings (TBCs) can also be used to give extra protection to this surface.

Tiled walls offer the following advantages:

1. The tiles can be cast from blade alloy materials having a much higher temperature capability (>100°C) than typical combustor alloys
2. Since the combustor shell remains at a uniform low temperature, relatively cheap alloys can be used
3. The low shell temperature minimizes thermal growth relative to the combustor casing
4. Maintenance time and cost is reduced because changing a tile is simpler than repairing a liner
5. Significant reduction in cooling-air requirement

The main drawbacks of tiled combustors are:

1. A substantial increase in weight
2. The various ports of entry for the combustion and dilution airstreams are difficult to modify during combustor development
3. Difficult to scale down the tile attachment features for application to small engines

8.12 Augmented Cold-Side Convection

The rate of convective heat transfer on the cool side of the liner can be increased by the use of fins, pedestals, and ribs, or any other form of secondary surface that increases the effective area for heat exchange. Secondary surfaces cannot be 100% efficient because a temperature gradient must exist along each protuberance, whatever its physical form, to allow heat to be conducted away from its base.

Pedestals have been used on heat shields in the dome area of RR RB211 combustors for many years, and the RR Tay combustor also features a pedestal head. Pedestals are also used to augment convective heat transfer on the cold sides of wall tiles. Finned outer surfaces have featured successfully in a number of industrial gas turbines, but only recently has their use become more widespread. Usually, the fins or ribs are arranged to run longitudinally, but circumferential fins are also used at the expense of a small increase in pressure loss. Dutta et al. [22] have described a catalytic combustor that employs continuous round wire welded to the outside of the liner. The design is based on data from Norris [23] and Evans and Noble [24], which show an average threefold augmentation in convective heat transfer when compared with similar geometries without the wires.

More detailed information on the use and performance of various extended-surface configurations, including ribs, fins, and pedestals, may be found in Gardner [25] and Lohmann and Jeroszko [26].

8.13 Thermal Barrier Coatings

One attractive approach to the problem of achieving satisfactory liner life is to coat the inside of the liner with a thin layer of refractory material. A suitable material of low emissivity and low thermal conductivity could reduce the wall temperature in two ways: (1) by reflecting a large part of the incident gas radiation; and (2) by providing a layer of thermal insulation between the hot gas and the wall, thereby reducing the temperature of the supporting base metal. A further benefit may be gained if an oxidation-resistant base coat is applied because it reduces the oxidation constraint on the choice of liner-wall material.

An ideal TBC would be chemically inert and have good mechanical strength, resilience to thermal shock, and resistance to wear and erosion. Above all, it would have a low thermal conductivity and a thermal expansion coefficient that is similar to that of the base metal. A typical TBC comprises a metallic base coat (such as 0.1 mm of Ni Cr AL Y), plus one or two layers of ceramic (such as partially yttria stabilized zirconia). Recent developments in the strain tolerance of TBCs have reduced the necessity for an intermediate coat, and two-layer coatings are now sometimes specified for improved mechanical integrity.

Plasma flame spraying is generally used to apply the ceramic and base coat layers because it is found to provide durable and reproducible coatings. A typical overall coating thickness is around 0.4–0.5 mm, which gives metal temperature reductions of the order of 40–70 K, depending on the heat flux through the liner wall. In this context, it should be noted that for a TBC to be fully effective, there must be adequate heat removal from the "cold" side of the liner wall. Inevitably, this means that liner geometries will become more complex as various features (such as fins, ribs, etc.) are added to augment the *J* convective heat transfer from the outer wall in order to derive full benefit from the TBC on the inner wall.

The reduction in wall temperature obtained from using a TBC may be calculated by adding another term to Equation 8.25 to give

$$R_1 + C_1 = R_2 + C_2 = K_{1-i} = K_{i-2}, \tag{8.46}$$

where

$$K_{1-i} = (k/t)_{\text{TBC}}(T_1 - T_\text{i}),$$

$$K_{i-2} = (k/t)_\text{w}(T_i - T_2),$$

where T_1 is the hot-side surface temperature of TBC, T_2 is the cold-side surface temperature of liner wall, and T_i is the temperature at interface between TBC and liner wall.

From the previous example we have

$$R_1 = 794{,}460 - 0.00321 T_1^{2.5}\ \text{W/m}^2,$$

$$C_1 = 1926(1176 - T_1)\,\text{W/m}^2,$$

$$R_2 = 2.29(T_2/100)^4 - 13{,}715\,\text{W/m}^2,$$

$$C_2 = 921T_2 - 810{,}400\,\text{W/m}^2.$$

Solution of these equations for no TBC gave

$$T_1 = 1283\text{K} \quad T_2 = 1265\text{K}.$$

The value of thermal conductivity used in the previous example for the Nimonic wall was 26 W/(m·K). The thermal conductivity of a TBC is typically an order of magnitude lower. Thus, for a coating of thickness 0.5 mm on a wall of 1.2 mm thickness we have

$$K_{1-i} = (2.6/0.0005)(T_1 - T_i),$$

and

$$K_{i-2} = (2.6/0.0012)(T_i - T_2).$$

Substitution of these expressions for $R_1, C_1, R_2, C_2, K_{1-i}$, and K_{i-2} into Equation 8.46 yields

$$T_1 = 1304\text{K} \quad T_i = 1236\text{K} \quad T_2 = 1220\text{K}$$

Thus, the TBC has reduced the peak metal temperature from 1283 to 1236 K. Note that the hot-side temperature of the TBC is 21 K higher than the metal surface without TBC.

8.14 Materials

Continuing efforts to improve engine performance and reduce fuel consumption rely heavily on the development of new combustor materials to withstand the harsher environmental conditions associated with operating at higher pressures and temperatures. During the past 30 years, materials and manufacturing processes have improved appreciably in terms of higher temperature capability and lower cost. With the continuing development of new materials and new methods of construction, this progress seems likely to continue for the foreseeable future. The material requirements for future aero engines have been reviewed in papers by Kirk [27] and Rosen and Facey [28].

Current production combustors are typically fabricated from sheets of nickel- or cobalt-based alloys. These conventional materials still have considerable development potential and will continue to dominate the aeronautical scene for some time to come. In the longer term, ceramics and ceramic composites offer major improvements provided a number of inherent drawbacks can be overcome.

The basic requirements for combustor materials can be listed as follows:

- High-temperature strength
- Resistance to oxidation and corrosion
- Low density
- Low thermal expansion
- Low Young's modulus
- Resistance to thermal fatigue
- Low cost
- Easy to fabricate
- High thermal conductivity

From inspection of this list, it is clear that the traditional requirements of good mechanical strength and oxidation resistance at high temperatures are by no means the only desirable attributes of combustor materials. The metal alloys now in common use are satisfactory for long-term operation at temperatures up to around 1100 K. Oxidation becomes rapid at temperatures above around 1300 K. However, developments in TBCs that incorporate an oxidation-resistant base coat have tended to diminish the importance of a material's oxidation resistance.

The requirement of low cost is just as important for the combustor as for other engine components, especially as most high-temperature materials tend to be high-cost materials also. Low density (i.e., low weight) is clearly of special importance for aero-engine combustors.

The steep temperature gradients that exist around features such as dilution holes and cooling slots result in high thermal stresses. Thus, good thermal fatigue strength is an important prerequisite for satisfactory liner life.

A high thermal conductivity is a desirable material property because it facilitates the dissipation of heat from local "hot spots" in the liner wall. It is especially beneficial in pseudoporous wall constructions, such as Lamilloy, Transply, and effusion cooling, which rely for their effectiveness on achieving a high rate of heat transfer from the liner wall to the cooling air flowing through it.

8.14.1 Metal Alloys

During the past 40 years, the nickel-based alloys Nimonic 75 and Hastelloy X have been widely used in the UK and the United States, respectively, as sheet materials for combustor liners. The success of these alloys is closely linked to their ease of fabrication by forming and welding. They are satisfactory for long-term operation at temperatures up to around 1100 K. Above this temperature, the strength of these materials falls to unacceptable levels.

Modern combustor liners make extensive use of Nimonic 263 and the cobalt-based Haynes 188 (HS 188), which is generally regarded as a replacement for Hastelloy X because of its exceptionally high strength at temperatures above 1070 K. Nimonic 263 has superior strength-temperature characteristics to Nimonic 75, is easier to fabricate, and is relatively cheap. Nimonic 86 has been developed as an alternative to Nimonic 263. It is highly oxidation resistant and is readily fabricated, but its mechanical strength is inferior to that of Nimonic 263 over most of its temperature range.

8.14.2 Ceramics

Although metal materials will continue to be of prime importance for the foreseeable future, with developments taking place both in the materials themselves and in manufacturing processes, for the long term, only ceramic materials have the capability of meeting future engine requirements.

Ceramics have good mechanical strength at high temperatures, low density, and are oxidatively stable at temperatures well beyond the capability of unprotected metals and alloys. These are clearly attractive qualities where the reduction of liner-wall cooling air is an important design requirement.

Silicon compounds are considered to be most promising and foremost among these are silicon carbide and silicon nitride. Monolithic silicon nitride and silicon carbide exhibit high strength and stiffness up to about 1680 and 1880 K, respectively [27]. The main drawback to these and other ceramics is that, although they are strong at high temperatures, they do not possess the toughness and ruggedness that engineers have become accustomed to with ductile metals. To some extent, this problem can be alleviated by the incorporation of particles or whiskers to deflect and arrest cracks [28]. An important asset of this ceramic matrix construction is that when failure occurs, it does so in a gradual and progressive manner instead of the catastrophic fracture that is normally associated with monolithic materials. Also, continuous ceramic filaments allow ceramic components to tolerate minor flaws and, generally, to mimic metallic behavior, but with a higher temperature capability. Ceramic composites that utilize continuous silicon carbon fibers to reinforce silicon carbide have been commercially available for some time [27].

A series of full-scale engine tests have been carried out at Solar on a Centaur 50 industrial gas turbine fitted with silicon carbide composite combustion liners. These liners functioned well over long periods of engine operation with no apparent problems [29]. This progress encourages the notion that developments in material properties, design methods, and manufacturing processes are now reaching a stage where the introduction of ceramic liners in certain engine applications merits serious consideration. However, a number of concerns still exist that are related to the inherent brittleness of ceramics such as, for example, damage arising due to the ingestion of foreign objects.

8.14.3 Mechanical Integrity

In designing a combustor, the mechanical integrity of all the components in the module is of vital importance. Predicting component life is an essential part of reliability and service warranties. There are several key elements involved in ensuring the mechanical integrity of casing: pressure containment, life, fan-blade-off, and shock loads.

The casing must neither buckle nor rupture under the most extreme pressure loadings seen by the engine, and pressure-vessel tests are performed to assess this. The casing must last the life of the modern engine, which is approximately: 15,000 hours for naval marine engine, 25,000 flights for a large civil aero engine, or 100,000 hours for an industrial engine. If a fan blade is lost during engine running, engine surges, or a bird is ingested, the casing must not buckle and flange integrity must be maintained despite the large vibration caused by the out-of-balance engine. In the event of a flame-out, there is still high compressor-air pressure outside the combustor wall, while the pressure inside the combustor rapidly collapses causing a buckling load on the outer combustor wall. The combustor should be able to preserve its mechanical integrity under these conditions. Finally, the casing should pass various high-cycle fatigue tests.

8.15 Liner Failure Modes

Inspection of combustion-chamber components during engine overhaul often reveals buckling and cracking of the liner. When such failure occurs after a relatively short period of operation, the cause is usually found to be errors in manufacture or design. Hot spots on the liner wall may be created by igniter plugs, liner support struts, or any other object whose presence in the annulus airflow causes a breakdown in convective cooling in the wake region immediately downstream. Even if these problems do not arise, after long periods of engine operation it is customary to observe signs of distress in the form of liner buckling and cracking. Usually, the cracks originate at geometrical discontinuities, such as cooling rings and air-admission holes, or at other points where residual stresses may be induced during manufacture.

High thermal stresses are usually due to a combination of temperature distribution and liner stiffness [1]. Cooling rings are not only stiff, but they also operate close to the cooling-air temperature. Downstream of the cooling ring, the metal temperature rises as the film-cooling air loses its effectiveness and a thermal gradient is created between the stiff ring and the relatively weak metal downstream. As the hot areas try to expand, they are subjected to compression by the surrounding cold metal, which causes them to yield. After engine shutdown, these same areas are forced into tension by the surrounding unyielded metal. After many cycles of engine startup and shutdown, distortion and/or cracking can occur.

In general, buckling results from long periods of operation under a combination of high temperature and high-temperature gradients. Loss of material thickness, or pitting of the surface due to oxidation or high-temperature corrosion, clearly facilitates buckling and must be circumvented by using materials with good oxidation resistance or by the application of TBCs. With aircraft engines, buckling loads are highest when flame blowout occurs at maximum power and low altitude. For marine and industrial engines running offshore, corrosion caused by salt ingestion and by the high sulfur content of the fuel are additional factors to consider in assessing the durability and failure of combustor liners.

Nomenclature

C_1 convection heat flux from combustion gas to liner, W/m^2
C_2 convection heat flux from liner to annulus air, W/m^2
C/H carbon/hydrogen mass ratio of fuel
D_L liner diameter (can) or height (annular), m
d_h hydraulic mean diameter (4 × flow area/wetted perimeter), m
h heat-transfer coefficient, W(m^2 K)
K conduction heat transfer along liner wall, W/m^2
K_{1-2} conduction heat flux through liner wall, W/m^2
k thermal conductivity, W/(m·K)
L luminosity factor
l_b mean beam length of radiation path, m
m mass velocity ratio $[(\rho U)_a/(\rho U)_g]$
$\dot{m}$ mass flow rate, kg/s
Nu Nusselt number
P total pressure, kPa
Q heat flux, W/m^2
q fuel/air ratio by mass
R_1 radiation heat flux from combustion gas to liner, W/m^2
R_2 radiation heat flux from liner to casing, W/m^2
Re Reynolds number
Re_s slot Reynolds number $(U_a\rho_a S/\mu_a)$

S $(x/ms)\ \mathrm{Re}_s^{0.25}$
s slot height, m
T absolute temperature, K
$T_{w,ad}$ adiabatic wall temperature, K
ΔT_{comb} temperature rise due to combustion, K
t slot lip thickness, m
t_w liner wall thickness, m
U velocity, m/s
x distance downstream of slot, m
α absorptivity
ε emissivity
η film-cooling effectiveness
η_c local combustion efficiency
μ dynamic viscosity, kg/(ms)
ρ density, kg/m^3
σ Stefan–Boltzmann constant, 5.67×10^{-8} $W/(m^2\,K^4)$

Subscripts

a air
an annulus
g gas
L liner
pz primary zone
ref reference value
s slot
w liner wall
1 flame side of liner wall
2 coolant side of liner wall
3 combustor inlet condition

References

1. Dodds, W. J., and Bahr, D. W., "Combustion System Design," in A. M. Mellor, ed., *Design of Modern Gas Turbine Combustors*, pp. 343–476, Academic Press, New York, NY, 1990.
2. Lefebvre, A. H., and Herbert, M. V., "Heat-Transfer Processes in Gas Turbine Combustion Chambers," *Proceedings of the Institution of Mechanical Engineers*, Vol. 174, No. 12, pp. 463–73, 1960.
3. Hottel, H. C., "Some Problems in Radiative Transport," in D. B. Olfe and S. S. Penner, eds., *International Developments in Heat Transfer*, ASME, New York, pp. 126–33, 1960.
4. Reeves, D., "Flame Radiation in an Industrial Gas Turbine Combustion Chamber," National Gas Turbine Establishment, UK, NGTE Memo M285, 1956.

5. Fishenden, M., and Saunders, O. A., *An Introduction to Heat Transfer*, Oxford University Press, New York, 1950.
6. Kretshmer, D., and Odgers, J., "A Simple Method for the Prediction of Wall Temperatures in a Gas Turbine Combustor," ASME Paper 78-GT-90, 1978.
7. Lefebvre, A. H., "Influence of Fuel Properties on Gas Turbine Combustion Performance," AFWAL-TR-84-2104, 1985.
8. Hottel, H. C., "Radiant Heat Transmission," in W. H. McAdams, ed., *Heat Transmission*, 3rd ed., Chap. 4, McGraw-Hill, New York, 1954.
9. Stollery, J. L., and El-Ehwany, A. A. M., "A Note on the Use of a Boundary-Layer Model for Correlating Film-Cooling Data," *International Journal of Heat and Mass Transfer*, Vol. 8, No. 1, pp. 55–65, 1965.
10. Ballal, D. R., and Lefebvre, A. H., "Film-Cooling Effectiveness in the Near Slot Region," *Journal of Heat Transfer*, Vol. 95, pp. 265–6, 1973.
11. Ballal, D. R., and Lefebvre, A. H., "A Proposed Method for Calculating Film-Cooled Wall Temperatures in Gas Turbine Combustion Chambers," ASME Paper 72-WA/HT-24, 1972.
12. Kacker, S. C., and Whitelaw, J. H., "An Experimental Investigation of the Influence of Slot Lip Thickness on the Impervious Wall Effectiveness of the Uniform Density, Two-Dimensional Wall Jet," *International Journal of Heat and Mass Transfer*, Vol. 12, pp. 1196–201, 1969.
13. Sturgess, G. J., "Correlation of Data and Prediction of Effectiveness from Film Cooling Injection Geometries of a Practical Nature," in E. R. Norster, ed., *Cranfield International Propulsion Symposium*, Vol. 11, pp. 229–50, Pergamon, Oxford, 1969.
14. Marek, C. J., and Juhasz, A. J., "Simultaneous Film and Convection Cooling of a Plate Inserted in the Exhaust Stream of a Gas Turbine Combustor," NASA TN D-7156, 1973.
15. Mularz, E. J., and Schultz, D. F., "Measurements of Liner Cooling Effectiveness within a Full-Scale Double-Annular Ram-Induction Combustor," NASA TN D-7689, 1974.
16. Nealy, D. A., "Combustor Cooling—Old Problems and New Approaches," in A. H. Lefebvre, ed., *Gas Turbine Combustor Design Problems*, pp. 151–85, Hemisphere, Washington, DC, 1980.
17. Burrus, D. L., Charour, C. A., Foltz, H. L., Sabla, P. E., Seto, S. P., and Taylor, J. R., "Energy Efficient Engine Combustor Test Hardware—Detailed Design Report," NASA CR-168301.
18. Wassell, A. B., and Banghu, J. K., "The Development and Application of Improved Combustor Wall Cooling Techniques," ASME Paper 80-GT-66, 1980.
19. Nealy, D. A., and Reider, S. B., "Evaluation of Laminated Porous Wall Materials for Combustor Liner Cooling," *Journal of Engineering for Power*, Vol. 102, pp. 268–76, 1980.
20. Dodds, W. J., and Ekstedt, E. E., "Broad Specification Fuel Combustion Technology Program," Phase II, Final Report, 1989.
21. Gritsch, M., Schulz, A., and Wittig, S., "Adiabatic Wall Effectiveness Measurements of Film-Cooling Holes with Expanded Exits," ASME Paper 97-GT-164, 1997.
22. Dutta, P., Cowell, L. H., Yee, D. K., and Dalla Betta, R. A., "Design and Evaluation of a Single-Can Full Scale Catalytic Combustion System for Ultra-Low Emissions Industrial Gas Turbines," ASME Paper 97-GT-292, 1997.

23. Norris, R. H., "Some Simple Approximate Heat Transfer Correlations for Turbulent Flow in Ducts with Rough Surfaces," *Augmentation of Convective Heat and Mass Transfer*, ASME, pp. 16–26, 1970.
24. Evans, D. M., and Noble, M., "Gas Turbine Combustor Cooling by Augmented Backside Cooling," ASME Paper 78-GT-33, 1978.
25. Gardner, K. A., "The Efficiency of Extended Surfaces," *Transactions of the American Society of Mechanical Engineers*, Vol. 67, pp. 621–31, 1945.
26. Lohmann, R. P., and Jeroszko, R. A., "Broad Specification Fuels Technology Program," Phase I Final Report, NASA CR-168180, NASA, Washington, DC, 1982.
27. Kirk, G. E., "Material Requirements for Future Aeroengines," *Ninth International Symposium on Air Breathing Engines*, Paper No. 89-7033, AIAA, Washington, DC, 1989.
28. Rosen, R., and Facey, J. R., "Civil Propulsion Technology for the Next Twenty-Five Years," *Eighth International Symposium on Air Breathing Engines*, Paper No. 87-7000, AIAA, Washington, DC, 1987.
29. Richerson, D. W., "Ceramics for Turbine Engines," *Mechanical Engineering*, Vol. 119, No. 9, pp. 80–3, 1997.

9

Emissions

9.1 Introduction

Pollutant emissions from combustion processes have become of great public concern due to their impact on health and the environment. The past decade has witnessed rapid changes both in the regulations for controlling gas turbine emissions and in the technologies used to meet these regulations. During this period, the consumption of fuel by civil aviation has increased to the extent that air transport is now perceived as one of the world's fastest growing energy-use sectors. At the same time, stationary gas turbines have become firmly established as prime movers in the gas and oil industry, and have acquired new ranges of application in combined cycle plants and in many areas of utility power generation. All these developments have been accompanied by continuous and increasing pressure on the combustion engineer to reduce pollutant emissions from all types of gas turbines.

The material presented in this chapter is divided into six main sections:

- A general overview of emissions concerns and the regulations that have been introduced to address these concerns
- The mechanisms of pollutant formation and the methods employed in alleviating pollutant emissions from conventional gas turbine combustors
- The use of variable geometry and staged combustion for emissions reduction by control of flame temperature
- Basic approaches to the design of "dry" low oxides of nitrogen (NO_x) and "ultralow" NO_x combustors
- Alternative methods for achieving ultralow NO_x emissions, including rich-burn, quick-quench, lean-burn, and catalytic combustors
- Correlations for NO_x and carbon monoxide (CO) emissions.

Emissions from new alternative and bio-fuels used in gas turbines are presented and discussed in Chapter 10.

9.2 Concerns

The exhaust from an aircraft gas turbine is composed of CO, carbon dioxide (CO_2), water vapor (H_2O), unburned hydrocarbons (UHC), particulate matter (mainly carbon), NO_x, and excess atmospheric oxygen and nitrogen. CO_2 and H_2O have not always been regarded as pollutants because they are the natural consequence of complete combustion of a hydrocarbon fuel. However, they both contribute to global warming and can only be reduced by burning less fuel. Thus, improvements in engine thermal efficiency not only reduce direct operating costs, but also reduce pollution.

The principal pollutants are listed in Table 9.1. CO reduces the capacity of the blood to absorb oxygen and, in high concentrations, can cause asphyxiation and even death. UHC are not only toxic, but they also combine with NO_x to form photochemical smog. Particulate matter (generally called soot or smoke) creates problems of exhaust visibility and soiling of the atmosphere. It is not normally considered to be toxic at the levels emitted, but recent studies by Seaton et al. [1] indicate a strong association between asthma and other respiratory diseases and atmospheric pollution by concentrations of small particles in the microgram range. Moreover, some smoke suppressants contain heavy metals, such as barium, which add another pollutant to the exhaust gases. NO_x ($NO + NO_2$), of which the predominant compound at high emission levels is NO, not only contribute to the production of photochemical smog at ground level, but also cause damage to plant life and add to the problem of acid rain.

Relative to other sources, aircraft engines are only minor contributors to the overall NO_x burden. For example, in the United States, NO_x emissions from aircraft engines account for only about 2% of the total emissions nationwide from all sources [2]. On a global basis, NO_x emissions from aircraft engines constitute less than 3% of all man-made NO_x emissions. However, of special concern is that these emissions lead to the formation of ozone in the troposphere—the region that extends from ground level to approximately 12 km above the earth's surface. This is the region in

TABLE 9.1

Principal Pollutants Emitted by Gas Turbines

Pollutant	Effect
Carbon monoxide (CO)	Toxic
Unburned hydrocarbons (UHC)	Toxic
Particulate matter (C)	Visible
Oxides of nitrogen (NO_x)	Toxic, precursor of chemical smog, depletion of ozone in stratosphere
Oxides of sulfur (SO_x)	Toxic, corrosive

which stationary gas turbines and subsonic aircraft operate. The relevant reaction mechanisms are

$$NO_2 = NO + O,$$

$$O + O_2 = O_3.$$

Measurements taken over a long period of time at altitudes from 1 to 3 km indicate that the level of ozone over western Europe is now approaching 50 ppb (parts per billion). Prolonged exposure to ozone concentrations around 100 ppb is associated with respiratory illnesses, impaired vision, headaches, and allergies. Ground-level ozone is especially important in regions where the topographical features prevent the local weather system from removing the ozone formed in combustion, and where strong sunshine can promote the photochemical reactions that lead to smog. Los Angeles is the classic example of such a region. It is hardly surprising, therefore, that the drive toward very stringent emissions legislation on NO_x first emanated from this city.

Similar studies indicate that NO_x emissions emitted at the extreme altitudes at which supersonic aircraft are required to operate can deplete the stratospheric ozone layer via the reactions

$$NO + O_3 = NO_2 + O_2,$$

$$NO_2 + O = NO + O_2.$$

Note that at the end of these reactions, the NO is liberated to produce more ozone.

Depletion of the ozone layer allows increased penetration of solar ultraviolet radiation, which produces a corresponding increase in the incidence of skin cancer.

With stationary engines burning residual fuels, an additional pollutant of concern is oxides of sulfur (SO_x), mainly SO_2 and SO_3, which are formed when sulfur-containing compounds in the fuel react with oxygen in the combustion air. They are toxic and corrosive and lead to the formation of sulfuric acid in the atmosphere. Since virtually all the sulfur in fuel is oxidized to SO_x, the only viable limitation strategy is to remove sulfur from fuel before combustion.

For all types of stationary gas turbines, the problems posed by exhaust gas emissions are no less challenging than for aircraft engines. World energy demand is forecast to grow over the next 30 years at around 1.8% per annum. This demand will be met predominantly by the combustion of fossil fuels [3]. Thus, the manufacturers and users of gas turbines for utility power generation now find themselves at the forefront in regard to responsibility for emissions issues.

9.3 Regulations

9.3.1 Aircraft Engines

The International Civil Aviation Organization (ICAO) has promulgated regulations for civil subsonic turbojet/turbofan engines with rated thrust levels above 26.7 kN (6000 pounds) for a defined landing-takeoff cycle (LTO), which is based on an operational cycle around airports. This LTO cycle is intended to be representative of operations performed by an aircraft as it descends from an altitude of 914 m (3000 ft) on its approach path, to the time it subsequently attains the same altitude after takeoff.

ICAO standards for gaseous emissions are presented in Table 9.2, in which π_{00} is the engine pressure ratio at takeoff. They are expressed in terms of a parameter that consists of the total mass in grams of any given gaseous pollutant emitted during the LTO cycle per kilonewton of rated thrust at sea level. We have

$$\underset{(\mathrm{g/kN})}{\text{Emission}} = \underset{(\mathrm{g/kg\ fuel})}{\text{Emission Index}} \times \underset{(\mathrm{kg\ fuel/hr\ kN})}{\text{Engine SFC}} \times \underset{(\mathrm{hr})}{\text{Time in mode}}. \tag{9.1}$$

This equation shows that two methods are available to the engine manufacturer for reducing NO_x. One is to make improvements to the combustor that reduce its emissions index (EI), and the other is to choose an engine cycle that yields a lower SFC. Since the CO and UHC levels of modern engines have been significantly reduced at all low-power conditions, and only NO_x is emitted in appreciable amounts at altitude cruise, in practice the emissions generated by aircraft engines consist primarily of NO_x. A typical example of the emissions mass distribution associated with the flight of a modern subsonic aircraft has been provided by Bahr [4] and is shown in Table 9.3. This table represents a flight of 500 nautical miles and shows that NO_x emissions predominate both in the vicinity of the airport and during altitude cruise.

TABLE 9.2

ICAO Gaseous Emissions Standards

Emission (g/kN)	Subsonic Turbojet/Turbofan Engines[a]	Supersonic Turbojet/Turbofan Engines
HC	19.6	$140\ (0.92)^{\pi_{00}}$
CO	118.0	$4550\ (\pi_{00})^{-1.03}$
NO_x	$32 + 1.6\pi_{00}$	$36 + 2.4\pi_{00}$
	$-1.04 + 2\pi_{00}$ (2007 + engines)	

Source: www.icao.org.

[a] Newly manufactured engines with rated takeoff thrust greater than 26.7 kN.

TABLE 9.3

Typical Distribution of Total Emission Mass Quantities Generated During a Flight of An Aircraft Equipped with Modern Engines

	Percent of Total Emission Mass		
Category	**During ICAO Landing-Takeoff Cycle**	**During Climbout Cruise/ Descent**	**Overall**
Smoke	–	0.1	0.1
HC	0.6	1.0	1.6
CO	5.4	7.0	12.4
NO_x	7.8	78.1	85.9
Total	13.8	86.2	100.0
	(56.5% NO_x)	(90.6% NO_x)	

Source: Bahr, D.W., ASME Paper 92-GT-415, 1992. With permission.
Note: Aircraft: twin-engine transport; range: 500 nautical miles.

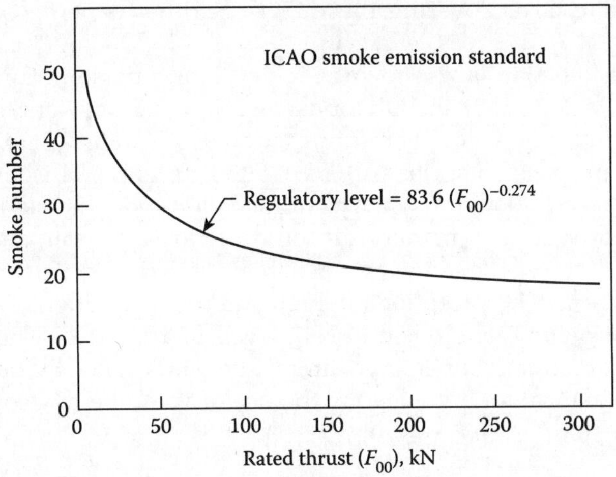

FIGURE 9.1
ICAO smoke emissions standards.

For a longer flight, NO_x emissions would account for an even larger fraction of the total emissions mass.

The ICAO standard for smoke measurement (engine year of manufacture 2000+) is expressed in terms of a smoke number (SN), which is related to the engine takeoff thrust (F_{00}) by the expression

$$SN = 83.6(F_{00})^{-0.274}. \tag{9.2}$$

This expression is shown graphically in Figure 9.1. The intention of this standard is to eliminate any visible smoke from the engine exhaust. As smoke visibility depends on both the smoke concentration, as indicated by the value of SN, and on the viewing path length, the allowable SN of a high thrust engine is lower than for a low thrust engine because of its larger exhaust diameter.

The situation in regard to compliance with ICAO regulations is generally satisfactory for subsonic aircraft engines, due mainly to the efforts of the engine manufacturers during the past 20 years in improving combustor design and in reducing engine-specific fuel consumption. However, there is continuing pressure to reduce NO_x emissions from all sources. In Sweden, a tax is now imposed on NO_x LTO emissions generated during domestic flights, whereas in Switzerland the charge bands are based on both NO_x and UHC LTO emissions [5]. Current ICAO regulations are restricted to operations at low altitudes in and around airports, but growing concerns regarding ozone depletion at high altitudes could lead to them being extended to other flight regimes, such as altitude cruise, where the bulk of NO_x emissions occur. The feasibility of introducing certification standards covering these flight regimes is being considered by ICAO [6].

The emissions standards shown in Table 9.2 for supersonic turbojet engines were set to ensure that the Olympus engine, which powers the Concorde, would be in compliance. For future supersonic transport (SST) engines, NASA has proposed a cruise NO_x EI of 5 g/kg fuel [7]. This target does not seem to be too challenging when examined alongside the cruise NO_x EI levels of 8–12 produced by modern subsonic aircraft. However, due to the large pressure rise across the sonic wave generated by SST aircraft, combustor inlet temperatures will be exceptionally high and the application of current combustor technology would yield EI NO_x levels of around 45. Thus, the future of second generation supersonic aircraft depends crucially on compliance with goals that can only be met by the use of yet-to-be-developed ultralow NO_x combustor designs.

9.3.2 Stationary Gas Turbines

Regulations governing emissions from stationary gas turbines tend to be highly complex because the legislation varies from one country to another and is supplemented by local or site-specific regulations and ordinances governing the size and usage of the plant under consideration and the type of fuel to be used. Detailed information on environmental legislation and regulations for stationary engines may be found in Schorr [8]. For the large number of engines burning natural gas, the emissions of UHC, particulate matter, and SO_x are negligibly small, and most of the drive toward more stringent regulations for stationary gas turbines has been directed at NO_x. In the United States, the Environmental Protection Agency (EPA) has promulgated emissions standards (Federal Register 71 FR 38482 dated July 6, 2006

Standards 40 CFR Part 60), which depend on the engine's input energy and intended use (utility or industrial).

- For new, electricity-producing, turbine-firing natural gas, the NO_x limits are: 42 ppmv below 3 MW (4000 HP); 25 ppmv (3–110 MW); and 15 ppmv (above 110 MW)
- For new electricity-producing turbines firing fuels other than natural gas, the NOx limits are: 96 ppmv below 3 MW (4000 HP); 74 ppmv (3–110 MW); and 42 ppmv (above 110 MW)
- For new mechanical drive turbines (below 3.5 MW), the NOx limits are: 100 ppmv for natural gas firing and 150 ppmv for fuels other than natural gas
- SO_2 emissions are limited to 110 ng/J gross energy output for turbines that are located in continental areas, and 780 ng/J gross energy output for turbines located in noncontinental areas.

Note that the above limits are expressed in parts per million by volume (ppmv), referenced to 15% oxygen on a dry basis. The purpose is partly to remove ambiguity when comparing different sets of experimental data, but also to indicate that combustors burning less fuel are expected to produce less NO_x.

The correction formula is

$$(NO_x)_{ref15\%oxygen} = (5.9)(NO_{xmeas}) / (20.9 - O_{2meas}), \tag{9.3}$$

where NO_x concentrations are expressed in ppmv (dry) and O_2 content is expressed in percentage by volume.

European and Japanese regulations are broadly in line with EPA standards. In some locations, notably Southern California and parts of Japan, growing public awareness of the contribution of NO_x to the production of smog has created pressures for increasingly stringent NO_x standards, and some local regulations now call for NO_x limits as low as 9 ppmv. In all countries, the published standards are considered to be minimum requirements and there is often a requirement to use the "best available control technology" (BACT) or the "lowest available emission rate" (LAER). This has led to concerns such as those expressed by Angello and Lowe [9], that if a new technology is developed that significantly improves the ability to reduce NO_x emissions, it effectively sets the emission standard that all subsequent plants must meet. Thus, as new technologies are developed to meet the ever increasingly restrictive emission limits, they become the standard by which the next round of emission regulations is guided.

Until the late 1980s, the BACT for achieving NO_x levels down to 25 ppmv was by water or steam injection into the combustion zone. The technology currently used to reduce NO_x concentrations to below 10 ppmv is a combination of diluent injection (water or steam), or lean, premixed combustion, supplemented by exhaust gas cleanup using selective catalytic reduction (SCR), as described below. This process is not only very expensive, but it also requires the use of additional control systems. Furthermore, it can exacerbate the overall pollution problem by releasing ammonia gas and increased CO concentrations into the atmosphere. This illustrates the difficulties involved in determining how all the various technical, economic, and environmental tradeoffs should be assessed.

At the present time, there are no EPA standards for CO and UHC emissions, but many local standards exist that are usually established on a site-specific basis. Typical CO limits range from 10 to 40 ppm.

The attainment of low CO levels has not usually presented any major difficulties in the past, due largely to the user's insistence on high combustion efficiencies to minimize fuel consumption. However, the continuing pressure to reduce NO_x emissions has resulted in copious amounts of water or steam being injected into the combustion zone and, more recently, to the adoption of lean, premix combustion. The success of these techniques relies on the lowering of flame temperature, which tends to promote the formation of CO. Thus, the control of CO levels is now posing a more difficult problem than when the emissions regulations for stationary engines were first formulated.

9.4 Mechanisms of Pollutant Formation

The concentration levels of pollutants in gas turbine exhausts can be related directly to the temperature, time, and concentration histories of the combustion process. These vary from one combustor to another and, for any given combustor, with changes in operating conditions. The nature of pollutant formation is such that the concentrations of CO and UHC are highest at low-power conditions and diminish with an increase in power. By contrast, NO_x and smoke are fairly insignificant at low-power settings and attain maximum values at the highest power condition. These characteristic trends are illustrated in Figure 9.2.

9.4.1 Carbon Monoxide

When a combustion zone is operating fuel-rich, large amounts of CO are formed owing to the lack of sufficient oxygen to complete the reaction to CO_2. If, however, the combustion zone mixture strength is stoichiometric or moderately fuel-lean, significant amounts of CO will also be present due to the dissociation of CO_2 (see Chapter 2). In practice, CO emissions are found to be much higher than predicted from equilibrium calculations and to be highest at low-power conditions, where burning rates and peak temperatures are relatively low. This is in conflict with the predictions of equilibrium theory,

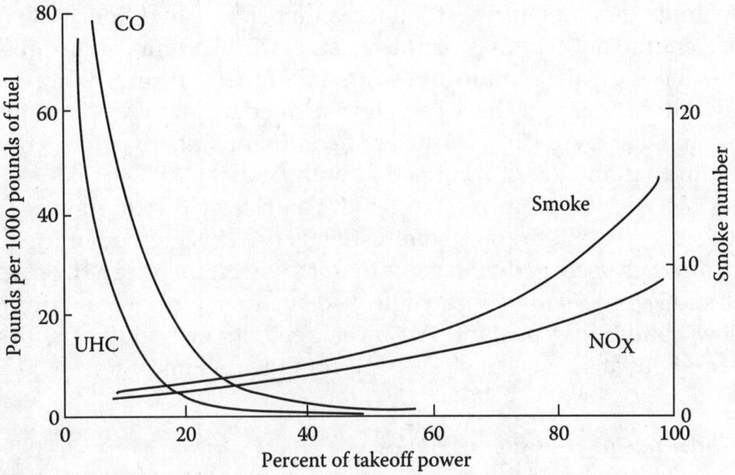

FIGURE 9.2
Emissions characteristics of gas turbine engines.

and it suggests that much of the CO arises from incomplete combustion of the fuel, caused by one or more of the following:

- Inadequate burning rates in the primary zone, due to a fuel/air ratio that is too low and/or insufficient residence time
- Inadequate mixing of fuel and air, which produces some regions in which the mixture strength is too weak to support combustion, and others in which over-rich combustion yields high local concentrations of CO
- Quenching of the postflame products by entrainment into the liner wall-cooling air, especially in the primary zone

In principle, it should be possible to reduce the CO formed in primary combustion to a very low level by the staged admission of additional air downstream to achieve a gradual reduction in burned gas temperature. However, once formed, CO is relatively resistant to oxidation, and in many practical systems its oxidation is rate determining with respect to the attainment of complete combustion. At high temperatures, the major reaction removing CO is

$$CO + OH = CO_2 + H.$$

This is a fast reaction over a broad temperature range. At lower temperatures, the reaction

$$CO + H_2O = CO_2 + H_2,$$

is important as a means of removing CO.

The main factors influencing combustion efficiency, and hence CO emissions, are engine and combustor inlet temperatures, combustion pressure, primary-zone equivalence ratio and, with liquid fuels, the mean drop size of the spray. All these aspects have been investigated by many workers, including Rink and Lefebvre [10], who used a continuous flow tubular combustor, 150 mm in diameter, in conjunction with an array of 36 equally spaced "microscopic" airblast atomizers, to achieve a uniform distribution of liquid fuel in the mixture entering the combustion zone. This method of fuel injection had another useful advantage in that it allowed the mean drop size in the fuel spray to be varied in a controlled manner while maintaining all other flow conditions constant. All measurements of pollutant emissions were carried out at a distance of 170 mm from the fuel injectors.

9.4.1.1 Influence of Equivalence Ratio

Some of the results obtained by Rink and Lefebvre [10] for a light diesel oil (DF 2) are presented in Figure 9.3, which shows the variation of CO emissions with an equivalence ratio for three values of inlet air pressure. All three curves exhibit the same general characteristics. They show that CO emissions diminish with an increase in the equivalence ratio, reaching minimum values at an equivalence ratio of around 0.8, above which any further increase

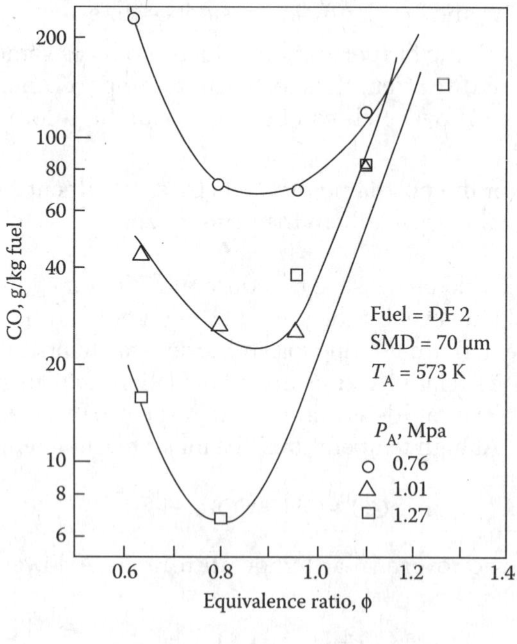

FIGURE 9.3
Influence of pressure and equivalence ratio on CO. (From Rink, K.K. and Lefebvre, A.H., *International Journal of Turbo and Jet Engines*, 6(2), 113–22, 1989. With permission.)

in the equivalence ratio causes CO emissions to rise. These trends are typical of those observed for other types of combustion systems. The high levels of CO at low equivalence ratios are due to the slow rates of oxidation associated with low combustion temperatures. An increase in the equivalence ratio raises the flame temperature, which accelerates the rate of oxidation so that CO emissions decline. However, at temperatures higher than around 1800 K, the production of CO by chemical dissociation of CO_2 starts to become significant. Thus, only in a fairly narrow range of equivalence ratios around 0.8 can low levels of CO be achieved.

9.4.1.2 Influence of Pressure

Figure 9.3 also demonstrates the beneficial effect of an increase in combustion pressure in reducing CO emissions. In recent high-pressure industrial combustor tests, performed at 9 and 14 bars, it was found that at low equivalence ratios, a 50% increase in pressure decreases CO by a factor of two and at high equivalence ratios, the same increase in combustion pressure virtually eliminated CO emissions by suppressing chemical dissociation.

9.4.1.3 Influence of Ambient Air Temperature

Hung and Agan [11] have examined the influence of ambient air temperature on the CO emissions from a 7 MW industrial engine supplied with natural gas fuel. A strong air temperature effect on measured CO was observed. CO emissions for an air temperature of 287 K were three to four times higher than the corresponding values at 298 K. A correlation of these data carried out by Hung [12] yielded the following expression for calculating the effect of ambient air temperature on CO. It is considered to be valid for temperatures up to 303 K.

$$CO_T / CO_{288} = 1 - 0.0634(T - 288), \qquad (9.4)$$

where CO_T is the emissions of CO in ppmv for 15% oxygen at ambient temperature T, and CO_{288} is the emissions of CO in ppmv for 15% oxygen at 288 K.

This equation should be used with caution because it is likely to be very engine specific. Nevertheless, it serves to highlight the strong dependence of CO emissions on ambient air temperature, and helps to explain some of the anomalies that are sometimes encountered when analyzing CO measurements obtained from repeat tests carried out over a period of time.

9.4.1.4 Influence of Wall-Cooling Air

An important factor influencing CO emissions is the amount of liner wall-cooling air employed in the primary combustion zone. CO formed in primary combustion can migrate toward the liner walls and become entrained in the wall-cooling air. The temperature of this air is so low that all chemical reactions are effectively frozen. Thus, the film-cooling air emanating from the primary zone normally contains significant quantities of CO. Unless this CO is subsequently entrained into the hot central core with sufficient time to react to completion, it will appear in the exhaust gas.

9.4.1.5 Influence of Fuel Atomization

The main effect of mean drop size on CO emissions stems from its strong influence on the volume required for fuel evaporation. At low-power operation, where these emissions attain their highest concentrations, a significant proportion of the total combustion volume is occupied in fuel evaporation. Consequently, less volume is available for chemical reaction.

9.4.2 Unburned Hydrocarbons

UHC include fuel that emerges from the combustor in the form of drops or vapor, as well as the products of the thermal degradation of the parent fuel into species of lower molecular weight. They are normally associated with poor atomization, inadequate burning rates, the chilling effects of film-cooling air, or any combination of these. The reaction kinetics of UHC formation are more complex than for CO formation, but it is generally found that those factors that influence CO emissions also influence UHC emissions and in much the same manner.

9.4.3 Smoke

Exhaust smoke is caused by the production of finely divided soot particles in fuel-rich regions of the flame that, in conventional combustors, are always close to the fuel spray. These are the regions in which recirculating burned products move upstream toward the fuel injector, and local pockets of fuel vapor become enveloped in oxygen-deficient gases at high temperature. In these fuel-rich zones, soot may be produced in considerable quantities.

Most of the soot produced in the primary zone is consumed in the high-temperature regions downstream. Thus, from a smoke viewpoint, a combustor may be considered to comprise two separate zones—the primary zone, which governs the rate of soot formation, and the intermediate zone (and, on modern high-temperature engines, the dilution zone also), which determines the rate of soot consumption. The soot concentration actually observed in the exhaust gas is the difference between two large numbers.

Analysis of the soot found in exhaust gases shows that it consists mostly of carbon (96%) and a mixture of hydrogen, oxygen, and other elements. Soot is not an equilibrium product of combustion except at mixture strengths far richer than those employed in the primary zones of gas turbines. Thus, it is impossible to predict its rate of formation and final concentration from kinetic or thermodynamic data. In practice, the rate of soot formation tends to be governed more by the physical processes of atomization and fuel–air mixing than by kinetics.

9.4.3.1 *Influence of Pressure*

Problems of soot and smoke are always most severe at high pressures. There are several reasons for this; some derive from chemical effects, whereas others stem from physical factors that affect spray characteristics and hence the distribution of mixture strength in the soot-forming regions of the flame. For premixed kerosine/air flames, it is found that no soot is formed at pressures below 0.6 MPa and equivalence ratios below 1.3.

One adverse effect of an increase in pressure is to extend the limits of flammability, so that soot is produced in regions that, at lower pressures, would be too rich to burn. Increased pressure also accelerates chemical reaction rates, so that combustion is initiated earlier and a larger proportion of the fuel is burned in the fuel-rich regions adjacent to the spray. With pressure atomizers, reduced spray penetration is one of the main causes of smoke at high pressures. At low pressures, the fuel is distributed across the entire combustion zone, but at high pressures it tends to concentrate in the soot-forming region just downstream of the fuel nozzle. Another adverse effect of an increase in pressure is to reduce the cone angle of the spray. This encourages soot formation, partly by increasing the mean fuel drop size, but mainly by raising the mixture strength in the soot-forming zone. The total effect of all these factors is that with pressure atomizers, smoke emission increases steeply with pressure.

With airblast atomizers, the influence of pressure on spray characteristics is much less pronounced. Recent experimental studies carried out by Zheng et al. [13] on a modern practical airblast atomizer, showed that spray cone angle and spray volume are largely independent of pressure, provided that the air/fuel ratio is kept constant, which corresponds to the normal engine situation at power settings above idle. It was also observed that changes in pressure have very little effect on the mean drop size in the spray. Thus, in contrast to pressure atomizers, the spray characteristics of gas turbine airblast atomizers are largely uninfluenced by variations in ambient air pressure. This is the main reason that combustors fitted with airblast atomizers exhibit only small increases in soot formation and smoke with an increase in combustion pressure.

9.4.3.2 *Influence of Fuel Type*

Fuel properties can influence smoke production in two ways; first by inducing the formation of local fuel-rich regions, and second, by exerting variable resistance to carbon formation. The former is controlled by physical properties, such as viscosity and volatility, which affect the mean drop size, penetration, and rate of evaporation of the fuel spray, whereas the latter relate to molecular

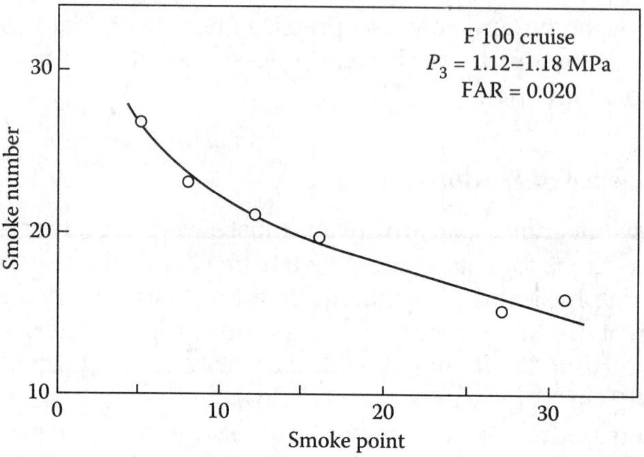

FIGURE 9.4
Correlation of smoke number with smoke point for an F100 combustor. (From Chin, J.S. and Lefebvre, A.H., ASME Paper 89-GT-261, 1989. With permission.)

structure. It is well established that smoking tendency increases with an increase in aromatic content of the conventional petroleum-based hydrocarbon fuel. This is because the polycyclic aromatic hydrocarbons (PAHs) form the nuclei for the growth of the soot particles. Therefore, as discussed in Chapter 10, if the fuel has no aromatic compounds (e.g., coal-derived fuels produced by Fischer–Tropsch synthesis), it will produce virtually no soot, as shown later in Figure 10.31. Also, hydrogen content is commonly used in correlating rig and engine test data on various soot-related parameters such as smoke emissions, flame radiation, and liner-wall temperature. However, Chin and Lefebvre [14] have shown that a better index of sooting tendency is the ASTM smoke point, which is obtained experimentally by burning the test fuel in a wick lamp and slowly increasing the height of the flame until it begins to smoke. The height of the flame in millimeters is the smoke point; the higher this is, the lower the tendency of the fuel to soot formation.

The correlation shown in Figure 9.4 was obtained from an analysis of measurements of SN carried out on a Pratt & Whitney F100 combustor by Russel [15]. The generally high quality of the data fit obtained with this and several other aircraft combustors led Chin and Lefebvre to conclude that smoke point is superior to hydrogen content as a correlating parameter for soot-related combustion phenomena.

9.4.3.3 Influence of Fuel Atomization

The influence of fuel drop size on soot formation has been investigated by Rink and Lefebvre [10] using the tubular combustor described above, supplied with a kerosine fuel. Their results for a combustion pressure of 1.52 MPa (15.5 atm) are shown in Figure 9.5. This figure shows that improvements

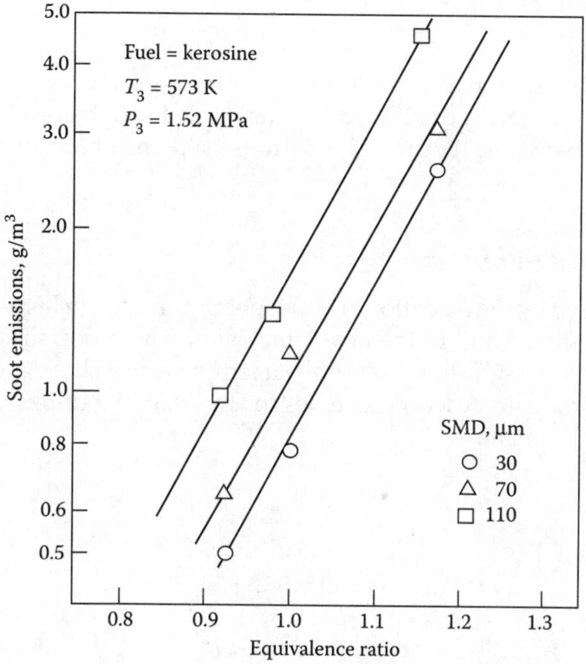

FIGURE 9.5
Influence of fuel mean drop size on soot formation. (From Rink, K.K. and Lefebvre, A.H., *International Journal of Turbo and Jet Engines*, 6(2), 113–22, 1989. With permission.)

in atomization quality inhibit soot formation. For example, at the highest equivalence ratios, reducing the mean drop size from 110 to 30 µm effectively halves the soot concentration. The importance of atomization quality to soot formation and smoke stems from the fact that, as the fuel spray approaches the flame front, heat transmitted from the flame starts to evaporate the drops. The smallest droplets in the spray have time to evaporate completely ahead of the flame front, and the resulting fuel vapors then mix with the combustion air and burn in the manner of a premixed flame. However, the largest drops in the spray do not have time to fully evaporate and mix completely with air before being consumed by the flame. In consequence, they burn in the mode of fuel-rich diffusion flames. Clearly, any increase in mean drop size will increase the proportion of large drops in the spray. This, in turn, will raise the proportion of fuel burned in diffusion-type combustion, as opposed to premixed combustion.

In general, exhaust smoke decreases with mean drop size, but if improved atomization should lead to a reduction in spray penetration, as occurs with all types of pressure atomizers, the smoke output may actually go up because of the local increase in fuel concentration. In fact, reduced spray penetration is one of the main causes of smoke on high-pressure ratio engines fitted with dual-orifice atomizers.

9.4.4 Oxides of Nitrogen

Most of the nitric oxide (NO) formed in combustion subsequently oxidizes to NO_2. For this reason, it is customary to lump NO and NO_2 together and express results in terms of NO_x, rather than NO. It can be produced by four different mechanisms: thermal NO, nitrous oxide mechanism, prompt NO, and fuel NO.

9.4.4.1 Thermal Nitric Oxide

This is produced by the oxidation of atmospheric nitrogen in high-temperature regions of the flame and in the postflame gases. The process is endothermic and it proceeds at a significant rate only at temperatures above around 1850 K. Most of the proposed reaction schemes for thermal NO utilize the extended Zeldovich mechanism:

$$O_2 = 2O,$$

$$N_2 + O = NO + N,$$

$$N + O_2 = NO + O,$$

$$N + OH = NO + H.$$

NO formation is found to peak on the fuel-lean side of stoichiometric. This is a consequence of the competition between fuel and nitrogen for the available oxygen. Although the combustion temperature is higher on the slightly rich side of stoichiometric, the available oxygen is then consumed preferentially by the fuel. The exponential dependence of thermal NO on flame temperature is demonstrated in Figure 9.6. This figure shows that NO production declines very rapidly as temperatures are reduced, particularly at normal combustor residence times of around 5 ms.

Figure 9.7 illustrates the exponential dependence of NO_x on flame temperature for both gaseous and liquid fuels. It is based on experimental data (not shown in the figure) obtained by Snyder et al. [16] in their studies on the combustion performance achieved when using a tangential entry lean-premixed fuel nozzle. Of special interest in this figure is that the well-known difference in NO_x emissions between liquid and gaseous fuels diminishes with an increase in flame temperature, becoming negligibly small at the highest levels of temperature. The reason for this is because when burning liquid fuels there is always the potential for near-stoichiometric combustion temperatures, and consequently high NO_x formation, in local regions adjacent to the fuel drops, although the average equivalence ratio throughout the combustion zone may be appreciably less than stoichiometric. With an increase

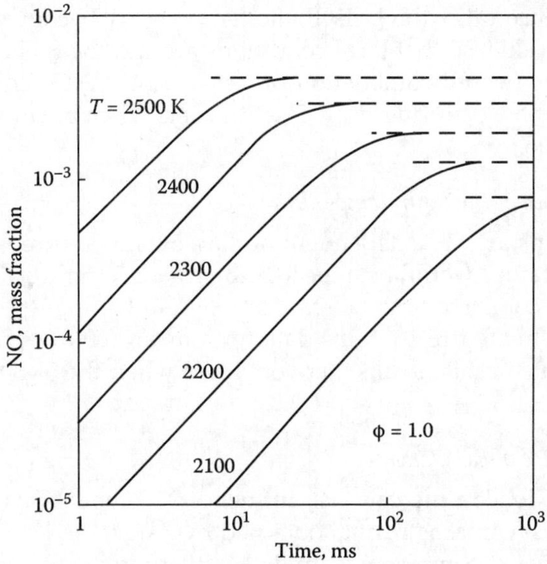

FIGURE 9.6
NO_x formation as a function of time and temperature; $P = 1$ MPa.

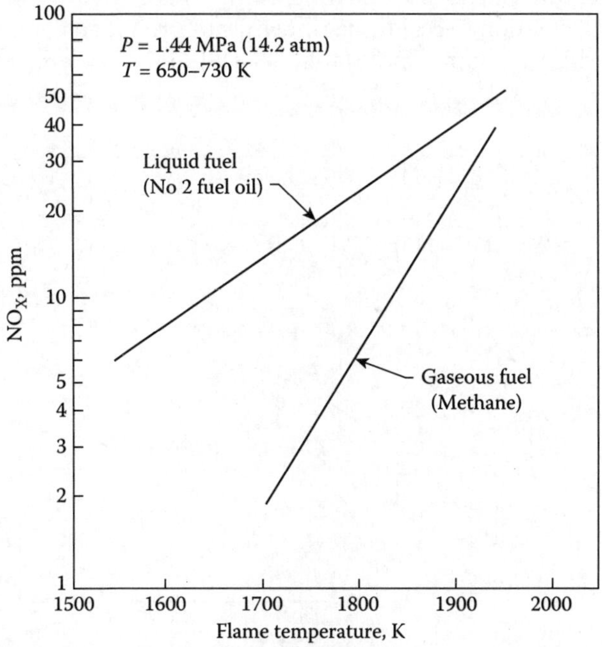

FIGURE 9.7
Dependence of NO_x on flame temperature for liquid and gaseous fuels. (From Snyder, T.S., Rosfjord, T.J., McVey, J.B., and Chiappetta, L.M., ASME Paper 94-GT-283, 1994. With permission.)

in the equivalence ratio, the bulk flame temperature becomes closer to the stoichiometric value, so that local conditions around the fuel drop have less influence on the overall combustion process and the NO_x emissions begin to approximate those produced by gaseous fuels when burning at the same equivalence ratio.

9.4.4.1.1 *Influence of Inlet Air Temperature*

As NO emissions are very dependent on flame temperature, an increase in inlet air temperature would be expected to produce a significant increase in NO, and this is confirmed by the results shown in Figure 9.8 from Rink and Lefebvre [17]. This figure contains data for a mean fuel drop size (SMD) of 110 microns, but similar results were obtained when the SMD was reduced to 30 microns.

9.4.4.1.2 *Influence of Residence Time*

Combustor residence time can also influence NO_x emissions, as shown in Figure 9.9, which contains results obtained by Anderson [18] when using a premix-prevaporize combustor supplied with premixed gaseous propane fuel. It shows that NO_x emissions increase with an increase in residence time, except for very lean mixtures ($\phi \cong 0.4$), for which the rate of formation is so low that it becomes fairly insensitive to time. Similar results showing the insensitivity of NO_x formation to residence time in lean-premixed combustion have been obtained by Leonard and Stegmaier [19] and Rizk and Mongia [20].

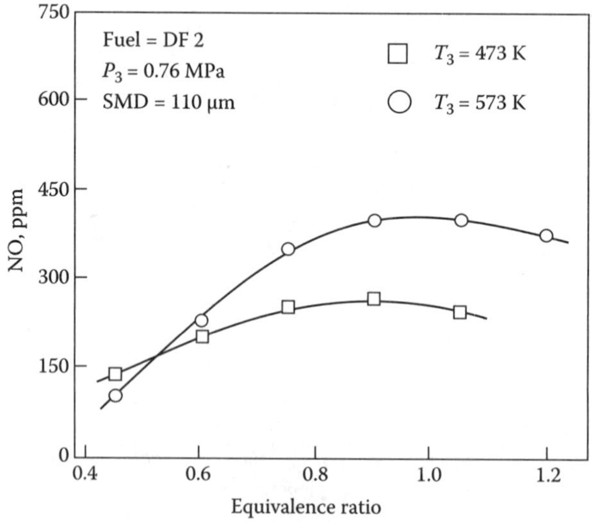

FIGURE 9.8
Influence of inlet air temperature on NO_x formation. (From Rink, K.K. and Lefebvre, A.H., *Combustion, Science and Technology*, 68, 1–14, 1989. With permission.)

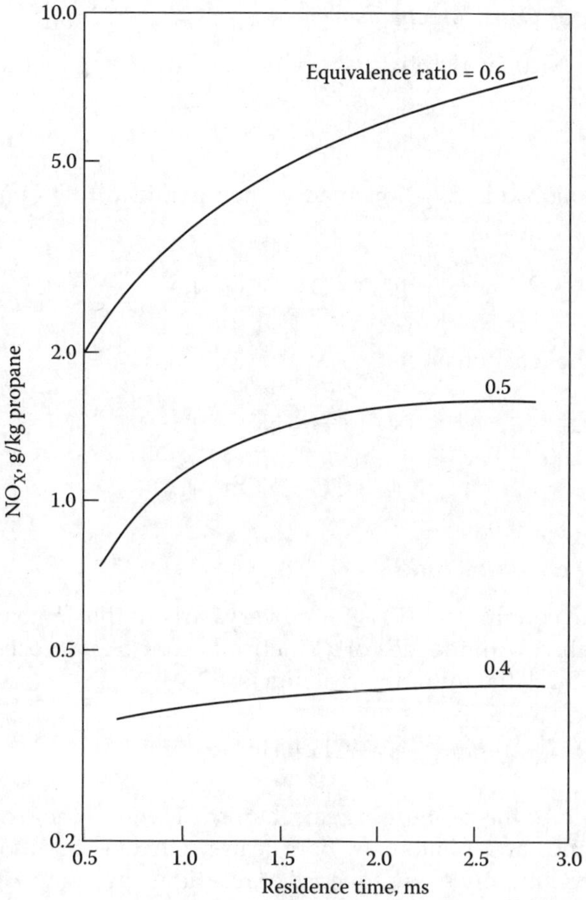

FIGURE 9.9
Effect of residence time on NO_x in a premixed fuel–air system. (From Anderson, D.N., ASME Paper 75-GT-69, 1975. With permission.)

These findings have important practical implications to the design of lean-premixed combustors.

The key points regarding thermal NO may be summarized as follows:

- Thermal NO formation is controlled largely by flame temperature
- Little NO is formed at temperatures below around 1850 K
- For conditions typical of those encountered in conventional gas turbine combustors (high temperatures for only a few milliseconds), NO increases linearly with time, but does not attain its equilibrium value
- For very lean-premixed combustors ($\phi < 0.5$), NO formation is largely independent of residence time

9.4.4.2 Nitrous Oxide Mechanism

According to Nicol et al. [21], this mechanism is initiated by the reaction

$$N_2 + O = N_2O,$$

and the nitrous oxide (N_2O) formed is then oxidized to NO mainly by the reaction

$$N_2O + O = NO + NO,$$

but also by the reactions

$$N_2O + H = NO + NH,$$

$$N_2O + CO = NO + NCO.$$

9.4.4.3 Prompt Nitric Oxide

Under certain conditions, NO is found very early in the flame region—a fact that is in conflict with the idea of a kinetically controlled process. According to Nicol et al. [21], the initiating reaction is

$$N_2 + CH = HCN + N.$$

The balance of the prompt NO mechanism involves the oxidation of the HCN molecules and N atoms. Under lean-premixed conditions, the HCN oxidizes to NO mainly by a sequence of reactions involving HCN → CN → NCO → NO. The N atom reacts mainly by the second Zeldovich reaction.

The influence of pressure is of special interest and importance because prompt NO can be a significant contributor to the NO emissions produced in lean premix (LPM) combustion [22]. Unfortunately, few data are available on this effect. Fennimore's [23] pioneering study of prompt NO in ethylene-air flames over a range of pressures from 1 to 3 atm concluded that prompt NO $\propto P^{0.5}$. Later work by Heberling [24] over a much wider range of pressures from 0.1 to 1.8 MPa showed that prompt NO was independent of pressure. Altermark and Knauber [25] also concluded that NO_x is independent of pressure for equivalence ratios below 0.6. The practical implications of these findings are discussed below.

9.4.4.4 Fuel Nitric Oxide

Light distillate fuels contain less than 0.06% of organically bonded nitrogen (usually known as fuel-bound nitrogen; FBN), but the heavy distillates may contain as much as 1.8%. During combustion, some of this nitrogen reacts to form the so-called "fuel NO." The fraction of nitrogen undergoing this change increases only slowly with increasing flame temperature. As far as gaseous fuels are concerned, natural gases contain little or no FBN, but some is found in certain processes and low-Btu gases. Depending on the degree of nitrogen conversion, fuel NO can represent a considerable proportion of the total NO [26].

Nicol et al. [21] analytically examined the relative contributions of the various mechanisms discussed above to the total NO_x emissions produced by a lean-premixed combustor burning methane fuel, for which the fuel NO is zero. The results of their study show that at relatively high temperatures of around 1900 K, and equivalence ratios of around 0.8, the contributions are about 60% thermal, 10% N_2O, and 30% prompt. With reductions in temperature and equivalence ratio, the contributions made by N_2O and prompt NO increase significantly until, at a temperature of 1500 K and an equivalence ratio of around 0.6, the relative contributions to the total NO_x emissions become 5% thermal, 30% N_2O, and 65% prompt. At the lowest equivalence ratios ($\phi = 0.5–0.6$), the major source of NO_x is that formed by the N_2O mechanism. These results clearly have great importance to the design of ultralow NO_x lean-premixed combustors.

9.4.5 Influence of Pressure on Oxides of Nitrogen Formation

Pressure effects on NO_x formation are of special importance due to the continual trend toward engines of higher pressure ratio to meet the need for lower fuel consumption. Combustor testing at high pressures is extremely expensive and it would, therefore, be highly convenient to carry out combustion tests at low levels of pressure and then extrapolate the results obtained to high levels of pressure where NO_x emissions attain their highest values. Such extrapolation could be carried out with confidence if the relationship between NO_x and pressure were accurately known. Unfortunately, the experimental data obtained on different combustor types are conflicting in this regard. They vary from no effect of pressure on NO_x, to quite significant increases in NO_x with an increase in pressure.

For conventional combustors, it is generally found that $NO_x \propto P^n$, where n has values ranging from around 0.5 to around 0.8. The results of Maughan et al. [27] from a well mixed combustor supplied with natural gas fuel showed an increase in n with an increase in exhaust gas temperature. For example, raising the combustor outlet temperature from 1227 to 1310 K caused n to increase from 0.38 to 0.51. Maughan et al. regard this result as evidence that the lowest NO_x levels result from the N_2O and prompt mechanisms, which dominate at low temperatures and are independent of pressure, whereas the higher NO_x levels associated with higher combustion temperatures are due primarily to thermal NO_x, which exhibits a square-root dependence on pressure.

These results and conclusions are fully consistent with those obtained by Correa et al. [22,28]. These workers studied turbulent premixed methane-air flames using an uncooled perforated plate burner operating at pressures from 1 to 10 atm, inlet air temperatures from 300 to 615 K, and equivalence ratios from 0.5 to 0.9. Their modeling featured a stirred reactor for flame stabilization followed by a plug flow reactor and a kinetic scheme that included thermal and prompt NO. The results confirmed that the low temperatures of lean flames preclude significant formation of NO by the thermal mechanism. At temperatures below 1800 K, the prompt mechanism appears to be dominant. The implications of these results to the effect of pressure on NO_x formation is well illustrated in Figure 9.10, which contains some of the experimental data from Correa et al. and highlights their conclusions in regard to the influence of flame temperature on the pressure dependence of NO_x formation. This figure shows that NO_x is independent of pressure in the leanest premixed flames. An increase in flame temperature, corresponding to an increase in equivalence ratio, causes the pressure exponent to increase until, in the near-stoichiometric region, it attains the value of 0.5, corresponding to NO formation by the thermal mechanism.

Additional evidence to support the argument that NO_x formation in well-mixed, low-temperature flames is largely independent of pressure has been provided by Leonard and Stegmaier [19] and Steele et al. [29]. Their experiments covered ranges of pressure from 0.1 to 3.0 MPa and 0.1 to 0.7 MPa, respectively. Leonard and Stegmaier found little or no effect of pressure, whereas the results obtained by Steele et al. using a lean, premixed, high-intensity

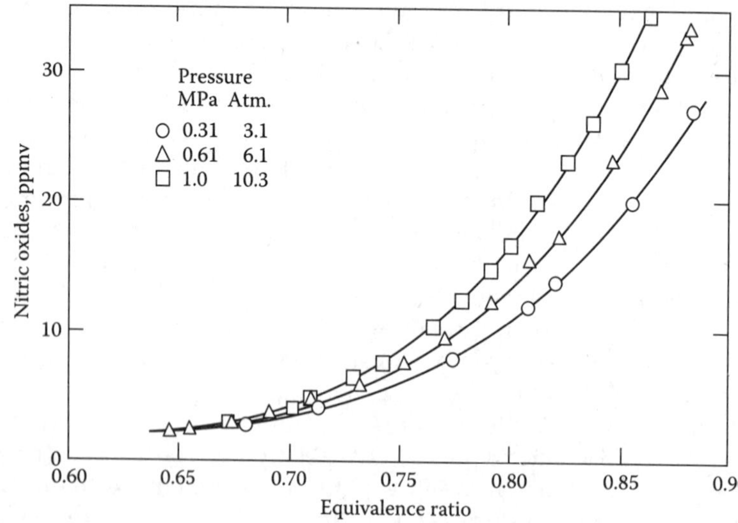

FIGURE 9.10
Data illustrating the effect of pressure on NO_x formation. (From Leonard, G.L. and Correa, S.M., *Second ASME Fossil Fuel Combustion Symposium*, PD-30, 69–74, 1990.)

combustor showed a neutral or even slightly negative effect of pressure on NO_x. Comparatively little is known about the pressure dependence of NO_x formation in fuel-rich flames. Rizk and Mongia [30] performed a three-dimensional analysis to examine the influences of pressure and residence time on NO_x formation in the rich zone of a rich/quench/lean (RQL) combustor. Their predictions indicate that the value of the pressure exponent n varies with rich-zone equivalence ratio according to the relationship:

$$n = 116.5 \exp - (\phi/0.222). \tag{9.5}$$

For a typical rich-zone equivalence ratio of 1.4, this equation gives a value for n of 0.21.

9.4.6 Influence of Fuel Atomization on Oxides of Nitrogen Formation

The manner and extent to which NO_x are influenced by the sizes of the fuel droplets in the spray is very dependent on equivalence ratio. This aspect was addressed in the experimental studies carried out by Rink and Lefebvre [10,17] using the continuous flow combustor referred to above in which the mean drop size could be varied and controlled independently of other operating variables. The data presented in Figure 9.11 show that NO emissions increase with an increase in mean drop size, especially at low equivalence ratios. At first sight this may seem surprising, because at the fairly high pressures employed in this study, evaporation rates are so fast that even for the larger drops, the time required for their evaporation is small in comparison with the total residence time of the combustion zone. However, an increase in SMD means that a larger proportion of the total number of fuel drops

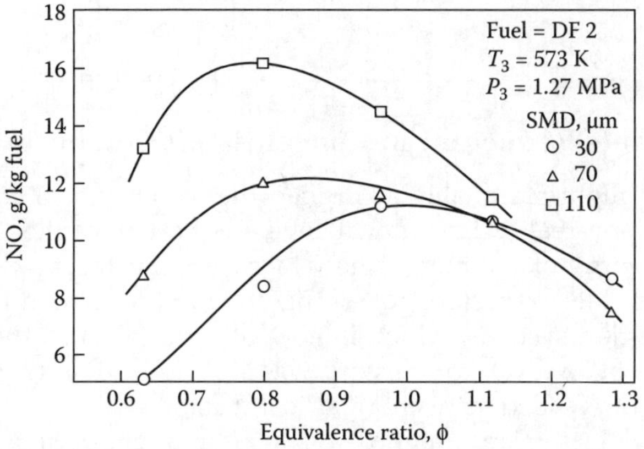

FIGURE 9.11
Influence of fuel atomization on NO emissions. (From Rink, K.K. and Lefebvre, A.H., *International Journal of Turbo and Jet Engines*, 6(2), 113–22, 1989. With permission.)

in the spray is capable of supporting "envelope" flames. These envelope flames, which surround the larger drops, burn in a diffusion mode at near-stoichiometric fuel/air ratios, giving rise to many local regions of high temperature in which NO_x is formed in appreciable quantities. Reduction in mean drop size impedes the formation of envelope flames, so that a larger proportion of the total combustion process occurs in what is essentially a premixed mode, thereby generating less NO_x.

Envelope flames are unlikely to occur in combustion zones supplied with light distillate fuels but, even if none are present, with increasing drop size a larger proportion of the fuel burns in the fuel-rich regions created in the wakes of the moving drops. Although, in theory, combustion within these localized regions can take place at any equivalence ratio within the flammability limits, it tends to occur preferentially at the stoichiometric value, i.e., at the maximum temperature, thereby producing high levels of NO_x. This hypothesis serves to explain why NO_x emissions increase with SMD for lean mixtures. However, as the overall equivalence ratio increases toward unity, the local fuel/air ratio adjacent to the fuel drops approaches the premixed value. According to this hypothesis, mean drop size should have no influence on NO_x emissions for stoichiometric mixtures, and this is generally confirmed by the results shown in Figure 9.11. This figure is important because it demonstrates that even at low equivalence ratios, where the average combustion temperature is so low that only negligible amounts of NO should, in theory, be formed, the presence of fuel drops in the combustion zone gives rise to conditions in which combustion can and does proceed at near-stoichiometric equivalence ratios, regardless of the average equivalence ratio in the combustion zone. This, of course, is the rationale for the various types of lean, premix, prevaporize (LPP) combustors whose success relies largely on the elimination of all fuel drops from the combustion zone.

9.5 Pollutants Reduction in Conventional Combustors

Although it might reasonably be argued that conventional combustors no longer pose any real technical challenge, they do, nevertheless, constitute the large majority of combustors now in service. Furthermore, most of our knowledge of the key factors governing pollutant formation in continuous flow combustion systems, which is now being applied to the design and development of low-NO_x combustors, was acquired from experience gained on what are now called "conventional" combustors.

In the previous section, attention was focused on the various mechanisms and processes involved in the formation of pollutant emissions. Of equal interest and importance is the application of this knowledge to the problems of alleviating pollutant emissions in practical combustion systems.

The main factors controlling emissions from conventional combustors may be considered in terms of:

- Primary-zone temperature and equivalence ratio
- Degree of homogeneity of the primary-zone combustion process
- Residence time in the primary zone
- Liner-wall quenching characteristics
- Fuel spray characteristics (with liquid fuels)

In reviewing practical design methods for pollutants reduction, a convenient approach is to consider each individual pollutant species in turn. It will become clear, however, that with conventional combustors a great deal of compromise is involved in design, not only between one species and another, but also among the many other performance requirements, such as lean blowout limits and pattern factor.

9.5.1 Carbon Monoxide and Unburned Hydrocarbons

The presence of these species in the exhaust gases is a manifestation of incomplete combustion. Thus, all approaches to CO and UHC reduction are based on a common philosophy, which is to raise the level of combustion efficiency. An effective method of achieving this is by redistributing the airflow to bring the primary-zone equivalence ratio closer to the optimum value of around 0.8. A higher equivalence ratio (up to around 1.05) would increase burning rates even further, but it would not yield lower emissions of CO and UHC because of lack of the oxygen that these species need in order to convert to CO_2 and H_2O. Good fuel–air mixing in the primary zone is also essential for low CO and UHC. Even when operating at the optimum equivalence ratio, poor mixing can produce local regions in which the mixture strength is either too fuel-lean to provide adequate burning rates or so fuel-rich that there is insufficient O_2 to convert all the CO produced into CO_2.

Another effective means of reducing CO and UHC is by using less liner wall-cooling air, especially in the primary zone. Figure 9.12 shows the effect of replacing a conventional film-cooled wall with a non-film-cooled wall in the primary zone of a Rolls Royce Industrial RB211 low-emissions combustor when operating at atmospheric pressure. CO is significantly reduced (at 1850 K from 1500 to 700 ppm), whereas the lean blowout temperature is lowered by 110 K [31]. Today, most modern combustors operate at over 35 bars with 99% combustion efficiency, a fine spray, and with one-half the film-cooling air requirement of their predecessors. Thus, the emissions of CO and UHC are minimal (in single digits) for aero engines and almost nonexistent for stationary industrial gas turbines. Clearly, the development of new materials and methods of liner-wall construction, which allow the liner to operate at higher metal temperatures, along with the development of new methods of wall

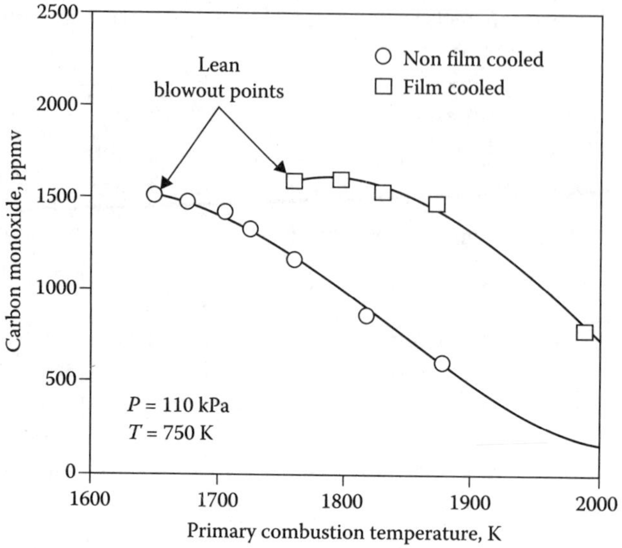

FIGURE 9.12
Effect of eliminating hot-side film cooling on CO emissions. (From Willis, J.D., Toon, I.J., Schweiger, T., and Owen, D.A., ASME Paper 93-GT-391, 1993. With permission.)

cooling that require much less cooling air, such as effusion and transpiration cooling, has made a very direct and significant contribution to the reduction of CO and UHC emissions. In summary, CO and UHC emissions are reduced by the following:

- Redistribution of the airflow to bring the primary-zone equivalence ratio closer to the optimum value of around 0.8
- Increase in primary-zone volume and/or residence time
- Reduction in liner wall-cooling air, especially in the primary zone
- Improved fuel atomization

Figure 9.13 illustrates the reductions in CO to be gained from improvements in atomization quality, whereas Figure 9.14 shows that UHC emissions are also greatly diminished by reductions in mean drop size [10]. Only at low equivalence ratios, where burning rates tend to be limited more by chemical reaction rates than by evaporation rates, is the influence of fuel drop size on emissions less pronounced.

9.5.2 Smoke

The main factors governing smoke emissions are combustor inlet air temperature, pressure, and fuel spray characteristics. The influence of inlet air temperature is complex because an increase in this parameter serves to accelerate

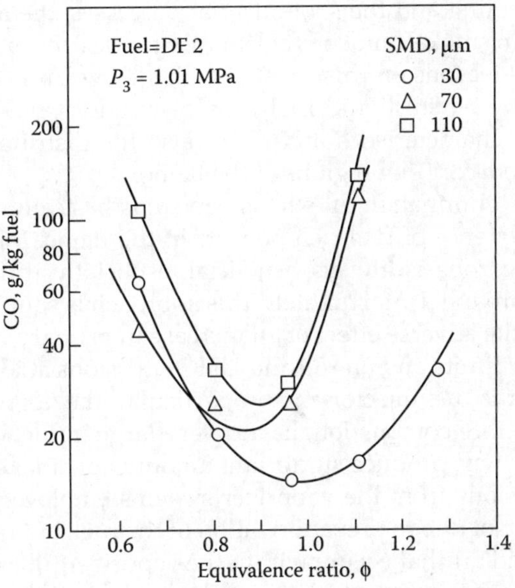

FIGURE 9.13
Influence of fuel atomization on CO emissions. (From Rink, K.K. and Lefebvre, A.H., *International Journal of Turbo and Jet Engines*, 6(2), 113–22, 1989. With permission.)

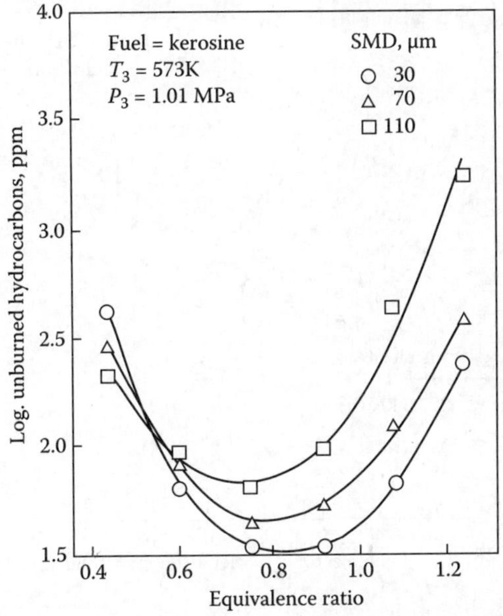

FIGURE 9.14
Influence of fuel atomization on UHC emissions. (From Rink, K.K. and Lefebvre, A.H., *International Journal of Turbo and Jet Engines*, 6(2), 113–22, 1989. With permission.)

both the soot-forming and the soot-burnout processes; the net result is usually a reduction in smoke. Smoke problems are most severe at high pressures. There are several reasons for this, most of which derive from chemical effects as discussed above. With liquid fuels, there are additional physical factors that affect spray characteristics and hence also the distribution of mixture strength in the soot-forming regions of the flame.

In practice, the elimination of exhaust smoke is basically a matter of preventing the occurrence of fuel-rich pockets in the flame. Injecting more air into the primary zone is always beneficial, especially if accompanied by more thorough mixing. Unfortunately, this approach is somewhat limited in scope, owing to the adverse effect of an increase in primary-zone air on ignition and stability limits and on CO and UHC emissions at idle.

The design of the fuel injector and, in particular, the degree of premixing of fuel and air before combustion, have a very large influence on whether a given combustor will produce significant amounts of smoke. The relatively low smoke emissions from the vaporizer systems employed on some Rolls Royce engines is not due to prevaporization of the fuel, but rather to the premixing of fuel and air that occurs within the vaporizer tubes.

Alleviating soot formation and smoke by fuel–air mixing is only fully effective if sufficient air is used. This is well illustrated in Figure 9.15 from Sturgess et al. [32], which shows how smoke was drastically reduced in a P&W JT9D-70 combustor when operating at takeoff conditions by the addition of

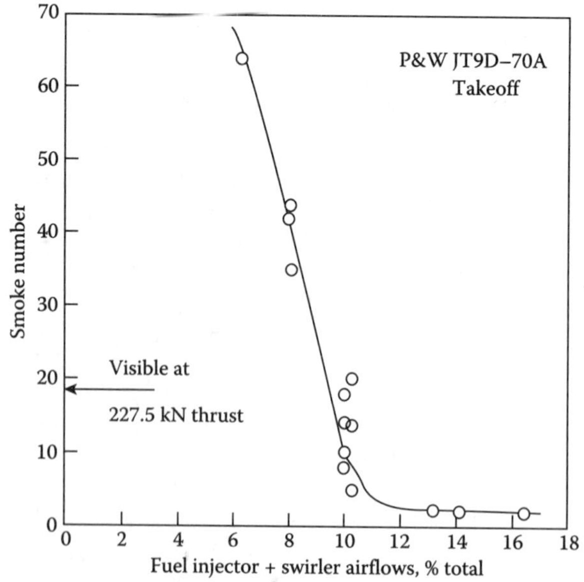

FIGURE 9.15
Control of exhaust smoke through atomizer and swirler airflows. (From Sturgess, G.J., McKinney, R., and Morford, S.A., *Journal of Engineering for Gas Turbines and Power*, 115(3), 570–80, 1993. With permission.)

more air through the fuel injector and air swirler. The injection of air through these components is particularly effective in reducing smoke because it all flows directly into the soot-forming zone.

The advantages of airblast atomizers over dual-orifice pressure atomizers in regard to smoke emissions are well established. It is not just a question of better atomization, although this is very significant at high combustion pressures where smoke levels attain their highest values, but because the airblast atomization process virtually guarantees good mixing of air and fuel drops prior to combustion. Another important asset of the airblast atomizer is that atomization quality is high over the entire operating range from idle to full power. This is also true for the piloted-airblast injector because there is no physical interference between the pilot and main sprays. With dual-orifice nozzles, owing to the interaction of the pilot and main sprays, there is always a range of fuel flows, starting at the point where the main fuel is first admitted, over which atomization quality is poor and CO and HC emissions are inevitably high.

9.5.3 Oxides of Nitrogen

In any attempt to reduce NO_x, the prime goal must be to lower the reaction temperature. The second objective should be to eliminate hot spots from the reaction zone, as there is little point in achieving a satisfactorily low average temperature if the reaction zone contains local regions of high temperature in which the rate of NO_x formation remains high. Finally, the time available for the formation of NO_x should be kept to a minimum.

Practical approaches to low NO_x in conventional combustors include the addition of more air into the primary combustion zone to lower the flame temperature, improved atomization (see Figure 9.11), increase in liner pressure drop to promote better mixing, thereby eliminating hot spots from the combustion zone, and reduction in combustor residence time. Unfortunately, reductions in flame temperature and residence time lead to increased output of both CO and UHC. In fact, as a generalization, it can be stated that any change in operating conditions or combustor configuration that reduces NO_x tends also to exacerbate the problems of CO and UHC, and vice versa.

Over the last two decades, a judicious application of the techniques above has yielded a significant decrease in NO_x emissions. For example, aircraft NO_x emissions decreased by 33%, NO_x emissions from liquid-fueled industrial gas turbines decreased by one-half, and from natural gas-fired industrial combustors decreased by a factor of six (see www.icao.org and www.epa.gov.).

9.5.3.1 Water Injection

As NO_x formation is exponentially dependent on temperature, an obvious way of reducing NO_x emissions is by lowering the temperature of the combustion zone. Additional air is effective, but can only be used sparingly because it raises the primary-zone velocity, which has an adverse effect on both ignition and stability performance. An alternative approach is to introduce a heat sink, such as water or steam, into the combustion zone. The technique is clearly inappropriate for aero engines, but is a practical proposition for large stationary engines, especially if large amounts of water or steam are available. It has been widely used to control NO_x emissions to the level required by EPA regulations. For example, Davis and Washam [33] have reported a 40% reduction in NO_x down to the 75 ppmv goal when using a water/oil ratio of 0.4. In some cases, the water or steam is injected directly into the flame, either through a number of separate nozzles located at the head end of the combustor or through holes that are integrated into the fuel nozzle [34]. Alternatively, the water injection may take place upstream of the combustor liner, usually into the airstream, which subsequently flows into the combustion zone through the main air swirler. This method ensures good atomization because the smaller droplets are carried by the airflow through the swirler into the combustion zone, whereas the larger drops impinge on the swirler vanes where they form a thin liquid film, which is airblast atomized as it flows over the downstream edge of the vane [35].

When steam is used to reduce NO_x emissions it may also be injected directly into the combustion zone or into air that subsequently flows into the combustion zone. In some installations, the steam is injected into the compressor discharge air. The method is simple, but inherently wasteful because only about 40% of the steam actually flows into the combustion zone. This may be only a minor consideration if excess steam is available [34].

The effectiveness of water and steam for reducing NO_x has been demonstrated by many workers. According to Hung [36], the relationship between NO_x reduction and water/fuel mass ratio, X, can be expressed as

$$\text{wet NO}_x/\text{dry NO}_x = \exp - (0.2X^2 + 1.41X). \tag{9.6}$$

This relationship was found to apply to both liquid and gaseous fuels. It shows, for example, that equal mass flow rates of water and fuel (for which $X = 1$) yields an 80% reduction in NO_x. Very similar results were obtained for both gaseous and liquid fuels by Claeys et al. [37] on the General Electric MS7001F gas turbine.

Equation 9.6 should not be regarded as having universal application. For example, Wilkes [38] has shown that water injection is much less effective with fuels containing FBN, whereas Toof [39] actually observed a slight increase in the yield of NO from fuel nitrogen. The main effect of water addition is to reduce thermal NO_x, although it does also slightly reduce prompt NO. This implies that water injection is most effective when combustion takes place at high pressures and temperatures where thermal NO_x production is high, and is less effective at low pressures and temperatures where a larger proportion of the total NO_x is formed via the prompt mechanism.

The key point is that water and/or steam injection always reduces NO_x, but the extent of the reduction depends on combustor operating conditions and fuel type.

Hilt and Waslo [34] have reported NO_x reductions of around 60% for a steam/fuel mass ratio of unity on two GE industrial engines burning natural gas. As steam is a less effective diluent than water, due to the latent heat of evaporation of water, the reductions in NO_x achieved with steam injection tend to be less dramatic than when water is used. According to Schorr [40], about 60% more steam than water is needed to achieve a given NO_x reduction.

Although both water and steam injection are very effective in reducing NO_x emissions and have been used on stationary engines that operate at near-constant load conditions since the early 1970s, they do have a number of drawbacks. White et al. [41] have reported an increase in capital cost of US$10–15 per kW and an increase in fuel consumption of 2–3%. This additional fuel is needed to heat the water to combustion temperature, although power output is enhanced due to the additional mass flow through the turbine. The water must be of high purity to prevent deposits and corrosion in the hot sections downstream of the combustor. The treatment of this water is expensive and requires a separate plant based on reverse osmosis and de-ionization. User experience with water injection has shown a significant increase in inspection and hardware maintenance. There are, therefore, practical limits to the amount of water or steam that can be injected into the combustor. The deterioration in combustion performance arising from water–steam injection is manifested as increases in the levels of CO and UHC emissions and by increases in combustor pressure oscillations. These oscillations can become amplified by coupling with the combustion process, and cause deterioration of combustor hardware.

The various penalties associated with water injection, as discussed above, may be summarized as:

- Higher capital cost
- Increase in fuel consumption
- High cost of water treatment
- Potential for corrosion of hot section components
- Higher maintenance costs
- Increase in CO and UHC emissions
- Increase in combustion pressure pulsations

These drawbacks of water and steam injection have encouraged the development of the so-called "dry low-NO_x" (DLN) combustors, i.e., combustors that can meet the emission goals without having to resort to diluent injection.

9.5.3.2 Selective Catalytic Reduction

This is a method for converting NO_x in a gas turbine exhaust stream into molecular nitrogen and H_2O by injecting ammonia into the stream in the presence of a catalyst. Exhaust gases first pass through an oxidation catalyst and are then mixed with ammonia before entering the SCR catalyst. The oxidation catalyst removes the CO and UHC emissions by oxidizing them to CO_2 and H_2O. To reduce NO_x emissions, ammonia is injected in a manner designed to achieve intimate mixing with the engine exhaust stream. After mixing, the exhaust gases pass over a catalyst (usually vanadium pentoxide), which results in the selective reduction of NO_x to form N_2 and H_2O. The principal reactions are:

$$6NO + 4NH_3 \rightarrow 5N_2 + 6H_2O,$$

$$6NO_2 + 8NH_3 \rightarrow 7N_2 + 12H_2O.$$

Water or steam injection is used first to reduce the NO_x level down to around 40 ppmv, leaving the SCR process to achieve a further reduction down to less than 10 ppmv [40]. SCR works best with natural gas fuel and is fairly intolerant to sulfur-bearing liquid fuels. It requires that the temperature of the exhaust stream be within a fairly narrow range from 560 to 670 K, and so is restricted to systems in which the exhaust gas flows into a heat recovery device, usually a steam generator [33]. A major problem with this method is the requirement for a control system that feeds the requisite amount of ammonia, and the need for a continuous monitoring system that can give the feedback to the ammonia supply mechanism under differing load conditions. Another problem is the size and weight of the equipment. According to Davis and Washam [33], for an 83 MW MS7000 gas turbine, an SCR designed to remove 90% of the NO_x from the exhaust stream has a volume of 175 m^3 and weighs 111 tons. Despite these drawbacks, the method is quite widely used.

9.5.3.3 Exhaust Gas Recirculation

The underlying principle of this approach is the reduction of flame temperature by recirculating cooled combustion products back into the primary zone, as illustrated in Figure 9.16. The practical feasibility of this method of NO_x reduction has been investigated by Wilkes and Gerhold [42], who found that significant reductions (50%) could be achieved with recirculation rates of 20% or less at base load conditions. The major thermal effect stems from the reduced concentration of oxygen in the inlet air, but there is also a secondary effect due to the higher heat capacity of this air with an increased H_2O and CO_2 content.

The main advantage of the method is that little or no combustor development is required and standard production combustors can be used. Its main drawback lies in the need for an intercooler between the exhaust and

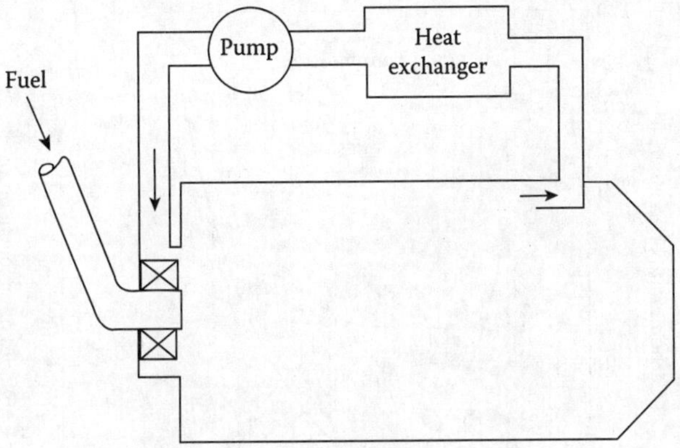

FIGURE 9.16
Schematic diagram to illustrate the principle of exhaust gas recirculation.

inlet. This virtually rules it out for simple gas turbines, but application to combined cycle plants offers more promise due to the substantially lower exhaust gas temperatures. Another drawback is that only very clean fuels can be used to avoid problems of fouling and contamination.

9.6 Pollutants Reduction by Control of Flame Temperature

Of all the factors influencing pollutant emissions from gas turbine combustors, the most important by far is the temperature of the combustion zone. With conventional combustors, this can range from 1000 K at low-power operation to 2500 K at high-power operation, as indicated in Figure 9.17. This figure also shows that too much CO is formed at temperatures below around 1670 K, whereas excessive amounts of NO_x are produced at temperatures higher than around 1900 K. Only in the fairly narrow band of temperatures between 1670 and 1900 K are the levels of CO and NO_x below 25 and 15 ppmv, respectively. The basic objective of all the various approaches toward low-emissions combustors described below is to maintain the combustion zone (or zones) within a fairly narrow band of temperatures over the entire power range of the engine.

9.6.1 Variable Geometry

An ideal variable-geometry system would be one in which large quantities of air are admitted at the upstream end of the combustion liner at maximum power conditions to lower the primary-zone temperature and provide adequate film-cooling air. With a reduction in engine power, an increasing

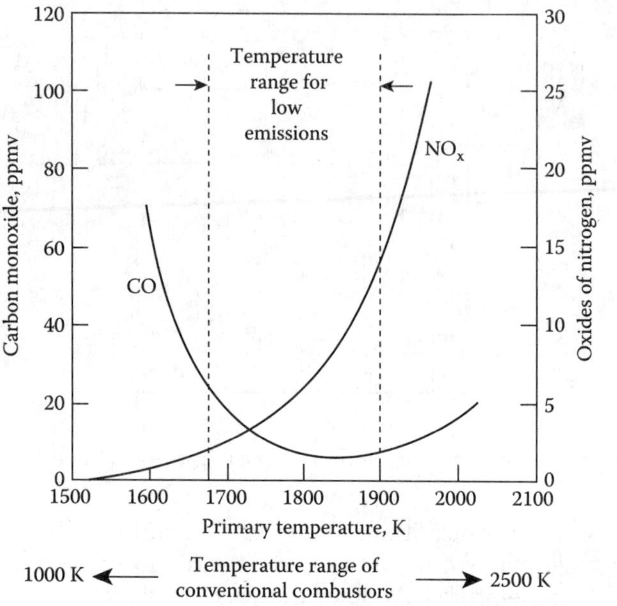

FIGURE 9.17
Influence of primary-zone temperature on CO and NO_x emissions.

proportion of this air is diverted to the dilution zone to maintain the primary-zone temperature within the low-emissions "window" shown in Figure 9.17. Practical ways of achieving some variation in airflow distribution include the use of variable-area swirlers to control the amount of air flowing into the combustion zone [43,44], variable air openings into the dilution zone [45,46], or a combination of these.

The drawbacks to all forms of variable-geometry systems include complex control and feedback mechanisms that tend to increase cost and weight and reduce reliability. Problems of achieving the desired temperature pattern in the combustor efflux gases could also be encountered, especially if the liner pressure drop is allowed to vary too much. The incentive for surmounting these practical problems is that variable geometry has the potential for simultaneously reducing all the main pollutant species without sacrificing other aspects of combustion performance. It also has several other advantages; for example, as the combustion temperature never falls below a certain minimum value of around 1670 K, chemical reaction rates are always relatively high. This enables the combustion zone to be made smaller, with consequent advantages in terms of reductions in combustor size and weight. For aircraft applications, variable geometry also has the potential for wide stability limits and improved altitude relight performance.

Ideally, variable geometry combustors should be used in conjunction with premix-prevaporize fuel-injection systems. Only in this way is it possible to avoid the local high-temperature, high NO_x-forming regions, created by the presence of fuel droplets in the combustion zone.

Although variable geometry has been used in some large industrial engines, there have been few successful applications of this technique in small-to-medium-size gas turbines due to size and cost limitations and also because of concerns regarding operational reliability [47].

9.6.2 Staged Combustion

With variable-geometry systems, the combustion temperature is controlled to within fairly narrow limits by switching air from one zone to another with changes in engine power setting. By contrast, the airflow distribution within staged combustors remains constant; the fuel flow is switched from one zone to another in order to maintain a fairly constant combustion temperature. One simple method of fuel staging is by "selective fuel injection," as described by Bahr [48]. With this technique, fuel is supplied only to selected combinations of fuel injectors at lightoff, relight, and engine idle conditions, as illustrated in Figure 9.18. Only at power settings above idle is the full

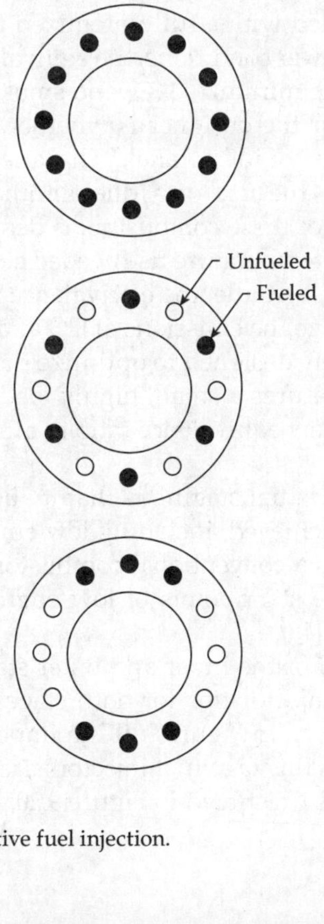

FIGURE 9.18
Illustration of the use of selective fuel injection.

complement of fuel injectors employed. The objective of this modulation technique is to raise the equivalence ratio and hence also the temperature of the localized combustion zones at low-power operation. This approach, which is now in common use, not only reduces CO and UHC emissions, but also has the added advantage of extending the lean blowout limit to lower equivalence ratios.

A major drawback of selective fuel injection is the "chilling" of chemical reactions that occurs at the outer edges of the individual combustion zones. This chilling lowers combustion efficiency, as discussed above, and increases the formation of CO and UHC. Furthermore, the circumferentially nonuniform exit temperature distribution results in loss of turbine efficiency. These limitations have led to the development of "staged" combustors in which no attempt is made to achieve all the performance objectives in a single combustion zone. Instead, two or more zones are employed, each of which is designed specifically to optimize certain aspects of combustion performance.

A typical staged combustor has a lightly loaded primary zone, which provides all the temperature rise needed to drive the engine at low-power conditions. It operates at an equivalence ratio of around 0.8 to achieve high combustion efficiency and low emissions of CO and UHC. At higher power settings, its main role is to act as a pilot source of heat for the main combustion zone, which is supplied with a fully premixed fuel–air mixture. When operating at maximum power conditions, the equivalence ratio in both zones is kept low at around 0.6 to minimize NO_x and smoke.

An important choice for the designer is whether the staged combustion should take place in "series" or in "parallel." The latter approach, often called "radial staging," features the use of a dual-annular combustor, as illustrated in Figure 9.19. One of these combustors is designed to operate lightly loaded and provide all the temperature rise needed at startup, altitude relight, and engine idle conditions. At idle, the equivalence ratio of the combustion zone is selected to minimize the emissions of CO and UHC. The other annular combustor is specifically designed to optimize the combustion process at high-power settings. It features a small, highly loaded combustion zone of short residence time and low equivalence ratio to minimize the formation of NO_x and smoke.

The main advantage of radial staging is that it allows all the combustion performance goals to be achieved, including low emissions, within roughly the same overall length as a conventional combustor. This short-length feature is attractive from the standpoints of low engine weight and reduced rotor dynamics problems [48].

If the combustor domes of the inner and outer stages are arranged to be radially in-line, the fuel injector tips for both stages can be mounted on a common feed arm, as shown in Figure 9.19. An important advantage of this arrangement is that the main stage fuel injectors are cooled by the continuously flowing pilot fuel, as illustrated in Figure 9.20. This prevents coking of

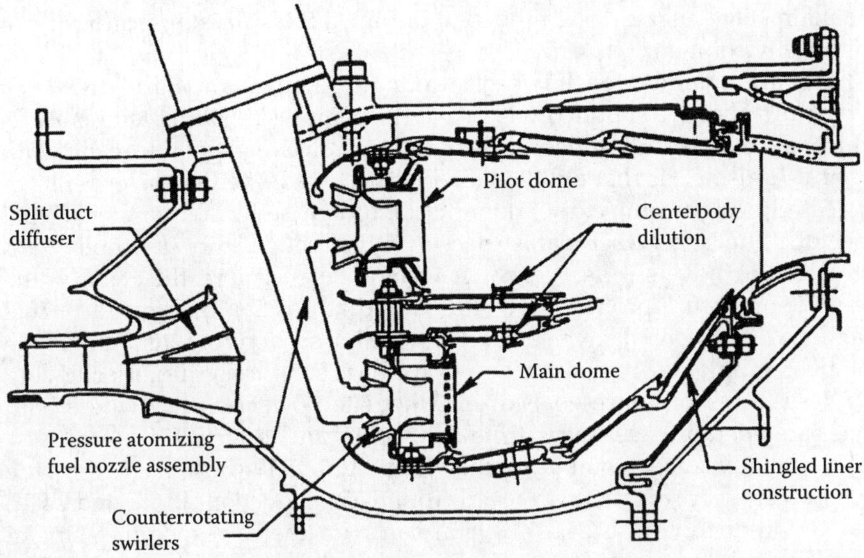

FIGURE 9.19
General Electric dual-annular combustor. (From Bahr, D.W., *Journal of Propulsion and Power*, 3(2), 179–86, 1987. Reprinted with permission from AIAA.)

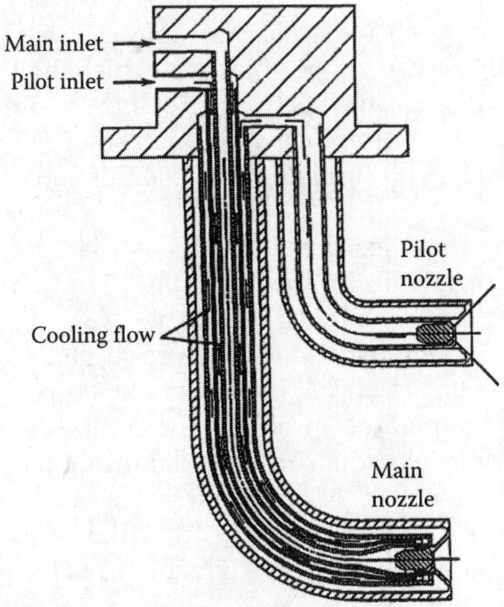

FIGURE 9.20
Fuel nozzle for GE dual-annular combustor. (Courtesy of Parker Hannifin Corporation.)

the main stage nozzles when they are unfueled but still exposed to the hot engine environment.

There are a number of drawbacks to radial staging. One basic drawback is that all zones are supplied with air at the compressor outlet temperature, which means that all zones have the same relatively poor lean blowout limit. It is also clear that pollutants reduction is achieved at the expense of increased design complexity and a marked increase in the number of fuel injectors. The larger liner wall surface area demands additional cooling air, which has an adverse effect on pattern factor. Furthermore, the peaks of the radial temperature profile could shift in radial position as a result of fuel staging, with potential adverse effects on the hot sections downstream of the combustor. Another basic problem with radial staging is that of achieving the desired performance goals at intermediate power settings where both zones are operating well away from their optimum design points.

The radially staged combustor shown in Figure 9.19 was designed by the General Electric company. It achieved around 35% reductions in CO and UHC, and 45% reduction in NO_x, in comparison with the corresponding single-annular combustor. The GE CFM56-5B engine, fitted with this dual-annular combustor, is now in service on Airbus Industrie A320 and A321 aircraft. The GE90 dual-annular combustor has also received flight certification.

With "series" or "axial" fuel staging, a portion of the fuel is injected into a fairly conventional primary combustion zone. Additional fuel, usually premixed with air, is injected downstream into a "secondary" or "main" combustion zone, which operates at low equivalence ratios to minimize the formation of NO_x and smoke. The primary combustion zone is used on engine startup and generates the temperature rise needed to raise the rotational speed up to engine idle conditions. At higher power settings, fuel is supplied to the secondary combustion zone and, as the engine power rises toward its maximum value, the function of the primary zone becomes increasingly one of providing the heat needed to initiate rapid combustion of the fuel supplied to the second stage.

Axial staging does have certain advantages over radial staging. Since the main stage is downstream of the pilot stage, ignition of the main stage directly from the pilot is both rapid and reliable. Also, the hot gas flow from the pilot into the main combustion zone ensures high combustion efficiency from the main stage, even at low equivalence ratios. According to Segalman et al. [49], the radial temperature profile at the combustor exit can be developed to a satisfactory level using conventional dilution hole trimming and, once developed, does not change significantly as a result of fuel staging.

The main drawback to axial staging is that the in-line arrangement of stages tends to create additional length, which makes the problem of retrofit difficult for some engines. In comparison with conventional combustors, the liner surface area that needs to be cooled is higher. The fuel injectors for the two combustion stages require separate feed arms, which involve two separate penetrations of the combustor casings. Furthermore, the pilot fuel

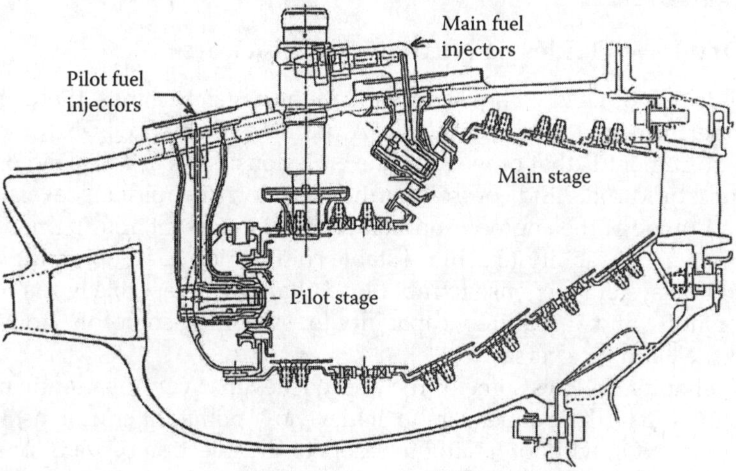

FIGURE 9.21
Pratt and Whitney axially staged combustor.

cannot be used to cool the main stage fuel as can be done quite conveniently with radial staging.

Figure 9.21 shows a cross-sectional view of an axially staged combustor developed by the Pratt and Whitney company [50]. For clarity, the main stage fuel injectors are shown rotated half an injector pitch to be in line with the pilot stage injectors. The engine centerline is at the bottom of the figure. This combustor has the benefits of the axially in-line stage arrangement without any length penalty and is designed to fit into the existing P&W V2500-AS engine. The pilot combustion zone is specifically designed to provide wide stability limits and high combustion efficiency (low CO and UHC). With an increase in engine power above idle, fuel is admitted to the main combustion zone, where combustion is initiated and sustained by the hot gas emanating from the pilot zone. The relative amounts of fuel supplied to the pilot and main zones is such that no thrust lag is created when fuel is first introduced into the main zone. In combination, the pilot and main zones maintain a low equivalence ratio that ensures low NO_x emissions at higher power settings.

Of special interest in Figure 9.21 is the inboard location of the pilot combustion zone. This greatly reduces the susceptibility to flame blowout in heavy rain because the compressor centrifuges the water to the outer portion of the airflow path. Another advantage of having the main zone outside the pilot is that the radial temperature profile at the combustor outlet peaks toward the outer radius of the turbine flowpath, a situation that is conducive to long turbine blade life.

In the longer term, it is possible that staging of the airflow by variable geometry in conjunction with fuel staging may become more of a design option.

9.7 Dry Low-Oxides of Nitrogen Combustors

In the design of DLN combustors for stationary gas turbines, there are two major performance criteria to be met. As pointed out by Davis [51], one obvious requirement is that of meeting the emissions goals at base load on both gas and liquid fuels and controlling the variation of emissions levels across the load range of the engine. Another, equally important, requirement is for high system operability to achieve stable combustion at all operating conditions, good system response to rapid load changes, acceptable levels of combustion noise and, if required, capability for switching smoothly from gas to liquid fuel, and vice versa.

This chapter reviews some of the approaches that various manufacturers are following in their endeavors to achieve low pollutant emissions, in particular low NO_x, without having to resort to the injection of water or steam. Combustors of this type are known as "dry low-NO_x" (DLN) or "dry low-emissions" (DLE) combustors.

9.7.1 Solar Dry Low-Emissions Concepts

Solar Turbines in San Diego has been among the pioneers in the development of DLE combustors for industrial gas turbines. The results of this company's efforts have appeared in a number of publications, e.g., White et al. [41], Roberts et al. [45], Smith et al. [52–54], and Etheridge [55]. Figure 9.22

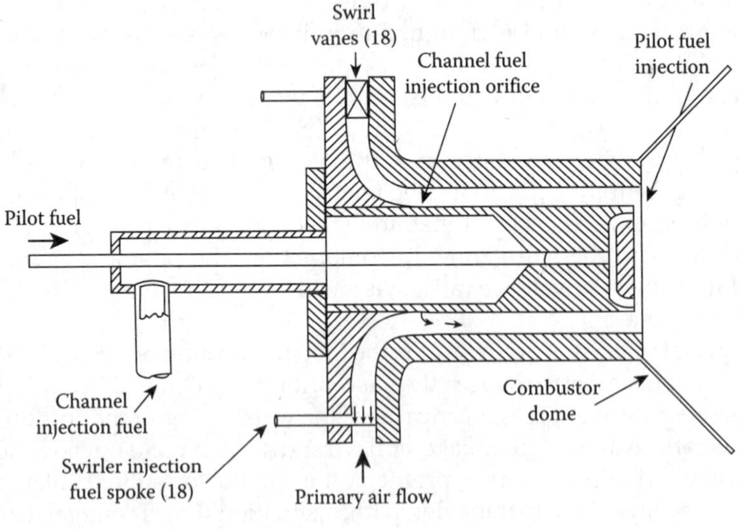

FIGURE 9.22
Solar low-NO_x fuel injector for natural gas. (From Smith, K.O., Angello, L.C., and Kurzynske, F.R., ASME Paper 86-GT-263, 1986. With permission.)

shows a cross-sectional view of a fuel injector designed for installation in multiple-can combustion systems for the Mars and Centaur engines. An 18-vane radial flow swirler is used to impart a high degree of rotation to the combustor primary air, which serves both to promote fuel–air mixing and to induce a recirculatory flow in the primary zone. The fuel injector/air swirler assembly permits three different modes of fuel injection, as indicated in Figure 9.22. Best mixing is achieved by injecting the gaseous fuel through 18 spokes, each spoke being located between a pair of swirl vanes. As each spoke contains six holes of 0.89 mm diameter, the total number of injection points is 108. Combustion tests carried out with this fuel injector assembly attached to a cylindrical combustion liner showed that the concept is capable of achieving NO_x emissions below 10 ppmv when burning natural gas at pressures up to 1.1 MPa, along with low values of CO and UHC [52].

The manner in which the fuel-injection system described above was adapted for liquid fuels by Smith and Cowell [54] is shown in Figure 9.23. The system employs two different modes of liquid fuel injection. The "inner filming" mode involves filming of the fuel on the cylindrical swirler centerbody. Fuel is delivered to the outer surface of the centerbody through eight holes located around the centerbody circumference. This fuel forms a film that is carried downstream by the swirling primary airflow. It vaporizes and mixes with air as the film progresses along the centerbody and into the primary zone. The "outer filming" mode operates in the same manner, but the film formation now takes place on the outer cylindrical surface of the air swirler channel.

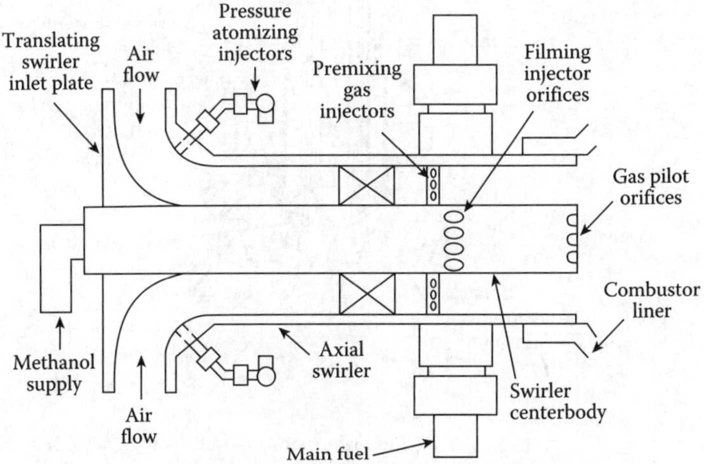

FIGURE 9.23
Solar low-NO_x fuel injector for liquid fuels. (From Smith, K.O. and Cowell, L.H., ASME Paper 89-GT-264, 1989. With permission.)

Combustion tests showed that NO_x emissions were lowest with either total inner fueling or combined inner and outer fueling. The combustor yielded around 12 ppmv NO_x at 0.6 MPa and 20 ppmv at 0.9 MPa. CO was always below 50 ppmv. In common with most well-mixed systems, low concentrations of both CO and NO_x were attainable only over a fairly narrow range of operating conditions. Potential improvements for this concept include increasing the number of fuel-injection holes used to deliver fuel to the filming surface to aid in the formation of a more uniform film, and lengthening the injector centerbody to allow a longer time for fuel evaporation and mixing [54].

9.7.2 Siemens Hybrid Burner

This burner was originally developed to operate on natural gas in either diffusion or premix modes. It has two separate air passages—an inner one, which features an axial swirler near its exit, and a concentric outer passage, which is designated in Figure 9.24 as "diagonal swirler" because it is tilted

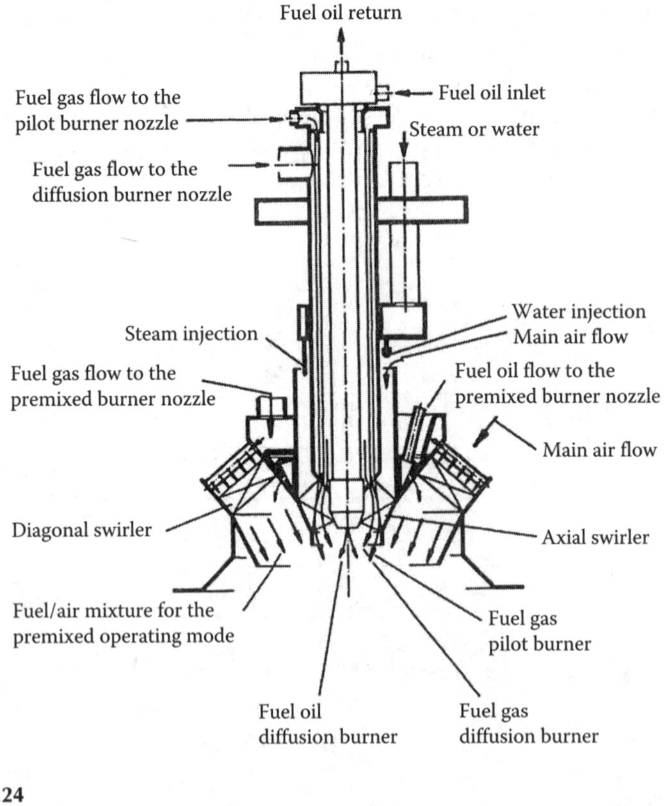

FIGURE 9.24
Siemens hybrid burner. (From Angello, L. and Lowe, P., ASME Paper 89-GT-254, 1989; Bonzani, F., Di Meglio, A., Pollarolo, G., Prade, B., Lauer, G., and Hoffmann, S., Presented at Power-Gen Europe '97, Madrid, June 1997. With permission.)

with respect to the burner axis. The inner passage, which carries about 10% of the total air, contains the gas-diffusion and pilot burners. During startup and low load operation, all the fuel is confined to this passage. As the load increases, a fuel/air ratio is eventually reached at which the burner switches from a diffusion to a premix mode. The gas is then injected into the outer air passage through small premixer tubes (one per diagonal swirler vane channel), each containing ten holes. This arrangement ensures a high level of radial and circumferential uniformity in the fuel–air mixture entering the premix combustion zone. Good stability is achieved by the presence of the pilot burner situated in the inner passage. Essentially, the system functions as a diffusion burner at low engine loads, and then operates as a premix burner in the upper load range.

For liquid fuels, the burner is equipped with a central oil burner lance. A further nozzle system for water or steam injection into the diffusion flame ensures that NO_x emissions can be reduced to low values over the entire engine operating range [56].

The Siemens hybrid burner is now fully established as a low-emissions system for engines in the 150 MW class, and has consistently achieved single-figure NO_x emissions levels (9 ppmv) when burning natural gas. Its flexibility has been demonstrated by its application by MAN GHH to its THM 1304 engine. This 9 MW class gas turbine is a two-shaft, heavy frame machine that features two tubular combustion chambers mounted on top of the engine casing. NO_x emissions are less than 5 ppmv between 75% and 100% load when operating on natural gas.

In the early Siemens silo combustors, all the hybrid burners were of the same design and size. Adaptation to different combustor sizes was accomplished by changing the number of burners. However, in the new hybrid burner ring (HBR) annular combustors, the number of burners is kept constant at 24 in order to achieve a satisfactory temperature pattern factor (see Chapter 4). This means that the size of the burner must be varied to suit the size of the combustor. However, the basic hybrid burner design remains unchanged [57].

The new annular combustor developed jointly by Siemens AG and Ansaldo Energia in Italy for the V64.3A engine features premixed operation for both natural gas and fuel oil. The gaseous fuel is injected into the diagonal swirlers in the manner described above; the liquid fuel is injected through plain-orifice atomizers into the crossflowing airstream issuing from these swirlers. Water or steam is not required to meet the emissions regulations.

9.7.3 General Electric DLN Combustor

The GE DLN combustor, shown schematically in Figure 9.25, is a two-stage concept designed for application to natural gas-fired, heavy-duty gas turbines, but capable of liquid fuel operation with diluent injection to control NO_x. It has been described in some detail in a number of publications, including

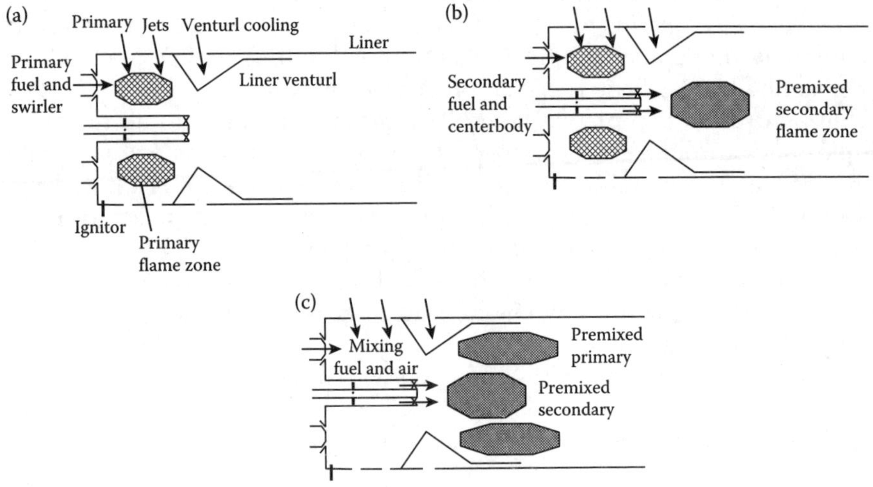

FIGURE 9.25
GE DLN combustor in various operational modes: (a) primary; (b) lean-lean; and (c) premix. (From Maughan, J.R., Luts, A., and Bautista, P.J., ASME Paper 94-GT-252, 1994. With permission.)

those of Maughan et al. [27], Davis and Washam [33], Hilt and Waslo [34], Schorr [40], Davis [51], and Washam [58]. The essence of this concept is the use of two-stage combustion to achieve low emissions and high operability over the entire load range. The combustion system consists of four main components: primary fuel nozzles, liner, venturi, and cap/centerbody assembly. These components are arranged to provide three main zones:

1. A primary zone that extends from the six primary nozzles mounted on the cap face to the end of the centerbody
2. A secondary zone that includes the volume from the centerbody exit to the plane of the dilution holes
3. A dilution zone that occupies the space from the dilution holes to the end of the liner

The combustor operates in four distinct modes, which are designated as primary, lean-lean, secondary, and premix as described below.

9.7.3.1 Primary

As illustrated in Figure 9.25a, lightoff is accomplished with fuel flowing through the primary nozzles located in the head end of the liner. Primary combustion air enters through swirlers surrounding each nozzle and through the primary air holes. This mode of operation is used to ignite, accelerate, and operate the machine at low-power settings up to around 40% full load.

9.7.3.2 Lean-Lean

As the engine load increases, fuel is supplied to the secondary zone from four radial stub pipes located in the centerbody, as shown in Figure 9.25b. This fuel mixes with air and then flows through a swirler at the centerbody exit to create a swirl-stabilized secondary combustion zone. The primary and secondary zones both operate at low equivalence ratios, hence the term lean-lean for this operating mode, which can raise turbine output to base load.

9.7.3.3 Secondary

This mode represents a transition between lean-lean and premix modes. The fuel supply to the primary zone is gradually reduced while increasing the fuel flow to the secondary zone. Eventually, the primary flame is extinguished, leaving flame only in the secondary zone.

9.7.3.4 Premix

Fuel is reintroduced through the primary nozzles and the primary combustion zone (now premixed) is shifted to a region downstream of the liner venturi, where it is ignited by the secondary flame zone (Figure 9.25c). The venturi consists of a converging–diverging section that accelerates the flow from the first stage to prevent flashback. It also creates a toroidal recirculation zone over its downstream conical surface to stabilize the primary combustion zone in this premix mode, which is attained at or near the engine design point. This mode corresponds to minimum pollutant emissions.

If required, both the primary and secondary fuel injectors can be dual-fuel nozzles, to permit automatic transfer from gas to oil throughout the load range. The system can achieve NO_x and CO levels of 9 and 25 ppmv, respectively, at base load when operating on natural gas [51]. With liquid fuels, the NO_x and CO emissions from the DLN combustor at loads less than 20% of the base load are similar to those obtained with the standard combustor. This result is hardly surprising because both systems feature diffusion flames in this range. The combustor operates in the lean-lean mode between 20 and 50% load and in the premix mode from 50 to 100% load. NO_x emissions are appreciably lower than for a standard combustor due to the premixing, but are considerably higher than the low levels achievable with gas in the fully premixed mode. With water injection, the combustor achieves NO_x and CO levels of 42 and 20 ppmv, respectively, at base load when operating on distillate oil fuel [51].

9.7.4 ABB EV Burner

The ABB company has developed a conical premix burner module, called the EV-burner, which has demonstrated good performance in a wide range of DLN combustion applications [59,60]. A cross-sectional view to illustrate

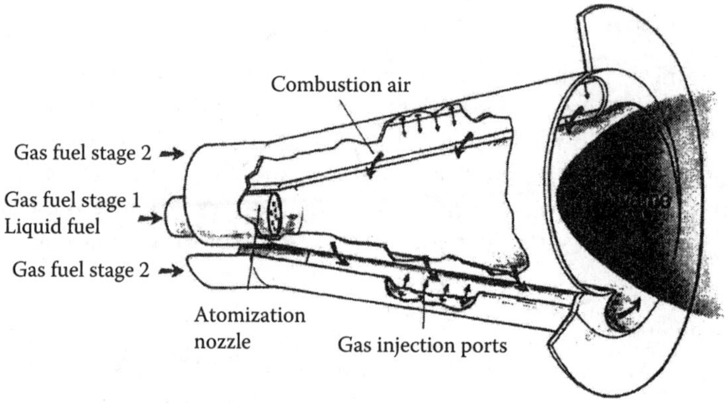

FIGURE 9.26
Operating principle of ABB EV conical premix burner. (From Sattelmayer, T., Felchin, M.P., Haumann, J., Hellat, J., and Styner, D., *Journal of Engineering for Gas Turbines and Power*, 114(1), 118–25, 1992. With permission.)

the operating principles of the burner is given in Figure 9.26. Each burner is formed by two offset half cones, which are shifted to form two diametrically opposed air inlet slots of constant width. Gaseous fuels are injected into the combustion air flowing into the slots by means of two fuel distribution tubes containing rows of small holes that inject the fuel across the airstream. Fairly complete mixing of fuel and air is obtained shortly after injection and the swirling mixture flows out of the cone and into the flame zone. A unique feature of this burner is that flame stabilization is achieved in free space near the burner outlet because of the sudden breakdown of a swirling flow.

The device can operate satisfactorily on both gaseous and liquid fuels. The latter are injected at the apex of the cone using a pressure or air-assist type of atomizer. The fuel is not fully evaporated by the time it reaches the stabilization zone and a diffusion-type flame penetrates a short distance upstream into the burner, which explains why no pilot stage is needed to achieve adequate flame stability. Also, because the flame never touches the walls, the cone body remains clean and metal temperatures are relatively low.

In February 1991, an ABB GT11N gas turbine was retrofitted with a new silo combustor of the type shown in Figure 9.27. This silo combustor is equipped with 37 EV burners, all of which operate in a pure premix mode [59]. For part load operation, fuel is supplied to only a fraction of the total number of burners. NO_x values of 13 ppmv have been reported by Aigner and Muller [60] for the base load conditions of 1.25 MPa and $T_3 = 643$ K.

When burning MBtu syngases, a different fuel-injection strategy is called for than for natural gas because of the very high flame speeds and fast reaction times of MBtu fuels [61]. Injection of the fuel along the inlet air slots is no longer appropriate. Instead, fuel injection is delayed until the burner exit, where it enters the swirling airstream through a number of plain holes that direct the fuel radially inward.

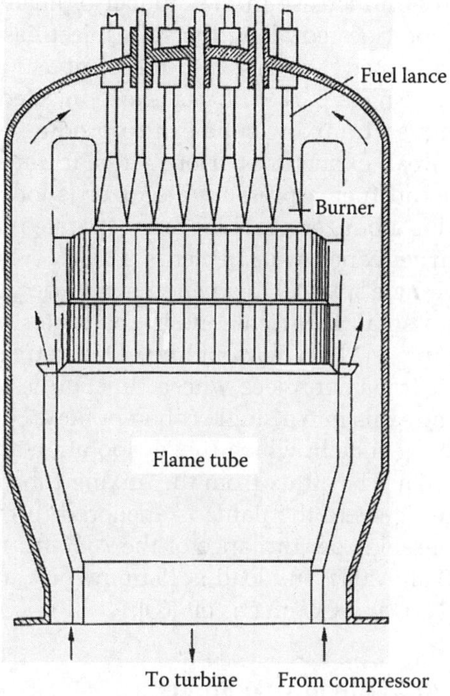

FIGURE 9.27
Silo burner fitted with ABB EV burners. (From Aigner, M. and Muller, G., *Journal of Engineering for Gas Turbines and Power,* 115(3), 533–6, 1993. With permission.)

The EV-burner technology has also been used in the design of annular combustors. The ABB GT10 (23 MW) combustor features a single row of 18 EV burners, whereas the heavy-duty ABB GT13E2 gas turbine (>150 MW) has 72 EV-burners that are arranged in two staggered circumferential rows within the annular combustor [62].

In 1994, the 17 MW GT35 7-can combustor was equipped with 3 EV burners per can. It operates on gas only and startup is achieved using one burner per can. Between idle and half load, a second burner is also used. At higher loads, all three burners are in operation. Burner staging has a surprisingly small effect on combustor pattern factor [63].

From the different sizes of engines and from the different types of combustor (can, annular, and silo) and the number of burners employed, it is clear that the burner modules cannot have the same size for all engines. Thus, a major asset of the EV burner is that it can be scaled with only minor modifications to suit a wide range of engine applications.

The latest engine from ABB—the GTX100—is a single shaft 43 MW machine that features an annular combustor containing 30 AEV burners, which represent the most recent development in EV burner technology. The

AEV burner features an increase in the number of air inlet slots from two to four. Gas feed pipes located along the slots inject gaseous fuel through "tuned" holes with "tuned" spacing into the combustion air flowing into the slots. When operating at part load, gas is also injected through six equispaced holes around each mixing tube exit to produce a ring of diffusion flames with good weak extinction performance. To meet the requirement for operation on liquid fuel, a pressure atomizer is located in the apex of the burner cone. This atomizer provides four separate fuel sprays—one for each slot in the burner cone. At its center is a small pilot atomizer, which produces a narrow-angle spray of low penetration. The purpose of this pilot nozzle is to create a small fuel-rich zone in the center of the mixing tube exit, thereby extending the lean blowout limit. Downstream of the four cone segments is a short transition piece whose function is to convert the four individual fuel-airstreams into a single coherent flow. Further downstream is a cylindrical mixing tube in which fuel evaporation and fuel–air mixing proceed to completion. The efflux from the mixing tube is then discharged into an annular liner, where the flame is anchored in free space as in the EV burner. The emissions performance of the AEV burner on the GTX100 engine between 50 and 100% full load is 15 ppmv NO_x and CO on gaseous fuel and 25 ppmv NO_x and CO on gas oil [63].

9.7.5 Rolls Royce RB211 Industrial Burner

Most of the operating experience gained in DLE technology has been with heavy-duty gas turbines whose applications call for extended periods of base load running. However, there are many other applications, such as mechanical drive for pipeline compressors, where considerable operational flexibility is required. Aeroderivative engines have much to offer in this regard, and there are a number of stationary engines in the small-to-medium-size category that have been derived from successful high-performance aero engines. One notable example is the Rolls Royce RB211 engine in which the annular aero-combustor has been replaced by nine radially positioned reverse flow combustors, as shown in Figure 9.28 from Willis et al. [31]. This arrangement results in an 80% increase in combustion volume. The primary zone is fed by two counter-rotating air swirlers, with several gaseous fuel-injection points located in each swirl passageway. The secondary mixing duct is wrapped around the primary combustor, but is separated from it by another annular duct that provides the wall-cooling air. Gaseous fuel is injected into the secondary duct from 36 equispaced axial spray bars, each containing six injection holes. This fuel bar arrangement was determined by trajectory calculations and an air velocity profile predicted by a CFD code. Fuel sampling and combustion tests showed uniformity of fuel–air mixing to within 4%. Combustion testing, carried out over a range of pressures from 0.1 to 2.0 MPa, demonstrated the ability of this axially staged, DLE combustor to achieve simultaneously low NO_x and low CO over wide ranges of power and ambient

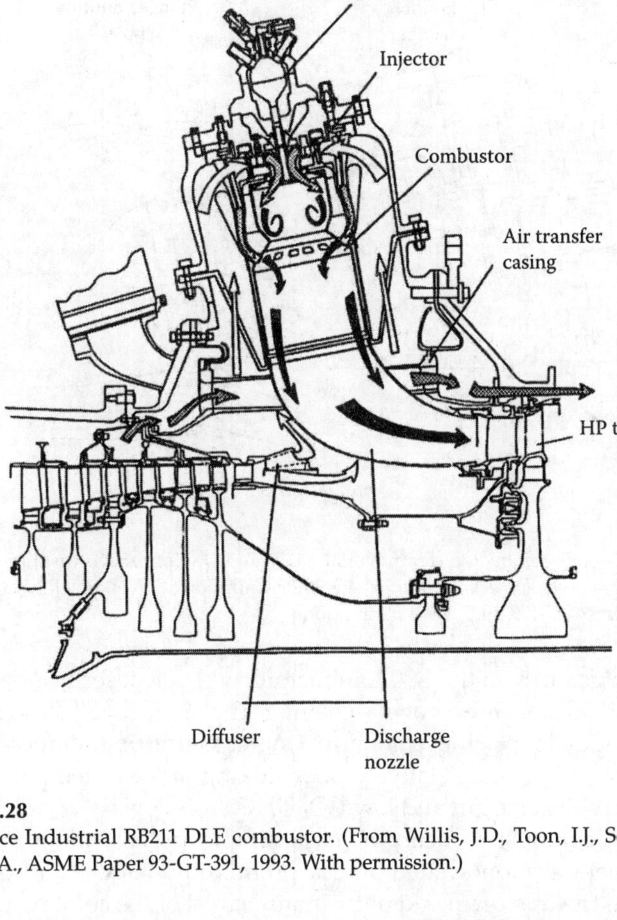

FIGURE 9.28
Rolls Royce Industrial RB211 DLE combustor. (From Willis, J.D., Toon, I.J., Schweiger, T., and Owen, D.A., ASME Paper 93-GT-391, 1993. With permission.)

temperature without resorting to either variable geometry or air bleeds. It was also demonstrated that a uniform fuel distribution before combustion is essential for achieving low emissions, especially at high pressures. Based on the test data obtained so far, at full base load conditions the predicted engine emissions are 17.4 ppmv NO_x, 5 ppmv CO, and zero UHC.

9.7.6 EGT DLN Combustor

The European Gas Turbine company (EGT) has adopted a simple fixed-geometry, partially premixed system for its G30 DLN combustor. The NO_x emission goal of 25 ppmv is achieved by partially premixing the fuel with half of the total combustor airflow. Also, the incorporation of an impingement-cooled, thermal barrier coated liner greatly reduces the wall-quenching effects associated with conventional film-cooled liners and limits CO and UHC emissions to below 50 and 20 ppmv, respectively.

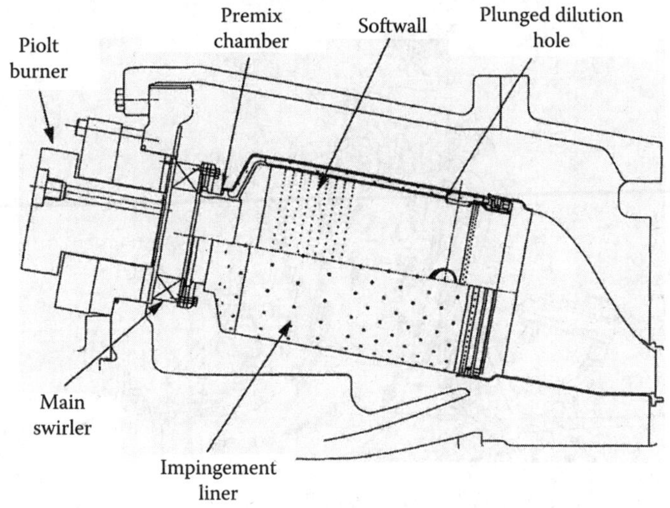

FIGURE 9.29
EGT dry low-emissions combustor. (From Norster, E.R. and DePietro, S.M., *Institution of Diesel and Gas Turbine Engineers, DEUA Publication* 495, 1996; Gallimore, S., Vickers, R.M, and Boyns, M.B., ASME Paper 97-GT-159, 1997. With permission.)

The essential features of the G30 combustor have been described by Norster and DePietro [64] and are shown schematically in Figure 9.29. The basic design philosophy is to achieve good mixing at high firing temperatures to limit NO_x production, and relatively poor mixing at lower temperatures to give a good stability margin and low CO/UHC emissions.

The tubular combustor incorporates a simple radial inflow swirler, a swirler slot fuel-injection system, and a premixing chamber, all of which are attached to the upstream end of the main barrel of the combustor. These components provide major control of combustion air, fuel injection, and mixing. Ignition and flame stabilization occurs within the vortex core of the prechamber. Gaseous fuel is injected at the entrance of each swirler slot through a metering jet that is sheltered below a step on the upstream side of the slot. At low fuel flows, corresponding to low firing temperatures, the injected fuel remains close to the rear wall of the slot and delivers a poorly mixed fuel–air mixture to the prechamber. With increasing fuel flow, the fuel penetrates further across the swirler slot and mixes more effectively with the swirler air, thereby providing a more uniform mixture to the burning zone.

During starting and engine acceleration, there is a need for a piloting flame of high stability and efficiency. This is provided by a pilot/igniter burner, which is centrally located in the head of the prechamber (see Figure 9.29). The amount of fuel supplied to the pilot under starting and acceleration conditions is adjusted automatically to achieve smooth and consistent starting. At full engine speed, with no load or low firing temperature, the pilot fuel proportion is fairly high (around 50%) to assist flame stability at these fuel-lean conditions. This proportion is gradually reduced with an increase in firing temperature and reaches a minimum value at full load.

The G30 combustor was designed and developed initially for the 4.9 MW Typhoon gas turbine. It is intended to be retrofitable across EGT's Typhoon/Tornado/Tempest range of small engines (<10 MW). Base load emissions of below 15 ppmv NO_x and 10 ppmv CO, along with zero UHC, have been achieved on the Tornado single-shaft industrial engine [65].

9.7.7 General Electric LM6000 Combustor

Another important aeroderivative gas turbine is General Electric's LM6000. Figure 9.30 gives a cross-sectional view of the DLN combustor designed for this engine, as described by Leonard and Stegmaier [19] and Joshi et al. [66]. This premix combustor employs about twice the volume of the conventional annular combustor it replaces in order to maintain low levels of CO and UHC while greatly reducing the emissions of NO_x. Part of the air used in combustion, which at maximum power is around 80% of the total combustor airflow, flows into the combustion zone through three annular rings of premixers, as shown in Figure 9.30. The two outer rings each have 30 fuel–air premixers, whereas the inner ring has 15. This arrangement of premixers facilitates fuel staging at part-load operation. The total of 75 fuel nozzles is formed by having 15 stems with three premixers on each stem, as shown in Figure 9.31, plus

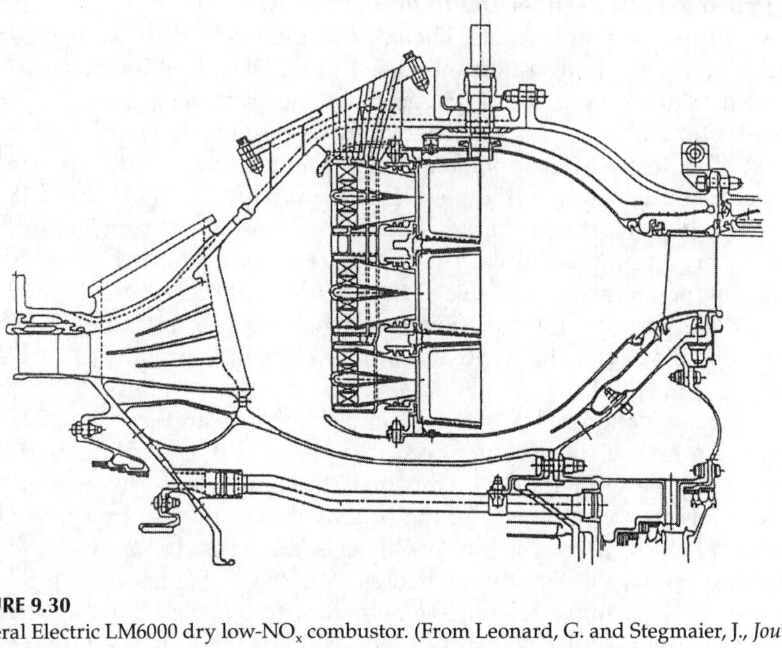

FIGURE 9.30
General Electric LM6000 dry low-NO_x combustor. (From Leonard, G. and Stegmaier, J., *Journal of Engineering for Gas Turbines and Power*, 116, 542–6, 1993. With permission.)

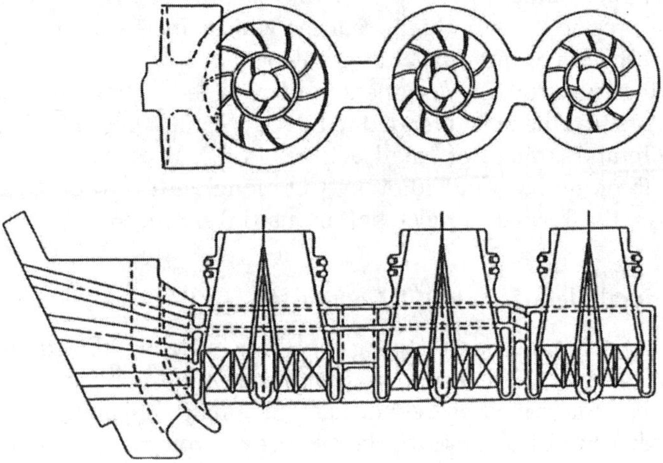

FIGURE 9.31
Fuel nozzle assembly for LM6000 dry low-NO_x combustor. (From Leonard, G. and Stegmaier, J., *Journal of Engineering for Gas Turbines and Power*, 116, 542–6, 1993. With permission.)

15 stems with two premixers on each stem. Each stem incorporates two or three separate fuel circuits for independently fueling the premixers.

A short annular liner was selected to minimize the amount of air needed for wall cooling. Only backside cooling is used, so a thermal barrier coating is applied to both the liner and in the dome area to keep the metal temperatures within acceptable limits. The use of a multipass diffuser also permits further reduction in overall combustor length. Of special importance to the attainment of low emissions is the design of the premixers. Figure 9.32 shows cross-sectional views of three different mixer designs that were subjected to combustion testing. The double annular counter rotating swirler (DACRS) was conceived to satisfy the restraints of autoignition and size. The duct diameter is reduced toward the exit in order to create an accelerating flow, thereby preventing flashback. The conical centerbody located along the centerline of the premixer can be used to supply liquid fuel to an atomizer at its tip, and gas passages for diffusion burning at low-power conditions [66].

The objective with this type of mixing device is to produce a completely homogeneous mixture of fuel and air at the premixer exit. As the total area of the fuel-injection holes is fixed by the flow rate and the available fuel-injection pressure, the design procedure is essentially one of finding the best compromise between the desire for small injection holes to give a large number of fuel-injection points, and the equally important requirement of large injection holes to allow the fuel jets to penetrate across the airstream.

In the premixer design designated as DACRS I in Figure 9.32a, the fuel is injected radially outward into the airstream from holes in the centerbody just downstream of the swirl vanes. This configuration suffered from unsatisfactory fuel jet penetration, so a modification was made by adding eight

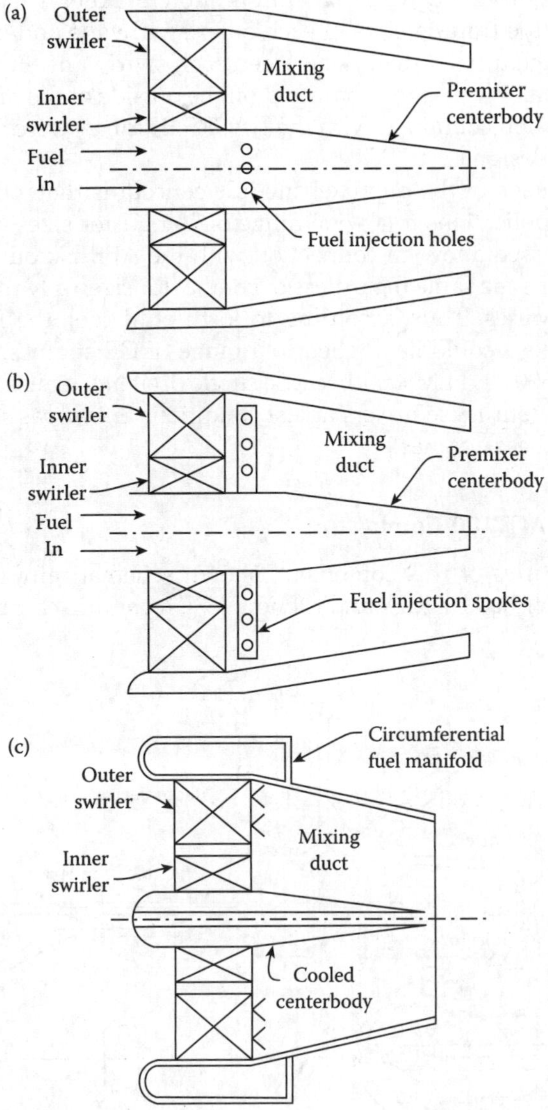

FIGURE 9.32
Cross-sectional views of three mixer designs. (From Joshi, N.D., Epstein, M.J., Durlak, S., Marakovits, S., and Sabla, P. E., ASME Paper 94-GT-253, 1994. With permission.)

radial spokes in the location of the holes in the centerbody, as shown in Figure 9.32b. Each spoke has three holes to inject gaseous fuel perpendicular to the flowing airstream. Combustion testing of this DACRS II mixer showed that single digit NO_x emissions are attainable with this concept. A further modification to the premixer designs described above was made by incorporating *fuel-injection holes into the swirl vanes of the outer swirler*, as illustrated in Figure 9.32c. In this DACRS III configuration, the fuel is injected through three holes in the trailing edge of each outer swirl vane and one hole in the outer wall of the mixing duct in between each swirl vane. Fuel is fed to the hollow outer vanes through a manifold on the outside of the premixing duct. The NO_x emissions obtained with the DACRS III mixer were very similar to the DACRS II design.

A big advantage of the premixer module concept is that, once developed, it has broad applications to a wide range of combustor sizes and configurations, as discussed above in connection with the ABB-EV burner. The basic module remains the same regardless of combustor size; only the number and arrangement varies. Thus, according to Joshi et al. [66], the DACRS II and DACRS III mixers could be applied to a range of GE engines, including the LM1600, LM2500, and LM6000, because single digit NO_x emissions have been attained with both these mixers at test conditions encompassing the operating ranges of these engines.

9.7.8 Allison AGT100 Combustor

The main features of this combustor, shown schematically in Figure 9.33, have been described by Rizk and Mongia [67]. It comprises a prechamber in

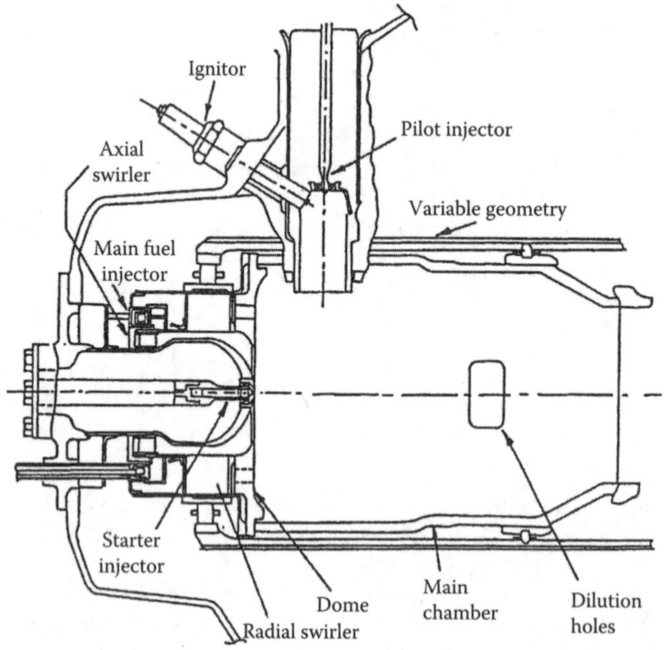

FIGURE 9.33
Allison AGT100 combustor. (From Rizk, N.K. and Mongia, H.C., *Twenty-Third Symposium (International) on Combustion*, 1063–70, The Combustion Institute, Pittsburgh, PA, 1990.)

which the fuel is vaporized and mixed with air, a pilot and ignition chamber, and the main cylindrical chamber. Variable geometry is employed to control the stoichiometry in the primary zone.

The prechamber contains a centerbody that houses both the main fuel injector and a pilot nozzle, which is employed only for lightup and acceleration to engine idle speed. The main fuel is introduced from a manifold surrounding the prechamber, just downstream of the prechamber axial swirler. Uniform filming of the fuel is achieved by spraying it through eight tangential holes onto the etched surface of the prechamber. The swirling air assists in the prefilming process. The high temperatures of the inlet air and the prechamber walls combine to promote rapid vaporization of the fuel within the prechamber. At power modes higher than idle, additional air is admitted into the prechamber through a radial swirler to merge and mix with the air flowing through the axial swirler.

Engine lightup is initiated in a small pilot chamber located on the side of the main combustion chamber. This piloting device also serves as a sustainer source when the combustor is operating at low inlet air temperatures or at conditions that lie outside the normal lean blowout limits.

The swirling vaporized fuel–air mixture flows into the main chamber through a round opening in the center of the dome. At high-power settings, additional air is injected into the main chamber through eight holes that are drilled in a manner designed to impart a swirling motion to the flowing air. Four simple rectangular dilution holes were chosen to ease fabrication of the ceramic liner. Variable geometry, in the form of sliding bands, is used to vary and control the flow areas of the dilution holes and the radial swirler in the prechamber. At low-power modes, most of the air flows through the dilution holes. As the fuel flow rate is increased above idle, the variable geometry is moved to increase the airflow through the radial swirler and to reduce, by a corresponding amount, the airflow through the dilution holes.

The use of variable geometry enabled the AGT100 combustor to meet the program goals of 5.0 and 37 g/kg fuel for NO_x and CO, respectively. Moreover, the experimental data acquired in the course of this investigation was used by Rizk and Mongia to develop a model for calculating NO_x formation in LPP combustors [68]. This model takes into account the effects of pressure, residence time, and air distribution between different combustion zones. It also provides useful insight into the contribution of the pilot chamber to the total NO_x emissions.

9.7.9 Developments in Japan

The strict NO_x regulations in Japan have promoted several developments in DLN combustion. Hosoi et al. [69] adopted a three-stage configuration in their design of a DLN combustor for application to a 2 MW gas-fired machine. The arrangement is shown schematically in Figure 9.34. The burner assembly consists of primary and secondary annular nozzles and a pilot nozzle at the

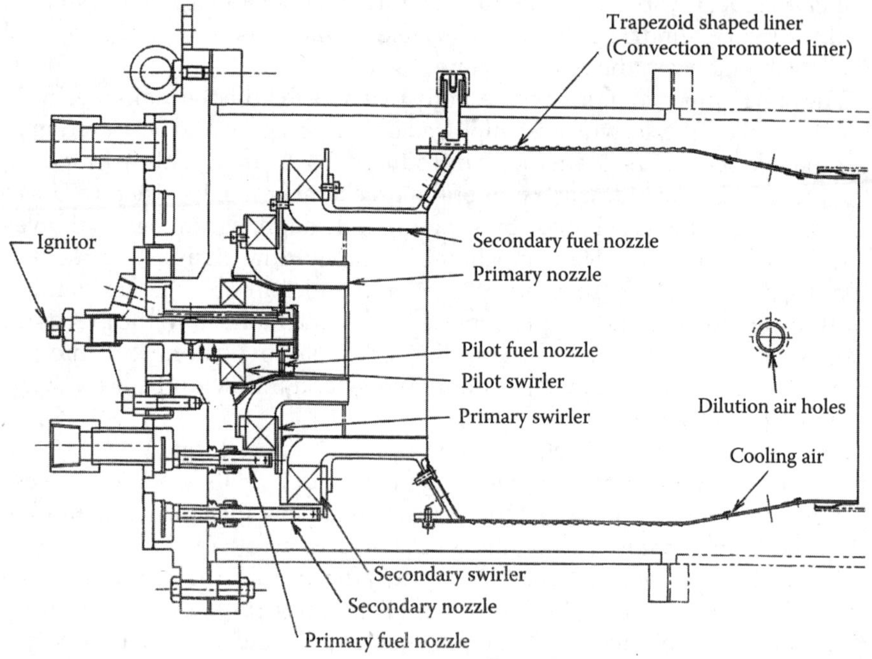

FIGURE 9.34
Dry low-NO_x combustor for a 2 MW class gas turbine. (From Hosoi, J., Watanabe, T., Toh, H., Mori, M., Sato, H., and Ishizuka, A., ASME Paper 96-GT-53, 1996. With permission.)

center. The fuel-injection schedule for the three coaxial burners is divided into three modes. Mode 1 employs both pilot and primary nozzles with equal fuel flow rates to each. It is used to sustain combustion from startup to 50% of the maximum engine rpm. At this point, the scheduling system switches to mode 2 in which the pilot fuel flow is held constant and the primary fuel flow increases with an increase in load up to 50% base load. In mode 3, the pilot and primary fuel flows both remain constant and the secondary fuel flow increases with an increase in load from 50 to 100% base load. Tests carried out on this combustor at a pressure of 1.18 MPa and an inlet air temperature of 643 K indicated NO_x levels of 10 ppmv (for 16% O_2) at base load, and combustion efficiencies above 99.8% at all loads above 50%.

In the small engine size range, the Japan Automobile Research Institute is collaborating with the Toyota Central Research and Development Laboratory in the development of a low-emissions combustor for a 100 kW automotive ceramic gas turbine. Some of the studies carried out in a premix-prevaporize system, operating both at atmospheric pressure and on the engine at high pressures, have been described by Kumakura et al. [70] and Ohkubo et al. [71]. The aim of this work is to provide quantitative data on the influence of fuel drop size and fuel distribution on the degree of vaporization achieved.

9.8 Lean Premix Prevaporize Combustion

A common feature of all the DLN combustors described above is that positive efforts are made to eliminate local regions of high temperature within the flame by mixing the fuel and air upstream of the combustion zone. The LPP concept represents the ultimate in this regard. Its underlying principle is to supply the combustion zone with a completely homogeneous mixture of fuel and air, and then to operate the combustion zone at an equivalence ratio that is very close to the lean blowout limit. The smaller the margin between stable combustion and flame blowout, the lower the output of NO_x.

A typical LPP combustor can be divided into three main regions. The first region is for fuel injection, fuel vaporization, and fuel–air mixing. Its function is to achieve complete evaporation and complete mixing of fuel and air before combustion. By eliminating droplet combustion and supplying the combustion zone with a homogeneous mixture of low equivalence ratio, the combustion process proceeds at a uniformly low temperature and very little NO_x is formed. In the second region, the flame is stabilized by the creation of one or more recirculation zones. Combustion is completed in this region and the resulting products then flow into region three, which may comprise a fairly conventional dilution zone.

A useful by-product of LPP combustion is that it is essentially free from carbon formation, especially when gaseous fuels are used, in which case the description "lean premixed" or "LPM" is more appropriate. The absence of carbon not only eliminates soot emissions, but also greatly reduces the amount of heat transferred to the liner walls by radiation, thereby reducing the amount of air needed for liner wall cooling. This is an important consideration because it means that more air is made available for lowering the temperature of the combustion zone and improving the combustor pattern factor.

Another important advantage of LPP systems is that for flames in which the temperature does not exceeds 1900 K, the amount of NO_x formed does not increase with an increase in residence time [18,19]. This means that LPP systems can be designed with long residence times to achieve low CO and UHC, while maintaining low NO_x levels. This finding is especially significant for industrial engines, where size is less important than for aero engines. As noted above, this approach leads to an LPM combustor volume that is approximately twice that of a conventional combustor [19].

The main problem with the LPP concept is that the long time required for fuel evaporation and fuel–air premixing upstream of the combustion zone may result in the occurrence of autoignition at the high inlet air temperatures and pressures associated with operation at high-power settings. Appropriate equations for calculating evaporation and autoignition delay times over wide ranges of mixture temperature, pressure, and equivalence ratio are given in Chapter 2. They may be used in the design of LPP combustors to ensure that at no operating condition does the sum of the fuel evaporation and mixing times exceed the autoignition delay time. Another problem associated with all well-mixed combustion systems is that of acoustic resonance, which occurs when the combustion process becomes coupled with the acoustics of the combustor. Lean-premixed systems are especially prone to this problem, as discussed in Chapter 7.

In summary, LPP combustion has considerable potential for ultralow NO_x emissions. NO_x levels below 10 ppmv have been reported by Poeschl et al. [72], even for flame temperatures higher than 2000 K. However, many formidable problems remain, the principal being that of achieving complete evaporation of the fuel and thorough mixing of fuel and air within the autoignition delay time and without risk of acoustic resonance or flashback.

9.8.1 Fuel–Air Premixing

Most types of ultra-low-emissions combustors rely on the attainment of near-perfect mixture homogeneity before combustion for their success. A homogeneous combustible mixture has the added advantage that it greatly reduces the possibility of autoignition. Although fuel-lean mixtures tend to have long autoignition delay times, imperfections in mixing result in local regions in which the equivalence ratio is higher than the average value, and ignition delay times are thereby greatly reduced. Thus, a high degree of mixture homogeneity is essential, not only for the attainment of low NO_x emissions, but also to alleviate the problem of autoignition.

The influence of mixture inhomogeneity on NO_x formation has been examined by several workers, both theoretically and experimentally. Lyons [73] used a multipoint fuel injector spraying Jet A fuel to achieve different equivalence ratio profiles across the diameter of the flametube. The results showed that spatial nonuniformity in the equivalence ratio resulted in increased NO_x emissions for equivalence ratios below 0.7 and decreased NO_x emissions for near-stoichiometric mixtures. Flanagan et al. [74] used a simple mixing tube fitted with a bluff-body flameholder at its exit. By changing the location of the natural gas fuel injector along the length of the tube, the degree of fuel–air mixing in the mixture approaching the stabilizer could be varied. When the system was operating at an equivalence ratio of 0.66, a nearly fivefold increase in NO_x emissions was recorded when going from well-mixed to incompletely mixed conditions. Fric [75] used an experimental apparatus very similar to that employed by Flanagan et al., to examine the NO_x emissions produced when burning natural gas in air at normal atmospheric pressure. He found that temporal fluctuations in the equivalence ratio can also raise NO_x emissions, in addition to spatial nonuniformities. For example, temporal fluctuations of 10% resulted in a doubling of NO_x.

Leonard and Stegmaier [19] used a gas-fired GE LM6000 combustor to examine the effects of premixing on NO_x formation. The results obtained are given in Figure 9.35, which shows NO_x as a function of average flame temperature for various degrees of premixing. Nonuniformities are the result of

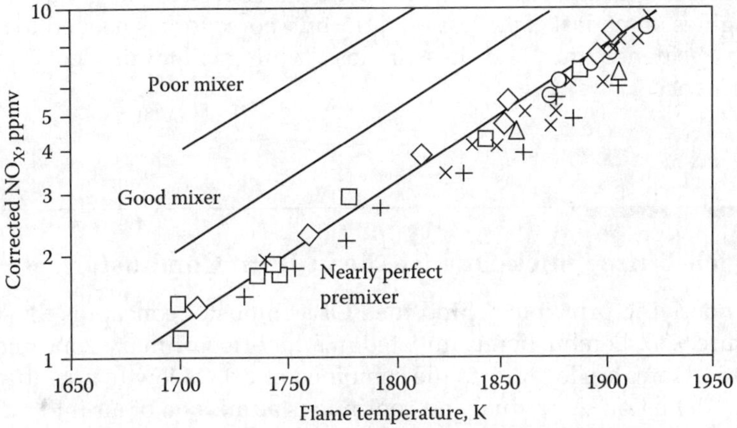

FIGURE 9.35
Effects of nonuniform fuel–air premixing on NO_x formation. (From Leonard, G. and Stegmaier, J., *Journal of Engineering for Gas Turbines and Power*, 116, 542–6, 1993. With permission.)

fluctuations in time as well as variations in space. Figure 9.35 contains data for a nearly perfect premixer, a well-designed premixer, and a nonoptimized premixer. It clearly illustrates the tremendous advantage to be gained from thorough mixing of air and fuel. Leonard and Stegmaier also noted that the amount of NO_x formed in a nonoptimized premixer increased with increasing pressure. They attributed this result to the fact that NO_x is formed in the hot spots (>2000 K) of poorly premixed flames by the thermal mechanism, which is pressure dependent.

The only exception to the general rule that better premixing yields less NO_x are the results obtained by Santavicca et al. [76]. These workers examined the effects of incomplete fuel–air mixing on the emissions characteristics of an LPP coaxial mixing tube combustor. Contrary to expectations, it was found that an improvement in fuel–air mixing resulted in comparable NO_x emissions for the same conditions of inlet temperature and equivalence ratio. However, the better mixed device was able to demonstrate lower NO_x emissions due to its ability to operate at lower equivalence ratios.

A number of studies have been carried out on the use of mechanical mixers for achieving a satisfactory degree of fuel–air premixing. Static mixers are widely used in process engineering for mixing of both gases and liquids, but they appear to have evoked little interest for combustion applications. Poeschl et al. [72] examined the mixing capability of a commercially available static mixer and observed excellent homogeneity, with a standard deviation of lower than 5% for a 2% loss in total pressure. According to these workers, flashback should not be a problem if the velocity through the mixer is higher than 20 m/s. Also, by flattening the velocity profile, the mixer eliminates the boundary layer along which flashback is most prone to occur. According to Valk [77], the static mixer system has good potential for engines of modest compression ratio, but more work is needed to reduce length, residence time, and pressure loss, while maintaining good mixing performance.

9.9 Rich-Burn, Quick-Quench, Lean-Burn Combustor

The underlying principle behind the RQL combustor concept is illustrated in Figure 9.36. Combustion is initiated in a fuel-rich primary zone and NO_x formation rates are low due to the combined effects of low temperature and oxygen depletion. A gradual and continuous admission of air into the combustion products emanating from the primary zone would raise both their temperature and oxygen content, thereby greatly accelerating the rate of NO_x formation, as indicated by the high NO_x route in Figure 9.36. If, however, the additional air required to complete the combustion process and reduce the gas temperature to the desired predilution zone level could be mixed uniformly and instantaneously with the primary-zone efflux, the combustion process would then follow the low NO_x route shown in Figure 9.36. This figure serves to demonstrate that the design of a rapid and effective quick-quench mixing section is of decisive importance to the success of the RQL concept.

A practical form of RQL combustor is shown schematically in Figure 9.37. Combustion is initiated in a fuel-rich primary zone operating at an equivalence ratio of between 1.2 and 1.6. A higher equivalence ratio would be even more efficacious in reducing NO_x, but could lead to excessive soot formation

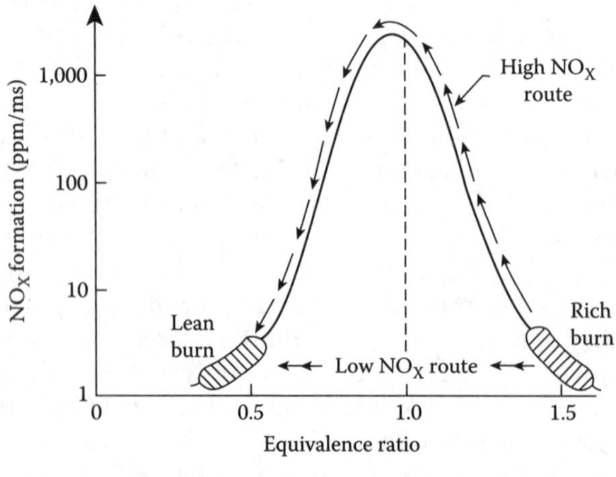

FIGURE 9.36
Graph to illustrate the principle of rich-burn, quick-quench, lean-burn (RQL) combustion.

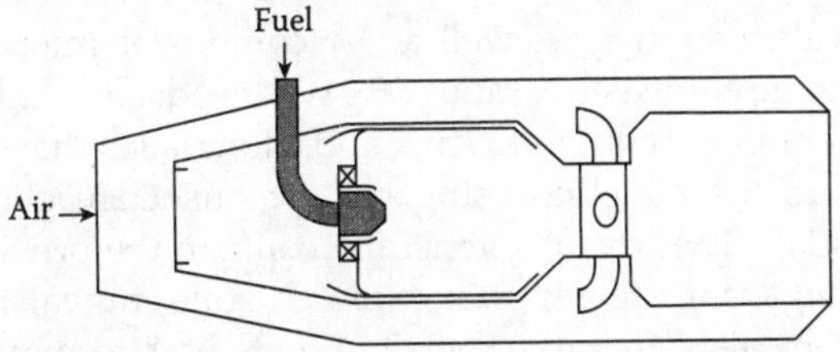

FIGURE 9.37
Schematic diagram of RQL combustor.

and smoke. For RQL combustion to be fully effective, the fuel must be finely atomized and uniformly distributed throughout the fuel-rich zone. Moreover, the primary-zone airflow pattern must be designed to prevent the occurrence of localized flow recirculation zones, which could increase residence times, thereby increasing the production of NO_x [44].

As well as reducing thermal NO_x, this initial fuel-rich combustion process also discourages NO_x formation from FBN by converting a large fraction of the FBN into nonreactive N_2 [78]. A further advantage of an initial fuel-rich stage in the combustion of low heating value (LHV) fuels containing ammonia (NH_3) is that it can greatly reduce the conversion of NH_3 into NO_x [79].

As the fuel-rich combustion products flow out of the primary zone, they encounter jets of air that rapidly reduce their temperature to a level at which NO_x formation is negligibly small. As mentioned above, this transition from a rich zone to a lean zone must take place quickly to prevent the formation of near-stoichiometric, high NO_x-forming streaks.

If the temperature of the lean-burn zone is too high, the production of thermal NO_x becomes excessive. On the other hand, the temperature must be high enough to consume any remaining CO, UHC, and soot. Thus, the equivalence ratio for the lean-burn zone must be carefully selected to satisfy all emissions requirements. Typically, lean-burn combustion occurs at equivalence ratios between 0.5 and 0.7 [80]. After the requirements of combustion and liner-wall cooling have been satisfied, any remaining air can be used as dilution air to tailor the exit temperature pattern for maximum turbine durability.

In some designs, the atomizing air is arranged to flow over the outside of the liner wall in the rich zone before entering the fuel nozzle. This regenerative backside convective cooling is an important design feature because conventional film cooling in the rich zone would create local near-stoichiometric mixtures that would produce high levels of NO_x.

Work has been in progress on RQL combustors since the late 1970s [81]. In early experimental studies by Novick et al. [82], NO_x emissions appeared to be controlled only by inlet temperature and rich-zone equivalence ratio, whereas CO and smoke emissions were influenced markedly by both rich-zone and lean-zone equivalence ratios, as well as by combustor inlet temperature. A minimum lean-zone equivalence ratio of 0.6 was needed to achieve satisfactory smoke levels. In more recent work, Rizk and Mongia [83] have applied three-dimensional emissions modeling, using well-established reaction mechanisms, to RQL combustion. Their results generally confirm the previous findings of Novick et al. in regard to the importance of rich-zone equivalence ratio to NO_x emissions, but they also stress the contribution to NO_x formation of residence time and combustion pressure.

In Japan, Nakata et al. [79,84] have designed an RQL combustor for a 150 MW class, stationary gas turbine with the double objective of maintaining stable combustion when burning LHV gas and reducing the NO_x emissions that are produced from the NH_3 in fuel. An interesting feature of the design is that strong swirl is imparted to the air and fuel as they enter the fuel-rich primary zone with an equivalence ratio of 1.6. Tests carried out at atmospheric pressure gave very satisfactory results in terms of good combustion stability, and low NO_x emissions (3 ppm for combustor exit temperatures up to 1500°C), albeit at the expense of fairly high CO emissions.

The GE RQL2 combustor for LHV gas also features a swirl-stabilized fuel-rich primary zone. This zone terminates in a converging section that serves both to prevent the swirling flow from drawing lean-stage gases back upstream into the primary zone and to reduce the flow area to a reasonable size for proper quenching [85]. Rapid quenching is achieved by injecting the air through holes of different sizes to obtain a uniform distribution of quench air across the hot gas stream. When operating at 1670 K combustor exit temperature, the NO_x and CO emissions were 50 and 5 ppmv, respectively. Perhaps of greater significance is that the conversion of NH_3 to NO_x was only 5%.

The RQL concept is being actively studied for aircraft applications by the Pratt and Whitney company and other laboratories in the United States as part of NASA's HSCT (High Speed Civil Transport) program. The aim of this program is to demonstrate the feasibility of attaining NO_x levels of 3–8 g/kg fuel (i.e., around 40–100 ppmv) at supersonic cruise conditions with kerosine fuel [86].

Most of the work carried out so far on the RQL concept has confirmed its potential for ultralow NO_x combustion and low conversion of FBN into NO_x. With LHV fuels, the conversion of NH_3 into NO_x is also greatly reduced. In comparison with conventional combustors, RQL combustors have inherently better ignition and lean blowout performance. In comparison with staged combustors, they have the important practical advantage of needing fewer fuel injectors. However, in order to fully exploit these assets, significant improvements in quench mixer design are needed. With liquid fuels, other potential problems include high soot formation in the rich primary zone, which could give rise to high flame radiation and exhaust smoke. These problems are exacerbated by long residence times, unstable recirculation patterns, and nonuniform mixing.

9.10 Catalytic Combustion

Catalytic combustion is a process that employs a catalyst to initiate and promote chemical reactions in a flowing premixed fuel–air mixture at leaner conditions than are possible in homogeneous gas-phase combustion. This allows stable combustion to proceed at equivalence ratios that are below the normal lean flammability limit of the fuel–air mixture. Combustion at such reduced temperatures can be expected to dramatically decrease the production of thermal NO_x.

The principle of catalytic combustion is shown schematically in Figure 9.38. Fuel is injected upstream of the reactor to vaporize and mix with the inlet air. The fuel–air mixture then flows into a catalyst bed, or reactor, which may consist of several stages, each made of a different kind of catalyst. For the first stage, it is desirable to use a catalyst that is active at low temperatures, whereas subsequent stages need to be selected for good oxidation efficiency. Downstream of the catalytic bed, a thermal reaction zone is usually provided to raise the gas temperature to the required turbine entry value and to reduce the concentrations of CO and UHC to acceptable levels.

The potential of catalytic reactors for very low pollutant emissions has been recognized since 1975 [87], but the harsh environment in a gas turbine combustor and its wide range of operating conditions pose formidable problems that must be overcome to secure the implementation of catalytic combustion in operational gas turbines [88]. It is difficult to design a catalyst that will ignite a fuel–air mixture at the low compressor exit temperatures corresponding to crank lighting and engine operation at low load. Moreover, combustor exit temperatures are usually in the range from 1450 to 1770 K, which are well above the stability limits of most catalyst substrate materials. Even ceramics that can withstand high combustion temperatures are susceptible to thermal shock failure during engine transients. The doubts concerning

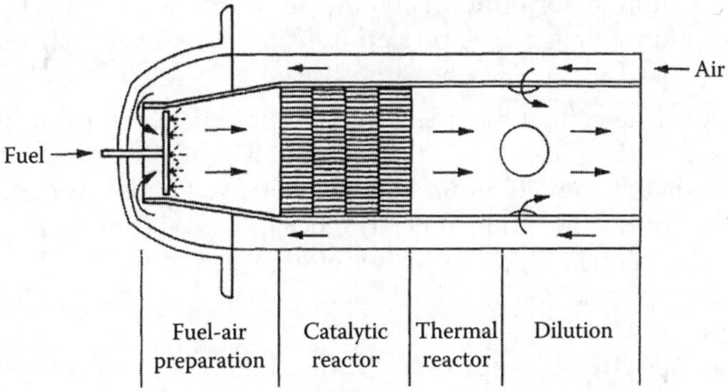

FIGURE 9.38
Schematic representation of catalytic combustor.

the long-term durability of catalyst substrates and sustained high catalytic activity for periods up to several thousand hours constitute a significant barrier to the development of viable catalytic combustors for gas turbines.

Current research and development activities in catalytic combustion for gas turbines are mainly in the form of small-scale laboratory tests and combustor rig tests. Many different concepts have been examined. Some of the designs tested are fairly complex, and require the use of a preburner along with fuel staging and/or variable geometry to achieve a satisfactory operating range. However, the results obtained from these various studies fully confirm the potential of catalytic combustion for ultralow NO_x (<5 ppmv) and provide the incentive to overcome the formidable challenges posed by durability issues and the limited temperature range capability of catalytic reactors.

9.10.1 Design Approaches

The various approaches to the design of catalytic combustors for gas turbines have been described in a number of publications [89–99]. They fall broadly into three main categories:

1. Traditional systems, as illustrated in Figure 9.38, in which the catalyst is fed with a fuel–air mixture whose adiabatic flame temperature is equal to the required combustor outlet temperature, T_4. This approach is satisfactory for low values of T_4 but, on modern gas turbines where T_4 could be as high as 1570 K, it would give rise to problems such as sintering and vaporization of the active catalyst components and thermal shock fracturing of ceramic supports.
2. Systems where only a part of the fuel is injected upstream of the catalyst to limit catalyst temperatures below 1270 K, and the rest of the fuel is injected downstream of the catalyst to achieve the desired combustor outlet temperature. Examples of this approach may be found in Cowell and Larkin [95], Ozawa et al. [96], and Fujii et al. [97].
3. Systems where all of the fuel is injected upstream of the catalyst, but is only partially reacted within the catalyst bed. Combustion then proceeds to completion via homogeneous gas-phase reactions in a postcatalyst combustion zone. The designs of Vortmeyer et al. [98], Dalla Betta et al. [89,91], Schlatter et al. [92], and Dutta et al. [93,94] use this approach.

The main objective of Approaches 2 and 3 is to keep substrate temperatures low to prevent problems of thermal sintering and catalyst deactivation. Limiting the maximum catalyst temperature in this way not only extends catalyst life, but also broadens the selection of suitable catalyst components.

9.10.2 Design Constraints

The problems facing the designer of catalytic combustors tend to focus on temperature and the need to reconcile the broad range of temperatures required by the engine with the relatively narrow range of temperatures over which the catalyst bed can satisfy the conflicting requirements of high catalytic activity and mechanical integrity.

Any given catalytic combustor has a certain range of operating conditions over which it can achieve stable combustion and low emission levels. This range of conditions, which is often referred to as an operating "window," is determined primarily with regard to three important temperatures [91–94].

1. The inlet mixture temperature, which must be high enough to activate the catalyst. Below a certain minimum temperature, known as the "lightoff' temperature, the exothermic oxidation reactions occurring on the reactor walls are too slow to generate the heat needed to sustain the reactions. In general, temperatures in excess of 700 K are necessary for catalyst lightoff. For example, noble metals, such as platinum and palladium, require 617–783 K lightoff temperatures, whereas metal oxide catalysts with active elements, such as nickel and cobalt, require 866–1367 K temperatures [99].
2. The gas temperature leaving the catalyst must be high enough to allow the catalytically initiated reactions to proceed to completion in the available residence time or, if additional fuel is injected downstream of the catalyst, to promote rapid combustion of this fuel, thereby reducing the emissions of CO and UHC to acceptable levels.
3. Catalyst wall temperatures must be low enough to provide the durability needed for stable long-term reactor operation.

9.10.3 Fuel Preparation

Successful operation of the catalyst bed is highly dependent on its entry flow conditions, which should be very uniform in terms of mixture strength, velocity, and temperature to ensure effective use of the entire catalyst area and prevent damage to the substrate because of local high temperatures.

Failure to achieve a completely uniform mixture will detract from performance in a number of ways that follow directly from the concept of an operating window, as discussed above. For any given inlet temperature, a low fuel/air ratio may result in low catalyst temperatures and high emissions of CO and UHC. On the other hand, too high a fuel/air ratio may cause the catalyst to overheat. Clearly, if there are local variations in temperature and fuel/air ratio in the mixture entering the reactor, some regions of the catalyst will experience conditions on the low temperature (high CO and UHC emissions) side of the operating window, whereas other regions will be exposed to potential damage to the substrate due to local high gas temperatures

[91,93]. Tests carried out by Dutta et al. [94] showed that inhomogeneities in fuel concentration greater than 10% (peak to peak) can lead to catalyst damage and high CO and UHC emissions. Nonuniformities in inlet conditions also make it more difficult to maintain the catalyst temperature within the operating window, which curtails the degree of flexibility in responding to changes in engine operating conditions [91].

Mechanical mixers are sometimes used to promote more uniform flow conditions at inlet to the catalyst bed [89,96–98]. The advantages and drawbacks of mixers in premixed combustion applications have already been discussed in the context of LPP combustors. Their main drawback is the obvious one—the introduction of additional pressure loss that penalizes engine performance. Also in common with LPP combustors, the residence time in the fuel-preparation zone must be short enough to avoid the risk of spontaneous ignition.

9.10.4 Catalyst Bed Construction

A common catalyst specification is one where the major active ingredient (usually a metal, such as palladium, platinum, or magnesium, or a metal oxide) is applied to a stabilized alumina washcoat on a honeycomb-type ceramic monolith, such as cordierite. The washcoat is porous, to provide a large surface area, and may contain a dispersion of active catalyst particles, such as aluminum oxide or zirconium oxide.

One problem with ceramic monoliths is that as they become larger, their mechanical and thermal reliability is reduced. This problem can be overcome by replacing a large catalyst bed with a number of small segments that are clustered together without cementing [96,97]. Although this method of construction eases the problem, the durability of ceramic monoliths is always a cause for concern, especially in heavy-duty gas turbines. A more robust form of construction is one whereby each stage of the catalyst reactor is formed by corrugating a strip of oxidation-resistant metal foil, and coiling the strip in such a way as to form a channeled monolith structure through which the fuel–air mixture can flow and react on the channel walls [91,92,100]. The active catalytic material is deposited as a coating on the foils.

The reactor employed by Beebe et al. [100] and Dalla Betta et al. [91] consisted of three such stages, each designed to deliver gas at the appropriate temperature to the next stage or to a final homogeneous combustion section. An interesting feature of the design [90] is the manner in which it takes advantage of the unique thermodynamics of palladium oxidation and reduction to control surface temperatures. Palladium oxide decomposes to palladium metal at temperatures between 1050 and 1200 K, depending on the pressure. This transition between oxide and metal can be exploited to limit the catalyst temperature, which allows a corrugated metal (Fecralloy) support to be used instead of the less durable ceramic substrate.

9.10.5 Postcatalyst Combustion

The main function of the postcatalyst combustion zone is to reduce the concentrations of CO and UHC to the required levels and to achieve the desired combustor exit temperature. Placing the zone of maximum gas temperature downstream of the catalyst allows the catalyst bed to operate at a relatively low temperature (<1270 K) with consequent advantages in terms of reliability and extended life. It is of interest to note that the temperature gradient in a catalytic combustor is in the opposite direction to that in a conventional gas turbine combustor. In the latter, maximum temperatures are attained in the primary combustion zone and gas temperatures then decline in the downstream direction as more air is injected into the hot gas stream to reduce its temperature to the required combustor exit value.

Dalla Betta et al. [89] have pointed out the following advantages of postcatalyst combustion:

- It allows the catalyst to operate at low temperatures that can reduce or eliminate many of the deactivation mechanisms that curtail catalyst life
- It permits the use of a wide variety of catalyst and substrate materials
- The problems associated with thermal shock fracture of substrates during startup, shut down, and transient operations are alleviated
- The combustor can be developed to higher outlet temperatures without changing the catalyst material

9.10.6 Design and Performance

Two different types of catalytic combustors for gas turbines are described below to illustrate the main features of current designs. Fujii et al. [97] used Approach 2 in designing a catalytic combustor to operate with an outlet temperature of 1570 K while keeping the catalyst bed temperature down to around 1270 K. The system, shown schematically in Figure 9.39, consists of an annular preburner, six fan-shaped catalyst segments, six premixing fuel nozzles, and a premix combustion section downstream of the catalyst. The inlet air is heated to 720 K by the preburner and is distributed to both the catalytic segments and the premixing nozzles, which are arranged alternately in the form of a circle, as shown in Figure 9.39. The major active ingredient of the catalysts is palladium, which is supported on a stabilized alumina washcoat on a honeycomb-type monolith made of cordierite [96]. At its downstream end, each nozzle is shaped to inject the premixed gas–air mixture flowing between the catalyst segments into the catalyst efflux at an angle close to 90°. The mixture is ignited by the hot combustion products emanating from the catalyst segments, and LPM combustion then occurs to raise the combustor outlet temperature to 1570 K.

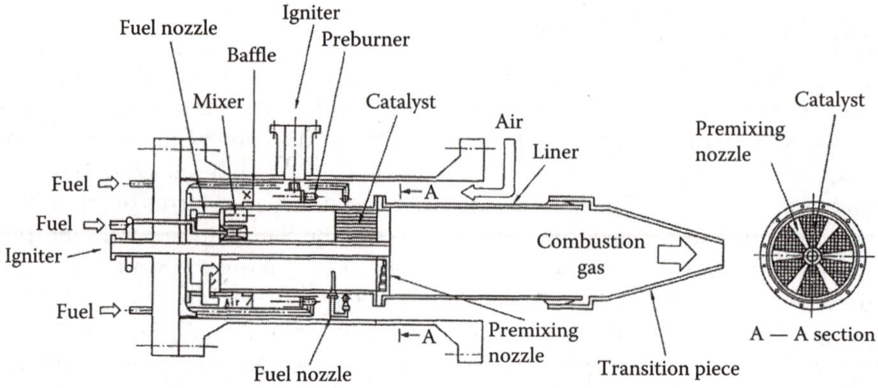

FIGURE 9.39
High-pressure catalytic combustor. (From Ozawa, Y., Hirano, J., Sato, M., Saiga, M., and Watanabe, S., *Journal of Engineering for Gas Turbines and Power,* 116, 511–16, 1994; Fujii, T., Ozawa, Y., Kikumoto, S., and Sato, M., ASME Paper 96-GT-382, 1996. With permission.)

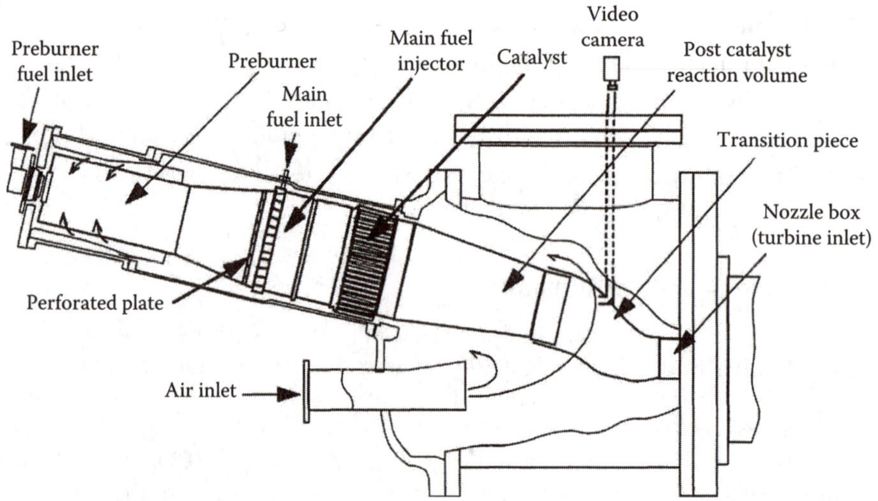

FIGURE 9.40
GE catalytic combustor in test stand. (From Schlatter, J.C., Dalla Betta, R.A., Nickolas, S.G., Cutrone, M.B., and Beebe, K.W., ASME Paper 97-GT-57, 1997. With permission.)

When operating on natural gas over a pressure range from 0.10 to 1.35 MPa (1–13.5 bar), NO_x emissions were always below 10 ppmv, of which a significant portion came from the preburner. The maximum overall combustor pressure loss was about 2.5%, which is well below the values normally associated with conventional combustors.

The alternative method (Approach 3) for avoiding exposure of the catalyst to excessive temperatures is illustrated schematically in Figure 9.40, which shows a catalytic combustor now under development by the General Electric Company for its natural gas-fired MS9001E gas turbine [92]. The technology used to design this combustor has been demonstrated in a number of sub-scale and full scale tests, e.g., References [89,91]. A basic feature of the concept is that all the fuel required to achieve the combustor exit temperature of 1380 K is supplied to the catalytic reactor, which comprises two or more separate stages. The first stage operates at a relatively low temperature, which makes possible the high catalytic activity needed for catalyst operation at the compressor outlet temperature. The last stage can operate successfully with a lower catalytic activity because its inlet temperature is higher, but its substrate temperature must also be high to provide the outlet gas temperatures needed to initiate the homogeneous combustion reactions downstream of the catalyst. Typically, about half of the fuel is reacted within the catalyst stages and the remainder is consumed in the postcatalyst combustion zone.

The catalytic reactor for the GE MS9001E consists of three individually supported stages, each 508 mm in diameter. The catalyst stages are formed by corrugating strips of very thin (50 μm) oxidation-resistant metal foil and then depositing the active catalytic material (palladium oxide) as a coating on the strips [92]. The strips are coiled in order to form channels through which the fuel–air mixture flows while reacting on the channel walls. The overall length of the catalyst container is 305 mm of which the catalyst occupies 230 mm.

The fuel-injection system consists of 93 individual venturi tubes arranged across the flow path, with four fuel-injection orifices at the throat of each venturi [101]. The objective is to deliver to the catalyst a fuel–air mixture that is uniform in composition, temperature, and velocity. The target for fuel-air uniformity is a maximum range of 10% between the highest and lowest concentration. By suitable tailoring of the individual fuel-injector orifices, a range of 12% has been achieved so far, which is close to the target value.

Results of tests carried out in June 1996 at the simulated base load operating point ($P_3 = 1.25$ MPa, $T_3 = 714$ K, $T_4 = 1465$ K), indicated NO_x levels below 5 ppmv and even lower concentrations of CO and UHC [94].

9.10.7 Use of Variable Geometry

Depending on the type of catalyst employed and the degree of homogeneity achieved in the fuel–air mixture entering the reactor, stable combustion can be sustained over a temperature range of only a few hundred degrees Kelvin without serious loss of performance. This corresponds to a range of overall fuel/air ratios of around 1.4–1, as opposed to the 5–1, which is readily achieved with conventional combustors. It is essential, therefore, that catalytic combustion be combined with variable geometry and/or fuel staging to maintain stable combustion over the broad range of conditions encountered in modern gas turbines.

One example of the use of variable geometry to broaden the operating range of an engine fitted with a catalytic combustor is the Solar design for the Advanced Turbine Systems (ATS) program. This program is aimed at the development of advanced recuperative gas turbines with pollutant emissions at single digit levels. The Solar catalytic combustion system is modular in design and includes a fuel–air mixer upstream of the reactor and a postcatalyst homogeneous gas-phase reaction zone downstream of the catalyst bed to complete the combustion process. Startup is accomplished using a conventional (LPM) low-emissions fuel injector. The system transitions to catalyst operation using a variable geometry valve that diverts airflow into the catalyst at loads greater than 50% of full load. This arrangement overcomes the limitations created by the narrow turndown ratio of the catalyst and allows the catalyst outlet temperature to remain fairly constant over the 50–100% load range [93].

9.10.8 Future

All the laboratory and combustor rig tests carried out on catalytic combustors for gas turbines have amply demonstrated their capabilities in regard to ultralow pollutant emissions. Considerable progress has also been made in increasing the life expectancy of catalysts under typical high-temperature conditions. Although most combustor rig testing to date has involved relatively short test periods of less than 100 hours, one test has been reported in which a catalyst system operated satisfactorily for 4500 hours on natural gas fuel at atmospheric pressure and an adiabatic combustion temperature of 1570 K without any apparent damage to the catalyst and its support structure [89].

Most of the work now in progress on catalytic combustion for gas turbines is directed toward stationary applications and is aimed at meeting the broad operating range and arduous operating conditions of gas turbines. Work is continuing on catalyst development to lower the required reactor inlet temperature and system development to enlarge the operating window. Engine tests over long periods are now called for to fully validate the technology. The current dearth of experience on stationary engines certainly eliminates catalytic combustion from serious consideration for aircraft applications at this time.

9.11 Correlation and Modeling of Oxides of Nitrogen and Carbon Monoxide Emissions

Many empirical and semi-empirical models are now in widespread use for correlating experimental data on pollutant emissions in terms of all the relevant parameters. These include combustor dimensions, design features, and operating conditions, as well as fuel type and fuel spray characteristics. Empirical models can play an important role in the design and development of low-emissions combustors. They serve to reduce the complex problems associated with emissions to forms that are more meaningful and tractable to the combustion engineer. They also permit more accurate correlations of emissions for any one specific combustor than can be achieved by more comprehensive numerical models.

As the chemical reactions governing the formation of UHC and smoke are highly complex, most of the empirical models developed so far have been confined to NO_x and CO. Lefebvre [102] assumed for both these species that their exhaust concentrations are dependent on three terms, which are selected to represent the following: (1) mean residence time in the combustion zone, (2) chemical reaction rates, and (3) mixing rates. Expressions for these three parameters were derived in terms of combustor size, liner pressure loss, airflow proportions, and operating conditions of inlet pressure, temperature, and air mass flow rate. This approach led to the development of semi-empirical expressions for NO_x and CO, which are presented below.

9.11.1 Oxides of Nitrogen Correlations

According to Lefebvre [102], we have

$$NO_x = 9 \times 10^{-8} P^{1.25} V_c \exp(0.01 T_{st}) / \dot{m}_A T_{pz} \text{ g/kg fuel.} \tag{9.7}$$

The values of the constant and exponents in this equation were obtained from analysis of experimental data on NO_x emissions from several different aero-engine combustors. Equation 9.7 takes account of the fact that in the combustion of heterogeneous fuel–air mixtures, it is the stoichiometric flame temperature that determines the formation of NO_x, not the average flame temperature. However, for the residence time in the combustion zone, which is also significant to NO_x formation, the appropriate temperature term is the average value T_{pz}, as indicated in the denominator of Equation 9.7. The excellent correlation of experimental data on NO_x provided by Equation 9.7 for GE J79-17A and F101 combustors is illustrated in Figures 9.41 and 9.42, respectively.

Equation 9.7 is suitable for conventional spray combustors only. For LPP combustors, in which the maximum attainable temperature is T_{pz}, it may still be used, provided that T_{pz} is substituted for T_{st}.

Many other semi-empirical models for predicting NO_x emissions have been derived. For a critical review of the models developed before 1980, reference should be made to Mellor [103]. Some of the more recent expressions for NO_x emissions include the following.

9.11.1.1 Odgers and Kretschmer [104]

$$NO_x = 29 \exp -(21{,}670/T_c) P^{0.66} \times [1 - \exp -(250\tau)] \text{g/kg fuel.} \tag{9.8}$$

Odgers and Kretschmer's recommended NO_x formation times for aircraft combustors are 0.8 ms (airblast atomizers) and 1.0 ms (pressure atomizers). Quoted formation times for industrial combustors burning liquid fuels range from 1.5 to 2.0 ms.

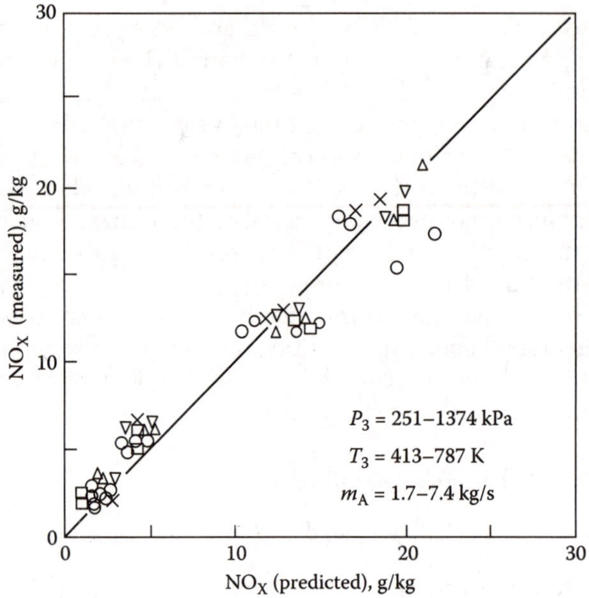

FIGURE 9.41
Comparison of measured and predicted values of NO_x for a GE J79-17A combustor.

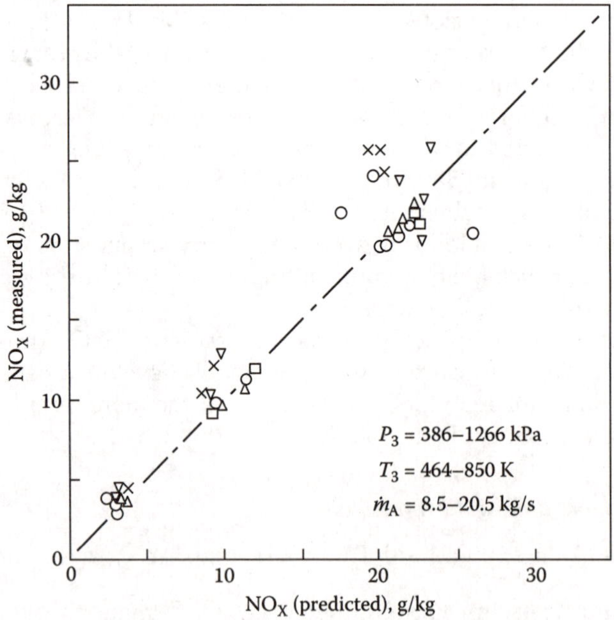

FIGURE 9.42
Comparison of measured and predicted values of NO_x for a GE F101 combustor.

9.11.1.2 Lewis [105]

$$NO_x = 3.32 \times 10^{-6} \exp(0.008T_c)P^{0.5} \text{ ppmv.} \quad (9.9)$$

This equation is intended to show the amount of NO_x formed in lean, homogeneous combustion. It suggests that NO_x formation depends only on the postcombustion temperature and pressure and is completely independent of the residence time of the gases in the combustor. According to Lewis [105], this is because the relevant time is not the residence time of the combustion products, but rather the relaxation time of the molecules involved, primarily the nitrogen molecule, and thus, is the same in all combustion systems using air. However, expressions for NO_x that take no account whatsoever of residence time can often provide a good prediction of experimental data because the residence time of all aero gas turbine combustors tends to be roughly the same at around a few milliseconds. It is only when expressions derived for industrial gas turbine combustors are applied to aero combustors, or vice-versa, that the lack of a term for residence time becomes important.

9.11.1.3 Rokke et al. [106]

$$NO_x = 18.1P^{1.42}\dot{m}_A^{0.3}q^{0.72} \text{ ppmv.} \quad (9.10)$$

This equation was found to very satisfactorily correlate measurements of NO_x emissions from five different natural gas-fired industrial machines operating in the power range from 1.5 to 34 MW. Although combustion temperature is conspicuous by its absence in Equation 9.10, its influence on NO_x emissions is acknowledged by the inclusion of a fuel/air ratio term.

9.11.1.4 Rizk and Mongia [107]

$$NO_x = 15.10^{14}(t - 0.5t_e)^{0.5} \exp -(71{,}100/T_{st})P^{-0.05} (\Delta P/P)^{-0.5} \text{ g/kg fuel.} \quad (9.11)$$

An interesting feature of this equation is that it includes a term, t_e, to account for the influence of fuel evaporation on NO_x emissions. According to Equation 9.11, a reduction in mean drop size should increase NO_x emissions by reducing the time required for fuel evaporation. However, if combustion takes place under conditions where the evaporation time is negligibly small in comparison with the total combustor residence time, for example, at high combustion pressures, NO_x emissions can actually go down with a reduction in mean drop size, as observed by Rink and Lefebvre [17].

For more information on semi-analytical equations for the estimation of NO_x emissions from gas turbines, reference should be made to Becker and Perkavec [108] and Nicol et al. [109].

Comparisons between the various published equations for predicting NO_x emissions from conventional combustors are prohibited by the fact that in some cases the units for NO_x are in parts per million by volume (ppmv) and in others grams per kilogram of fuel (EI). Conversion from one set of units to the other cannot be undertaken unless the equivalence ratio is known but, as a rough guide, one EI is roughly equivalent to around 12 ppmv. In general, most expressions for NO_x provide an excellent fit to the experimental data employed in their derivation. As NO_x formation in conventional combustors occurs primarily via the thermal mechanism, which is almost solely dependent on reaction temperature, this is not too surprising.

9.11.2 Carbon Monoxide Correlations

Similar correlations to those presented above for NO_x have been developed for CO. One important difference stems from the fact that the formation of CO in the primary combustion zone takes appreciably longer than the time required to produce NO_x. In consequence, the relevant temperature is not the local peak value adjacent to the evaporating fuel drops, but the average value throughout the primary zone, T_{pz}. Also, because CO emissions are most important at low pressure conditions, where evaporation rates are relatively slow, it is necessary to reduce the combustion volume, V_c, by the volume occupied in fuel evaporation, V_e. Lefebvre's approach [102] gives

$$CO = 86\dot{m}_A T_{pz} \exp-(0.00345T_{pz})/(V_c - V_e)(\Delta P/P)^{0.5} P^{1.5} \text{g/kg fuel}, \quad (9.12)$$

where V_e, the volume employed in fuel evaporation, is obtained as

$$V_e = 0.55\dot{m}_{pz} D_0^2 / \rho_{pz} \lambda_{eff}. \quad (9.13)$$

In this equation, it is of interest to note that V_e is proportional to the square of the initial mean drop size. This highlights the importance of good atomization to the attainment of low CO. The ability of Equation 9.12 to predict CO emissions from a P&W F100 and a GE F101 combustor is illustrated in Figures 9.43 and 9.44, respectively.

A similar form of expression to Equation 9.12 has been derived by Rizk and Mongia [107] as

$$CO = 0.18 \times 10^9 \exp(7800/T_{pz})/P^2(t - 0.4t_e)(\Delta P/P)^{0.5} \text{ g/kg}. \quad (9.14)$$

This equation yields a slightly lower dependence on combustion temperature and a slightly higher dependence on pressure than Equation 9.12.

The most recent study on the prediction and correlation of CO emissions from stationary gas turbines is that of Connors et al. [110]. These workers found that Mellor's characteristic time model [103] could be used to satisfactorily predict CO emissions from two heavy-duty power-generation units operating on natural gas or No. 2 fuel oil without inert injection.

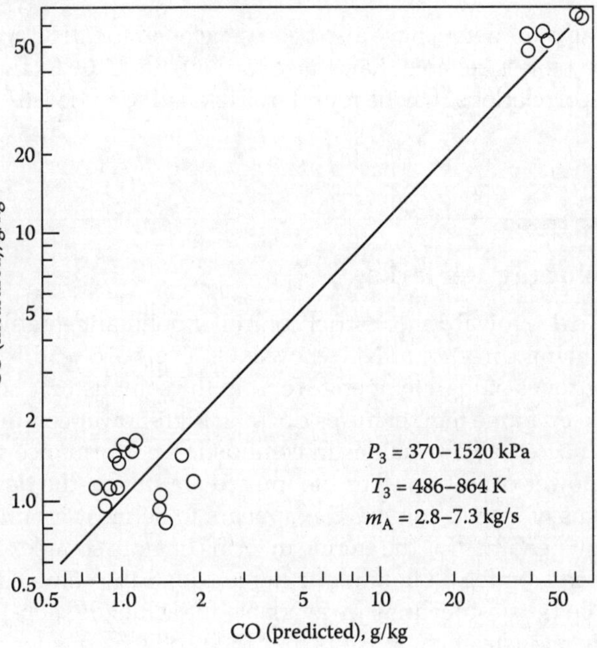

FIGURE 9.43
Comparison of measured and predicted values of CO for a P&W F100 combustor.

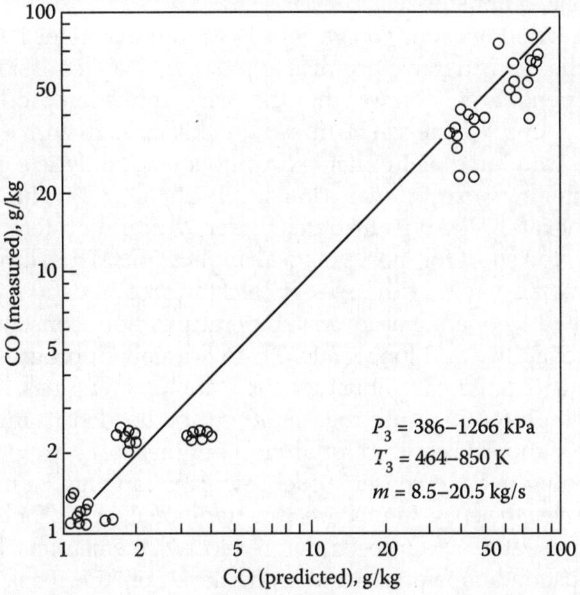

FIGURE 9.44
Comparison of measured and predicted values of CO for a GE F101 combustor.

Correlating parameters have also been developed for UHC and smoke, but they tend to be less reliable than those for NO_x and CO. Examples of UHC and smoke correlations may be found in Rizk and Mongia [107,111].

9.12 Concluding Remarks

Most of the drive toward more strict control of pollutant emissions from gas turbines is being directed at NO_x. Low NO_x levels are readily achieved by eliminating zones of high temperature from the combustor. The challenge is broadly to keep flame temperatures down at high-power conditions without incurring unacceptable penalties in combustion performance when operating at low-power conditions. For the immediate future, the development of various forms of staged combustion appears to be most promising, despite the attendant penalties for the engine in terms of more complex fuel scheduling and control strategy. Looking further ahead, lean-premixed combustion appears to be the best technology available for achieving sub-10 ppmv NO_x emissions in industrial combustors, but the problems of spontaneous ignition and flashback need to be fully addressed before lean-premixed combustion can be applied with confidence to aircraft engines. Problems of acoustic resonance could also be of great importance to the future development of lean-premixed combustors.

The work carried out so far on various RQL configurations has shown that this concept has considerable promise for very low NO_x emissions. Its future prospects depend largely on whether the rich combustion products emanating from the primary zone can be mixed quickly enough with the remaining combustion air to largely eliminate NO_x and soot. Catalytic combustors are capable of achieving exceptionally low levels of NO_x. Unfortunately, the life expectancy of catalysts and substrates under typical high-temperature conditions is still too short for most practical applications. This drawback can be alleviated to some extent by the use of postcatalyst combustion, which allows the catalyst bed to operate at lower temperatures with consequent benefits in terms of reliability and longer life. The work now in progress is directed mainly toward catalytic combustors for stationary engines. Considerable advances in high-temperature materials will be needed to raise their reliability to the standard required for aircraft engines.

The continuing need to conserve fuel resources can only be met by raising the engine cycle efficiency. In practice, this traditionally calls for an increase in engine pressure ratio, an approach that reduces CO_2 emissions, but results in higher combustion temperatures and higher levels of NO_x. Thus, the desire to burn less fuel, thereby generating less CO_2, is in direct conflict with the equally important need to reduce NO_x. For the foreseeable future, it seems likely that engine pressure ratios will rise to a maximum value of around 60 and yet designers of future combustors will be called on to reduce NO_x emissions by another 50%, and virtually eliminate soot emissions and combustion instability. These will be the research and development challenges of the future.

Nomenclature

D_o	Sauter mean diameter, m
L	liner length, m
$\dot{m}_A$	combustor airflow rate, kg/s
$\dot{m}_{pz}$	primary-zone airflow rate, kg/s
P	pressure (kPa in Equations 9.7, 9.11, 9.12, and 9.14, Pa in Equation 9.8, and atm in Equations 9.9 and 9.10)
$\Delta P/P$	nondimensional pressure drop
q	fuel/air ratio by mass
SMD	Sauter mean diameter, m
T	temperature, K
T_c	combustion temperature, K
T_{st}	stoichiometric flame temperature, K
t	residence time in combustion zone, s
t_e	evaporation time, s
τ	NO_x formation time, s
U	liner flow velocity (average), m/s
V_c	combustion volume, m^3
V_e	volume occupied in evaporation, m^3
ϕ	equivalence ratio
λ_{eff}	evaporation constant, m^2/s
ρ	density, kg/m^3

Subscripts

pz	primary zone value
st	stoichiometric value
3	combustor inlet value

References

1. Seaton, A., MacNee, W., Donaldson, K., and Godden, E., "Particulate Air Pollution and Acute Health Effects," *Lancet*, Vol. 345, pp. 176–8, 1995.
2. Bahr, D. W., "Aircraft Engines NO_x Emissions—Abatement Progress and Prospects," *Tenth International Symposium on Air-Breathing Engines*, ISABE 91-7064, AIAA, Washington, DC, pp. 229–38, 1991.

3. Anon, "International Energy Outlook," Energy Information Adminstration, U.S. Department of Energy, Washington DC, May 2009.
4. Bahr, D. W., "Aircraft Turbine Engine NO_x Emission Limits—Status and Trends," ASME Paper 92-GT-415, 1992.
5. Bahr, D. W., Lecture notes for short course on gas turbine combustion; University of California, Irvine, March 13–17, 1995.
6. http://www.icao.org.
7. Wesocky, H. L., and Prather, M. J., "Atmospheric Effects of Stratospheric Aircraft: A Status Report from NASA's High-Speed Research Program," *Tenth International Symposium on Air Breathing Engines,* ISABE 91-7020, AIAA, Washington, DC, pp. 211–20, 1991.
8. Schorr, M. M., "NO_x Emission Control for Gas Turbines: A 1991 Update on Regulations and Technology" (Part I), *Turbomachinery International*, Sept./Oct., pp. 24–30, 1991.
9. Angello, L., and Lowe, P., "Dry Low NO_x Combustion Development for Electric Utility Gas Turbine Applications; A Status Report," ASME Paper 89-GT-254, 1989.
10. Rink, K. K., and Lefebvre, A. H., "Influence of Fuel Drop Size and Combustor Operating Conditions on Pollutant Emissions," *International Journal of Turbo and Jet Engines*, Vol. 6, No. 2, pp. 113–22, 1989.
11. Hung, W. S. Y., and Agan, D. D., "The Control of NO_x and CO Emissions from 7-MW Gas Turbines with Water Injection as Influenced by Ambient Conditions," ASME Paper 85-GT-50, 1985.
12. Hung, W. S. Y., "Carbon Monoxide Emissions from Gas Turbines as Influenced by Ambient Temperature and Turbine Load," *Journal of Engineering for Gas Turbines and Power*, Vol. 115, No. 3, pp. 588–93, 1993.
13. Zheng, Q. P., Jasuja, A. K., and Lefebvre, A. H., "Influence of Air and Fuel Flows on Gas Turbine Sprays at High Pressures," *Twenty-Sixth (International) Symposium on Combustion,* pp. 2757–62, The Combustion Institute, Pittsburgh, PA, 1996.
14. Chin, J. S., and Lefebvre, A. H., "Influence of Fuel Chemical Properties on Soot Emissions from Gas Turbine Combustors," ASME Paper 89-GT-261, 1989.
15. Russel, P. L., "Fuel Mainburner/Turbine Effects," AFWAL-TR-81-21000, May 1982.
16. Snyder, T. S., Rosfjord, T. J., McVey, J. B., and Chiappetta, L. M., "Comparison of Liquid Fuel/Air Mixing and NO_x Emissions for a Tangential Entry Nozzle," ASME Paper 94-GT-283, 1994.
17. Rink, K. K., and Lefebvre, A. H., "The Influence of Fuel Composition and Spray Characteristics on Nitric Oxide Formation," *Combustion, Science and Technology,* Vol. 68, pp. 1–14, 1989.
18. Anderson, D. N., "Effects of Equivalence Ratio and Dwell Time on Exhaust Emissions from an Experimental Premixing Prevaporizing Burner," ASME Paper 75-GT-69, 1975.
19. Leonard, G., and Stegmaier, J., "Development of an Aeroderivative Gas Turbine Dry Low Emissions Combustion System," *Journal of Engineering for Gas Turbines and Power,* Vol. 116, pp. 542–6, 1993.
20. Rizk, N. K., and Mongia, H. C., "Three-Dimensional NO_x Model for Rich-Lean Combustor," AIAA Paper 93-0251, 1993.
21. Nicol, D., Malte, P. C., Lai, J., Marinov, N. N., and Pratt, D. T., "NO_x Sensitivities for Gas Turbine Engines Operated on Lean-Premixed Combustion and Conventional Diffusion Flames," ASME Paper 92-GT-115, 1992.

22. Correa, S. M., "Lean Premixed Combustion for Gas Turbines: Review and Required Research," *Fossil Fuel Combustion*, ASME PD-Vol. 33, 1991.
23. Fennimore, C. P., "Formation of Nitric Oxide in Premixed Hydrocarbon Flames," *Thirteenth Symposium (International) on Combustion*, pp. 373–80, The Combustion Institute, Pittsburgh, PA, 1971.
24. Heberling, P. V., "Prompt NO Measurements at High Pressures," *Sixteenth Symposium (International) on Combustion*, pp. 159–68, The Combustion Institute, Pittsburgh, PA, 1976.
25. Altemark, D., and Knauber, R., "Results of the Study of a Premixed Burner Under Pressure with Extremely Low NOx Emissions," *VDI Berichte*, No. 645, pp. 299–311, 1987.
26. Merryman, E. L., and Levy, A., "Nitric Oxide Formation in Flames: The Role of NO_2 and Fuel Nitrogen," *Fifteenth Symposium (International) on Combustion*, pp. 1073–83, The Combustion Institute, Pittsburgh, PA, 1975.
27. Maughan, J. R., Luts, A., and Bautista, P. J., "A Dry Low NO_x Combustor for the MS3002 Regenerative Gas Turbine," ASME Paper 94-GT-252, 1994.
28. Leonard, G. L., and Correa, S. M., "NO_x Formation in Lean Premixed High-Pressure Methane Flames," *Second ASME Fossil Fuel Combustion Symposium*, PD-30, ASME, New York, NY, pp. 69–74, 1990.
29. Steele, R. C., Jarrett, A. C., Malte, P. C., Tonouchi, J. H., and Nicol, D. G., "Variables Affecting NO_x Formation in Lean-Premixed Combustion," ASME Paper 95-GT-107, 1995.
30. Rizk, N. K., and Mongia, H. C., "A Three-Dimensional Analysis of Gas Turbine Combustors," *Journal of Propulsion and Power*, Vol. 7, No. 3, pp. 445–51, 1991.
31. Willis, J. D., Toon, I. J., Schweiger, T., and Owen, D. A., "Industrial RB211 Dry Low Emission Combustion System," ASME Paper 93-GT-391, 1993.
32. Sturgess, G. J., McKinney, R., and Morford, S. A., "Modification of Combustor Stoichiometry Distribution for Reduced NO_x Emission from Aircraft Engines," *Journal of Engineering for Gas Turbines and Power*, Vol. 115, No. 3, pp. 570–80, 1993.
33. Davis, L. B., and Washam, R. M., "Development of Dry Low NO_x Combustor," ASME Paper 89-GT-255, 1989.
34. Hilt, M. B., and Waslo, J., "Evolution of NO_x Abatement Techniques Through Combustor Design for Heavy-Duty Gas Turbines," *Journal of Engineering for Gas Turbines and Power*, Vol. 106, pp. 825–32, 1984.
35. McNight, D., "Development of a Compact Gas Turbine Combustor to Give Extended Life and Acceptable Exhaust Emissions," *Journal of Engineering for Power*, Vol. 101, No. 3, pp. 349–57, 1979.
36. Hung, W. S. Y, "Accurate Method of Predicting the Effect of Humidity or Injected Water on NO_x Emissions from Industrial Gas Turbines," ASME Paper 74-WA/GT-6, 1974.
37. Claeys, J. P., Elward, K. M., Mick, W. J., and Symonds, R. A., "Combustion System Performance and Field Test Results of the MS7001F Gas Turbine," *Journal of Engineering for Gas Turbines and Power*, Vol. 115, No. 3, pp. 537–46, 1993.
38. Wilkes, C., "Residual Fuel Combustion in Industrial Gas Turbines," in A. H. Lefebvre, ed., *Gas Turbine Combustor Design Problems*, pp. 87–110, Hemisphere, New York, 1980.
39. Toof, J. L., "A Model for the Prediction of Thermal, Prompt, and Fuel NO_x Emissions from Combustion Turbines," *Journal of Engineering for Gas Turbines and Power*, Vol. 108, No. 2, pp. 340–7, 1986.

40. Schorr, M. M., "NO_x Emission Control for Gas Turbines: A 1991 Update on Regulations and Technology," (Part II), *Turbomachinery International*, Nov./Dec., pp. 28–36, 1991.
41. White, D. J., Batakis, A., Le Cren, R. T., and Yacabucci, H. G., "Low NO_x Combustion Systems for Burning Heavy Residual Fuels and High Fuel-Bound Nitrogen Fuels," *Journal of Engineering for Gas Turbines and Power*, Vol. 104, pp. 377–85, 1982.
42. Wilkes, C., and Gerhold, B., "NO_x Reduction from a Gas Turbine Using Exhaust Gas Recirculation," ASME Paper 80-JPGC/GT-5, 1980.
43. Bayle-Laboure, G., "Pollutant Emissions from Aircraft Engines: A Situation Under Control," *Revue Scientifique SNECMA*, Paris, France, 2nd ed., 1991.
44. Micklow, G. J., Roychoudhury, S., Nguyen, H., and Cline, M. C., "Emissions Reduction by Varying the Swirler Airflow Split in Advanced Gas Turbine Combustors," *Journal of Engineering for Gas Turbines and Power*, Vol. 115, No. 3, pp. 563–69, 1993.
45. Roberts, P. B., Kubasco, A. J., and Sekas, N. J., "Development of a Low NO_x Lean Premixed Annular Combustor," *Journal of Engineering for Power*, Vol. 104, pp. 28–35, 1982.
46. Sasaki, M., Kumakura, H., and Suzuki, D., "Low NO_x Combustor for Automotive Ceramic Gas Turbine—Conceptual Design," ASME Paper 91-GT-369, 1991.
47. Aoyama, K., and Mandai, S., "Development of a Dry Low NO_x Combustor for a 120 MW Gas Turbine," *Journal of Engineering for Gas Turbines and Power*, Vol. 106, pp. 795–800, 1984.
48. Bahr, D. W., "Technology for the Design of High Temperature Rise Combustors," *Journal of Propulsion and Power*, Vol. 3, No. 2, pp. 179–86, 1987.
49. Segalman, I., McKinney, R. G., Sturgess, G. J., and Huang, L. M., "Reduction of NO_x by Fuel-Staging in Gas Turbine Engines—A Commitment to the Future," *AGARD Conference Proceedings 536*, Paris, France, pp. 29/1–17, 1993.
50. Koff, B. L., "Aircraft Gas Turbine Emissions Challenge," ASME Paper 93-GT-422, 1993.
51. Davis, L. B., "Dry Low NO_x Combustion Systems for GE Heavy-Duty Gas Turbines," ASME Paper 96-GT-27, 1996.
52. Smith, K. O., Angello, L. C., and Kurzynske, F. R., "Design and Testing of an Ultra-Low NO_x Gas Turbine Combustor," ASME Paper 86-GT-263, 1986.
53. Smith, K. O., Holsapple, A. C., Mak, H. H., and Watkins, L., "Development of a Natural Gas-Fired, Ultra-Low NO_x Can Combustor for an 800 kW Gas Turbine," ASME Paper 91-GT-303, 1991.
54. Smith, K. O., and Cowell, L. H., "Experimental Evaluation of a Liquid-Fueled, Lean-Premixed Gas Turbine Combustor," ASME Paper 89-GT-264, 1989.
55. Etheridge, C. J., "Mars SoLoNO$_x$ Lean Premix Combustion Technology in Production," ASME Paper 94-GT-255, 1994.
56. Jeffs, E., "Siemens Completes Testing of Model V84. 3," *Turbomachinery International*, Vol. 35, No. 2, pp. 36–9, 1993.
57. Bonzani, F., Di Meglio, A., Pollarolo, G., Prade, B., Lauer, G., and Hoffmann, S., "Test Results of the V64. 3A Gas Turbine Premix Burner," Presented at Power-Gen Europe '97, Madrid, June 1997.
58. Washam, R. M., "Dry Low NO_x Combustion System for Utility Gas Turbine," ASME Paper 83-JPGC-GT-13, 1983.

59. Sattelmayer, T., Felchin, M. P., Haumann, J., Hellat, J., and Styner, D., "Second Generation Low-Emission Combustors for ABB Gas Turbines: Burner Development and Tests at Atmospheric Pressure," *Journal of Engineering for Gas Turbines and Power*, Vol. 114, No. 1, pp. 118–25, 1992.
60. Aigner, M., and Muller, G., "Second Generation Low-Emission Combustors for ABB Gas Turbines: Field Measurements with GT11N-EV," *Journal of Engineering for Gas Turbines and Power*, Vol. 115, No. 3, pp. 533–6, 1993.
61. Dobbeling, K., Knopfel, H. P., Polifke, W., Winkler, D., Steinbach, C., and Sattelmayer, T., "Low NO_x Premixed Combustion of MBTU Fuels Using the ABB Double Cone Burner (EV Burner)," ASME Paper 94-GT-394, 1994.
62. Senior, P., Lutum, E., Polifke, W., and Sattelmayer, T., "Combustion Technology of the ABB GT13E2 Annular Combustor," *20th International Congress on Combustion Engines*, CIMAC, London, Paper No. G22, 1993.
63. Anderson, L., ABB STAL AB, Finspong, Sweden, private communication, August 1997.
64. Norster, E. R., and DePietro, S. M., "Dry Low Emissions Combustion System for EGT Small Gas Turbines," *Institution of Diesel and Gas Turbine Engineers, DEUA Publication 495*, London, UK, 1996.
65. Gallimore, S., Vickers, R. M, and Boyns, M. B., "The Design Modifications of the EGT Tornado Industrial Gas Turbine to Incorporate a Dry Low Emissions Combustion System," ASME Paper 97-GT-159, 1997.
66. Joshi, N. D., Epstein, M. J., Durlak, S., Marakovits, S., and Sabla, P. E., "Development of a Fuel-Air Premixer for Aero-Derivative Dry Low Emissions Combustors," ASME Paper 94-GT-253, 1994.
67. Rizk, N. K., and Mongia, H. C., "Lean Low NO_x Combustion Concept Evaluation," *Twenty-Third Symposium (International) on Combustion*, pp. 1063–70, The Combustion Institute, Pittsburgh, PA, 1990.
68. Rizk, N. K., and Mongia, H. C., "NO_x Model for Lean Combustion Concept," *Journal of Propulsion and Power*, Vol. 11, No. 1, pp. 161–9, 1995.
69. Hosoi, J., Watanabe, T., Toh, H., Mori, M., Sato, H., and Ishizuka, A., "Development of a Dry Low NO_x Combustor for 2 MW Class Gas Turbine," ASME Paper 96-GT-53, 1996.
70. Kumakura, H., Sasaki, M., Suzuki, D., and Ichikawa, H., "Development of a Low-Emission Combustor for a 100-kW Automotive Ceramic Gas Turbine," *Journal of Engineering for Gas Turbines and Power*, Vol. 118, No. 1, pp. 167–72, 1996.
71. Ohkubo, Y., Idota, Y., and Nomura, Y., "Evaporation Characteristics of Spray in a Lean Premixed-Prevaporization Combustor for a 100 kW Automotive Ceramic Gas Turbine," ASME Paper 94-GT-401, 1994.
72. Poeschl, G., Ruhkamp, W., and Pfost, H., "Combustion with Low Pollutant Emissions of Liquid Fuels in Gas Turbines by Premixing and Prevaporization," ASME Paper 94-GT-443, 1994.
73. Lyons, V. J., "Fuel-Air Non-Uniformity Effect on Nitric Oxide Emissions," AIAA Paper 81-0327, 1981.
74. Flanagan, P., Gretsingir, K., Abbasi, H. A., and Cygan, D., "Factors Influencing Low Emissions Combustion," ASME PD-Vol. 39, *Fossil Fuels Combustion*, 1992.
75. Fric, T. F., "Effects of Fuel-Air Unmixedness on NO_x Emissions," AIAA Paper 92-3345, 1992.

76. Santavicca, D. A., Steinberger, D. L., Gibbons, K. A., Citeno, J. V., and Mills, S., "The Effect of Incomplete Fuel-Air Mixing on the Lean Limit and Emissions Characteristics of a Lean Prevaporized Premixed (LPP) Combustor," *AGARD Conference Proceedings 536*, AGARD, Paris-France, pp. 22/1–12, 1993.
77. Valk, M., private communication, 1994.
78. Martin, F. J., and Dederick, P. K. J., "NO_x from Fuel Nitrogen in Two-Stage Combustion," *Sixteenth Symposium (International) on Combustion*, pp. 191–98, The Combustion Institute, Pittsburgh, PA, 1976.
79. Nakata, T., Sato, M., Ninomiya, T., and Hasegawa, T., "A Study on Low NO_x Combustion in LBG-Fueled 1500°C-Class Gas Turbine," ASME Paper 94-GT-218, 1994.
80. Talpallikar, M. V., Smith, C. E., Lai, M. C., and Holderman, J. D., "CFD Analysis of Jet Mixing in Low NO_x Flametube Combustors," ASME Paper 91-GT-217, 1991.
81. Mosier, S. A., and Pierce, R. M., "Advanced Combustion Systems for Stationary Gas Turbine Engines," EPA Contract 68-02-2136, 1980.
82. Novick, A. S., Troth, D. L., and Yacobucci, H. G., "Design and Preliminary Results of a Fuel Flexible Industrial Gas Turbine Combustor," *Journal of Engineering for Power*, Vol. 104, pp. 368–76, 1982.
83. Rizk, N. K., and Mongia, H. C., "Low NO_x Rich-Lean Combustion Concept Application," AIAA Paper 91-1962, 1991.
84. Nakata, T., Sato, M., Ninomiya, T., Yoshine, T., and Yamada, M. "Design and Test of a Low-NO_x Advanced Rich-Lean Combustor for LBG Fueled 1300°C-Class Gas Turbine," ASME Paper 92-GT-234, 1992.
85. Feitelberg, A. S., and Lacey, M. A., "The GE Rich-Quench-Lean Gas Turbine Combustor," ASME Paper 97-GT-127, 1997.
86. Shaw, R. J., "Propulsion Challenges for a 21st Century Economically Viable, Environmentally Compatible High-Speed Civil Transport," *Tenth International Symposium on Air-Breathing Engines*, ISABE 91-7008, AIAA, Washington, DC, pp. 93–103, 1991.
87. Pfefferle, W. C., "Catalytically Supported Thermal Combustion," U.S. Patent 3, 928, 961, 1975.
88. Kolaczkowski, S. T., "Catalytic Stationary Gas Turbine Combustors: A Review of the Challenges Faced to Clear the Next Set of Hurdles," *Transactions of the Institution of Chemical Engineers*, Vol. 73, Part A, 1995.
89. Dalla Betta, R. A., Schlatter, J. C., Nickolas, S. G., Yee, D. K., and Shoji, T., "New Catalytic Combustion Technology for Very Low Emissions Gas Turbines," ASME Paper 94-GT-260, 1994.
90. Dalla Betta, R. A., Schlatter, J. C., Nickolas, S. G., Razdan, M. K., and Smith, D. A., "Application of Catalytic Combustion Technology to Industrial Gas Turbines for Ultra-Low NO_x Emissions," ASME Paper 95-GT-65, 1995.
91. Dalla Betta, R. A., Schlatter, J. C, Nickolas, S. G., Cutrone, M. B., Beebe, K. W., Furuse, Y., and Tsuchiya, T., "Development of a Catalytic Combustor for a Heavy-Duty Utility Gas Turbine," ASME Paper 96-GT-485, 1996.
92. Schlatter, J. C., Dalla Betta, R. A., Nickolas, S. G., Cutrone, M. B., and Beebe, K. W., "Single-Digit Emissions in a Full Scale Catalytic Combustor," ASME Paper 97-GT-57, 1997.
93. Dutta, P., Cowell, L. H., Yee, D. K., and Dalla Betta, R. A., "Design and Evaluation of a Single-Can Full Scale Catalytic Combustion System for Ultra-Low Emissions Industrial Gas Turbines," ASME Paper 97-GT-292, 1997.

94. Dutta, P., Yee, D. K., and Dalla Betta, R. A., "Catalytic Combustor Development for Ultra-Low Emissions Industrial Gas Turbines," ASME Paper 97-GT-497, 1997.
95. Cowell, L. H., and Larkin, M. P., "Development of a Catalytic Combustor for Industrial Gas Turbines," ASME Paper 94-GT-254, 1994.
96. Ozawa, Y., Hirano, J., Sato, M., Saiga, M., and Watanabe, S., "Test Results of Low NO_x Catalytic Combustors for Gas Turbines," *Journal of Engineering for Gas Turbines and Power*, Vol. 116, pp. 511–16, 1994.
97. Fujii, T., Ozawa, Y., Kikumoto, S., and Sato, M., "High Pressure Test Results of a Catalytic Combustor for Gas Turbine," ASME Paper 96-GT-382, 1996.
98. Vortmeyer, N., Valk, M., and Kappler, G., "A Catalytic Combustor for High-Temperature Gas Turbines," *Journal of Engineering for Gas Turbines and Power*, Vol. 118, No. 1, pp. 61–4, 1996.
99. Anderson, S. J., Friedman, M. A., Krill, W. V., and Kesselring, J. P., "Development of a Small-Scale Catalytic Gas Turbine Combustor," *Journal of Engineering for Power*, Vol. 104, pp. 52–7, 1982.
100. Beebe, K. W., Cutrone, M. B., Mathews, R. N., Dalla Betta, R. A., Schlatter, J. C., Furuse, Y., and Tsuchiya, T., "Design and Test of a Catalytic Combustor for a Heavy-Duty Industrial Gas Turbine," ASME Paper 95-GT-137, 1995.
101. Beebe, K. W., Ohkoshi, A., Radak, L., and Weir, A., "Design and Test of Catalytic Combustor Fuel-Air Preparation System," Paper 51, Presented at 1987 Tokyo International Gas Turbine Congress, Japan, October 26–31, 1987.
102. Lefebvre, A. H., "Fuel Effects on Gas Turbine Combustion—Liner Temperature, Pattern Factor, and Pollutant Emissions," *Journal of Aircraft*, Vol. 21, No. 11, pp. 887–98, 1984.
103. Mellor, A. M., "Semi-Empirical Correlations for Gas Turbine Emissions, Ignition, and Flame Stabilization," *Progress in Energy and Combustion Science*, Vol. 6, No. 4, pp. 347–58, 1981.
104. Odgers, J., and Kretschmer, D., "The Prediction of Thermal NO_x in Gas Turbines," ASME Paper 85-1GT-126, 1985.
105. Lewis, G. D., "A New Understanding of NO_x Formation," *Tenth International Symposium on Air-Breathing Engines*, ISABE 91-7064, Nottingham, UK, AIAA, Washington, DC, pp. 625–9, 1991.
106. Rokke, N. A., Hustad, J. E., and Berg, S., "Pollutant Emissions from Gas Fired Turbine Engines in Offshore Practice—Measurements and Scaling," ASME Paper 93-GT-170, 1993.
107. Rizk, N. K., and Mongia, H. C., "Emissions Predictions of Different Gas Turbine Combustors," AIAA Paper 94-0118, 1994.
108. Becker, T., and Perkavec, M. A., "The Capability of Different Semianalytical Equations for Estimation of NO_x Emissions of Gas Turbines," ASME Paper 94-GT-282, 1994.
109. Nicol, D. G., Make, P. C., and Steele, R. C., "Simplified Models for NO_x Production Rates in Lean-Premixed Combustion," ASME Paper 94-GT-432, 1994.
110. Connors, C. S., Barnes, J. C., and Mellor, A. M., "Semiempirical Predictions and Correlations of CO Emissions from Utility Combustion Turbines," *Journal of Propulsion and Power*, Vol. 12, No. 5, pp. 926–32, 1996.
111. Rizk, N. K., and Mongia, H. C., "Semianalytical Correlations for NO_x, CO and UHC Emissions," *Journal of Engineering for Gas Turbines and Power*, Vol. 115, No. 3, pp. 612–19, 1993.

10

Alternative Fuels

10.1 Introduction

In the early days of gas turbine development, the view was widely held that gas turbines are "omnivorous" machines, capable of operating efficiently on a wide variety of cheap fuels—solid, liquid, and gaseous. The choice of fuel is dictated on the grounds of cost, availability, and ease of handling. This is true even today, provided one keeps in mind the important and restrictive effects of aircraft engine requirements on the physical and chemical properties of fuel.

Today, the ever-rising cost of petroleum fuel is prompting research into developing alternative liquid fuels based on coal, biomass, and other feedstock. The terrorist activities in the Middle East are leading various industrialized and developing countries to develop domestic sources of fuel and energy. Finally, these domestically produced alternative fuels have to be capable of using the available infrastructure of fuel refining, transportation, distribution, and consumption.

In general, for any given aircraft application, the optimal fuel is the one that represents the best compromise solution to the various problems confronting the fuel companies, the engine and aircraft manufacturers, and the operators. For civil aircraft, the main requirements are safety, reliability, low cost, and ease of handling. For military aircraft, fuel cost is of secondary importance compared with availability, supply logistics, and the need for trouble-free operation over a wide range of conditions. Industrial and marine gas turbines can compete effectively with the diesel engine only through the use of very cheap fuels, such as residual oil or surplus gas. As a rule, gaseous fuels present no special problems, but residual oils produce a highly destructive ash and sometimes copious amounts of exhaust smoke. Attempts to burn pulverized solid fuels in open-cycle gas turbines have generally proved unsuccessful. However, developments in integrated gasification combined cycle (IGCC) plants employing coal gasifier and fluidized-bed combustion are dramatically changing that situation.

The alternative synthetic liquid fuels of major interest will be largely derived from biomass (carbohydrates, algae, and vegetation), coal, oil shale, tar sands, and heavy oil. For longer term, methane and hydrogen are among the candidate fuels now being considered. All these fuels must be compatible with engine and fuel-system requirements and, where appropriate, with aircraft design features and operational procedures. The main impact of trends and developments relative to combustor design will be felt in fuel-nozzle design for multifuel capability and in fuel and air management for minimum soot and gaseous emissions. The future of the alternative fuel industry depends on the following key factors:

- Right fuel properties and handling for the engines and infrastructure already in place,
- Environmental impact that includes competition with food, water, and land,
- CO_2 life cycle analysis and carbon footprint issues,
- Economics of return on investment, production, and sustainability.

This chapter presents the physical and chemical properties of conventional (petroleum-based) liquid and gaseous fuels for gas turbines. Next, alternative (synthetic) fuels and conventional-alternative fuel blends are reviewed. The influence of these different fuels and their blends on combustor performance and design are investigated. Reference is made to the special requirements of aircraft fuels and the problems encountered with industrial fuels. Important aspects of oil production and supplies are discussed in recent publications [1–7].

10.2 Types of Hydrocarbons

Pure hydrocarbon fuels are compounds of two elements only, carbon and hydrogen. They may be gaseous, liquid, or solid at normal pressure and temperature, depending on the number of carbon atoms and their molecular structure. Those with up to four carbon atoms are gaseous; those with twenty or more are solid, and those in between are liquid [8]. It is usual to classify the hydrocarbons present in petroleum fuel into four main groups: paraffinic, olefinic, naphthenic, and aromatic. The proportions in which these groups are present largely define the character of the fuel.

10.2.1 Paraffins

Paraffinic oils are found mainly in the United States, North Africa, and Nigeria. They have the general formula C_nH_{2n+2}. Thus, the simplest hydrocarbon, methane, is in this class; its molecule can be represented as:

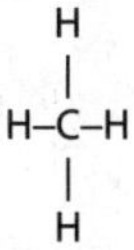

Methane, CH_4

The remaining *normal* paraffins are built up from methane as straight chains, e.g.,

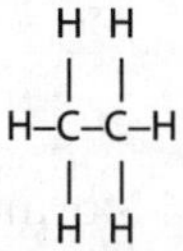

Ethane, C_2H_6

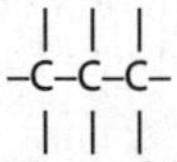

Propane, C_3H_8

Alternative paraffin configurations, or *isoparaffins,* are in the form of branched chains, such as

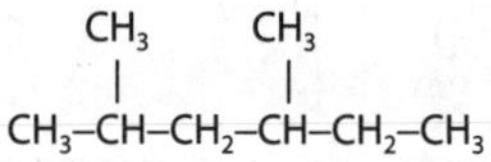

Dimethyl hexane, $C_6H_{12}(CH_3)_2$

Current aviation fuels contain an average of 60% paraffins, depending on the source of the crude oil and the distillation process. In general, paraffins tend to have a higher hydrogen/carbon ratio, lower density, lower freeze point, and high gravimetric calorific value than other types of hydrocarbon fuels. They possess high thermal stability, and their combustion is characterized by freedom from coke deposition and exhaust smoke.

10.2.2 Olefins

Olefins conform to the general formula C_nH_{2n}. They do not normally exist in crude oil, but are produced by conversion processes in the refinery. As olefins are unsaturated, i.e., their molecules contain less than the maximum possible number of hydrogen atoms, they are very active chemically and readily react with a great many compounds to form resinous gums and rubberlike materials. For this reason, olefins are very undesirable in gas turbine fuels, and are found only in trace quantities.

Olefin molecules must contain at least two carbon molecules, and the lightest molecule is, therefore, C_2H_4, ethylene. More complex molecules of the olefin series are obtained by adding CH_2, as in

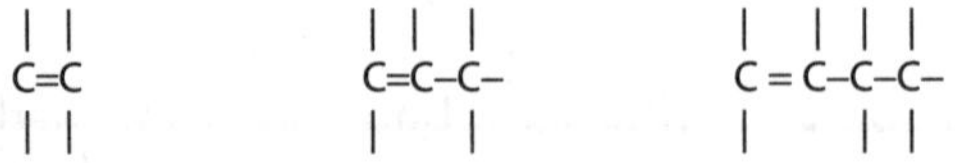

Ethylene, C_2H_4 Propylene, C_3H_6 Butylene, C_4H_8

10.2.3 Naphthenes

Naphthenes, which have the general formula $(CH_2)_n$, are saturated hydrocarbons in which the carbon atoms are linked to form rings instead of chains as in the case of paraffins. Naphthenes bear names identical to those of the paraffins that have the same number of carbon atoms, with the addition of the prefix "cyclo," e.g.,

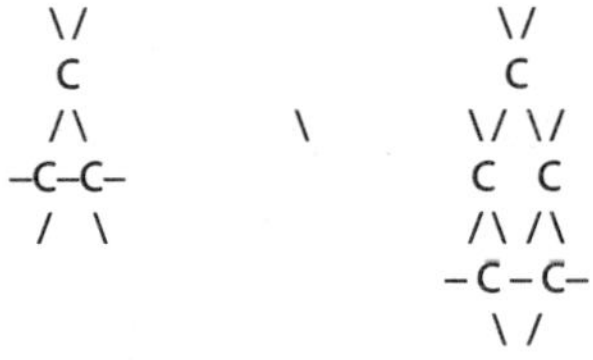

Cyclopropane, C_3H_6 Cyclopentane, C_5H_{10}

Naphthenes are major constituents of jet fuel, i.e., about 25–35%. They closely resemble paraffins in their chemical stability, high gravimetric heat of combustion, and low soot-forming tendencies.

10.2.4 Aromatics

Aromatics are ring compounds containing one or more six-member rings with the equivalent of three double bonds. Although similar in structure to the naphthenes, they contain less hydrogen and, in consequence, their specific energy is appreciably lower. Aromatic compounds in fuel cause swelling of o-ring and this helps seal the high pressure aircraft fuel system. The disadvantages of aromatic compounds include a marked tendency to soot formation and a high hygroscopicity that can lead to precipitation of ice crystals when the fuel is subjected to low temperatures. Aromatics also have a strong solvent action on rubber that can cause trouble in fuel systems and on aircraft fitted with soft-rubber-lined fuel tanks.

The characteristic formula for the aromatics is C_nH_{2n-6}. The simplest member is benzene, in which each carbon atom carries only one hydrogen atom:

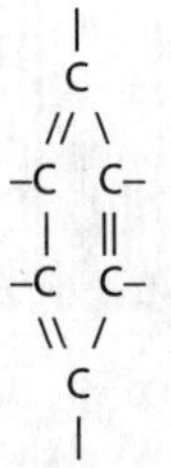

Benzene, C_6H_6

More complex molecules of the aromatic group are obtained either by replacing one or more of the hydrogen atoms with hydrocarbon groups or by "condensing" one or more rings [8]. Another example is:

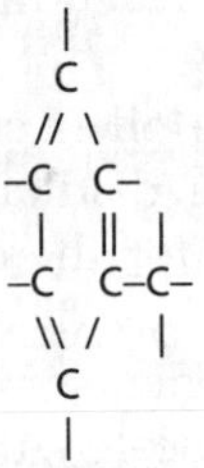

Toluene, $C_6H_5CH_3$

Although aromatics are undesirable in gas turbine fuels, the proportion of aromatics normally present in straight-run distillates is too high to justify the expense of their removal. Current practice with aircraft fuels is to limit the aromatic content to a maximum of 20% by volume [9].

In summary, the conventional petroleum-based jet fuel is a mixture of many different hydrocarbons. Jet fuel has an average composition of 60% paraffins, 20% naphthenes, 20% aromatics and contains about 500 ppm of sulfur. A broad specification allows significant variation in composition. For example, wide-cut jet fuel has a formula of $C_{10}H_{20}$ (MW = 140, density = 0.78), aviation kerosine is $C_{12}H_{24}$ (MW = 168, density = 0.8) and average formula is $C_{11}H_{21}$ (MW = 153, density = 0.79). Finally, commercial and military fuel specifications are very similar, except for slight variations. For example, Jet A (U.S.) has a higher freeze point (–51°C) than Jet A-1 (international, –47°C) and JP-8 (U.S. Air Force) has extra additives than Jet A-1. Figure 10.1 shows a gas chromatograph of the composition of jet fuel.

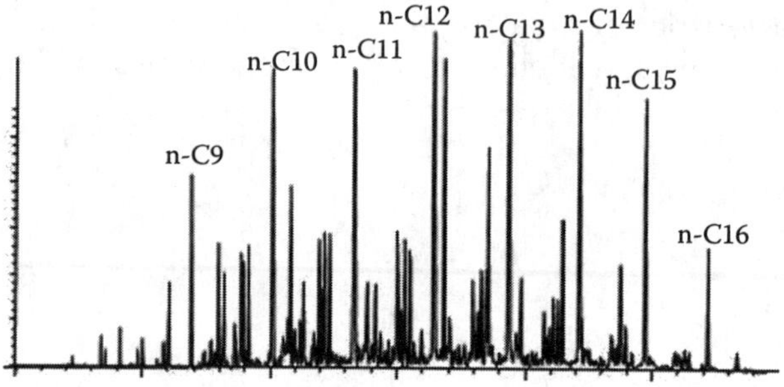

FIGURE 10.1
Gas chromatograph of the composition of jet fuel (Jet-A/JP-8/JP-5).

10.3 Production of Liquid Fuels

The production process can be divided into three basic categories:

1. **Separation process**: Crude oil is separated into its primary fractions, consisting of gasoline, distillate fuels, and fuel oil; these then provide the basic material for the desired range of fuel products. Separation is accomplished by a distillation process that exploits the fact that the various components in crude oil have different boiling points. When a crude oil is heated, the first gases evolved are chiefly methane, ethane, propane, and butane. Next, vapors are released that condense to form light distillates and then gasoline. As boiling proceeds, kerosine emerges, followed by the middle distillates used in gas oil and diesel fuel. Finally, a residue is left that is used in the manufacture of lubricating oils, wax, and bitumen.
2. **Upgrading process**: These processes improve the quality by using chemical reactions to remove any compounds present in trace quantities. Commonly used upgrading processes are sweetening, hydrotreating, and clay treatment.
3. **Conversion process**: These processes change the molecular structure of the feedstock, usually by "cracking" large molecules into small ones, e.g., catalytic cracking and hydrocracking.

10.3.1 Removal of Sulfur Compounds

The sulfur compounds of most concern are hydrogen sulfide, mercaptans, and free sulfur. Sulfur may account for up to about 5% by mass in crude oil. During combustion, sulfur is oxidized to SO_2 and then to acids, which are corrosive. Therefore, in low concentrations, hydrogen sulfide is easily removed by washing with a strong alkali; when the concentration is high, its removal is best effected by the Girbotol process in which the hydrogen sulfide is absorbed in liquid diethanolamine.

Mercaptans are undesirable because they are corrosive and have an offensive odor. Mercaptans may be removed by washing with caustic soda or caustic potash, but usually it is sufficient to convert them into sulfur compounds of a harmless nature that can remain in the fuel. Sometimes, a hydrogen treatment is employed whereby the sulfur compounds are removed by reaction with hydrogen in the presence of a catalyst. The hydrogen sulfide formed is then either removed or absorbed. Although the elimination of sulfur is generally desirable, some sulfur compounds provide boundary lubricity, thus their complete removal could lead to problems of fuel-pump seizure.

The corrosion tendency of sulfur compounds and free sulfur is controlled by requiring the fuel to pass a copper-strip corrosion test. This consists of a visual estimation of the change in color of a copper strip that has been immersed for a prescribed time in the test fuel, which is maintained at a specified temperature.

Clay treatment is used to remove surfactants and improve color and stability of fuel. Surfactants can degrade the ability of filter/separators to remove water from fuel, and free water in jet fuel can freeze at high altitudes. In this treatment, fuel is heated and then brought into contact with clay contained in a canister or a cloth bag. After vigorous mixing, the spent clay is removed by filtration.

10.3.2 Contaminants

In addition to sulfur and sulfur compounds, distillate fuels contain impurities, such as water, gum, and trace quantities of metal particles. Residual and crude oils contain larger quantities of these, as well as some metallic compounds that oxidize to form ash.

10.3.2.1 Asphaltenes

Asphaltenes are high-boiling-point hydrocarbons that are found in crude oils and their residual products. They encourage the formation of low-temperature sludge that can block fuel-filtration equipment.

10.3.2.2 Gum

All practical fuels contain minute quantities of inorganic elements, such as sulfur, nitrogen, oxygen, and various metals. Under the action of temperature, light, and the catalytic effect of suspended metal particles, fuels can react with atmospheric and/or dissolved oxygen to form gums. Such products may clog fuel filters and deposit on the surfaces of aircraft fuel systems and fuel injectors, restricting fuel flow and disrupting the spray pattern inside the combustor. Such deleterious changes in fuel properties with time are referred to as: the *fuel storage stability* and exposure to high temperature in the engine is the *fuel thermal stability.*

Antioxidants, such as hindered phenols, may be added to the fuel to improve storage stability. Antioxidants work by interrupting the chain of oxidation reactions, preventing the formation of peroxides, soluble gums, or insoluble particulates. Peroxides can attack elastomeric fuel system parts, gums produce engine deposits, and particulates can plug fuel filters.

Jet fuel is used to remove heat from engine oil, hydraulic fluid, and air-conditioning equipment. The resulting heating of fuel accelerates the reactions that lead to gum and particulate formation. A threshold temperature (usually over 160°C) above which such reactions begin is a measure of fuel "thermal stability." An apparatus used to quantify thermal stability is called a "Jet Fuel Thermal Oxidation Tester (JFTOT)" described in ASTM D 3241.

Gum formation can be minimized by excluding olefins, removing mercaptans and water, and eliminating trace metal particles from fuel.

10.3.2.3 Sediment

Sediment tends to be found in the heavier grades of residual fuel. It consists of finely divided solids that are produced either from carbonaceous materials through the action of air on unstable components, or from inorganic matter, including drilling mud, sand, metals, and chlorides from evaporated brine droplets. Sediment gives rise to deposits in storage tanks, to blockage in filters and flow lines, and to wear in the fuel injector. These problems are minimized by filtration and careful handling.

10.3.2.4 Ash

Ash is comprised of various metallic substances that undergo chemical change during combustion to form nonvolatile inorganic compounds of varying melting points and differing corrosion properties. These compounds tend to adhere to and foul the nozzle guide vanes and turbine blades at temperatures around 925 K and tend to promote corrosion at higher temperatures. The problem is extremely complex, and the extent of the damage caused is very dependent on the temperature history of the overall combustion process.

The ash content of distillate fuel is usually quite small and presents no problems. In the heavier fuel oils, where the ash content may exceed 0.1%, the problem is more serious. It may be alleviated by treatment during the refinery stage, but at a significant penalty in cost and availability. Insoluble ash can be removed by centrifuging.

10.3.2.5 Water

Water can occur in jet fuel in three different forms: dissolved water, free water, and fuel–water emulsion. Free water and/or fuel–water emulsion are potentially hazardous and must be avoided.

10.3.2.5.1 Dissolved Water

A typical water-saturated kerosine-type fuel contains between 40 and 80 ppm of dissolved water at 21°C (70°F). Figure 10.2 shows the water solubility. All distillate fuels contain dissolved water in amounts depending on their composition, temperature, and storage history. For example, fuels with high aromatic content can absorb more water into solution than paraffinic types, and more water will dissolve in warm than in cold fuel. Provided it remains dissolved, water does not have any adverse effects on engine operation [4].

10.3.2.5.2 Free Water

Free water exists as a separate liquid phase. When water-saturated fuel cools, free water separates out, in the form of numerous small droplets. These suspended droplets (dispersed water) give the fuel a hazy appearance. At subzero fuel temperatures, these water droplets form ice crystals and slush, which can collect and build up on fuel filters to cause blockage and malfunctioning.

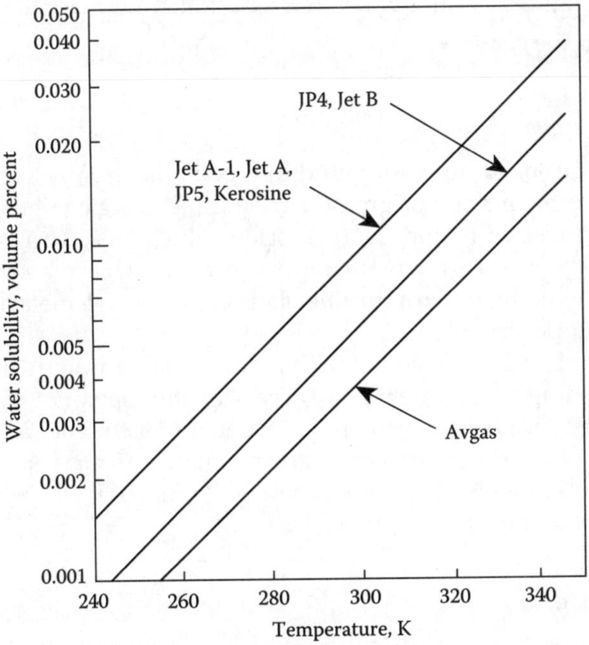

FIGURE 10.2
Solubility of water in aviation fuels.

The problem of engine-filter icing can be overcome by heating the fuel. This, however, incurs penalties in terms of extra weight and a small increase in engine fuel consumption. The problem may also be overcome through the use of fuel-soluble icing inhibitors, as described in the next section.

Another major problem with water-contaminated fuels is that of bacterial or fungal growth, leading to the corrosion of aluminum fuel tanks and fuel-distributing systems [10]. This problem is especially acute if the fuel contains water that is saline or brackish. In addition to providing the initial source of bacteria, this extraneous water usually contains the mineral salts that accelerate their development and growth within the aircraft fuel tank and also cause metal corrosion. The best approach to microbial contamination is prevention, i.e., keeping the amount of free water in the fuel to a minimum.

When microorganisms reach problem levels, approved biocides have been used. Even if biocides effectively stop biogrowth, it is still necessary to remove the accumulated biomass to avoid filter plugging. Also biocides are toxic, and water or fuel containing them must be disposed of appropriately.

10.3.2.5.3 Emulsion

This is a mixture in which very small droplets (less than 100 μm diameter) of one liquid are dispersed in the continuous phase of another liquid. The emulsified mixture is stabilized by surfactants that congregate on the surface of the droplets, preventing them from coalescing. Such surfactants can impair the ability of a filter/separator to remove free water from fuel; i.e., they "disarm the coalescer."

10.3.2.6 Sodium

Trace quantities of sodium are found in crude oils, usually as sodium chloride, owing to the marine origin of petroleum. During combustion, sodium chloride is converted to hydrogen chloride, which has a corrosive effect on the hot metal sections downstream. It can also attack the chromium-rich oxide layer on the surfaces of turbine blades, leading to a metal deterioration known as "black plague" [1].

Sodium can be removed, along with potassium and water-soluble calcium salts, by washing with water. The success of this operation depends on a working difference between the relative density of the fuel and that of the water. Fuels with relative densities greater than 0.97 may not be washable unless they are first blended with a compatible fuel of lower relative density or the density of the water is increased.

10.3.2.7 Vanadium

At temperatures above about 922 K, the trace quantities of vanadium that appear in some fuel oils produce molten compounds that solidify on contact with turbine blades and give rise to fouling and corrosion. The corrosive

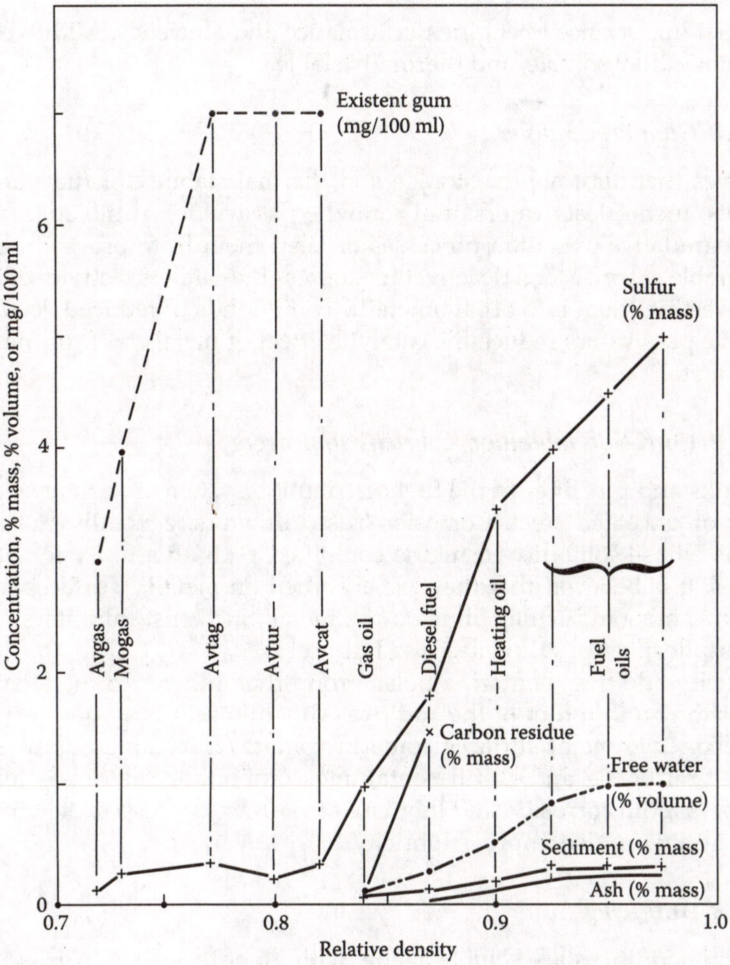

FIGURE 10.3
Concentrations of contaminants in petroleum fuels.

effects of vanadium can be combated through the addition of a suitable magnesium compound whose purpose is to raise the melting point of the ash above the turbine inlet temperature. The concentrations of contaminants present in various fuels of interest have been studied by Goodger [1]; his findings are shown in Figure 10.3. More detailed information is also contained in References [10–12].

10.3.3 Additives

Additives are fuel-soluble chemicals in the parts per million concentration range, which are blended with fuel to enhance fuel properties and handling, of crucial importance to engine performance and efficiency. Additives must be nontoxic, fuel soluble, and thermally stable.

10.3.3.1 Gum Prevention

Additives that improve the storage and thermal stability of fuels are antioxidants, metal deactivators, and catalytic passivators. Antioxidants interrupt degradative oxidation processes or keep them from proceeding to an undesirable extent. Metal deactivators depress the catalytic activity of metals by converting them into organometallic compounds of reduced activity [3]. Catalytic passivators reduce the catalytic effect of metals by forming a film on the metal.

10.3.3.2 Corrosion Inhibition/Lubricity Improvers

The tanks and pipelines of the fuel distribution system are constructed primarily of uncoated steel. Corrosion/rust inhibitors, generally, are hydrocarbons whose solubility in fuel is combined with an affinity for metallic surfaces. It is believed that they are absorbed on metallic surfaces to form protective monomolecular films on the metal. Such rust inhibitors include amines, phosphates, sulfonates, and fatty acids.

Lubricity additives contain a polar group that adheres to metal surfaces [13], forming a thin film of the additive. This film acts as a boundary lubricant when two metal surfaces come in contact. Trace amounts of certain oxygen-, nitrogen-, and sulfur-containing compounds can act as lubricity improvers. Both corrosion and lubricity are surface processes, therefore corrosion inhibitors also improve lubricity.

10.3.3.3 Anti-Icing

Anti-icing additives are soluble agents with a nonreversible hygroscopicity that allows them to combine with moisture in the fuel, thereby preventing its separation and the formation of ice crystals. A well-known anti-icing additive is di-ethylene glycol monomethyl ether (di-EGME). Fuel system icing inhibitors (FSII) work by combining with any free water that forms and lowering the freeze point of the mixture so that no ice crystals are formed. If a fuel containing FSII comes into contact with free water, the additive forms a thick, gelatinous-type paste. For this reason, FSII is not added at a refinery, but added during fuel delivery to the military aircraft. Commercial carriers do not use FSII, but have heaters on their main fuel filters to melt any ice formed.

10.3.3.4 Antistatic–Static Dissipators

During the fueling of an aircraft, static electricity charges can build up to such an extent that the ensuing spark discharge is capable of initiating fires and explosions [14,15]. Charging rates are highest when low-conductivity fuels are pumped at high flow rates, and for some time, the oil companies have taken precautions to reduce the ignition hazard by imposing flow-rate restrictions during product handling [15].

Another practical solution is to increase the electrical conductivity of the fuel so that the charges can dissipate as rapidly as they are generated. When the additive is used, the fuel conductivity must be between 50 and 450 CU at the point of delivery into the aircraft [4,8]. The only additive currently approved for use is Stadis 450 (ASTM D 2624)—a proprietary composition.

10.3.3.5 Metal Deactivators

When the fuel is contaminated with metals, like copper and zinc, its thermal stability is degraded. Metal deactivators are chelating agents [3] that form stable complexes with specific metal ions and inhibit their catalytic activity, thereby mitigating fuel oxidation that causes deterioration in thermal stability. Metal deactivators are not normally employed in commercial aviation.

10.3.3.6 Antismoke

Considerable interest has been show in the use of fuel additives to reduce exhaust smoke. The most successful of these are organometallic compounds of barium, manganese, and iron [16–19]. Tests on all three compounds by Hersh et al. [19] demonstrated very substantial reductions in exhaust emissions index when they were added to No. 2 fuel oil for an electric-utility gas turbine. The optimal injection rate was found to be approximately 25 ppm by mass of metallic base element. The results obtained with the manganese-based additive are shown in Figure 10.4, which demonstrates that the additive reduces the exhaust-plume opacity by reducing the carbonaceous emissions. From measurements carried out at base-load operation, it was found that exhaust-particle size distributions do not change significantly with additive use, the mean particle size being 0.5 μm.

There is an obvious drawback to the use of metal-based fuel additives: during the combustion process, they might form oxides that could deposit on the turbine blades. Conceivably, they could also create an extra pollutant in the engine exhaust gas, one that is less acceptable environmentally than smoke. For these reasons, improving fuel–air mixing and other parameters in combustor design are really the satisfactory solution to the smoke problem.

Synthetic alternative fuels offer the real opportunity of either completely eliminating or significantly reducing the aromatics content of fuel. As shown later, this leads to a corresponding decrease in soot/smoke formation.

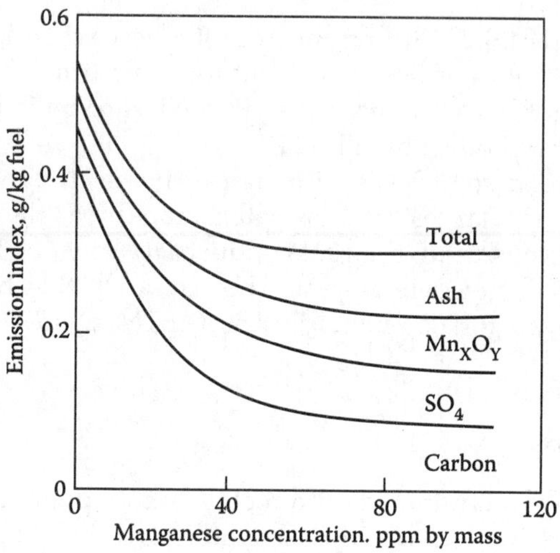

FIGURE 10.4
Exhaust particulate emissions index vs. manganese concentration. (From Hersh, S., Carr, R.C., and Hurley, J.F., ASME Paper 78-GT-64, Gas Turbine Conference, London, 1978.)

10.4 Fuel Properties

In a combustion chamber, the fuel must be injected, vaporized, and mixed with air before combustion can occur. The extent to which these processes are limiting to combustion depends, to a very great extent, on the physical properties of the fuel.

10.4.1 Relative Density

Fuel density is very important because it defines the energy content in the fuel tank with a fixed volume. For a military aircraft, fuel with a high volumetric energy content (MJ/L or Btu/gal) maximizes the energy stored in a fixed volume of the fuel tank, thus providing the longest flight range. For commercial aircrafts that take on just enough fuel to reach their intended destination safely, a less dense fuel with a higher gravimetric energy content (MJ/kg or Btu/lb) may minimize fuel weight. For example, for kerosine (density = 0.81 g/mL at 15°C), the volumetric energy content is maximum 35.06 MJ/L; whereas for a wide-cut jet fuel (density = 0.762 g/mL at 15°C), the gravimetric energy content is maximum at 43.54 MJ/kg.

The relative density d_r (formerly called specific gravity) of a fuel is related to its average boiling point and chemical composition. In general, it is highest for aromatics and lowest for paraffins, with naphthenes and olefins lying in between. The relative density is fairly easy to determine, and it provides

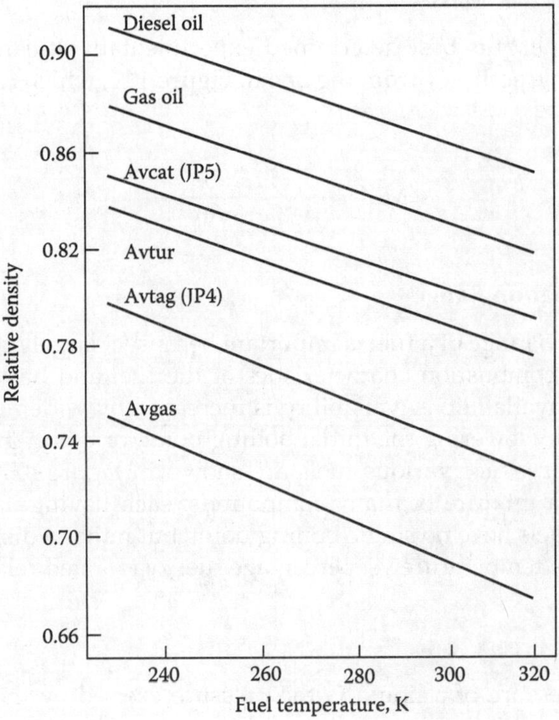

FIGURE 10.5
Relative densities of typical gas turbine fuels.

a useful indication of hydrogen/carbon ratio, calorific value, and tendency toward carbon formation. Typical values for liquid petroleum fuels range from about 0.72 for gasoline up to 0.97 for heavy fuel oil.

The relative density at T_1/T_2 of a fuel is usually defined as the ratio of the mass of a given volume of fuel at a temperature T_1 to the mass of an equal volume of pure water at a temperature T_2. A reference temperature of $T_1 = 289$ K is often used, with T_2 set at 277 K, where the density of water is highest. The relationship between relative density and fuel temperature for some typical aviation fuels is shown in Figure 10.5.

10.4.1.1 API Gravity

In the United States, fuel density is often specified in terms of the API gravity. It is calculated with the formula

$$\text{API gravity} = \frac{141.5}{d_r\left(\dfrac{\text{at } 289K}{289K}\right)} - 131.5.$$

10.4.1.2 Molecular Mass

This fuel property is best determined experimentally. It can be estimated with reasonable accuracy from the line in Figure 10.6, which conforms to the relationship:

$$MM = \frac{11{,}280}{(\text{API gravity})^{1.1}}.$$

10.4.2 Distillation Range

The distillation range of a fuel is important because it largely determines the physical and combustion characteristics of the fuel and has a dominating influence on availability. Availability is increased by widening the boiling range, either by lowering the initial boiling point, or both. The ASTM D-86 distillation curves for various fuels are shown in Figure 10.7. Since practical fuels are a mixture of many compounds, each having its own boiling point, these fuels have no single boiling point, but rather a distillation range defined by the temperature vs. percentage fuel evaporated relationship.

10.4.3 Vapor Pressure

The vapor pressure of a liquid is the pressure exerted by the vapor above its surface at a given temperature. A high vapor pressure is desirable from a

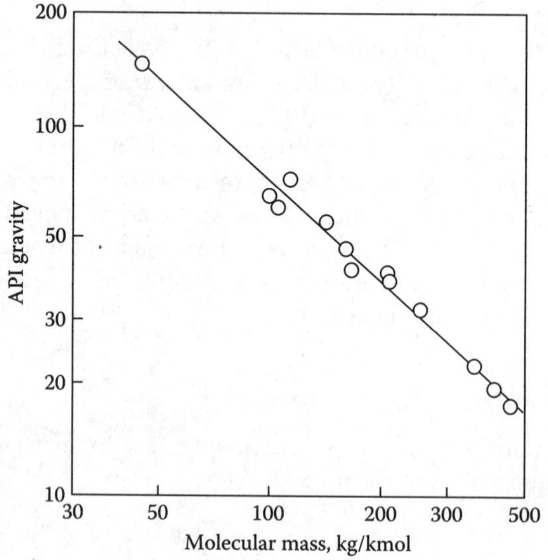

FIGURE 10.6
Relationship between API gravity and molecular mass.

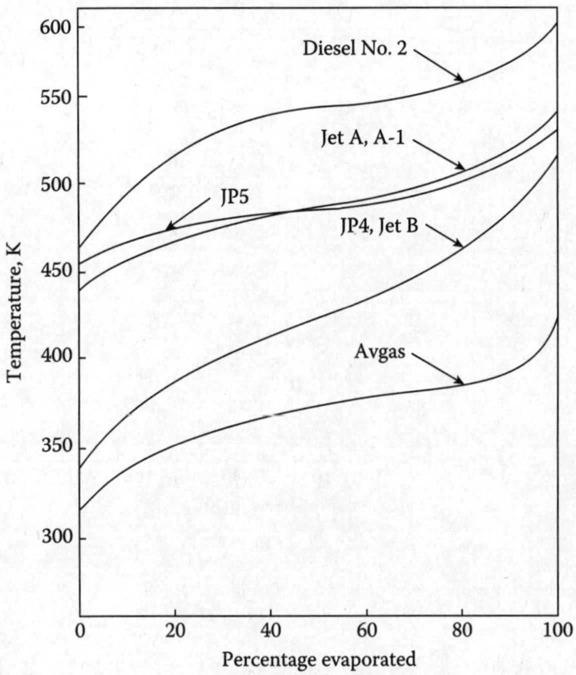

FIGURE 10.7
ASTM D-86 distillation curves for various fuels.

combustion viewpoint, because it ensures rapid evaporation of fuel in the primary combustion zone. On the other hand, low vapor pressure has advantages in terms of reduced pressure in unvented fuel tanks, lower fuel losses due to evaporation at high altitudes in vented tanks, and reduced fire hazard.

The vapor pressures of some typical kerosine fuels are shown in Figure 10.8 as a function of temperature. Note that the vapor pressure increases steeply with temperature. This is why fuels with low vapor pressure are essential for high-supersonic flight speeds.

10.4.4 Flash Point

The flash point is defined as the lowest temperature at which a fuel gives off sufficient vapor to form a flammable mixture with air. As might be expected, the flash point is directly related to vapor pressure; the higher the vapor pressure, the lower the flash point. In general, the flash point of kerosine provides a measure of its flammability, whereas that of heavier oils is an indication of their volatility. It is also useful for classifying fuels from the viewpoint of fire risk.

The significance of the flash point in the selection of aircraft fuels has been discussed by Johnson et al. [20]. The variation of flash point with relative

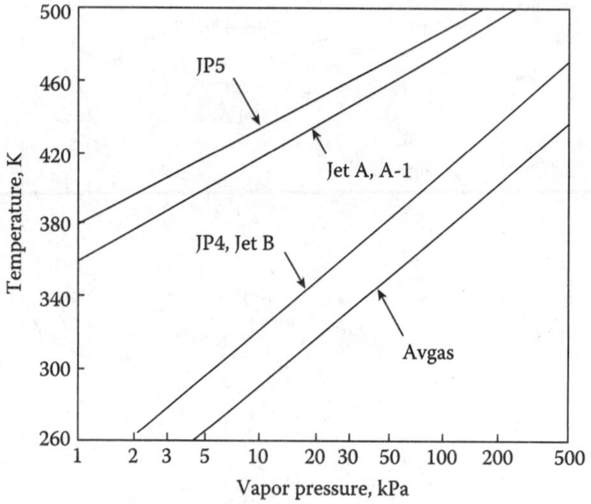

FIGURE 10.8
Effect of temperature on true vapor pressures of aviation fuels.

density is shown for petroleum fuels in Figure 10.9. The minimum flash point of Jet A kerosine-type fuel is 38°C (100°F). The US Navy uses a high flash-point (60°C) kerosine-type JP-5 fuel on aircraft carrier because of safety considerations.

10.4.5 Volatility Point

The volatility of a fuel may be assessed from a knowledge of its distillation range, vapor pressure, and flash point. Increased volatility affects combustion performance by providing easier lightup, improved stability, and higher combustion efficiency. These advantages are most apparent when combustion performance is limited by poor fuel atomization. The disadvantages of high volatility, as touched on above in connection with vapor pressure, are excessive fuel evaporation losses at high altitude and increased fire hazard. The correlation between volatility and flash point is shown in Figure 10.10.

10.4.6 Viscosity

Viscosity is a physical property that depends mainly on the chemical composition of the hydrocarbons contained in the fuel [1]. Apart from affecting the power required to pump the fuel through the fuel system, viscosity has marked effect on the formation of a well-atomized spray and hence on the rates of fuel evaporation and combustion. The higher the viscosity of a fuel, the poorer the quality of atomization. Modern fuel-injection systems are

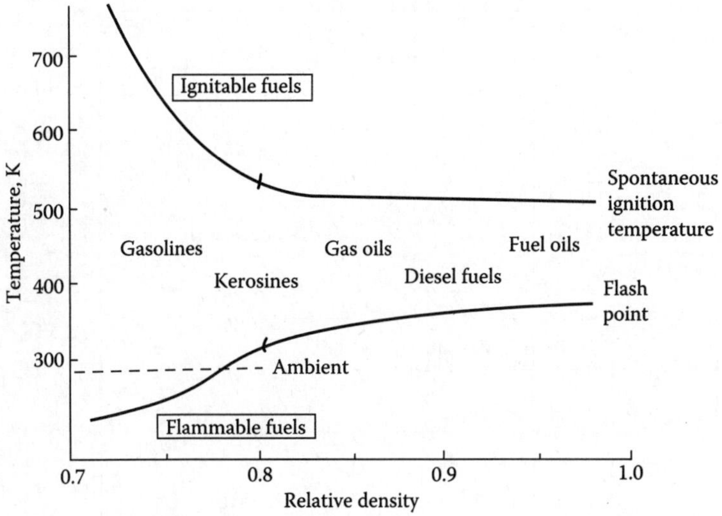

FIGURE 10.9
Ignition temperatures for petroleum fuels.

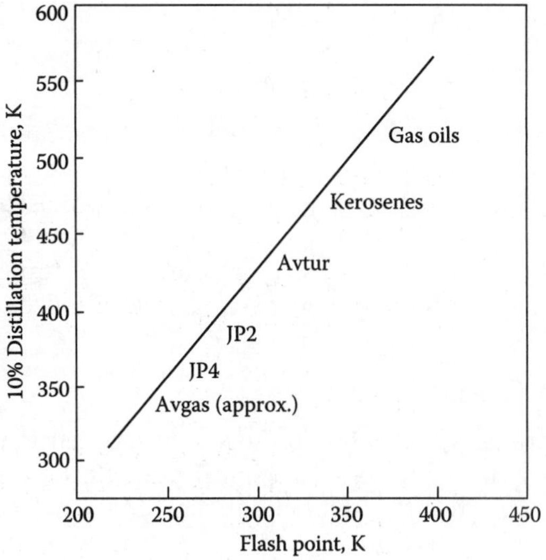

FIGURE 10.10
Relationship between flash point and volatility.

capable of providing satisfactory pressure atomization with fuel viscosities up to around 15×10^{-6} m^2/s, depending on the design of the atomizer and the available fuel-injection pressure. The performance of airblast atomizers is fairly insensitive to fuel viscosity, a significant advantage over atomizers of the pressure type [21].

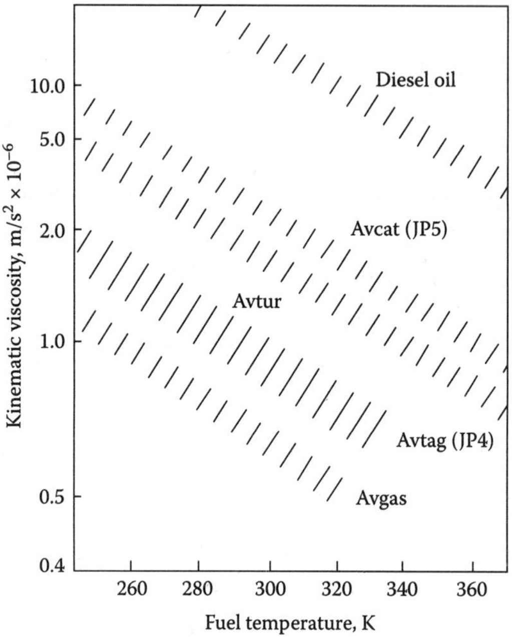

FIGURE 10.11
Viscosity-temperature characteristics of typical fuels.

The viscosity of gas turbine fuels depends greatly on temperature, as illustrated in Figure 10.11. Thus, heavy oils and residual fuels must usually be heated. Also, for aviation fuels, the increased viscosity at low temperatures (currently 12×10^{-6} m^2/s for starting) dictates the minimum operating temperature at which a satisfactory spray can be obtained for starting and/or altitude relight.

10.4.7 Surface Tension

Usually there are no specification limits on surface tension for gas turbine fuels. However, surface tension has a significant effect on fuel atomization, since, for both pressure and airblast atomizers SMD $\propto \sigma_F^{0.6}$. Some typical values of surface tension and their variation with temperature are shown in Figure 10.12.

10.4.8 Freezing Point

Aircraft operate most economically at high altitude where ambient temperature may be as low as 193 K. Because jet fuel is a complex mixture of many individual hydrocarbons, it does not freeze at one temperature like water. Hydrocarbons with highest freeze point solidify first, forming wax crystals; later, lower freeze point components solidify. The freezing point of the jet

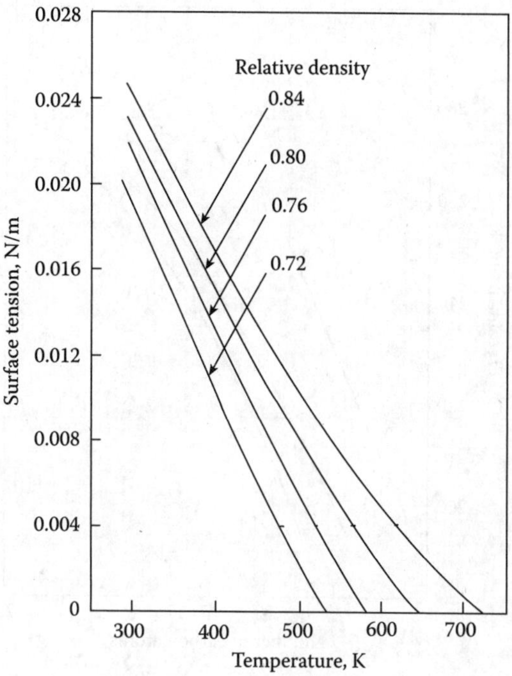

FIGURE 10.12
Surface tension-temperature relations for various hydrocarbon fuels.

fuel is defined as the temperature at which the last wax crystals melt when warming the solidified fuel. The primary engineering criterion is fuel pumpability, which is the ability to move highly viscous fuel from fuel tank to engine. Industry uses freezing point as an indicator of fuel low-temperature pumpability, which is approximately 5°C to 15°C below its freezing point.

The temperature at which a fuel first shows the presence of solid particles or wax crystals (as its temperature is decreased) is called its freezing point. Aircraft fuel specifications stipulate a maximum freezing point of 227 K (–51°F) or 215 K (–73°F), according to the type of service.

In general, the presence of iso-paraffin compounds in lieu of normal paraffins decrease freeze point. Thus, the refining process should be tailored to increase the yield of iso-paraffinic kerosine (IPK) in jet fuel. Figure 10.13 shows the relationship between freezing point and the number of carbon atoms. The relationship between freezing point and boiling point [22] is illustrated in Figure 10.14.

10.4.9 Specific Heat

In modern aircraft, fuel may be used as a heat sink to absorb heat from the airframe and engine. The specific heat is thus an important property

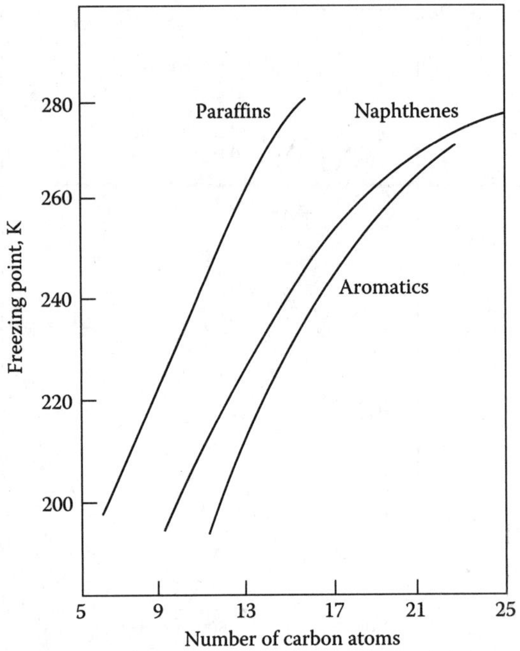

FIGURE 10.13
Relationship between freezing point and number of carbon atoms. (From Hager, J.A., *SAE J.*, 74, 38, 1966.)

of the fuels burned in these aircraft. Paraffinic fuels are the most attractive in this regard because they have considerably higher specific heats than either naphthenic or aromatic fuels [23–25]. Values of specific heat for kerosine-type fuels are plotted in Figure 10.15. For other liquid petroleum fuels, the specific heat may be estimated with reasonable accuracy from the expression

$$c_p = \frac{0.76 + 0.00335T_f}{d_r^{0.5}} \text{kJ/(kg}\cdot\text{K) or } c_p = \frac{0.388 + 0.00045T_f}{d_r^{0.5}} \text{Btu/lb.F.}$$

For fuel vapor:

$$c_p = (0.136 + 0.0012T_f)(4 - d_r)\text{kJ/(kg}\cdot\text{K) or}$$
$$= (0.105 + 0.00016T_f)(4 - d_r)\text{Btu/(lb}\cdot\text{F).}$$

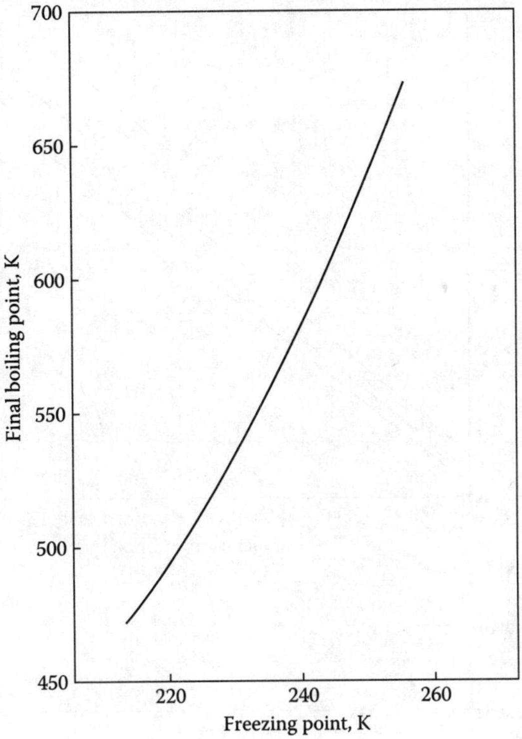

FIGURE 10.14
Effect of freezing point on final boiling point.

10.4.10 Latent Heat

The latent heat of evaporation at the average boiling point of the fuel may be related to relative density and fuel temperature by the expression

$$L = \frac{360 - 0.39T_{\mathrm{f}}}{d_{\mathrm{r}}}\,\mathrm{kJ/kg} \text{ or } L = \frac{115 - 0.1T_{\mathrm{f}}}{d_{\mathrm{r}}}\,\mathrm{Btu/lb}.$$

10.4.11 Thermal Conductivity

The thermal conductivity of a liquid petroleum fuel is given by:

$$k = \frac{0.134 - 0.000063T_{\mathrm{f}}}{d_{\mathrm{r}}}\,\mathrm{W/(m \cdot K)} \text{ or } k = \frac{0.065 - 0.0000206T_{\mathrm{f}}}{d_{\mathrm{r}}}\,\mathrm{Btu/(h \cdot ft \cdot F)}.$$

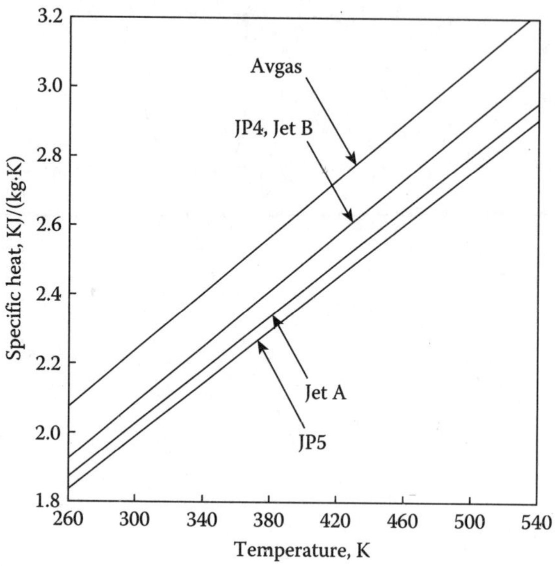

FIGURE 10.15
Influence of temperature on specific heat.

10.5 Combustion Properties of Fuels

The combustion properties of most interest are those that govern the flame temperature, the rate of chemical reaction, the burning range, and the tendency to form soot.

10.5.1 Calorific Value

The calorific value of a fuel is a measure of the heat librated when it is burned to completion under standard conditions. For gas turbine fuels, the lower calorific value is most relevant, since it pertains to complete combustion to CO_2 and H_2O vapor. On a gravimetric basis, calorific value is now described as specific energy; and on a volumetric basis, as energy density. For most hydrocarbon fuels, the specific energy can be related to the relative density, as shown in Figure 10.16. In general, the heaviest fuels have the highest energy density, while the lightest fuels have the highest specific energy. The choice of fuel for aircraft applications will depend on whether tankage space is limited, as in fighter aircraft, or if range is limited by all-up weight, as in most freight aircraft.

The specific energy of a fuel depends entirely on its chemical composition and, in particular, on the hydrogen/carbon ratio. Several empirical formulas have been derived for calculating the heat of combustion of a fuel from an

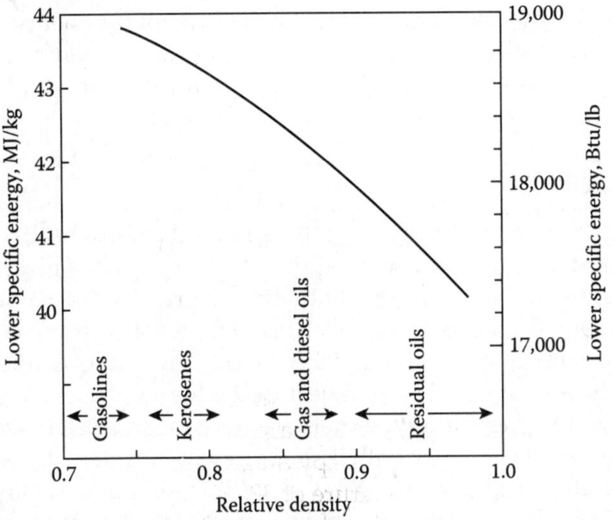

FIGURE 10.16
Variation of specific energy with relative density of fuels.

analysis of its constituents. Ragozin [26] recommends the following formula, which is attributed to Mendeleyev:

$$\text{LSE} = 0.339\text{C} + 1.03\text{H} - 0.109(\text{O} - \text{S}) - 0.025\text{W},$$

where LSE is the lower specific energy of the fuel in megajoules per kilogram, and C, H, O, S, and W are the mass percentages of carbon, hydrogen, oxygen, sulfur, and water, respectively, in the fuel.

If the chemical composition of the fuel is unknown, the heat of combustion can be estimated with less precision from the relative density alone, using the equation

$$\text{LSE} = 51.5 - \frac{5.95}{1.53 - d_r}\,\text{MJ/kg}.$$

The energy density of gaseous fuels is usually expressed either in megajoules per cubic meter or Btu per cubic foot. Since the amount of heat produced by a gas burner does not depend solely on the volumetric heat release, a quantity known as the *Wobbe index* is often used in comparing gaseous fuels and in preliminary assessment of the design requirements for the fuel system. It is defined as

$$\text{Wobbe index} = \frac{\text{Energy density}}{\text{Relative density}}.$$

Essentially, the Wobbe index provides a measure of the energy throughput for a given size of gas discharge orifice. Normally, it should not be allowed to vary by more than about 5% from the nominal value for which the system was designed.

10.5.2 Enthalpy

The enthalpy of a fuel is a measure of its capacity to absorb heat; the enthalpy also indicates the amount of heat required to accomplish a given change in the temperature or state of a fuel. Enthalpy diagrams of the type shown in Figure 10.17 for Jet A fuel, may be constructed for any fuel from its latent heat, distillation characteristics, and specific heats. The curve marked "liquid" represents the amount of absorbed heat in the liquid phase alone. Enthalpy diagrams for other kerosine-type fuels are contained in Reference [27,28].

As an example of the use of enthalpy diagrams, consider a Jet A fuel in the liquid phase at an initial temperature of 300 K. Its initial enthalpy is 250 kJ/kg, as shown by the saturated-liquid curve at 300 K. Assume that 800 kJ/kg is to be added to this fuel, giving a final enthalpy of 1050 kJ/kg. This heat can be added in several ways:

1. The fuel can be retained in the liquid phase. The final temperature will then have to be 600 K, as indicated by the temperature of saturated liquid at an enthalpy of 1050 kJ/kg. A pressure in excess of 1.8 MPa would be required to keep the fuel liquid.

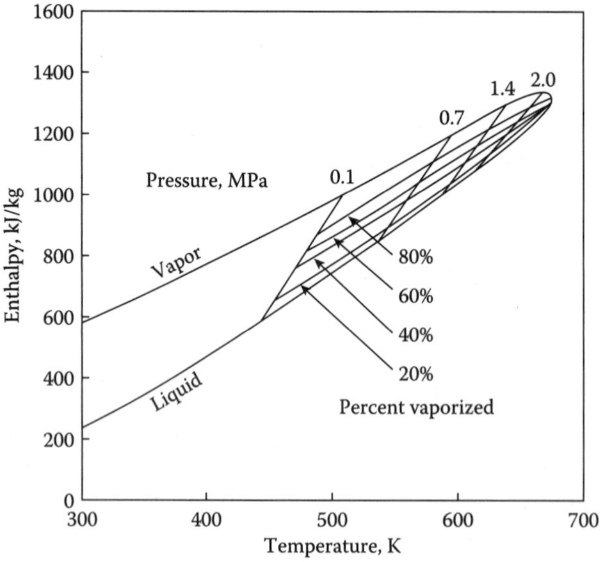

FIGURE 10.17
Enthalpy of Jet A fuel.

2. The fuel can be completely vaporized. The final temperature of the vapor will then have to be 522 K, as shown by the saturated-vapor curve at an enthalpy of 1050 kJ/kg.
3. The fuel can be partially vaporized at a temperature between 522 and 600 K, depending on the pressure. For example, at a pressure of 0.7 MPa, partial vaporization would yield a liquid-fuel-vapor mixture at a temperature of 567 K, with about half the total fuel in the form of vapor.

10.5.3 Spontaneous-Ignition Temperature

An important factor affecting aircraft fire and explosion hazards is the tendency of the fuel toward spontaneous ignition. When an aircraft crashes, a common cause of fire is the spontaneous ignition of fuel coming into contact with hot surfaces, especially near engines. The problem is especially important for high-speed aircraft because of kinetic heating of the aircraft structure. Above altitudes of 11 km, at about Mach 2.7, the leading edge of an aircraft attains a temperature that just exceeds the spontaneous-ignition temperature (SIT) of the fuel.

The SIT depends greatly on molecular structure, since the oxidation reactions leading to spontaneous ignition can occur only when the chemical bonds within the fuel are broken by thermal agitation. Consequently, the SIT is lower for fuels with higher relative density, whose molecules are less compact and more vulnerable to cracking [1]. Straight-chain paraffins are more easily ignited than either branched-chain paraffins or cycloparaffins containing the same number of carbon atoms. Very highly branched compounds and aromatics are highly resistant to spontaneous ignition [29].

The SIT is usually measured by injecting a small sample of the fuel into a heated crucible or flash and noting the delay until the onset of ignition. The test is repeated at lower temperatures (with associated longer delays), until the minimum ignition temperature is found [1]. Some typical results are shown in Figure 10.9.

Generally, for a hydrocarbon fuel, the SIT increases with decreasing pressure. Tests carried out on a JP-4 fuel [30] showed that decreasing the pressure to one-third its original value increased the SIT by a factor of 2.

A number of workers have investigated the effect of additives on the SIT of aviation fuels. Thomas [31] examined a number of additives, the most effective being tetraethyl lead. In a concentration of 3 mL per imperial gallon, this additive increased the minimum pressure needed to produce spontaneous ignition from 27.6 to 40.7 kPa. Tetraethyl lead is added as part of a mixture that also contains ethylene dibromide and dye. Ethylene dibromide acts as a scavenger for lead oxide and reacts to produce a mixture of lead bromide and lead oxybromides. These compounds are volatile and are exhausted from the engine; in this manner, deposits of lead oxide are prevented from collecting on valves and spark plugs. Of the number of additives investigated by

Kurtovich and Hayes [32], tetraisopropyl titanate was found to be the most promising. Although the results so far achieved are encouraging, no additive for aircraft fuels has yet been found that, when used in small concentrations, can effectively increase the SIT.

10.5.4 Limits of Flammability

Combustible gases and vapors are capable of burning in air only within closely defined limits of concentration. Within these limits, a flame, once initiated, can spread any distance away from the source of ignition. It is customary to define *rich* and *lean* flammability limits, which respectively represent the maximum and minimum fuel concentrations that produce combustible mixtures. The lean limit of flammability is closely related to the flash point. For kerosine-type fuels at room temperature, the range of flammability lies between fuel/air ratios of approximately 0.035–0.28 by mass, the exact value depending on the size and shape of the vapor space in the fuel tank. Limiting concentrations for a number of other fuels in air at standard pressure and temperature are listed in Reference [1].

Barnett and Hibbard [28] have shown that lean-limit mixtures of all hydrocarbons contain about the same heat of combustion per unit volume of fuel–air mixture. On this basis, the fuel concentrations for lean-limit mixtures at 1 atm are estimated as

$$L = \frac{4350}{\text{LSE} \times \text{MM}} \text{MJ/kg or } L = \frac{1.87 \times 10^6}{\text{LSE} \times \text{MM}} \text{Btu/lb.}$$

where L is the lean-limit concentration, percent by volume and MM is the molecular mass.

The limits of flammability do not depend greatly on pressure, except at very low pressures below 7 KPa. Generally, they tend to spread slightly with an increase in pressure. The influence of temperature is more pronounced. At low temperatures, the fuel may release insufficient vapor, so that the mixture strength lies below the lean limit of flammability. At high temperatures, the production of excessive amounts of vapor could move the mixture strength above the rich limit. Thus, flammable mixtures are produced only within a certain range of fuel temperatures. For an average kerosine and a typical wide-cut fuel (JP-4), the flammability *temperature* limits at sea level are 254–288 K and 316–338 K, respectively.

Fuel vaporization increases at high altitudes; thus, at altitude, flammable mixtures are formed at much lower temperatures than on the ground.

Of special interest in aviation is the possibility of forming flammable mixtures in the ullage space of fuel tanks at high altitude. Liquid hydrocarbon fuels absorb atmospheric gases; the solubility of oxygen being greater than that of nitrogen. At high altitudes, the dissolved oxygen in the fuel comes out of solution and can raise the oxygen content of ullage air from 21 to as high as 33% (= 10 K increase in flammability temperature). Also, fuel boiling can promote a significant loss of fuel through tank vents and produce vapor lock in pipelines and fuel pump cavitation. Consideration is given to "ullage purging" by nitrogen from a ground source prior to takeoff.

10.5.5 Smoke Point

The influence of fuel type and composition on soot formation is discussed in Chapter 9. Probably, the most widely used index of soot-forming tendency is the *smoke point* (ASTM D 1322). It is determined experimentally by burning the fuel under test in a special wick lamp and slowly increasing the height of the flame until it begins to smoke. The maximum height of smokeless flame in millimeters is the smoke point; the higher this is, the tendency of the fuel to soot formation is lower. For many aircraft fuels, the height of smokeless flame is specified to be not less than 20 mm.

Although the smoke point is not a fundamental property of a fuel, it provides a very satisfactory indication of the propensity to form soot. Naegeli and Moses [33] used a high-pressure research combustor, operating over a wide range of burner inlet conditions, to determine the effect of fuel molecular mixture on soot formation. Their results showed an excellent correlation of flame radiation with smoke point, as illustrated in Figure 10.18. In this figure, the average index R is defined as

$$R = \frac{\text{Flame radiation of test fuel}}{\text{Flame radiation of reference (Jet A) fuel}}.$$

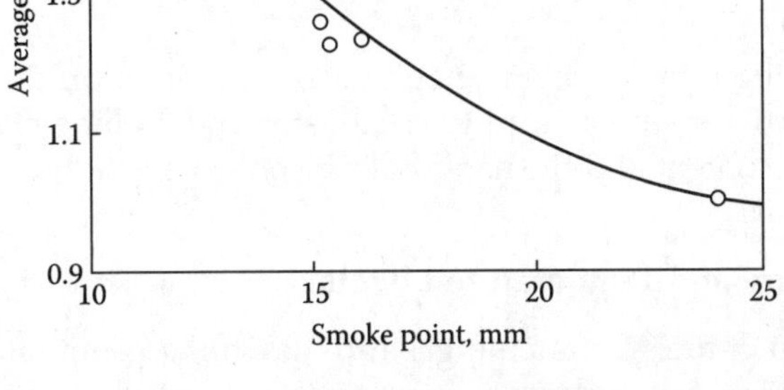

FIGURE 10.18
Effect of smoke point on average index of flame radiation. (From Naegeli, D.W. and Moses, C.A., ASME Paper 80-GT-62, Gas Turbine Conference, New Orleans, LA, 1980.)

It is of interest to note that all the fuels used in these tests had virtually the same hydrogen content. Thus, a correlation for soot-forming tendency based on hydrogen content alone could not account for the significant differences that are apparent in Figure 10.18. The differences are attributed by Naegeli and Moses [33] to variations in the polycyclic aromatic content of the fuel. The data suggest that the smoke point provides a better index of soot-forming tendency than a simple hydrogen correlation.

10.5.5.1 Luminometer Number

This index is measured on a modified smoke lamp (ASTM D1740) that includes thermocouples above and below the flame and a device for measuring flame radiation in the green-yellow band of the visible spectrum [1]. At a standardized level of radiation, the temperature rise, ΔT, across the flame is measured for the test fuel and compared with the corresponding values for isooctane (a paraffin) and tetralin (an aromatic), which are assigned values of 100 and 0, respectively. The luminometer number of the test fuel is then expressed as

$$\text{Luminometer number} = 100\frac{\Delta T_{\text{test fuel}} - \Delta T_{\text{tetralin}}}{\Delta T_{\text{isooctane}} - \Delta T_{\text{tetralin}}}.$$

The concept of luminometer number was introduced in the late 1950s in the hope that it would provide a more accurate description of the radiation signature of a fuel than is given by the smoke point. These early expectations were not fully realized in practice, and the apparatus and test method are no longer in common use.

10.5.5.2 Smoke Volatility Index

Sometimes used in preference to smoke point, this index is:

$$\text{SVI} = \text{smoke point (mm)} + 0.42$$

$$\text{(volume percent of total fractions boiling below 478 K)}.$$

Fuels with high aromatic and naphthalene contents produce more soot. Since soot particles are harmful to combustor and turbine blades, both the total aromatic content and the naphthalene content are tightly controlled.

10.5.6 Pressure and Temperature Effects

In the selection of fuels for aircraft gas turbines, full account must be taken of the very wide ranges of pressure and temperature to which the fuel may be subjected. At an altitude of 12 km, the ambient air pressure is only a quarter of the pressure at sea level. Under extreme conditions, air temperatures at altitude may be as low as 193 K, whereas very high fuel tank temperatures may be attained when an aircraft on the ground is exposed to tropical sun. Therefore, it is necessary to study the effects of pressure and temperature on fuel properties.

10.5.6.1 Subatmospheric Pressure

As the aircraft gains altitude, the air dissolved in the fuel begins to be released and may escape through the tank vents, carrying fuel vapor along with it. Additionally, "vapor lock" could occur. Also, when the fuel vapor pressure equals the high-altitude ambient pressure, fuel boiling may result. To minimize boiling, evaporation loss, and vapor lock, most civil aircraft have pressurized fuel tanks.

10.5.6.2 Low Temperature

Reference has already been made to the emergence of ice crystals at fuel temperatures below 273 K and the formation of wax crystals at the freezing point of the fuel. The main problem arising with both types of crystal formation is clogging of the fuel filters. Most commercial aircraft have heaters on their main fuel filters to melt any ice that is collected. Currently, a FSII approved for Jet A, Jet A-1 and JP-8 fuels is di-ethylene glycol monomethylether (di-EGME), which works by combining with any free water that forms and lowering mixture freeze point.

10.5.6.3 High Temperature

During flight at high supersonic speed, the bulk liquid fuel in the tank experiences a steady rise in temperature as the kinetic heat imparted to the aircraft structure is transferred to a continually diminishing volume of fuel. Other sources of heat include heat exchangers for engine lubricating oil and hydraulic fluid, and more heat is generated in the engine fuel pump. As a consequence, the temperature of the fuel entering the combustion chamber may reach 473 and 533 K at flight Mach numbers of 2.2 and 3.0, respectively [34]. When the fuel flow is reduced to start the descent, a steep rise in fuel temperature occurs in the fuel system through soakage of heat from the engines. This leads to fuel oxidation and deposition of gum and other insoluble solid particles on heat exchanger elements, filters, and fuel nozzles.

In the early 1990s, the U.S. Air Force initiated a program to develop a fuel with improved thermal stability. It led to the introduction of an additive package [35] to improve the thermal stability of fuel by about 100°F (60°C) from 325°F to 425°F. This additive became known as the " + 100 (plus one hundred)" and the new fuel was called JP-8 + 100. This new fuel is designed to alleviate the need to develop expensive specialty fuels, such as JP-7 and JP-TS, for future advanced aircraft and will also decrease maintenance costs for current aircraft. The initial JP-8 + 100 additive package is based on a detergent/dispersant, packaged with an anti-oxidant and MDA at a concentration of 256 mg/L. This additive has been shown to produce a 50–95% reduction in deposits with a wide variety of JP-8/Jet A fuels.

10.6 Classification of Liquid Fuels

Table 10.1 lists the typical properties of conventional liquid fuels for gas turbines. Light distillates such as gasoline and kerosine are used for aircraft, and heavy distillates are used for industrial gas turbines.

TABLE 10.1

Typical Properties of Liquid Fuels

	True Distillates		Ash-Bearing Fuels	
Fuel Type	**Kerosine**	**No. 2 Distillate**	**Blended Residuals and Crudes**	**Heavy Residuals**
Relative density at 311 K	0.793	0.82–0.88	0.80–0.92	0.92–1.05
Viscosity at 311 K (m^2/s)	1.4×10^{-6}	2×10^{-6}–4×10^{-6}	2×10^{-6}–10^{-4}	10^{-4}–18×10^{-4}
Viscosity at 311 K (cSt)	1.4	2.0–4.0	2–100	100–1800
Flash point (K)	311–344	339–367	283–367	353–403
Flash point (°F)	100–160	150–200	50–200	175–265
Pour point (K)	228	253–273	263–318	263–308
Pour point (°F)	–50	–10–30	15–110	15–95
Lower specific energy (MJ/kg)	42.8	42–43	42–43	40–41
Lower specific energy (Btu/lb)	18,400	18,060–18,490	18,060–18,490	17,200–17,630
Sulfur (% by mass)	0.01–0.1	0.1–0.8	0.2–3	0.5–4
Nitrogen (% by mass)	0.002–0.01	0.005–0.06	0.06–0.2	0.05–0.9
Hydrogen (% by mass)	12.8–14.5	12.2–13.2	12.0–13.2	10–12.5
Ash (fuel as delivered) (ppm)	1–5	2–50	25–200	100–1000
Ash (inhibited) (ppm)			25–250	100–7000
Trace metal contaminants (untreated)				
Sodium plus potassium (ppm)	0–0.5	0–1	1–100	1–350
Vanadium (ppm)	0–0.1	0–0.1	0.1–80	5–400
Lead (ppm)	0–0.5	0–1	0–1	0–25
Calcium (ppm)	0–1	0–2	0–10	0–50

Source: Goodger, E. M., *Transport Fuels Technology*, Landfall Press, Norwich, UK, 2000. With permission.

Gasolines (including naphtha) are excellent fuels of high burning quality. However, their low viscosity may result in poor lubricity, while their low flash point and high volatility may require special attention to safety. Aviation gasolines have a typical distillation range of 300–495 K. They include Avgas (UK military), JP-4 (US military), and Jet B (US civil) fuels. The kerosines consist of refined hydrocarbons from the distillation of crude petroleum, or blends of the latter with suitable cracked products. They include the normal aviation kerosines, such as Jet A-1, Jet A, and Navy's high-flash-point JP-5.

Light to heavy distillates include No. 2-GT gas turbine fuel, No. 2 burner fuel, diesel oil, and marine gas oil. Diesel fuels have additional requirements of cetane number, and the primary fuel in this group is the 2-D diesel fuel.

Blended heavy distillates include 3-GT gas turbine fuel, 4-D diesel fuel, and marine diesel fuel. Their vanadium content is less than 5 ppm, but their relatively high wax content may necessitate heating for pumping and filtering. These fuels lie between blended heavy distillates and heavy residuals. They are produced by blending to provide specific maximum sulfur levels. Their vanadium content is fairly high, normally between 5 and 100 ppm. Blended residuals include light residual oil and light furnace oil.

Heavy residual fuels represent what is left after all the various distillation processes are complete. These fuels have high viscosity and high content of asphaltenes, sulfur, vanadium, and sodium. However, after various treatments and with heating to facilitate pumping and atomization, they can be utilized in heavy-duty gas turbines.

10.6.1 Aircraft Gas Turbine Fuels

The fuel specifications for aircraft engines are more strict than those for all other types of gas turbines. The main requirements due to the airframe, engine fuel system, and combustion chamber are listed below.

10.6.1.1 Airframe

1. Low fire risk. This implies low vapor pressure, low volatility, high flash point, and high conductivity to minimize the buildup of static electricity during fueling.
2. High heat content for maximum range and/or payload, i.e., high calorific value on a weight or volume basis, depending on whether the aircraft is "weight limited" or "volume limited."
3. High thermal stability, to avoid filter plugging, sticking of control valves, etc.
4. Low vapor pressure, to minimize evaporation losses at high altitude.
5. High specific heat, to provide effective heat absorption on high-speed aircraft.

10.6.1.2 Engine Fuel System

1. Pumpability. That is, the fuel must remain liquid and flow freely to the atomizer. This is essentially a requirement for low viscosity.
2. Freedom from filter clogging by ice or wax crystals. Ice formation is eliminated by the use of additives or by fuel heating. Wax formation is related to gum content and thermal stability.
3. Freedom from vapor locking. This is achieved by the use of low-vapor-pressure fuels.
4. High lubricity for minimum pump wear. This is obtained through the presence or addition of highly polar compounds. (See Section 10.3.3.2 dealing with lubricity.)

10.6.1.3 Combustion Chamber

1. Freedom from contaminants that cause blockage of small passages in fuel nozzles.
2. Good atomization. Atomization quality is most strongly affected by viscosity.
3. Rapid evaporation. Evaporation rates are dependent on fuel volatility, and on atomization quality that determines the surface area of the atomized fuel. Maximum rates of evaporation are achieved with fuels of low viscosity and high volatility.
4. Minimum carbon formation, low flame radiation and low coke deposition.

10.6.2 Aircraft Fuel Specifications

Table 10.2 shows the typical aviation fuel properties of past and current aircraft fuels. The most widely used aviation fuel specifications are those issued by the U.S. ASTM D1655 and U.S. MIL-DTL-83133E, and the UK Ministry of Defense (MOD) DEF STAN 91-91. A number of other countries have similar agencies (Canadian General Standards Board CGSB-3.22, Russian GOST 10227, and International Air Transport Association (IATA) publication). Finally, the U.S. military maintains specifications of JP-8 fuel widely used by the Air Force and JP-5 used by the Navy. Two kerosine-type commercial jet fuels are widely used today: Jet A-1 (freeze point –51°C) on international flights and Jet A (freeze point –45°C) on almost all domestic flights in the United States.

10.6.3 Industrial Gas Turbine Fuels

For industrial and marine applications, the choice of fuel may be governed by economy and availability. In practice, this usually means residual

TABLE 10.2
Typical Aviation Fuel Properties

Property	Avgas	JP-4	JP-5	JP-7	JP-8 (Jet A/A-l)	RP-1
Approximate formula[a]	C_7H_{15}	$C_{8.5}H_{17}$	$C_{12}H_{22}$	$C_{12}H_{25}$	$C_{11}H_{21}$	$C_{12}H_{23}$
H/C ratio	2.09	2.00	1.92	2.07	1.91	1.95
Boiling range, °F (°C)	115–295 (46–145)	140–460 (60–240)	360–495 (180–260)	370–480 (190–250)	330–510 (165–265)	350–525 (175–275)
Freeze point, °F (°C)[b]		–80 (–62)	–57 (–49)	–47 (–44)	JP-8/Jet A-l: –60 (–51) Jet A: –50 (–45)	–55 (–48)
Flash point, °F (°C)		–10 (–23)	147 (64)	140 (60)	127 (53)	134 (57)
Net heating value Btu/lb (kJ/kg)		18,700 (43,490)	18,500 (43,025)	18,875 (43,895)	18,550 (43,140)	18,650 (43,370)
Specific gravity, 16°C (60°F)	0.72	0.76	0.81	0.79	0.81	0.81
Critical temperature, °F (°C)		620 (325)	750 (400)	750 (400)	770 (410)	770 (410)
Critical pressure, psia (atm)		450 (30.5)	290 (19.5)	305 (20.5)	340 (23)	315 (21.5)
Average composition						
Aromatics (vol%)	25	10	19	3	18	3
Naphthenes		29	34	32	35	58
Paraffins		59	45	65	45	39
Olefins	10	2	2		2	
Sulfur (ppm)		370	470	60	490	20

Source: Edwards, J.T., *AIAA Journal of Propulsion Power*, 19, 1089–1107, 2003. With permission.
[a] For illustration of average carbon number, not designed to give accurate H/C ratios.
[b] Typical.

oils or surplus gas, although sometimes a compromise is made (e.g., for the automobile gas turbine, gas oil is preferred to residual fuel, to avoid the more complex fuel system needed for the heavier fuel). Since industrial fuels must necessarily be cheap, they are often impure. Also, industrial handling and storage practices are not up to aeronautical standards, and contamination of liquid fuel by water, salt, and sand is commonplace. Thus, to make the heavier fuels suitable for reliable gas turbine use, it is usually necessary to provide the following treatments [1].

1. Washing to remove trace metals, such as sodium, potassium, and inorganic particulate matter
2. Inhibition of vanadium in the fuel by the addition of magnesium compounds
3. Filtration to remove solid oxides, silicates, and other compounds that could clog fuel pumps, flow dividers, and fuel nozzles

10.7 Classification of Gaseous Fuels

By far the most common gaseous fuel for industrial gas turbines is natural gas. However, the diminishing supply of natural gas has led to increased interest in other gaseous fuels, including by-products from industrial processes, low-energy gas from coke or oil, and coke-oven gas. All gaseous fuels are advantageous in terms of high thermal stability and clean (soot- and ash-free) combustion. Table 10.3 lists the typical properties of common gaseous fuels.

Natural gas consists mainly of methane, along with minor amounts of other gaseous hydrocarbons, such as butane, ethane, and propane. Some natural gases contain up to 15% of nitrogen and carbon dioxide. If the sulfur content is negligibly small, the gas is described as "sweet." However, if hydrogen sulfide is present in significant amounts, the gas is termed "sour" and must be purified prior to combustion.

Coal gas is produced by the carbonization of bituminous coals in gas retorts or coke ovens. Its composition and heating value vary with the type of coal and the temperature of carbonization. Fuels of high heat content tend to be rich in hydrogen and methane, with a nitrogen content of less than 11%. However, fuels of low heating value may contain as much as 55% nitrogen. The major impurity of concern is sulfur. Sulfur alone is not necessarily harmful, but if trace metal compounds, particularly sodium and potassium, are present along with sulfur, then turbine blade corrosion and erosion can occur.

The coke-oven gas contains substantial amounts of hydrogen and methane and has a high energy density, in the 20–24 MJ/m^3 range. Scrubbing

TABLE 10.3

Properties of Common Gaseous Fuels

Property	Natural Gas	Coal Gas (Low Btu)	Coal Gas (High Btu)	Coke-Oven Gas	Blast-Furnace Gas	Producer Gas
Energy density (Btu/ft^3)	950–1150	110–165	500–700	525–650	90–100	120–140
Energy density (MJ/m^3)	35–43	4.1–6.1	19–26	20–24	3.4–3.7	4.5–5.2
Relative density (air = 1.0)	0.58–0.72	0.80–0.92	0.41–0.48	0.40–0.45	0.95–1.05	0.86
Composition (vol%)						
Methane	75–97	0.5–4.5	20–35	28–32		1.2
CH–other hydrocarbons	2–20		2–4	2–4		
Hydrogen		12–16	40–55	50–55	1–4	16.5
Carbon monoxide		2–32	5–15	5–7	25–30	24
Nitrogen	1–16	30–55	4–11	1–6	55–60	50.8
Carbon dioxide	0.1	0.5–10	2–4	2–3	8–16	7.5

is normally needed to remove particulate matter. Since coke-oven gas is generally produced at normal atmospheric pressure, the gas turbine fuel-pressure requirements may demand special consideration.

Producer gas, which is obtained by the partial combustion of coal or coke in air, has a fairly low energy density, between 4.5 and 5.2 MJ/m^3. The energy density of blast-furnace gas, which is produced in fairly copious amounts in iron works, is even lower—of the order of 3.78 MJ/m^3. For this reason, it is not considered suitable as a gas turbine fuel.

Perhaps the most important difference between gaseous and liquid fuels is the wide energy-density range of gaseous fuels, as compared with liquid fuels. Most gaseous fuels can be accommodated in industrial engines by suitable modifications to the turbine control system and fuel-handling equipment. Gaseous fuels of very low energy density, less than around 6 MJ/m^3, present special problems due to the large volume of gas needed to sustain combustion; in addition, their low flame speed can give rise to combustion instability. Nevertheless, gas turbines have been operated successfully with gaseous fuels having energy densities as low as 4.1 MJ/m^3.

10.7.1 Gaseous Fuel Impurities

The problems arising from impurities in low-energy liquid and gaseous fuels are deposition, corrosion, and pollution. Of these, ash deposition has proved to be the most persistent, in some cases causing unacceptable losses in power output after only a few hundred hours of running [36–38]. In addition to sulfur, which may be present in fuel in concentrations up to 5% by mass, five trace metals are of most concern: calcium, lead, potassium, sodium, and vanadium. If they are present in the fuel in significant amounts, the last four can cause turbine-blade erosion, while all five can cause deposits. The two elements most commonly found in petroleum fuels are sodium and vanadium. Both can only be tolerated in small amounts, owing to their ability to form complex compounds of low melting point that are semi-molten and corrosive at metal temperatures as low as 894 K [37,38]. Clearly, turbine operation at such a low inlet temperature would severely limit power output and thermal efficiency. This is why limits must be placed on the acceptable levels of trace metals in fuels for modern heavy-duty gas turbines.

For heavy fuels, the normal practice is to employ water washing to reduce sodium and potassium to a specified level, and to introduce a magnesium-base additive to counteract the corrosive effect of vanadium. A magnesium/vanadium mass ratio of 3 to 1 is recommended [1]. This additive produces a solid ash, of which a small fraction adheres to the turbine blades and gradually reduces the power output of the machine. However, the ash is dry and noncorrosive, as opposed to the molten product that results from the use of untreated fuel.

The power lost through blade fouling can be restored by various methods, including nutshell injection, turbine shutdown, and shutdown plus washing. The latter is the most effective and involves injecting water into the combustion chamber after the machine has been shut down and cooled. The cleaning process is conducted at cranking speed, and it removes virtually all the deposits from the hot sections. Turbine shutdown for several hours, without washing, allows ash deposits formed at temperatures below 1172 K to flake and spall, so that they are blown away through the exhaust when the machine is started up. Nutshell injection is the least effective method, since it restores less than half the power lost via ash deposition. However, it has an advantage in that it can be performed while the machine is still running.

Further information on the treatment, handling, and combustion of fuels for industrial engines is contained in References [39–49].

10.8 Alternative Fuels

A fuel that either augments or replaces the conventional fuel on a potentially permanent basis with no adverse effects on engine performance, maintenance, or operational life may be defined as an alternative fuel. Given the current and future strong emphasis on fuel efficiency and emissions, alternative fuels range from highest quality fuels, such as hydrogen and methane, to low-grade liquid and gaseous fuels that remain deficient in many aspects, even after extensive refining. Other fossil fuels that can be processed to produce gaseous and liquid hydrocarbons include tar sands, oil shale, biomass, and coal.

10.8.1 Pure Compounds

These types of fuels include liquid hydrogen, liquid methane, liquid propane, liquid ammonia, and alcohols/oxygenates. Tables 10.4 and 10.5 list the properties of these fuels of interest for aircraft application, alongside the properties of kerosine for comparison.

10.8.1.1 Hydrogen

From a combustion viewpoint, hydrogen is probably the nearest thing to an ideal fuel. It is characterized by high flame speeds, wide burning limits, easy ignition, and freedom from soot formation. Moreover, liquid hydrogen has a cooling capacity far superior to that of any other fuel. The main drawbacks of hydrogen lie in its very low density and low boiling point, which necessitate the use of large, heavily insulated storage tanks on the aircraft. It is also quite costly to produce. Currently, industrial quantities of hydrogen gas are most economically derived from fossil sources using steam reforming of natural gas ($CH_4 + H_2O = CO + 3H_2$), partial oxidation of methane

TABLE 10.4

Properties of some Alternative Liquid Fuels

Property	Kerosine (Avtur)	Liquid Hydrogen	Liquid Methane	Liquid Propane	Liquid Ammonia
Lower specific energy (MJ/kg)	42.8	116	49	46	17.2
Lower specific energy (Btu/lb)	18,400	49,900	21,060	19,800	7380
Cooling capacity (MJ/kg)	0.38–0.85	20.2	2.55	1.20	3.39
Cooling capacity (Btu/lb)	164–364	8694	1098	518	1458
Relative density (289–289 K)	0.80	0.071 (b.p.)	0.424 (b.p.)	0.585 (b.p.)	0.682 (b.p.)
Specific heat (kJ/(kg·K))	1.97	7.32	3.43		
Specific heat (Btu/(lb·°F))	0.47	1.75	0.82		
Boiling point (K)	423–573	21	111	231	240
Boiling point (°F)	301–571	–422	–260	–44	–28
Freezing point (K)	223	13	91	91	195
Freezing point (°F)	–58.6	–437	–296	–296	–109
Flame speed (m/s)	0.39	2.67	0.37	0.43	0.30
Flame speed (ft/s)	1.28	8.76	1.21	1.41	0.98

TABLE 10.5
Properties of Alcohol Fuels

Property Chemical Formula	Kerosine (Avtur) $C_{12}H_{26}$	Methanol CH_3OH	Ethanol C_2H_5OH
Relative density at 15.5°C	0.80	0.797	0.794
Lower specific energy (MJ/kg)	42.8	19.9	26.8
Lower specific energy (Btu/lb)	18,400	8555	11,522
Molecular mass	170.3	32.04	46.068
Boiling point, K (°F)	423–573 (301–571)	338 (148)	351 (172)
Freezing point, K (°F)	223 (−59)	178 (−139)	156 (−180)
Stoichiometric f/a ratio (mass)	0.0676	0.155	0.111
Surface tension (N/m)	0.02767	0.0226	0.0223
Viscosity at 293 K (m^2/s)	1.65×10^{-6}	0.75×10^{-6}	1.51×10^{-6}
Viscosity at 293 K (cSt)	1.65	0.75	1.51

($2CH_4 + O_2 = 2CO + 4H_2$), or coal gasification ($C + H_2O = CO + H_2$), because electrolysis is very inefficient and expensive. Brewer [50] has shown that hydrogen may be more economical than synthetic aviation-grade kerosine when applied to a 400-passenger subsonic aircraft over a flight range of around 10,000 km. Thus, in spite of formidable design and logistics problems, liquid hydrogen is worthy of consideration as an alternative fuel.

10.8.1.2 Methane

Liquid methane has a specific energy of about 49 MJ/kg, as compared with about 42.8 MJ/kg for kerosine. Its cooling capacity is not as great as that of hydrogen, but is still very large, owing to the very low temperature (112 K) of the liquefied gas. This low temperature offers considerable cooling potential for supersonic aircraft, as well as the possibility of designing highly cooled turbine blades to permit the use of higher turbine inlet temperatures [1]. Other advantages of methane include good thermal stability and clean combustion.

The main problems arising with methane stem from its low density and low boiling point. Methane requires about 70% more storage space than current kerosine fuels (although significantly less space than hydrogen); this could prove a major problem with aircraft configurations having thin or variable-geometry wings [51,52]. Other problems include the condensation of atmospheric moisture, leading to ice formation on aircraft wings and loss of fuel by boil-off during climb.

10.8.1.3 Propane

From inspection of Table 10.4, it is clear that the characteristics of propane are similar to those of methane, and wherever methane has potential application, propane usually also merits consideration. Compared with methane, it has a lower specific energy and a lower cooling capacity. However, its higher boiling point implies easier handling; in particular, it may be stored as a liquid at ambient temperatures by modest pressurization of the fuel tank.

10.8.1.4 Ammonia

Ammonia has a low heat of combustion—only 40% that of kerosine—and is of interest mainly on account of its great potential as a heat sink. Because of its low heat release, it is unlikely to be used on aircraft as main fuel, but it could find application as a secondary fuel in situations where its high cooling capacity can be exploited to advantage. Like propane, it is storable on the ground as a pressurized liquid at ambient temperatures.

10.8.1.5 Alcohols

These fuels comprise hydrocarbons that contain one or more oxygen atoms within the molecular structure. There are two types:

1. The most common are the alkanes (paraffins) incorporating a hydroxyl radical to become $C_n\ H_{2n+1}\ OH$. These types include methanol (CH_3OH) and ethanol (C_2H_5OH).
2. These are ethers, comprising hydrocarbons with an internal oxygen atom to give fuels such as:
   a. CH_3–O–CH_3 (methyl ether)
   b. CH_3–O–C_4H_9 (methyl tertiary butylether MTBE).

As a commercial product, bioethanol generally comprises 98.5% ethanol plus water together with methanol. Both the United States and Brazil are producing bioethanol at less than the cost of gasoline.

Alcohols are not practical as fuels for long-range aircraft, owing to their high oxygen content and correspondingly low calorific value. For example, half the molecular mass of methanol (CH_3OH) is comprised of oxygen. The lighter alcohols are considered safer to handle than gasoline because of their higher flash point and the fact that alcohol fires can be extinguished with water. They do, however, tend to be corrosive to some metals, and special precautions would be required to avoid this problem.

Methanol, when available, is an attractive fuel for industrial gas turbines. It is ash-free and has minimal soot-forming tendency. It burns with a low-luminosity blue flame and a minimum of exhaust smoke, and it has wide flammability limits. Moreover, the low flame temperature ensures relatively low emissions of nitric oxides. Ethanol is now being used as an automotive fuel in a 10% mixture with gasoline in Brazil and in parts of the United States.

Methanol may be produced from the destructive distillation of biomass, direct oxidation of natural gas, or by coal gasification. Ethanol is being produced from fermentation of cellulosic carbohydrate materials from wood, corn, and grain. The main advantages of alcohol (oxygenate) fuels are:

1. Carbon neutrality due to their production from vegetable matter
2. Lower carbon content and lower freeze point
3. Higher flash point, latent heat of vaporization and octane rating
4. Reduced combustion particulates, carbon monoxide (CO) and oxides of nitrogen (NO_x)
5. Lean mixture operation due to higher flame speed

The main disadvantages are:

1. Toxicity of methanol
2. Lower specific energy and energy density
3. Highly corrosive, poor lubricity in pumps and injectors
4. Lower vapor pressure impedes cold starting, part load, and transient operation
5. Methanol can produce spark knock, has lower cetane rating
6. Generates aldehyde emissions of ozone pollution

10.8.2 Supplemental Fuels

These include alternative nonpetroleum fossil fuel sources, such as tars and shale oil, which occur naturally in the earth, but are not readily accessible because of their cohesion with rock or sand. Blazowski and Maggitti [9] report that a kerosine-type fuel derived from Canadian Athabasca tar sand was virtually indistinguishable in its combustion characteristics from a high-quality, petroleum-derived JP-5.

When oil shale is subjected to heating, its resinous content decomposes into an oily liquid from which a crude oil (syncrude) may be derived. Subsequent refining to reduce the content of nitrogen, oxygen, and sulfur can yield a product fairly close to that of Jet A, but with high aromatic content. Combustor tests on such fuels have resulted in higher than normal emissions of NO_x due to the higher content of fuel-bound nitrogen [42,47].

10.8.3 Slurry Fuels

The slurry fuels of interest for aircraft applications are suspensions of powdered metals, such as beryllium, boron, aluminum, and magnesium, in gasoline or kerosine. They offer the possibility of greater flight range or higher thrust than can be obtained with conventional hydrocarbons.

Pirans et al. [53] investigated the use of slurries consisting of 50% or more boron or magnesium in liquid hydrocarbon fuels for afterburners and ramjet engines. Tests carried out on various combustor designs showed that magnesium slurry burned readily, even under conditions in which the liquid hydrocarbon fuel would not burn. Boron slurries proved more difficult to burn than conventional jet fuel and gave rise to objectionable deposits in the combustor. Other slurry fuel problems include preparation, storage, and abrasion of pumps and fuel systems.

10.9 Synthetic Fuels

The term "synthetic" is used to describe fuels derived from nonpetroleum feedstock, such as coal and biomass. Two processes for producing liquid hydrocarbons from coal are: direct coal liquefaction and coal gasification. In the coal liquefaction process, a main objective is to increase hydrogen-carbon ratio. A small increase in this ratio produces a fairly heavy liquid similar to petroleum-based residual fuel oil. The more expensive degree of hydrogenation produces lighter liquid fuel comparable to gasoline.

The direct liquefaction of coal (or hydroliquefaction) is a two-stage process called the Bergius process [54]. The process involves the catalytic conversion of coal (slurried with heavy oil) in the presence of hydrogen and an iron oxide catalyst, at 450–500°C and 200–690 atm. The products are usually separated into light oils, middle distillates (BP = 300–750°F) and residuum (BP = 600–1000°F). In the second stage of the process, the middle distillates are catalytically cracked to produce heating oil, jet fuel, and kerosine. The residence time for this catalytic conversion is typically 80–85 minutes and the hydrogen consumption is also significant at about 11% by mass of dry ash-face coal.

The past evidence on fuels produced by coal liquefaction is somewhat contradictory. A JP5-type fuel refined from crude coal liquids failed to meet the specification requirements in several respects, in spite of intensive hydrogen treatment [9,55]. The thermal stability was poor, the heat of combustion marginal, the density too high, and the smoke point too low. Cohn et al. [56] and Singh et al. [57] carried out a series of tests on 12 different types of coal-derived liquid fuels and three oil-shale fuels, using both subscale and full-scale industrial-type combustors. Emissions measurements were made of nitric oxide, smoke, CO, and unburned hydrocarbons. Liner wall temperatures were also measured. The conclusion drawn from this test program was that liquid fuels derived from the coal and oil shale show no significant differences in combustion characteristic as compared with petroleum-derived fuels.

10.9.1 Fuels Produced by Fischer–Tropsch Synthesis of Coal/Biomass

Coal gasification is a two-stage process that involves the production of syngas ($CO + H_2$) via the gasification of coal and the conversion of that syngas to light hydrocarbons via Fischer–Tropsch (FT) synthesis. The FT process is currently being operated commercially by Sasol Corporation of South Africa, producing 40,000 barrels/day of liquid fuel. Inexpensive iron catalysts are used for the FT process. Other FT catalysts are cobalt, nickel, ruthenium, and molybdenum. Modern gasifiers produce syngas with low (0.6–0.7) H_2/CO ratio. Iron is known as a good water gas shift catalyst because syngas with $(H_2/CO) > 2$ is required to synthesize paraffins.

In 1985, Shell announced its SMDS (Shell Middle-Distillate Synthesis) process for the production of kerosine and gas oil from natural gas [54]. This two-stage process involves the slurry bubble column FT reactor, using a precipitated iron catalyst to produce long-chair hydrocarbon waxes and subsequent hydro conversion and fractionation into naphtha, kerosine, and gas oil.

The modern coal gasification process, using the proven FT synthesis, produces clean, high quality liquid fuels such as diesel, IPK jet fuel, and fuel oil. The FT process can also yield quantities of naphtha, ammonia, and methanol. IGCC technology can be incorporated in FT plant design to generate significant quantities of electricity for plant use or fed into the power grid. The FT technology generates significant quantities of CO_2 ($1.8 \times$ petroleum refining), albeit in a concentrated form that can be captured, compressed, and sequestered. By using biomass feedstock (up to 20% by mass of wood chips, switch grass, or corn stover) in the coal gasifier, CO_2 emissions can be decreased by 20% over equivalent fuel derived from petroleum refining.

Currently, significant efforts are underway to evaluate the greenhouse gas (GHG) footprint of FT and conventional fuels. The main benefits of FT transportation fuels are: large, secure domestic supply and clean burning fuel (with very low nitrogen, aromatics, and sulfur).

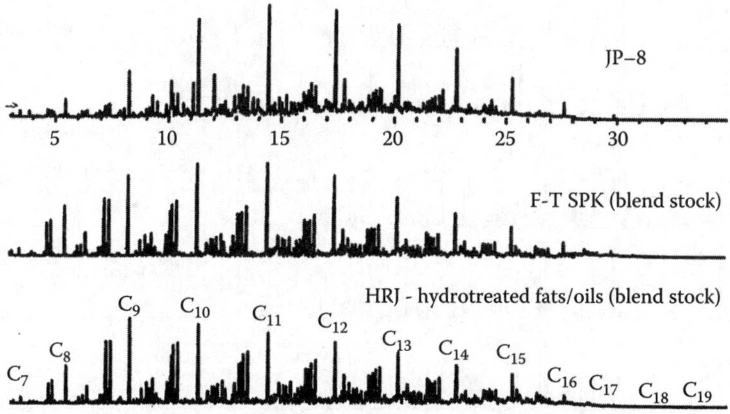

FIGURE 10.19
Gas chromatographs of JP-8, generic FT synthetic paraffinic kerosine 50/50 blend, and hydrotreated renewable jet (HRJ) 50/50 biofuel.

Figure 10.19 shows the gas chromatographs of conventional JP-8, generic FT 50/50 blend, and hydrotreated renewable jet (HRJ) 50/50 biofuels.

10.9.2 Biofuels

Gas turbines are fuel flexible energy converters. Biofuels are potential newcomers. There are 10 qualified biofuels in the European Union [1]. Of these, the five most promising are: biodiesel, bioethanol, biomethanol, vegetable oils (VO), and biodimethylether (bio-DME).

The heating value per unit of biofuel is typically 5–8 kBtu/lb, which is 30–50% lower than typical coal. However, fuel from biomass has very little or no sulfur or ash content. Also, it has zero net CO_2 footprint, because CO_2 is consumed from renewable generation of biomass.

Biofuels have a complex molecular structure, often in the form of carbohydrates, i.e., $C_m(H_2O)_n$. Figure 10.20 shows the gas chromatograph of the composition of various biofuels. The combustible oils from vegetable matter comprise natural esters of the trihydric alcohol, glycerol ($HOCH_2CHOHCH_2OH$) with the long straight-chain fatty acids (RCOOH), where the hydrocarbon radical R varies from about $C_{15}H_{31}$ to $C_{17}H_{35}$.

Vegetable oils and biodiesels have organic salts as contaminants (N_a + K 10–50 ppm, Ca 5–40 ppm, and viscosity of 20–300 CST). The main attractions of alternative biofuels over conventional petroleum-based fuels are:

1. Carbon neutrality and lower carbon content
2. Higher flash point giving greater fire safety

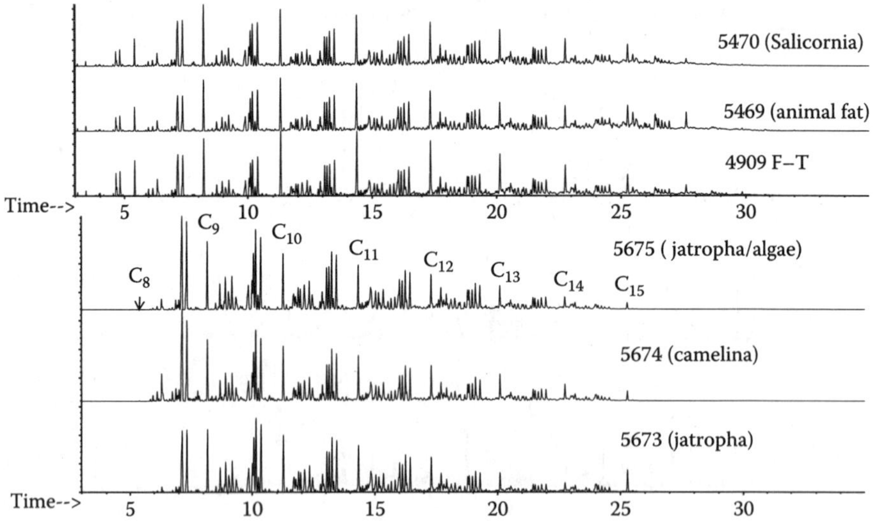

FIGURE 10.20
Gas chromatographs showing the FT fuel and composition of various biofuels.

3. Lower sulfur
4. Higher cetane number of rapeseed methyl ester (RME) and lower emissions of hydrocarbons and particulates

The main drawbacks are:

1. Higher viscosity and cold filter plugging point, hence need for gum removal
2. Lower specific energy and energy density, hence higher fuel consumption
3. Higher SIT, hence lower cetane number
4. Greater corrosivity with tendencies to carbon deposits and injector coking

10.9.3 Alternative Fuel Properties

In 2002, DEF STAN 91-91/Issue 4 approved the use of synthetic IPK in concentrations up to 50%, providing the fuel has adequate lubricity and at least 8% aromatics in the final blend with petroleum-derived kerosine. Moses and Stavinoha [58] performed the qualification of Sasol semisynthetic Jet A-1 as commercial jet fuel. A 50/50 mixture of petroleum-derived Jet A-1 and IPK derived from coal demonstrated that the mixture properties fell well within the Jet A-1 fuel specification range and should have no impact on engine operations. In 2009, Moses and Roets [59] reported on the Sasol fully synthetic fuel as fit-for-purpose jet fuel for civilian application. These authors provided results of both the properties and characteristics of the fuel as well as performance characteristics in engine and combustor tests. In 2006, the U.S. Air Force initiated "OSD Assured Fuels Initiative—Military Fuels Produced From Coal." Harrison and Zabarnick [60] measured the properties, characteristics, and behavior of sample synthetic FT-based fuels. Also, extensive studies of blending of conventional JP-8 (Jet A) and FT fuels were performed.

Table 10.6 lists the fuel properties of JP-8, 100% FT, and 50/50 FT fuel blended with JP-8. The 50/50 blending ratio was selected to meet the minimum density specification limit of 0.775 kg/L for JP-8. At this 50/50 ratio, the aromatics content in the blend is around 7% by volume. In general, all relevant specification properties showed linear dependence with blending. As seen in Table 10.6, the FT fuel used for blending contained no aromatics and sulfur and had an (iso/normal) paraffinic kerosine content ratio of 4.8 and a similar molecular weight range as JP-8. The elastomer property tests for fuel compatibility showed that fluorosilicone and fluorocarbon o-rings were relatively insensitive to various fuels. The nitrile tests, however, showed the expected increase in volume swell that was directly related to the aromatic content, and the swell produced was both acceptable and typical. Other bulk

TABLE 10.6

Properties of Conventional JP-8, Synthetic FT, and JP-8/FT (50/50) Blend Fuels

Property	JP-8 Min	JP-8 Max	FT[a] (100%)	FT[a]/JP-8 (50/50)
Aromatics (vol %)	--	25.0	0.0	6.8
Total sulfur (wt%)	--	0.30	0.0	0.046
Flash point (°C)	38.0	--	45.0	55.0
Freeze point (°C)	---	–47.0	–51.0	–50.0
Viscosity at –20°C (cSt)	--	8.0	4.9	4.8
Heat of comb (Btu/lb)	18,400	--	18,978	18,686
Hydrogen content (wt%)	13.4	--	15.4	14.6
Density (kg/L at 15C)	0.775	0.840	0.756	0.781

JP-8 specification update

Note: Definition to include coal, biomass, and natural gas via Fischer–Tropsch; -- blend instructions for petroleum and nonpetroleum; -- reference MIL-handbook for "fit-for-purpose" properties. Effect of these property nonconformities under evaluation.

[a] Syntroleum synthetic JP-8 (S-8).

physical properties, such as viscosity, bulk modulus, specific heat, and thermal conductivity, were typical of conventional jet fuels.

Harrison and Zabarnick [60] performed the thermal stability analysis and Figure 10.21 shows the results using a quartz crystal microbalance. The dynamic viscosity of conventional and synthetic fuels at low temperature is shown in Figure 10.22. Finally, the volume elastomer swell for nitrile rubber o-rings (Parker N-602-214) was also measured and plotted in Figure 10.23.

Table 10.7 shows the properties of bio-based synthetic paraffinic kerosine. Essentially, the properties of HRJ biofuels are indistinguishable from FT synthetic paraffinic kerosine. Therefore, the blending issues for HRJ (50/50) biofuels are also similar, namely, density, aromatic content, and GHG footprint. At present, not much information is available on the fit-for-purpose properties (lubricity, dielectric strength, cetane number), combustion operability and emissions, material compatibility, and blending issues.

Goodger [1] has plotted composite plots of the fuel properties of conventional fuels, biofuels, and alcohols. Figure 10.24 shows fuel density and Figure 10.25 shows kinematic viscosity as a function of fuel temperature. Finally, Figure 10.26 shows the variation of net specific energy with fuel density.

10.9.4 Combustion and Emissions Performance

10.9.4.1 Fischer–Tropsch Fuels

Moses and Roets [59] report engine and combustor performance tests as follows:

- Engine endurance
- Low-temperature atomization

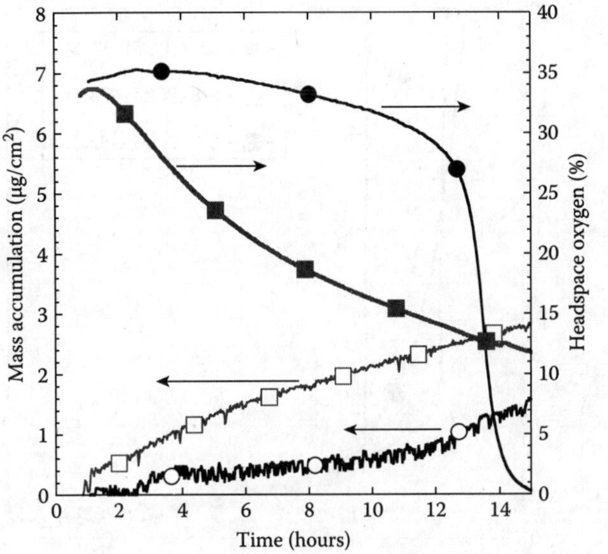

FIGURE 10.21
Quartz crystal microbalance (QCM) deposition at 140°C (open markers) and oxygen profiles (filled markers) for a JP-8 fuel (□) and synthetic syntroleum S-8 fuel (•). (From Harrison, W.E. and Zabarnick, S., Presented at DoE Clean Coal Conference, Clearwater, FL, 2007.)

- Cold-day ignition and altitude relight
- Lean blowout
- Exhaust emissions

The engine endurance, atomization and ignition tests showed no significant differences between the performance limits of fully synthetic jet fuels (FSJF) and Jet A-1 fuels.

As shown in Figure 10.27, the Sasol FSJF had no adverse effect on lean blowout limit. At higher altitudes, some limited fuel effects were noted, with FSJF between the Jet A and JP-5 results. Since the lean blowout is considered to be evaporation-limited, this result may be due to differences in boiling point distribution of the two fuels and not due to fuel chemistry.

When an engine is certified, it must meet emissions standards recommended by ICAO/CAEP (www.icao.org). Figures 10.28 through 10.30 show the results of NO_x, CO, and smoke number (SN), respectively. In the LTO cycle, the Sasol synthetic fuel produced 4% less NO_x and 19% less CO. These emissions levels for both fuels were lower than the limits for CAEP/6. At all other test conditions, the NO_x emissions were unchanged and the CO emissions were lower than that for the conventional fuel, presumably due to the lower viscosity and finer spray produced by the Sasol fuel. The differences in CO_2 and combustion efficiency were found to be negligible.

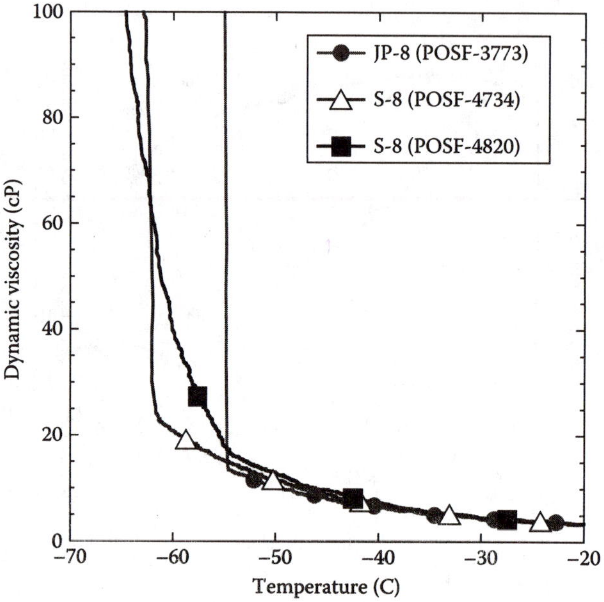

FIGURE 10.22
Dynamic viscosities of two synthetic (S-8) fuels compared with JP-8 fuel. (From Harrison, W.E. and Zabarnick, S., Presented at DoE Clean Coal Conference, Clearwater, FL, 2007.)

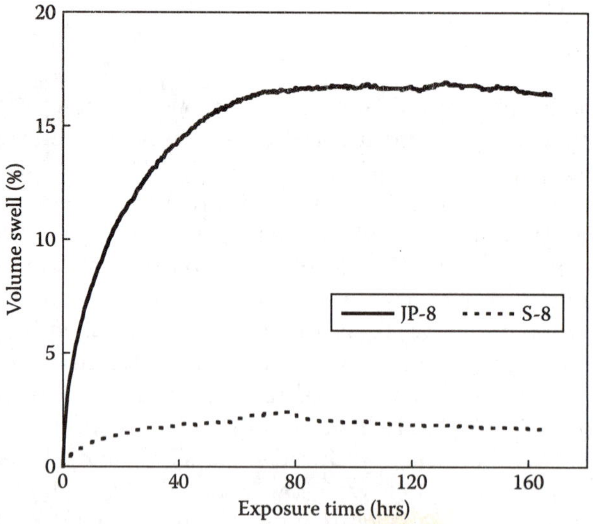

FIGURE 10.23
Volume swell vs. time for nitrile rubber o-ring samples. (From Harrison, W.E. and Zabarnick, S., Presented at DoE Clean Coal Conference, Clearwater, FL, 2007.)

TABLE 10.7
Properties of Bio-Based Synthetic Paraffinic Kerosine Fuels

Property		SPK	ASTM Test Method	Composition: Jatropha	Coconut	Soybean/Canola	Soybean	Palm
Paraffin (vol%)	Min	99.8	D2425					
Cycloparaffin (vol%)	Max	5	D2425					
Paraffin (vol%)[a]			Difference from D 1319	99.3	99.5	99.5	100	98.4
1. Aromatics (vol%)	Max	0.05	D 1319	0	0	0	0	0.8
2. Aromatics (vol%)	Max	0.053	D 6379	0	0	0	0	0.8
Sulfur (total max %)	Max	0.015	D 1266, D 2622, D 4294, or D5453	0.00009	0.0003	0.001	0.0008	< 0.0001
1. Physical distillation			D86					
Distillation temp (°C)								
10% recovered (temp (T10))	Max	205		172	188	189	186	197
50% recovered (temp (T50))		Report		192	200	214	226	236
90% recovered (temp (T90))		Report		223	231	248	280	254
Final boiling point (temp)	Max	300		243	263	261	286	259
T90–T10 (°C)	Min	25		51	43	59	94	57
Distillation residue (%)	Max	1.5		1.2	1.3	1.2	1.4	1.2
Distillation loss (%)	Max	1.5		0.4	0.5	0.8	1.0	0.7
2. Simulated distillation			D 2887					
Distillation temp (°C)								
10% recovered (temp)	Max	185		151.6	162	168	166.8	169
50% recovered (temp)		Report		195	190.8	218.6	226.6	249

(*Continued*)

TABLE 10.7

Properties of Bio-Based Synthetic Paraffinic Kerosine Fuels (Continued)

				Composition				
Property		SPK	ASTM Test Method	Jatropha	Coconut	Soybean/ Canola	Soybean	Palm
90% recovered (temp)		Report		237.6	238	267.2	297	269
Final boiling point (temp)	Max	340		273.8	299	284.4	310.6	280
Flash point (°C)	Min	38	D 56 or D 3828	50	64	62	60	64
Density at 15°C (kg/m^3)		751–840	D 1298 or D 4052	751	755	763	766	769
Fluidity								
Freezing point (°C)	Max	–47 Jet A-1	D 5872, D 7153, D 7154, or D 2386	–63	–56	–52	–36	–44
Viscosity –20°C (mm^2/s)	Max	8.0	D 445					
Combustion								
Net heat of combustion (MJ/kg)	Min	42.8	D 4529, D 3338, or D 4809	44.4	44.2	43.5	43.8	44.2
Metal content			D 7111					
Copper (ppb)	Max	100		< 0.01 ppm	< 0.01 ppm	< 0.01 ppm	< 0.03 ppm	< 0.02 ppm
Iron (ppb)	Max	100		< 0.01 ppm	< 0.01 ppm	0.04 ppm	< 0.03 ppm	0.02 ppm
Zinc (ppb)	Max	100		< 0.01 ppm	< 0.01 ppm	< 0.01 ppm	< 0.03 ppm	< 0.02 ppm
Vanadium (ppb)	Max	100		< 0.01 ppm	< 0.01 ppm	< 0.01 ppm	< 0.03 ppm	< 0.02 ppm
Thermal stability								
JFTOT (2.5 h at control temp of 280°C min)								
Filter pressure drop (mmHg)	Max	25	D 3241	< 0.1	25	0		0.0
Tube deposits less than		3		< 1	< 1	1		< 1

Source: From J. Holmgren, Private Communication, reproduced with permission, November 2008.

[a] Balance of composition is olefins.

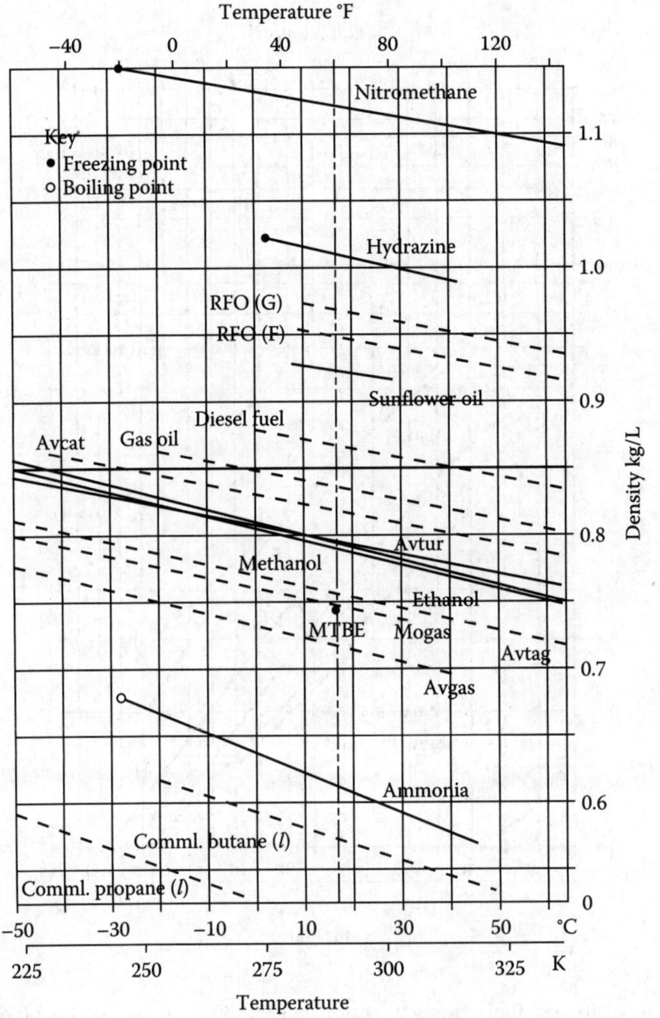

FIGURE 10.24
Fuel density vs. fuel temperature for conventional, bio, and alcohol fuels. (From Goodger, E.M., *Transport Fuels Technology*, Landfall Press, Norwich, UK, 2000. With permission.)

Figure 10.30 shows that the Sasol synthetic fuel produced lower smoke than the Jet A fuel as expected because of its lower aromatic content. Harrison and Zabarnick [60] have reported emissions from FT fuels using the T63-A-700 turboshaft engine. Figure 10.31 shows the percent change in particulate number density (PND) with percent volume of synthetic FT jet fuel in JP-8 for cruise and idle conditions. These results demonstrate reduced (smoke) particulate emissions on addition of the synthetic fuel. Clearly, this reduction is attributed largely to the reduced aromatic concentration in the resultant fuel blend. Reduction of the aromatics decreases the concentration of "seed"

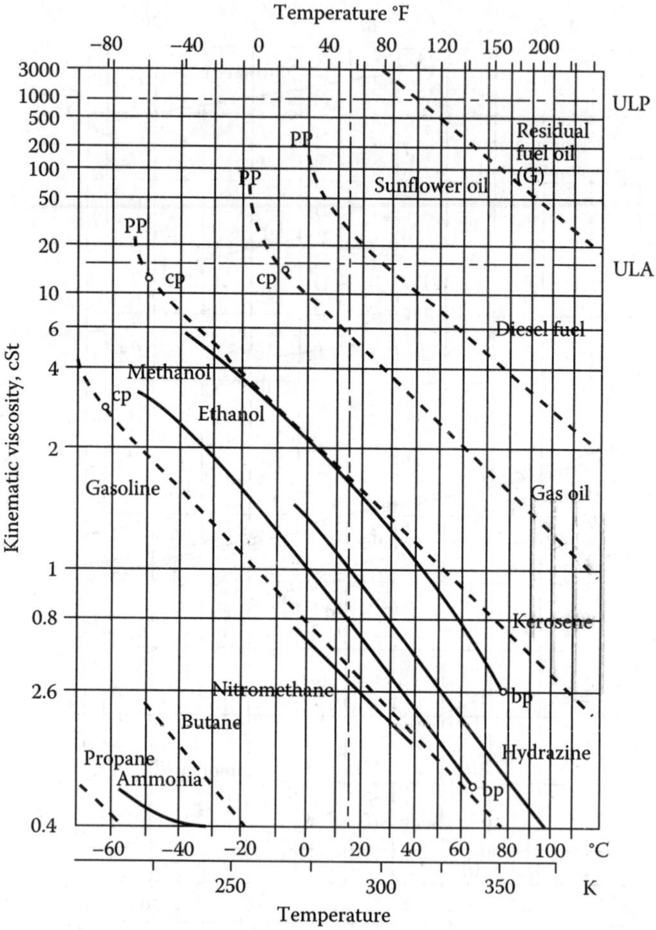

FIGURE 10.25
Kinematic viscosity vs. fuel temperature for conventional, bio, and alcohol fuels. (From Goodger, E.M., *Transport Fuels Technology*, Landfall Press, Norwich, UK, 2000. With permission.)

molecules that contribute to the formation of soot nuclei, thereby decreasing the production of fine soot particulate. Thus, blending with FT fuel significantly reduces the mean diameter of the particulate exhaust.

Lieuwen et al. [61] reviewed the impact of fuel composition on the operability of lean premixed gas turbine combustors. These workers investigated fuel blends of CH_4/H_2 (LNG), CH_4/CO_2 (landfill gas), and $CO/H_2/N_2$ (syngas). They concluded that the behavior of fuel mixtures can be markedly different than that of individual constituents. Moreover, properties such as turbulent burning velocity and ignition delay time may behave in a highly nonlinear fashion as fuel mixture changes.

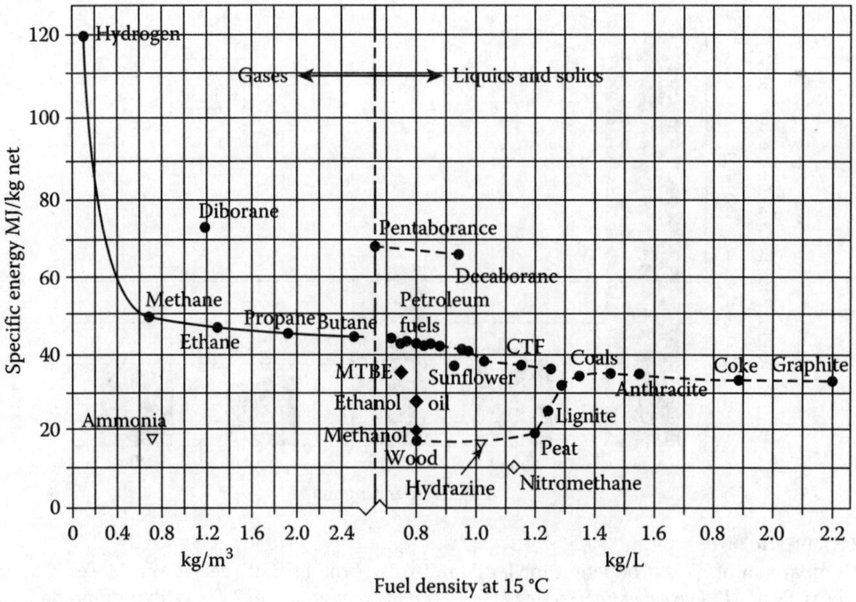

FIGURE 10.26
Specific energy vs. fuel density for conventional, bio, and alcohol fuels. (From Goodger, E.M., *Transport Fuels Technology*, Landfall Press, Norwich, UK, 2000. With permission.)

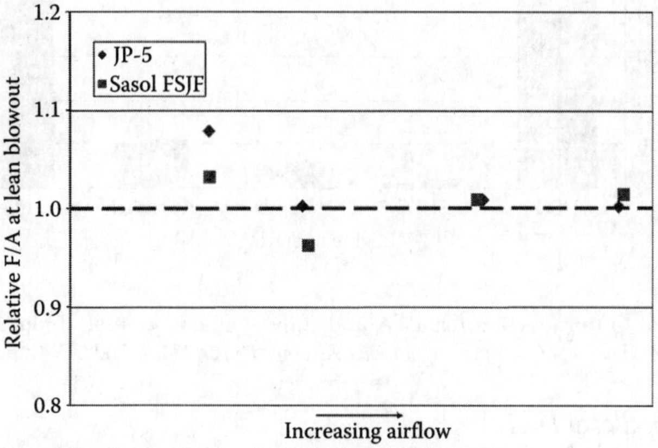

FIGURE 10.27
Lean blowout limits for JP-8 and Sasol fully synthetic jet fuel (FSJF). (From Moses, C. and Roets, P., *ASME Journal of Engineering for Gas Turbines Power*, 131, 4, 2009. With permission.)

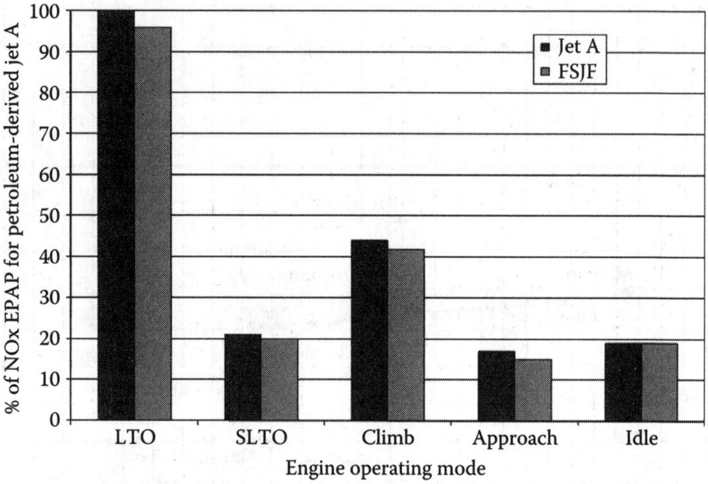

FIGURE 10.28
Comparison of NO_x production for Jet A and fully synthetic jet fuel. (From Moses, C. and Roets, P., *ASME Journal of Engineering for Gas Turbines Power*, 131, 4, 2009. With permission.)

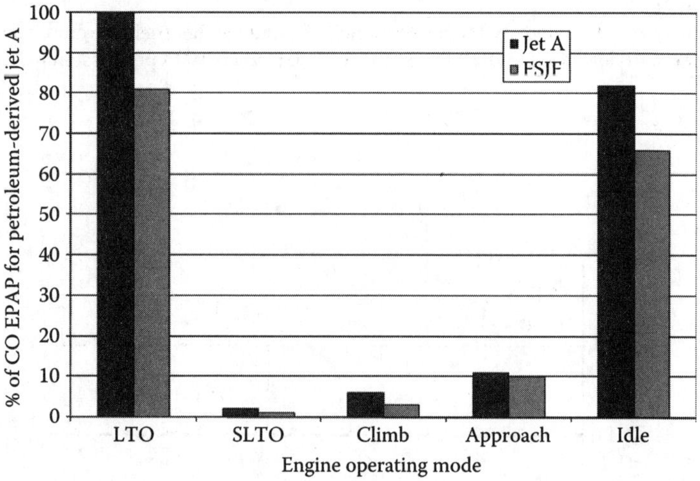

FIGURE 10.29
Comparison of CO production for Jet A and fully synthetic jet fuel. (From Moses, C. and Roets, P., *ASME Journal of Engineering for Gas Turbines Power*, 131, 4, 2009. With permission.)

10.9.4.2 Biodiesel Fuels

Recently, researchers [62,63] have analyzed the influence of biodiesel fuel on combustor blowout, flashback, auto-ignition, and combustion stability.

Panchasara et al. [62] studied the combustion performance of biodiesel and diesel–VO blends in a simulated gas turbine combustor. These experiments were performed at atmospheric pressure with air-assist injector and swirling

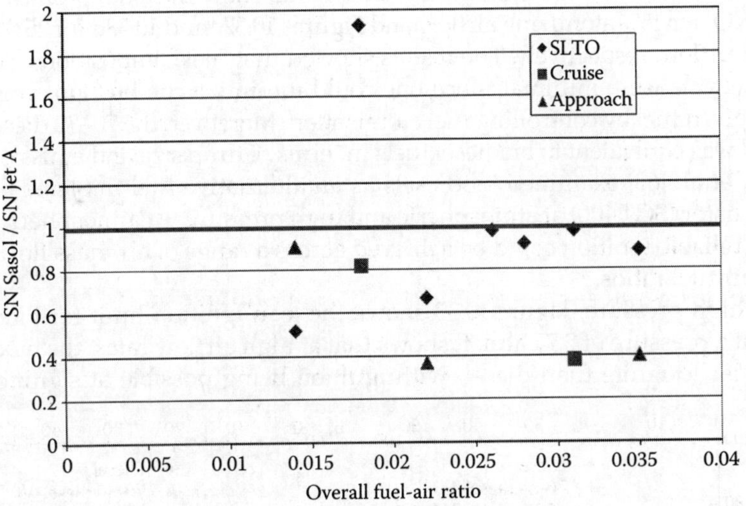

FIGURE 10.30
Comparison of smoke number for Jet A and fully synthetic jet fuel. (From Moses, C. and Roets, P., *ASME Journal of Engineering for Gas Turbines Power*, 131, 4, 2009. With permission.)

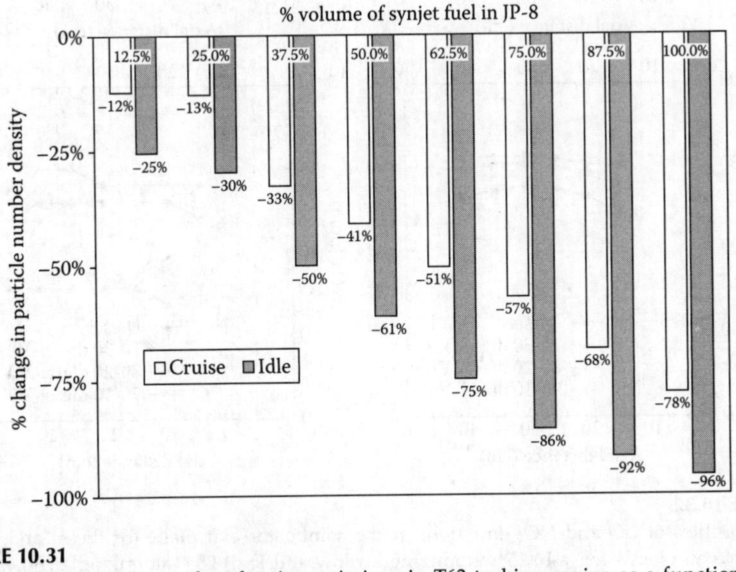

FIGURE 10.31
Reductions in particle number density emissions in T63 turbine engine as a function of the concentration of the FT fuel in JP-8 blends. (From Harrison, W.E. and Zabarnick, S., Presented at DoE Clean Coal Conference, Clearwater, FL, 2007.)

primary air around it. Figures 10.32a and 10.32b show the axial profiles of CO and NO_x for 15% atomizing airflow and Figures 10.32c and 10.32d for 25% atomizing airflow, respectively. The results showed that fuel composition or chemistry effects were minimal; since one could minimize combustion emissions for a given fuel by controlling fuel atomization. Moreover, the 70–30 diesel–VO blend was equivalent to biodiesel fuels in terms of drop size and emissions.

Liu et al. [63] examined biodiesels as an alternative fuel in Siemens DLE combustor (SGT-100) at atmospheric and high pressure. At atmospheric pressure, reliable ignition could be achieved across a range of air mass flow rates and air fuel ratios.

At high pressure, Figure 10.33 shows the lean ignition limit for biodiesel fuel at a pressure of 2.72 atm. It shows that at high airflow rates, the biodiesel is easier to ignite than diesel, with ignition being possible at significantly

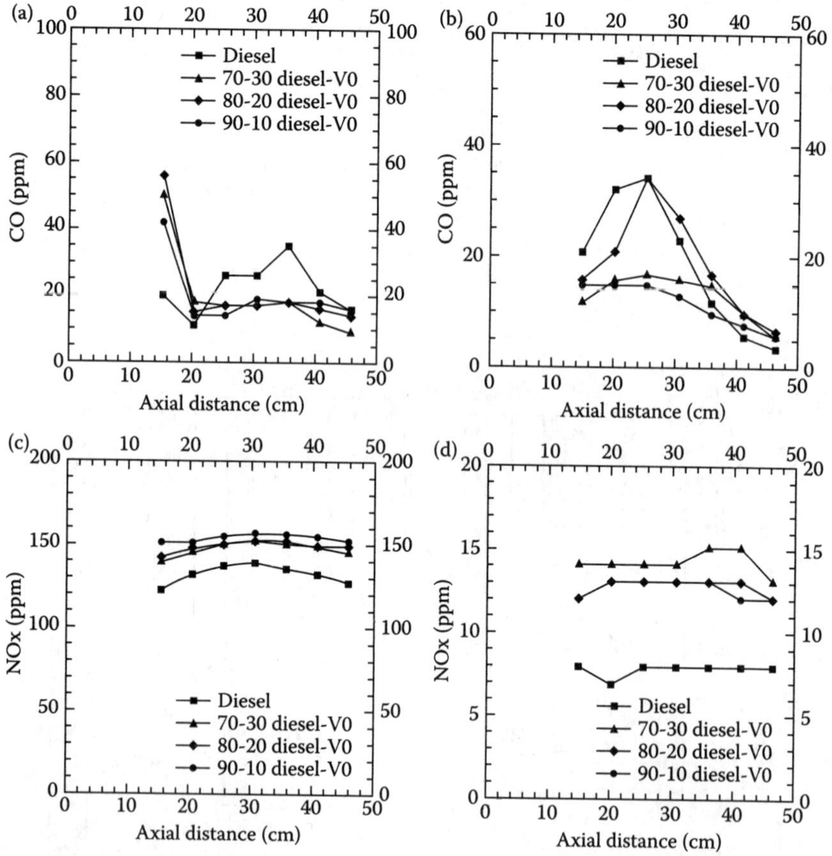

FIGURE 10.32
Axial profiles of CO and NO_x emissions at the combustor exit plane for diesel and diesel-vegetable oil blends for (a-b) 15% atomizing airflow and (c-d) 25% atomizing airflow. (From Panchasara, H., Simmons, B., Agrawal, A., Spear, S., and Doly, D., *ASME Journal of Engineering for Gas Turbines Power*, 131, 031503, 2009. With permission.)

higher air fuel ratios. At the lower flow rates, the opposite is true, i.e., diesel ignites at higher values of AFR. Also, it was found that reliable ignition can be achieved for the SGT-100 combustor operating with biodiesel.

Figure 10.34 shows the lean extinction limit burning biodiesel in air at 2.72 atm. This result shows that lean extinction occurred for both air temperatures at the same conditions, which were similar to those observed for the lean ignition.

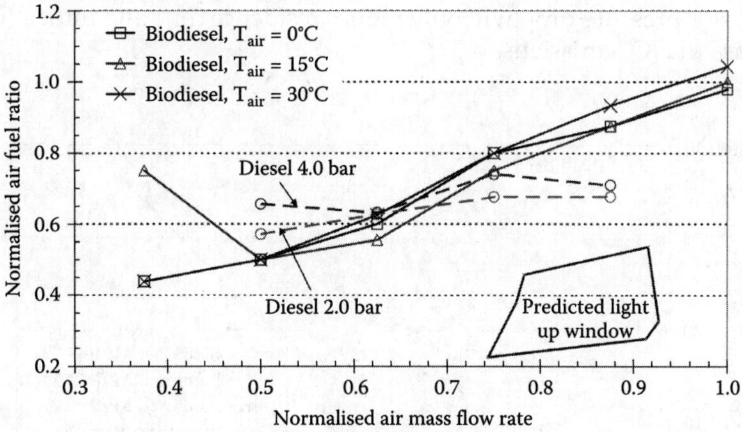

FIGURE 10.33
Lean ignition limit for dry low-emissions (SGT-100) combustor burning diesel and biodiesel fuels at air-assist injector pressure of 2.72 atm. (From Liu, K., Wood, J.P., Buchanan, E.R., Martin, P., and Sanderson, V., *ASME Journal of Engineering for Gas Turbines Power*, 2010. With permission.)

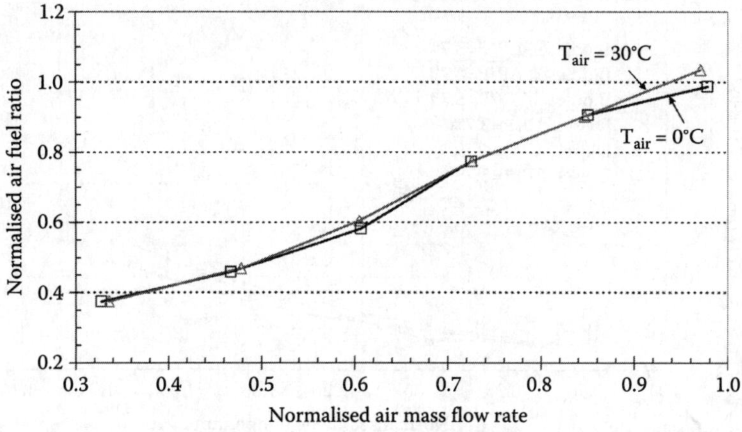

FIGURE 10.34
Effect of ambient air temperature on dry low emissions (SGT-100) combustor lean extinction limit burning biodiesel fuel at air-assist injector pressure of 2.72 atm. (From Liu, K., Wood, J.P., Buchanan, E.R., Martin, P., and Sanderson, V., *ASME Journal of Engineering for Gas Turbines Power*, 2010. With permission.)

Figures 10.35 (a–d) show the effect of flame temperature and combustor pressure drop on NO_x emissions at 14 and 9 bars. In these figures, flame temperature is normalized such that the value of 1:0 represents full load condition. At high pressure, diesel fuel has higher NO_x emissions than biodiesel fuel. Also, higher combustor pressure drop causes a strong reduction in NO_x emissions due to: (1) a decrease in residence time, (2) better mixing due to high turbulence, and (3) improved mixing due to downstream displacement of the flame. However, once perfect mixing is achieved, a further increase in combustor pressure drop will only reduce residence time and limit a further decrease in NO_x emissions.

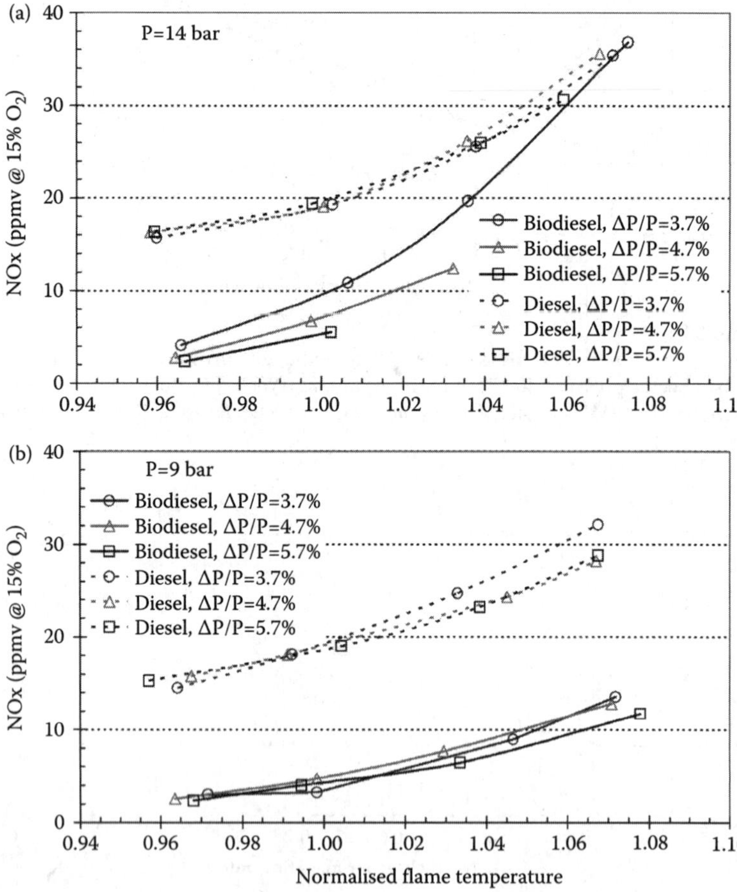

FIGURE 10.35
Effect of flame temperature and combustor pressure drop on NO_x emissions from diesel and biodiesel fuels for two different combustor pressures. (a–b) measured NOx, (c–d) measured NOx corrected for fuel-bound nitrogen (FBN). (From Liu, K., Wood, J.P., Buchanan, E.R., Martin, P., and Sanderson, V., *ASME Journal of Engineering for Gas Turbines Power*, 2010. With permission.)

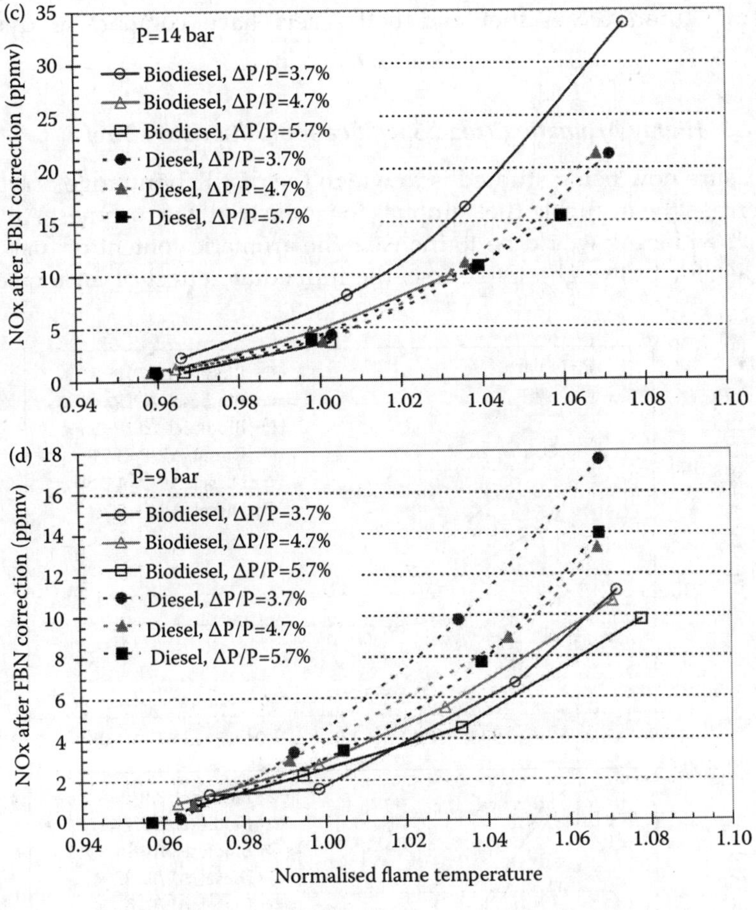

FIGURE 10.35 (Continued)

Campbell et al. [64] have demonstrated that biodiesel fuels have comparable NO_x emissions to diesel fuels. However, this could be because their biodiesel and diesel fuels had comparable nitrogen content. Figures 10.35c and d show NO_x emissions after correcting for fuel-bound nitrogen. It is observed that biodiesel fuels produce slightly higher NO_x emissions than diesel fuels.

Figure 10.36a through d show comparisons of CO and unburned hydrocarbon emissions, respectively, and reveal that biodiesels have higher CO emissions, especially low load conditions due to low flame temperature. Also, it is observed that biodiesels have lower unburned hydrocarbon (UHC) emissions than diesel fuel.

The frequency and amplitude of combustion dynamic pressure were measured by a pressure transducer mounted at the pilot burner face. Figure 10.37

shows these results. It is seen that biodiesels have generally lower dynamic pressures than diesel fuel and both fuels have comparable dynamic frequencies.

10.9.4.3 Highly Aromatic (Broad Specification) Alternative Fuels

A measure now being studied is to widen the distillation range of jet fuels to increase the available fuel supplies for industrial gas turbines. An effect of such widening would be to increase the aromatic content of fuels to as high as 35% by weight. Of course, the immediate effect of an increase in

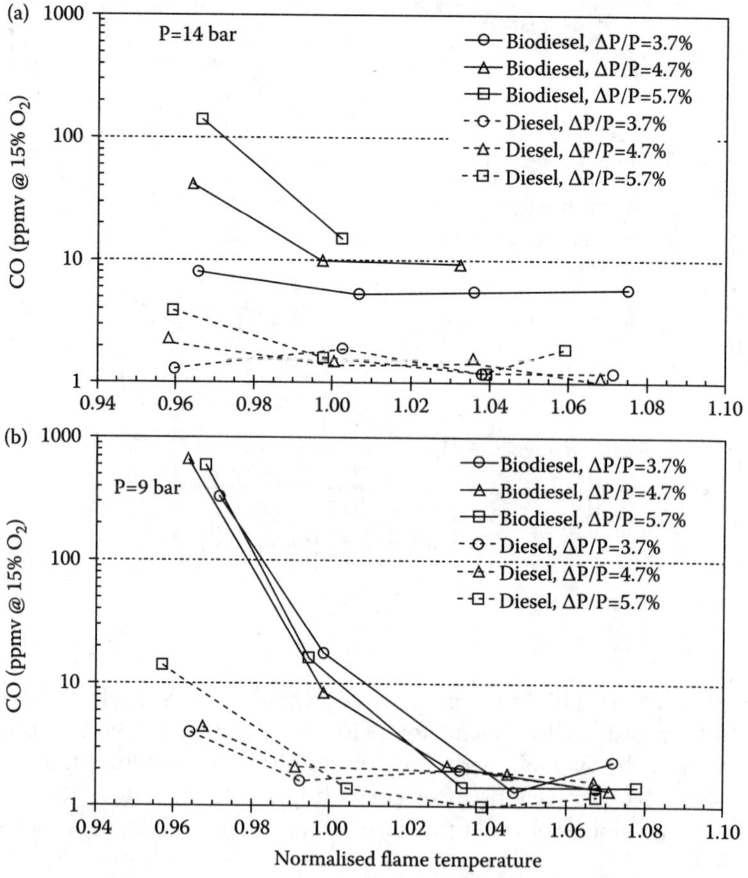

FIGURE 10.36
Effect of flame temperature and combustor pressure drop on emissions from diesel and biodiesel fuels for two different combustor pressures: (a–b) CO emission, (c–d) unburned hydrocarbon (UHC) emissions. (From Liu, K., Wood, J.P., Buchanan, E.R., Martin, P., and Sanderson, V., *ASME Journal of Engineering for Gas Turbines Power*, 2010. With permission.)

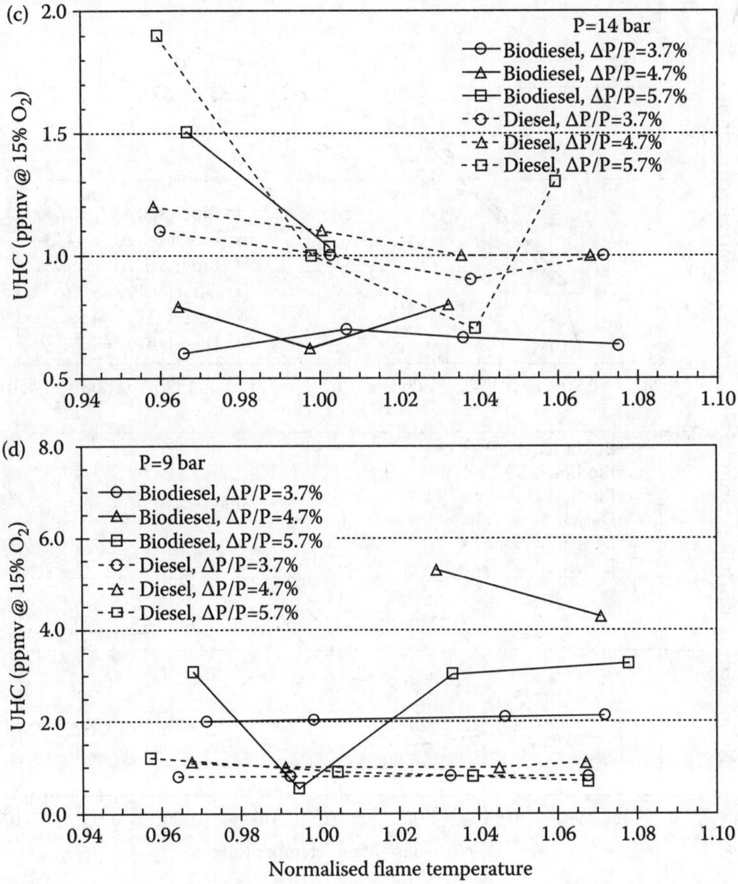

FIGURE 10.36 (Continued)

aromatic content on burning quality is to enhance the propensity toward soot formation in the flame. This is demonstrated in Figure 10.38, which shows, for a variety of fuels, how the smoke point varies with aromatic content. In practice, an increase in aromatic content is synonymous with a decrease in hydrogen content. A typical correlation between hydrogen content and liner wall temperature is shown in Figure 10.39. Figure 10.40 shows a similar correlation for hydrogen content and SN. Both sets of data were obtained with a single-can JT8D combustor at simulated takeoff and cruise conditions [65,66]. Based on measured values of the liner wall temperature, the relative liner life was estimated to fall by 53% for the F101 combustor and 65% for the J79 combustor when the fuel hydrogen content was reduced from 14.5 to 12%.

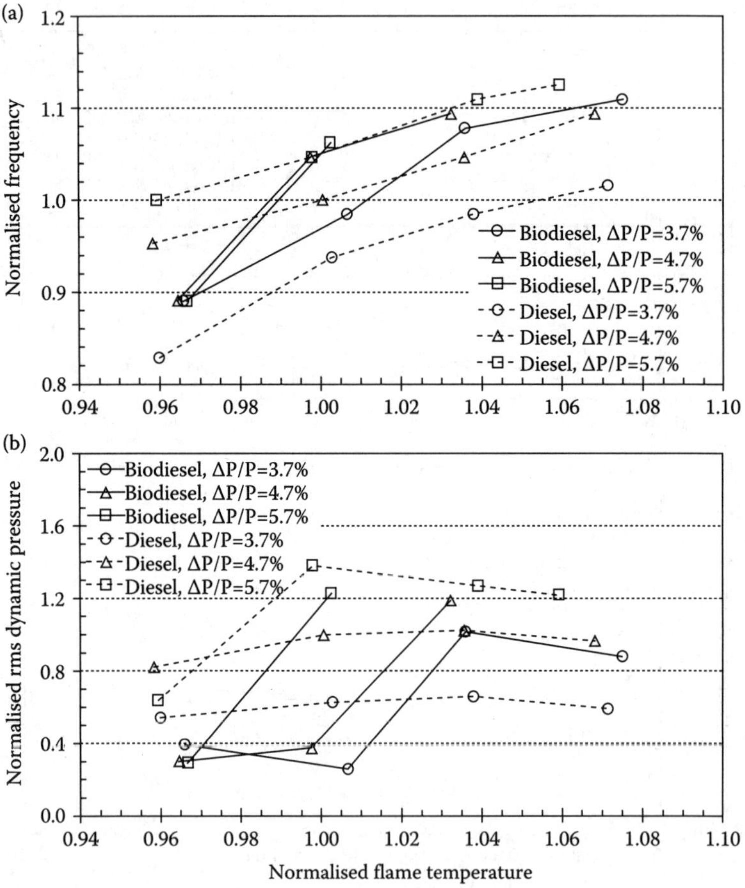

FIGURE 10.37
Comparison of combustion dynamics produced due to the burning of diesel and biodiesel fuel at an inlet air pressure of 14 atm. (a) Frequency, (b) RMS dynamic pressure. (From Liu, K., Wood, J.P., Buchanan, E.R., Martin, P., and Sanderson, V., *ASME Journal of Engineering for Gas Turbines Power*, 2010. With permission.)

Moliere and Geiger [67] performed a study of the combustion of highly aromatic fuels in industrial gas turbines. They examined a 77% aromatic content Benzene Heart Cuts (BHC) fuel, which results from the removal of mono-aromatics from gasoline and a poly-aromatic fuel (PArF with 58% aromatic content) rich in di- and tri-aromatics, such as light cycle oil (LCO). A GE frame 6B gas turbine was used. These field tests used diesel oil #2 as reference fuel.

Table 10.8 shows the results of these tests. For dry combustion, the NO_x emissions for BHC (at 290 ppm) and LCO (at 230 ppm) showed an increase of 30% and 22%, respectively, over diesel oil #2. With water injection, NO_x

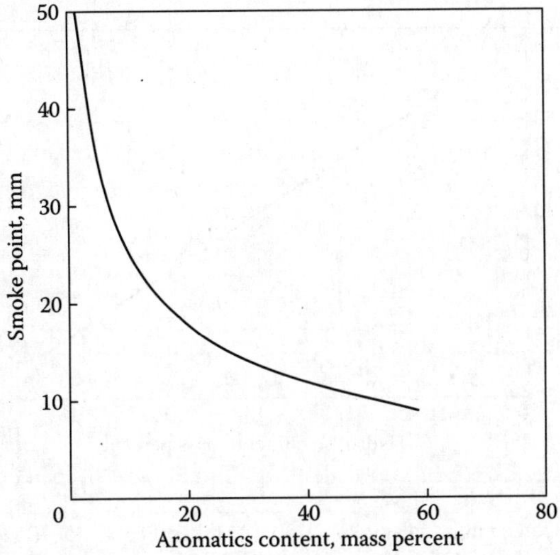

FIGURE 10.38
Relationship between smoke point vs. fuel aromatic content. (From Singh, P.P., Bazarian, E.R., Malik, P.R., Bauserman, G.W., Cohn, A., and Stein, T.R., ASME Paper 80-GT-64, Gas Turbine Conference, New Orleans, LA, 1980.)

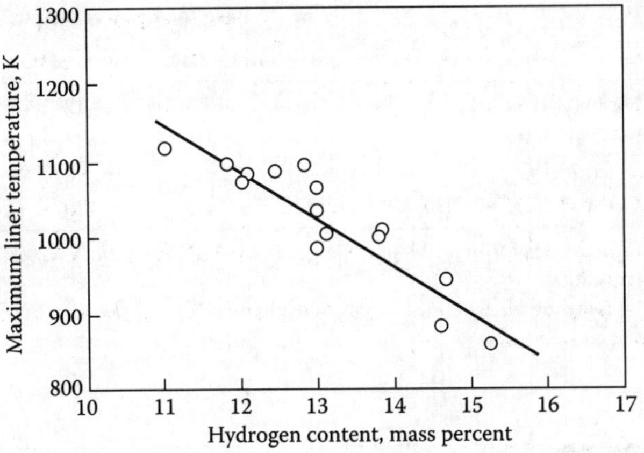

FIGURE 10.39
Effect of fuel hydrogen content on maximum combustor liner wall temperature. (From Gleason, C.C. and Bahr, D.W., ASME Paper 80-GT-55, Gas Turbine Conference, New Orleans, LA, 1980.)

emissions dropped to around 200 ppm VD at 15% O_2. These fuels also displayed higher CO and smoke emissions. In summary, the authors concluded that heavy-duty industrial gas turbines have the ability to clearly burn BHC- and LCO-type fuels. Such fuels can emerge as cheap alternatives to conventional fuels with aromatic content below 30%.

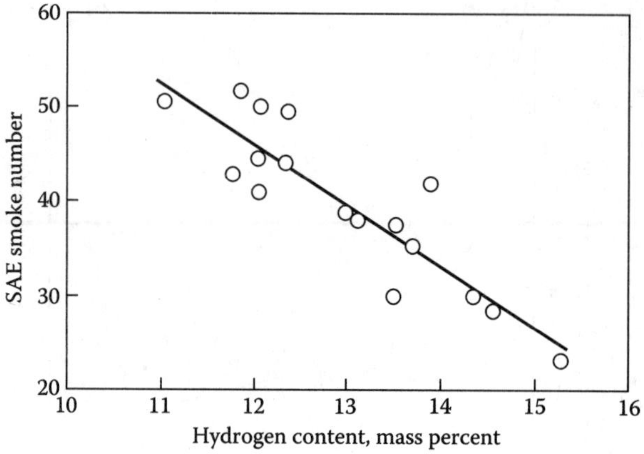

FIGURE 10.40
Effect of fuel hydrogen content on smoke number at takeoff condition. (From Rudey, R.A. and Grobman, J.S., Section 6, AGARD Lecture Series 96, 1978.)

TABLE 10.8

Emissions of NO_x and CO from burning highly aromatic alternative fuels

	Base load power output = 34 MW		
	100% DO	100% BHC	100% LCO
VOC (mg/Nm³ wet)	13	12	–
CO (mg/Nm³ D at 15% O_2)	1	4	< 2
NO_x (ppm VD at 15% O_2)	222	290	230
NO_x (ppm VD at 15% O_2 with $DeNO_x$ water)	114	136	121

Source: Adapted from Moliere, M. and Geiger F., ASME Paper No. GT2004-53272. With permission.

Note: BHC = Benzene Heart Cut, LC = light cycle oil, DO = #2 diesel oil; VOC = volatile organic compounds.

References

1. Goodger, E. M., *Transport Fuels Technology*, Landfall Press, Norwich, UK, 2000.
2. Goodger, E. M., *Aerospace Fuels*, Landfall Press, Norwich, UK, 2007.
3. Edwards, J. T., "Liquid Fuels and Propellants for Aerospace Propulsion, 1903–2003," *AIAA Journal of Propulsion and Power*, Vol. 19, pp. 1089–107, 2003.
4. *Aviation Fuels Technical Review*, Chevron Products Company, San Ramon, CA, 2000.
5. Longwell, J. P., "Synthetic Fuels and Combustion," *Sixteenth Symposium (International) on Combustion*, The Combustion Institute, Pittsburgh, PA, pp. 1–15, 1977.

6. Longwell, J. P., "Alternative Fuels and Combustion Problems," *Progress in Astronautics and Aeronautics*, Vol. 62, pp. 3–20, 1978.
7. Rudey, R. A., and Grobman, J. S., "Impact of Future Fuel Properties on Aircraft Engines and Fuel Systems," Section 6, AGARD Lecture Series 96, 1978.
8. *The Aeroshell Book*, 18th ed., Shell Aviation Ltd., London, 2003.
9. Blazowski, W. S., and Maggitti, L., "Future Fuels in Gas Turbine Engines," *Progress in Astronautics and Aeronautics*, Vol. 62, pp. 21–73, 1978.
10. Digman, W. J., "Effects of Fuel Contamination on Corrosion of Aircraft Fuel Systems," *Esso Air World*, Vol. 15, No. 4, pp. 135–43, 1963.
11. Felex, P. C., "Problems and Operating Experiences with Gas Turbines Burning Residual and Crude Oils," ASME Paper 78-GT-103, presented at Gas Turbine Conference, London, 1978.
12. Krulls, G. E., "Pretreat Liquid Fuels for Gas Turbines," *Power*, Vol. 118, No. 7, pp. 77–9, 1974.
13. Vere, R. A., "Lubricity of Aviation Turbine Fuels," SAE Paper 690667, 1969.
14. Strawson, H., and Lewis, A., "Electrostatic Charging in the Handling of Aviation Fuels," AGARD Conference 84 on Aircraft Fuels, Lubricants and Fire Safety, Hague, The Netherlands, 1971.
15. Kirtley, J. G., and Lewis, A., "Ten Years Experience of Anti-Static Additives in Aviation Fuels," International Conference on Electrical Phenomena, Las Vegas, NV, 1972.
16. Pagni, P. J., Hughes, L., and Novakov, T., "Smoke Suppressant Additive Effects on Particulate Emissions form Gas Turbine Combustion," Paper 28, AGARD Conference Proceedings 125, 1973.
17. Marx, S. D., and Klein, R. W., "Use of Chemical Additive (Manganese-Barium) at Consolidated Edison to Control Gas Turbine Emissions," *Combustion and Flame*, Vol. 47, No. 5, pp. 24–6, 1975.
18. Giammar, R. D., Krause, H. H., and Locklin, D. W., "The Effect of Additives in Reducing Particulate Emissions from Residual Oil Combustion," ASME Paper 75-WA/CD-7, 1975.
19. Hersh, S., Carr, R. C., and Hurley, J. F., "The Effects of Smoke Suppressants and Corrosion Inhibiting Fuel Additives on the Particulate and Gaseous Emissions from a Utility Gas Turbine," ASME Paper 78-GT-64, Gas Turbine Conference, London, 1978.
20. Johnson, C. E., Kennedy, E. R., and Little, W. W., "Selection of Fuels for Commercial Turbine Powered Aircraft," *Shell Aviation News*, 1954.
21. Lefebvre, A. H., "Airblast Atomization," *Progress in Energy and Combustion Science*, Vol. 6, pp. 1–29, 1980.
22. Smith, M., and O'Farrell, M., "Aviation Fuels," in I.M. Smallwood, ed., *Modern Petroleum Technology*, 2nd ed., pp. 134–88, Institute of Petroleum, London, 1954.
23. Hager, J. A., "High Temperature Fuels Pose Problem in Subsonic Environment," *SAE Journal*, Vol. 74, No. 5, p. 38, 1966.
24. Droegemueller, E. A., "Fuel Requirements for Supersonic Transport," *Esso Air World*, Vol. 16, No. 2, 1963.
25. Maxwell, J. B., *Data Book on Hydrocarbons*, p. 89, Macmillan, London, 1955.
26. Ragozin, N. A., *Jet Propulsion Fuels*, Pergamon, Oxford, 1962.
27. Szetela, E. J., and Chiapetta, L., "External Fuel Vaporization Study, Phase 1 Report," NASA CR 159850, 1980.

28. Barnett, H. G., and Hibbard, R. R., "Properties of Aircraft Fuels," NACA TN 3276, 1956.
29. Smith, M., *Aviation Fuels*, Foulis, Henley, UK, 1970.
30. Zabetakin, M. G., Furno, A. L., and Jones, G. W., "Minimum Spontaneous Ignition Temperatures of Combustibles in Air," *Industrial and Engineering Chemistry*, Vol. 46, p. 2173, 1954.
31. Thomas, A., Shell Report M.225, Thornton Research Center, Chester, UK, 1955.
32. Kurtovich, D. D., and Hayes, G. E., "Spontaneous Ignition and Supersonic Flight," *Esso Air World*, Vol. 14, No. 5, pp. 187–96, 1962.
33. Naegeli, D. W., and Moses, C. A., "Effect of Fuel Molecular Structure on Soot Formation in Gas Turbine Engines," ASME Paper 80-GT-62, Gas Turbine Conference, New Orleans, LA, 1980.
34. Lewis, A., "Fuels for Aircraft Gas Turbine Engines," in E. R. Norster, ed., *Cranfield International Symposium Series*, pp. 309–25, Pergamon, Oxford, 1971.
35. Heneghan, S. P., Ballal, D. R., and Harrison, W. E. III, "JP-8 + 100: The Development of High Thermal Stability Jet Fuel," *Transactions of ASME, Journal of Energy Resources Technology*, Vol. 118, pp. 170–9, 1996.
36. Bird, R. J., and Small, N. H., "The Combustion of Heavy Fuel Oil," *Journal of the Institute of Petroleum London*, Vol. 51, No. 494, pp. 494–503, 1965.
37. Rowling, H., "Ash Deposition—Petroleum Fuels," Chap. 19 in Sir Harold Roxbee Cox, ed., *Gas Turbine Principles and Practice*, Newnes, London, 1955.
38. Ekstedt, E. E., and Macauley, R. W., "Combustion of Heavy Distillate Fuel for the LM2500 Gas Turbine," ASME Paper 72-GT-24, 1972.
39. Del Bueno, R., "Gas Turbines Fire Heavy Oils, Power," *Power*, Vol. 116, No. 5, pp. 60–1, 1972.
40. White, A. O., "20 Years Experience Burning Heavy Fuels in Heavy Duty Gas Turbines," ASME Paper 73-GT-22, 1973.
41. Foster, A. D., "Gas Turbine Fuels," *Combustion and Flame*, Vol. 44, No. 10, pp. 4–14, 1973.
42. Frieder, A. J., and Felix, P. C., "Experiences with Gas Turbines Burning Non-Refined Fuel Oils and Related Theoretical Investigations," ASME Paper 74-GT-12, 1974.
43. Frieder, A. J., Felix, P. C., and Hess, H. J., "Gas Turbines Burning Non-Refined Fuel Oils," *Combustion and Flame*, Vol. 45, No. 11, pp. 18–29, 1974.
44. G. E. Krulls, "Design of Treatment Systems for Industrial Gas-Turbine Fuels-Liquids and Gaseous," *Power*, Vol. 118, No. 10, pp. 65–70, 1974.
45. Patterson, J. R., "Operating and Maintenance Experience for Base Load Gas Turbines Using Heavy Fuels—A Case Study," ASME Paper 75-GT-74, 1975.
46. Byam, J. W., Dreyman, S. S., and Pitrolo, A. A., "Residual Fuel Treating and Handing for Base Load Bas Turbines," ASME Paper 75-GT-72, 1975.
47. Bammert, K., and Sandstede, H., "Gas Turbines Burning Dirty Fuels," ASME Paper 75-GT-66, 1975.
48. Cauley, S., "Utilization of Petroleum Crude Oils as Boiler and Combustion Turbine Fuels," ASME Paper 76-JPGC-FU-2, 1976.
49. DeGreef, J. L., Maes, P., and Johnson, K. W., "Operating Experience on Residual Fuel Oil with a W251 Combustion Turbine," ASME Paper 78-GT-104, Gas Turbine Conference, London, 1978.
50. Brewer, G. D., "Liquid Hydrogen: Future Aircraft Fuel," *Automotive Engineering*, Vol. 88, No. 8, pp. 242–51, 1980.

51. Yaffee, M. L., "Methane Studied as Fuel for SST," *Aviation Week*, pp. 72–3, 1969.
52. Whitlow, J. B., Eisenberg, J. D., and Shovlin, M. D., "Potential of Liquid-Methane Fuel for Mach 3 Commercial Supersonic Transport," NASA TN-D3471, 1966.
53. Pirans, M. L., Olsen, W. T., Barnett, H. C., and Breitwieser, R., "NACA Research on Slurry Fuels," NACA Report 1388, 1958.
54. Lee, S., Speight, J. G., and Loyalka, S. K., *Handbook of Alternative Fuel Technologies*, CRC Press, Boca Raton, FL, 2007.
55. Nowack, C. J., Solash, J., and Delfosse, R. J., "Evaluation of Coal-Derived JP-5 Fuels," *CEP Technical Manual*, Vol. 3, pp. 122–6, 1977.
56. Cohn, A., Bauserman, G. W., Bazarian, E. R., Mulik, P. R., Singh, P. P., and Stein, T. R., "Combustion Effects of Coal Liquid and other Synthetic Fuels in Gas Turbine Combustors," ASME Papers 80-GT-67 and 80-GT-68, Gas Turbine Conference, New Orleans, 1980.
57. Singh, P. P., Bazarian, E. R., Mulik, P. R., Bauserman, G. W., Cohn, A., and Stein, T. R., "Comparative Testing of Petroleum Surrogate Fuels with Coal-Derived Liquids in a Combustion Turbine Burner," ASME Paper 80-GT-64, Gas Turbine Conference, New Orleans, LA, 1980.
58. Moses, C. A., and Stavinoha, L. L., "Qualification of Sasol Semi-Synthetic Jet A-1 as Commercial Jet Fuel," Report No. 8531, Southwest Research Institute, San Antonio, TX, November 1997.
59. Moses, C., and Roets, P., "Properties, Characteristics and Combustion Performance of Sasol Fully Synthetic Jet Fuel," *ASME Journal of Engineering for Gas Turbines and Power*, Vol. 131, No. 4, 041502, 2009.
60. Harrison, W. E., and Zabarnick, S., "The OSD Assured Fuels Initiative—Military Fuels Produced from Coal," presented at DoE Clean Coal Conference, Clearwater, FL, June 2007.
61. Lieuwen, T., McDonell, V., Petersen, E., and Santavicca, D., "Fuel Flexibility Influences on Premixed Combustor Blowout, Flashback, Autoignition, and Stability," *ASME Journal of Engineering for Gas Turbines and Power*, Vol. 130, 011506, 2008.
62. Panchasara, H., Simmons, B., Agrawal, A., Spear, S., and Doly, D., "Combustion Performance of Biodiesel and Diesel Vegetable Oil Blends in a Simulated Gas Turbine Burner," *ASME Journal of Engineering for Gas Turbines and Power*, Vol. 131, 031503, May 2009.
63. Liu, K., Wood, J. P., Buchanan, E. R., Martin, P., and Sanderson, V., "Biodiesel as An Alternative Fuel in Siemens DLE Combustors: Atmospheric and High Pressure Rig Testing," *ASME Journal of Engineering for Gas Turbines and Power*, Vol. 132, No. 1, 011501, 2010.
64. Campbell, A., Goldmeer, J., Healy, T., Washam, R., Moliere, M., and Citeno, J., "Heavy Duty Gas Turbines Fuel Flexibility," ASME Paper No. GT2008-51368.
65. Butze, H.F. and Ehlers, R.C. "Effects of Fuel Properties on Performance of a Single Aircraft Turbojet Combustor," NASA TMX-71789, 1975.
66. Gleason, C. C., and Bahr, D. W., "Fuel Property Effects on Life Characteristics of Aircraft Turbine Engine Combustors," ASME Paper 80-GT-55, Gas Turbine Conference, New Orleans, LA, 1980.
67. Moliere, M., and Geiger, F., "Gas Turbines in Alternative Fuel Applications: The Utilization of Highly Aromatic Fuels in Power Generation," ASME Paper No. GT2004-53272.

Author Index

A

Abramovich, G. N., 273
Adkins, R. C., 89, 91, 100–103, 105–106
Agan, D. D., 369
Aggarwal, S. K., 54
Agrawal, D. D., 500
Ahmadi, M., 236
Aigner, M., 404–405
Altermark, D., 378
Anderson, D. N., 306
Anderson, L., 376
Anderson, S. J., 377
Anderson, W., 377
Ando, Y., 107
Angello, L., 365, 398, 400
Arai, M., 274
Avery, W. H., 159

B

Bahr, D. W., 141–142, 146–147, 316, 362, 393, 395, 507
Bain, D. B., 137
Ballal, D. R., 44–46, 48, 55, 57–63, 173–175, 177–178, 180–182, 200, 203–204, 208, 295, 334–335
Banghu.J. K., 344
Barnett, C. M., 470
Barrere, M, 173, 176
Bautista, P. J., 402
Baxter, M. R., 175–176
Beck, J. E., 263
Becker, T., 431
Beebe, K. W., 424, 426
Beer, J. M., 141, 143
Benjamin, M. A., 246–247
Bergwerk, W., 269
Bhatia, J. C., 236
Binks, D., 105
Bloxsidge, G. J., 308
Bonzani, F, 400
Boyns, M. B., 408
Bracco, F. V., 269, 273
Bragg, S.L., 159, 294
Brodkey, R. O., 223
Brouwer, J., 309
Brustowski, T. A., 236
Buckner, H. N., 263
Burgoyne, J. H., 45

C

Candel, S. M., 306
Carr, E., 456
Carrotte, J. F., 90, 93, 98, 124
Cekalin, E. K., 45
Chen, S. K., 278–283
Cheng, S. L., 309
Chiappetta, L. M, 375
Chigier, N. A., 141, 143, 242, 269
Childs, J. H., 159
Chin, J. S., 52–55, 162–163, 199, 235, 245–246, 372
Choudhury, P. R., 308
Claeys, J. P., 388
Coban, A., 192
Cocanower, A. B., 87
Cochran, D. L., 99
Cockrell, D. J., 87
Cohen, J. D., 45
Cohen, L., 45
Connors, C. S., 432
Correa, S., 380
Cousin, J., 236
Cowell, L. H., 64–66, 399, 422
Crocco, L., 309
Culick, F. E. C., 310
Custer, J. R., 280
Cutrone, M. B., 426

D

Dalla Betta, R. A., 422, 424–426
Damkohler, G., 43–44
Das, S., 45
Davis, L. B., 388, 390, 398, 402
De Zubay, E. A., 173

DePietro, S. M., 408
Di Meglio, A., 400
Dodds, W. J., 141–142, 146–147, 316, 346
Dombrowski, N., 227
Ducourneau, F., 67
Dugger, G. L., 42
Durlak, S., 411
Dutta, P., 349, 422, 424

E

Eckerle, W. A., 282–283
Eisenklam, P., 227
Eissa, A. F. M., 64
Ekstedt, E. E., 346
El-Ehwany, A. A. M., 333–334
Elkotb, M. M., 258
El-Shanawany, M. S. M. R., 262
Epstein, M. J., 411
Etheridge, C. J., 398
Evans, D. M., 234
Evans, H. D., 349

F

Facey, J. R., 351
Felchin, M. P., 404
Fishenden, C. R., 97–98
Fishenden, M., 97–98
Flanagan, P., 416
Foster, H. H., 202–203
Fox, R. W., 83
Fraser, R. P., 226–227
Freeman, B. C., 64–65
Freeman, G., 121–122
Fric, T. F., 416
Fujii, T., 422, 425–426

G

Gallagher, D., 198
Gallimore, S., 408
Gardner, K. A., 349
Gaydon, A. G., 41
Gemmen, R. S., 309
Gerhold, B., 390
Giffen, E., 270–271, 275
Gleason, C. C., 507
Godsave, G. A. E., 51
Goodger, E. M, 64, 453, 474, 490, 495–497
Goris, N. H., 264–265
Greber, I., 128
Greenhalgh, D., 198
Greenhough, V. W., 155, 159–160
Grobman, J. S., 508
Guangshi, H., 108
Gupta, A. K., 141
Gupta, C. P., 141
Gutmark, E., 306

H

Halls, G. A., 156
Han, Z., 233
Harding, C. F., 271
Harrje, D. T, 306
Hart, R. W., 159
Hasson, D., 227
Hatch, M. S., 130–131
Haumann, J., 404
Heberling, P. V., 378
Heimel, S., 42
Hellat, J., 404
Henseler, H. J., 82
Herbert, M. V, 180, 320, 328
Hestermann, R., 91, 93, 98, 107–108
Hilt, M. B., 389, 402
Hinze, J. O., 223
Hirano, J., 426
Hiroyasu, H., 274
Hoffman, J. A., 92
Hoffman, S., 400
Holdeman, J. D., 128–129, 131–133, 137
Honami, S., 98
Hosoi, J., 413–414
Hottel, H. C., 319
Howard, J. H. G., 82–83, 89
Hoyt, J. W., 225
Hung, W. S. Y., 369, 388

I

Ibrahim, A. R. A. F., 303
Ibrahim, M., 303
Ishizuka, A., 414

J

Janus, M. C., 299–300, 304, 310
Jasuja, A. K., 253, 259, 262
Jeroszko, R. A., 349
Johnson, K., 459
Johnston, J. P., 89
Jones, W. P., 107
Joshi, N. D., 409, 411–412
Juhasz, A. J., 101

K

Kaddah, K. S., 121, 123
Kamotani, Y., 128
Karchmer, A., 295–296
Karki, K. C., 108
Keller, J. O., 299, 304
Kikumoto, S., 426
Kilik, E., 141, 144–145
Kim, S., 107
Kingdon, R. G., 197
Kirk, G. E., 351
Klein, A., 88–90, 94, 98, 103–104, 106, 108
Kline, S. J., 82, 89, 92, 99
Klomp, E. D., 81, 87–90
Knauber, R., 378
Kneeling, D., 107
Kors, D. L., 131–132
Koutmos, P., 107
Kretschmer, D., 11, 320, 429
Kumakura, H., 414
Kunz, H. R., 89
Kurzynske, F. R., 398

L

Lakshminarayana, B., 90
Langhorne, P. J., 308
Larkin, M. P., 422
Lauer, G., 400
Leonard, G., 376, 380, 409–410, 416–417
Lewis, B., 55, 137, 248
Lewis, G. D., 431
Li, J., 236
Li, X, 236
Lilley, D. G., 141
Little, A. R., 49, 89, 377
Lohmann, R. P., 90, 93, 349
Longwell, J. P., 38–39, 159, 173, 179–181
Lorenzetto, G. E., 262
Low, H. C., 253
Lowe, P., 365, 400
Lundberg, K., 64
Luts, A., 402
Lyons, V. J., 416

M

Mahan, J. R., 295–296
Maier, G., 248
Manners, A. P., 89, 107
Marakovits, S., 411
Marble, F. E., 173, 180
Marioka, T., 98
Markland, E., 87
Martin, C. A., 501–502
Martin, F. J., 504, 506
Matharu, D. S., 101–103
Mathur, M. L., 141, 146
Matsuoka, S., 274
Maughan, J. R., 379, 402
Mayer, D. W., 107
McCaldon, K., 255
McDonald, A. T., 83
McGuirk, J. J., 107
McKinney, R., 386
McManus, K., 306–307, 309
McVey, J. B., 375
Mehta, J. M., 299
Mellor, A. M., 70, 173, 429
Menon, S., 310
Mestre, A., 173, 176
Miller, D. S., 99
Mizutani, Y., 45
Mongia, H. C., 376, 381, 412–413, 420, 431–432, 434
Moore, C. A., 92
Morford, S. A., 386
Mori, M., 414
Moses, C. A., 164–165, 471–472, 489–490, 497–499
Mugele, R., 234
Muller.G., 404–405
Mullinger, P. J., 242
Mullins, B. P., 64, 67
Muraszew, A., 270

Muthukrishnan, M, 296
Myers, G. D., 45, 47–48, 255

N

Nakajima, A., 45
Nakata, T., 420
Nealy, D., 345
Nickolas, S. G., 426
Nicol, D., 378–379, 431
Nishimoto, T., 45
Noble, M., 349
Norris, R. H., 349
Norster, E. R., 125–126, 128, 130, 172, 408
Novick, A. S., 419–420
Nukiyama, S., 230, 244, 261–262

O

Odgers, J., 11, 159, 192, 320, 429
Ohkubo, Y., 414
Ohrn, T.R., 269, 274
Opdyke, G., 196
Ortman, J., 277
Owen, D. A., 384, 407
Ozawa, Y, 422, 426

P

Park, S. H., 223
Parnell, E. C., 253
Peng, F, 54
Perkavec, M. A., 431
Plee, S. L., 70, 173
Poeschl, G., 416–417
Pollarolo, G., 400
Polymeropoulos, C. E., 45
Prade, B., 400
Pratt, D. T., 6–7, 296, 331, 372, 397, 420
Putnam, A. A., 310

R

Radcliffe, A., 259, 270
Rammler, E., 230–236, 249
Rao, K. V. L., 56, 68, 199, 201–202, 205–206
Rayleigh, Lord, 224, 226–227, 297, 309
Razdan, M. K., 31, 246
Reardon, F. H., 306
Reeves, D., 319
Reid, R., 45, 207
Reitz, R. D., 269, 273
Reneau, L. R., 83, 88–90
Reynolds, B., 90–91, 107–108, 257–258, 266, 268–270, 322, 334
Richards, G. A., 304–306, 308–310
Rink, K. K., 368, 372–373, 376, 381, 385, 431
Rizk, N. K., 233, 246, 262, 270, 272, 275–276, 280, 376, 381, 412–413, 420, 431–432, 434
Roberts, P. B., 398
Rokke, N. A., 431
Rollbuhler, J. R., 251, 278–279, 281–283
Rosen, R., 351
Rosfjord, T. J., 282–283, 375
Rosin, P., 230–236, 249
Ruiz, R, 269
Russel, P., 283, 372

S

Sabla, P. E., 411
Saiga, M, 426
Saintsbury, J. A., 196
Santavicca, D. A., 417
Sato, H., 414
Sato, M., 426
Sato, Y., 426
Sattelmayer, T., 70, 263, 404
Scalzo, A. J., 301
Schadow, K. C., 308
Schlatter, J. C., 422, 426
Schorr, M. M., 364, 389, 402
Schweiger, T., 384, 407
Seaton, A., 360
Segalman, I., 396
Sellens, R. W., 236
Seto, S. P., 298
Sevik, M., 223
Shyy, W., 106–107
Simmons, H. C., 271, 500
Singh, R., 486, 507
Sivasegaram, S., 308
Smith, C.E., 101
Smith, D. A., 398
Smith, J. M., 399
Snyder, T. S., 374–375

Sojka, P. E., 263
Sotheran, A., 253
Sovran, G., 81, 87–90
Spadaccini, L. J., 64, 67
Spalding, D. B., 51, 173
Sridhara, K., 126–130
Srinivasan, R., 98, 131–132, 137
Steele, R. C., 380
Stegmaier, J., 376, 380, 409–410, 416–417
Stengele, J., 54
Stevens, S. J., 89–90, 93, 97–98, 107, 124
Stickles, R. W., 255
Stollery, J. L., 333–334
Strahle, W. C., 295–296
Straight, D. M., 82, 469
Sturgess, G. J., 172, 336–337, 386
Styner, D., 404
Subba Rao, H. N., 56, 201, 205–206
Suyari, M, 247–248, 271–272
Swett, C. C., 55, 200
Syred, N., 141

T

Tacina, R. R., 67
Takehira, A., 89–90
Talpallikar, M. V., 130
Tanasawa, Y., 230, 244, 261–262
Tankin, R. S., 236
Taylor, G. I., 172
Taylor, J. J., 225
Te Velde, J. A., 64, 67
Thomas, A., 295, 469
Thornton-Trump, A. B., 82
Toh, H., 414
Toof, J. L., 388
Toon, I. J., 384, 407
True, B., 474

V

Valk, M., 418
Vandsburger, U., 299
Verdouw, A. J., 101
Vickers, R. M., 4, 408
Vortmeyer, N., 422

W

Walker, R. B., 131
Walker, R. E., 132
Wang, X. F., 260–261, 283
Washam, R. M., 388, 390, 402
Waslo, J., 389, 402
Wassell, A. B., 344
Watanabe, T., 414, 426
Watson, E. A., 192, 200
Weber, C., 222–223, 225, 228, 256–258, 262–263, 266
Weinberg, F. J., 197
Weir, A., 264
Weiss, M. A., 38–39, 159
West, H. E., 194
White, C. D., 389
White, D. J., 398
Whitelaw, J. H., 308
Whitlow, J. D., 250–251
Whittle, F, 3–4, 6, 11, 16
Wilkes, C., 388, 390
Williams, G. J., 89, 93
Williams, G. T, 242, 253
Williams, M. R., 295
Willis, J. D., 384, 406–407
Winterfeld, G., 256
Wittig, S., 93, 263
Wolfer, H. H., 64
Wolfhard, H. G., 41
Wray, A. P., 98

X

Xiong, T. Y., 199
Xu, T.-H., 236

Y

Yang, V., 306, 310
Yokota, K., 274
Yost, J. O., 101–103

Z

Zheng, Q. P., 265, 279, 371
Zhiben, H., 108
Zhu, G., 131, 137
Zukowski, E. E., 173, 180

Subject Index

A

Absorptivity (gas), 318–319
Acoustic oscillations
  descriptions
    growl, 297–298
    howl, 298
Active control noise
  closed-loop system, 307
  examples, 307–308
  open-loop system, 307
Additives, 453
  anti-icing additives
    di-EGME and FSII, 454
  antismoke
    manganese concentration, 456
    optimal injection rate, 455
  antistatic–static dissipators, 454
    charging rates, 455
  corrosion inhibition/lubricity improvers, 454
  gum prevention
    antioxidants and metal deactivators, 454
  metal deactivators, 455
Adiabatic flame temperature
  factors influencing
    fuel/air ratio, 71
    initial air temperature, 71–72
    inlet-air vitiation, 72
    pressure, 72
AEC technique, *see* Angled effusion cooling (AEC) technique
Aero-derived GE LM6000 and RR211 DLE industrial engines, 31
Aerodynamic diffusers, 14–15
Aerodynamic processes, 113
  aerodynamic instabilities, 300–302
  axial swirlers, 142
    exit geometry, influence of, 146
    flow reversal, 145–146
    notation for, 143
    recirculation zone size, 144–145
    swirl number, axial flow, 143
  dilution zone design, 135
    Cranfield design method, 136–137
    NASA design method, 137
  dimensionless pressure-loss parameters
    cold pressure loss, 116–117
    combustor reference velocity, 116
    drag coefficient, 114
    effective flow area, 115
    hot loss, 114
    overall pressure loss, 114
    pressure drop across, 115
    pressure-loss factor, 114–115
    reference dynamic pressure, 114
  flat vanes versus curved vanes, 147
  flow in annulus, 118–120
  jet-mixing
    annular ducts, 133
    cylindrical ducts, 129–131
    influence of density ratio, 132
    rate of, 129
    rectangular ducts, 131
  jet trajectories
    configurations, 126–128
    in crossflow representation, 125
    penetration data, 125–126
    single jets, experiments on, 124–126
  liner holes, flow through
    discharge coefficient, 120–122
    initial jet angle, 123–124
  pattern factor data, correlation of
    for annular combustors, 138–139
    experimental data, 138
    for tubular and tuboannular combustors, 138–139
  radial swirlers
    airflow, 147
    notation for, 146
  reference quantities, 114
  rig testing for pattern factor, 140
  size and pressure loss, relationship between, 117–118

swirler, 140
flow recirculation, 141
flows and practical applications, 141
types, 142
vortex breakdown, 141
temperature traverse quality, 133
exit-temperature profile parameters, 135
pattern factor, 134
profile factor, 134
turbine profile factor, 135
Aero-engine combustor, 169
Air-assist atomizers, 242–243
Airblast atomizers, 19
advantages, 20
airblast simplex, 246–249
circumferential fuel distribution, 283–284
disadvantages, 20
double-swirler, 141–142
fuel distribution, airflow pattern, 20
piloted airblast, 245–246
plain-jet airblast, 243–244
prefilming airblast, 244–245
prompt mechanism, 221, 228, 264–266
Rayleigh mechanism, 226–227
Sauter mean diameter, 165
soot formation, 20
Airbus Industrie A320 and A321 aircraft, dual-annular combustor in, 396
Aircraft combustor, stability performance, 170
Air/fuel ratio (AFR), 36
Air swirlers, 141–142
Allison AGT100, 412–413
Allison T63 engine, 26
Alternative fuels
ambient air temperature, effect, 501
combustion dynamics, comparison, 506
combustion properties of
calorific value, 466–468
enthalpy, 468–469
flammability, limits, 470–471
pressure and temperature effects, 472–474
smoke point, 471–472
spontaneous-ignition temperature, 469–470
flame temperature and combustor pressure
effect of, 502–505
fuel-nozzle design for, 444
gaseous fuels, classification
coal gas, 478
coke-oven gas, 478, 480
common gaseous fuels, properties of, 479
impurities, 480–481
natural gas, 478
hydrocarbons, types
aromatics, 447–448
naphthenes, 446
olefins, 445–446
paraffinic oils, 444–445
IGCC plants, 443
lean ignition limit for dry low-emissions, 501
liquid fuels, classification
aircraft fuel specifications, 476
aircraft gas turbine fuels, 475–476
diesel fuels, 475
gasolines, 475
heavy residual fuels, 475
industrial gas turbine fuels, 476, 478
properties of, 474
liquid fuels, production
additives, 453–456
contaminants, 449–453
conversion process, 448
fuel properties, 456–466
separation process, 448
sulfur compounds removal of, 448–449
upgrading process, 448
maximum combustor, hydrogen effect, 507
NO_x and CO, emissions, 508
axial profiles of, 500
petroleum fuel, ever-rising cost, 443
pure compounds
alcohol fuels, properties, 483
alcohols, 484–485
ammonia, 484

hydrogen, 481, 483
methane, 483
propane, 483–484
properties of, 482
slurry fuels, 485–486
smoke number, hydrogen effect, 508
smoke point and fuel aromatic content, 507
supplemental fuels, 485
synthetic fuels
biodiesel fuel, 499–504
biofuels, 488–489
combustion and emissions performance, 490–491, 495–496
Fischer–Tropsch synthesis of coal/biomass, 486–488
fuel properties, 489–490
highly aromatic, 504–507
Angled effusion cooling (AEC) technique, 22, 346–347
drawback of, 23
Annular combustors, 12–13
altitude relight limits of, 190
pattern factor correlation, 139
Annular diffusers, 15, 89
characteristic
dimension for, 81
correlation of data, 138
geometries, 90
performance charts, 90
Annular reverse-flow combustor, 26–27
Aromatics
disadvantages, 446
formula for, 447
petroleum-based jet fuel, 447
gas chromatograph of, 448
Ash, distillate fuel, 450
Asphaltenes, 449
Atomization
air-assist atomizers, 242–243
airblast atomizers
piloted, 245–246
plain-jet, 243–244
prefilming, 244–245
simplex, 246–249
basic processes, 221
classical mechanism, 221
drops breakup, 222
in turbulent flow fields, 223
drop-size distributions
frequency distribution curves, 229
fuel-injection pressure on, 230
graphical representation of, 227–230
histogram, 229
prediction of, 236
effervescent atomizers
advantages of, 249
drawback, 249
hollow-cone spray, 250
plain-orifice, 250
pressure and air/liquid ratio on mean drop size, 251
wide-angled spray, 250
influence of
ambient air pressure, 277
liquid viscosity, 261, 278
jet and sheet breakup, 223
fuel jets, 223–226
fuel sheets, 226–227
mathematical distribution functions, 230–231
mean diameters, 234
modes of, 226
Ohnesorge number, 223, 258, 262
pressure atomizers, 237
dual-orifice, 239–240
plain orifice, 238
simplex, 238–239
spill return, 240–241
prompt, 227
representative diameter, 234–236
Rosin–Rammler relationship, 231–233
modified, 233
rotary atomizer, 241–242
Atomizers
air-assist, 242–243
airblast (*see* Airblast atomizers)
design, 18–19
effervescent atomizers
advantages of, 249
drawback, 249
hollow-cone spray, 250
plain-orifice, 250
pressure and air/liquid ratio on mean drop size, 251
wide-angled spray, 250

pressure atomizers, 237
dual-orifice, 239–240
plain orifice, 238
simplex, 238–239
spill return, 240–241
requirements, 237
rotary atomizer, 241–242
Augmented cold-side convection, 349
Autoignition, *see* Spontaneous ignition
Avon aero engines, 31
Axial-flow compressors, 79
Axial staging, 24, 396
Axial swirlers, 142
exit geometry, influence of, 146
flow reversal, 145–146
notation for, 143
recirculation zone size
experimental data, 145
stagnation point, 144
in swirling flow field, 144
velocity profiles, 144
swirl number
axial flow, 143

B

Backstop, 119
Beam length for radiation, 319–320
Best available control technology (BACT), 365
Blackbody radiation, 318
Bluff-body flameholders, stabilization
heterogeneous mixtures
chemical reaction rates, 179
fuel evaporation rates, 179
weak extinction limits, 177–178
homogeneous mixtures
cross-sectional area, 173
gaseous propane and air, 173
influence of size, 173
weak extinction limits, 174–176
BMW 003 annular combustor, 6
Boundary-layer mechanism, 69–70
Breakup of fuel drops, 222
influence of turbulence, 223
Breakup of fuel jets, 223–226
Breakup of fuel sheets, 226–227
Burner, ABB EV, 403–406
Burning liquid fuels, 36
Burning velocity
influence of equivalence ratio
fuel type, 41
initial temperature, 42
pressure, 42
turbulence, 43
model
chamber design, 156
combustion efficiency, 155
experimental data, 156–157
laminar burning velocity and turbulence intensity, 156
parameter, 156

C

Can-annular combustor, *see* Tuboannular combustors
Can-type combustor, sound power radiated by, 295–296
Carbon monoxide (CO), 383–384
correlations, 432–434
emissions standards, 362, 491
formation, 366–368
influence of
ambient air temperature, 369
equivalence ratio, 368–369
fuel atomization, 370
pressure, 369
wall-cooling air, 369–370
prediction, 366–367
Casing, 7, 10–13, 21
Casing area, 114
Catalytic combustion
bed construction, 424–425
design
approaches, 422
constraints, 422–423
and performance, 425–427
fuel preparation, 423–424
postcatalyst combustion, 425
variable geometry, use, 427–428
CF6-50 annular combustor, 13
Chemical ignition, 198–199
Chemical processes, 37
Circumferential fuel distribution
airblast atomizers, 283–284
pressure-swirl atomizers, 280–282

CIVB, *see* Combustion-induced vortex breakdown (CIVB)
Closed-loop system, 307
Co-flowing
  plain-jet airblast atomizer, 263
  prefilming airblast atomizer, 264
Coking of fuel nozzles, 254
Cold-side convection, 349
Combustion chambers
  flame stabilization in, 183
    drop-size term, 188
    experimental data correlation, 185
  measured and predicted values
    for J 79 combustor, 187
    for J 85 combustor, 187
    for TF 41 combustor, 186
  mode of fuel injection, influence of, 184–185
  stability loop, 168
Combustion efficiency, 153
  aircraft combustor, data for, 157
  combustion (predilution) temperature, 165–166
  evaporation and reaction efficiency, 165
  evaporation-controlled systems, 161–165
  experimental data on, 165
  fuel type, influence of, 164
  measured and predicted values
    for J 85 combustor, 167
    for TF 41 combustor, 166
  mixing-controlled systems, 160–161
  process, 154
  reaction-controlled systems, 154
    burning velocity model, 155–159
    stirred reactor model, 159–160
  Sauter mean diameter of fuel spray, 165
  for two different combustion designs, 158
Combustion-induced vortex breakdown (CIVB), 69
Combustion instabilities
  acoustic oscillations, descriptions
    growl, 297–298
    howl, 298
  active control, 306
    closed-loop system, 307
    examples, 307–308
    open-loop system, 307
  aerodynamic instabilities, 300–302
  ambient conditions, influence of, 300
  combustor operating conditions, influence of, 299–300
  compressor-induced oscillations, 303–304
  control signal frequency, influence, 309
  fuel-injector instabilities, 302–303
  fuel type, influence, 299
  LPM combustor noise, 304
  modeling of, 310
    Rayleigh criterion, 309
    τ–n analysis, 309
  passive control, 305–306
  test rig simulations, 304–305
  times, characteristic, 298–299
Combustors
  cold loss, 14
  components of, 18
  configuration, 8
  design chart for, 158
  developments, 2
    in Britain, 3–5
    in Germany, 5–6
    in United States, 6–8
  early designs
    BMW 003, 5
    GE J33, 6
    Jumo 004 tubular combustor, 5
    Metrovick annular combustor, 5
    P&W J57, 6
    Westinghouse J30, 6
    Whittle atomizer combustor, 3
    Whittle vaporizer combustor, 3
  geometry of, 1
  ignition loop for, 189
  inlet and outlet temperature, 14
  for low emissions, 23
    geometry, 24
    LPP combustor, 25
    staged combustion, 24
  operating conditions, influence, 299–300
  pressure
    drop across, 14
    losses in, 116

primary-zone airflow, 3
requirements, 9–10
"reverse-flow" type, 3–4
for small engines
annular reverse-flow combustor, 26–27
disadvantage, 27
"straight-through," 4
types, 10
aeroderivative, 31
annular, 12–13
catalytic, 25–26, 349
industrial, 29
lean premix (LPM), 64, 304
lean, premix-prevaporize (LPP), 376
reverse flow, 26
small, 26
staged, 393–397
tuboannular, 11–12
tubular, 11
variable geometry, 392
Compressor-generated turbulence, 93
Compressor-induced oscillations, 303–304
Cone angle of fuel sprays, 278–279
Conical diffuser
flow regimes, 82
performance, 87
Conical premix burner module, 31
Conical-sheet effervescent atomizer, 250
Control signal frequency, influence, 309
Convection
augmented, 349
cold-side, 349
external, 323–324
internal, 322–323
Cranfield design method, 136–137
Curved-vane swirlers, 142, 145
Cylindrical ducts, 129–131

D

Deflagration, 35
De Havilland Goblin, 4
Detonation, 35–36
Di-ethylene glycol monomethyl ether (di-EGME), 454
Diffractive light-scattering technique, 201
Diffusers, 14, 79
angle, 79
effect on pressure loss, 80
calculation methods for, 106–108
conical, performance chart for, 87
design considerations
dump diffusers, 97
faired diffusers, 93–97
hybrid diffuser, 101–104
incorporating splitter vanes, 99
tubular and tuboannular combustors, 104–105
vortex-controlled diffuser, 100–101
divergence angle influence on pressure loss, 80
flow regimes, 82–83
geometry of
conical, 81
two-dimensional, 81
ideal diffuser, design of, 80
inlet flow conditions effect
Mach number, 91–92
radial distribution of velocity, 91
Reynolds number, 91
subsequent development of flow within, 90
swirl, 93
turbulence, 92–93
performance charts, 90
performance criteria
continuity equation, 83
dynamic pressure, 83
energy conversion, 83
ideal pressure-recovery coefficient, 84–85
kinetic-energy coefficient, 86
loss coefficient, 85
one-dimensional incompressible flow, 84
overall effectiveness, 85
pressure loss in, 83
pressure-recovery coefficient, 84
static pressure, rise in, 84
testing of, 105–106
types
aerodynamic, 14–15
annular, 15

conical, 82
dump, 97
faired, 15
hybrid, 102
ideal, 80
merits, 104
two-dimensional diffuser, 88–89
VCD, 101
Diffusion flames, 36
Diffusion mechanism, 41
Dilution zone
design, 135
Cranfield design method, 136–137
NASA design method, 137
function, 17–18
Direct combustion noise
core noise prediction methods, 296
radiated sound power, 294
theory, 294–295
thermoacoustic efficiency, 294
Discharge coefficient
film thickness, 270–273
hole type and pressure-drop coefficient, 122
influence of hole shape on, 121
liner holes in, 120–121
plain-orifice atomizers, 268–269
pressure atomizers, 268
pressure-swirl atomizers, 270
Droplet
lifetime, 51
and spray evaporation, 49
vaporization modeling, 54
Drop-size distributions; *see also* Atomization
frequency distribution curves, 229
fuel-injection pressure on, 230
graphical representation of, 227–230
histogram, 229
prediction of, 236
Dry low emissions (DLE)
combustors, 146
versions, 29
Dry low-oxides of nitrogen (DLN) combustors
ABB EV Burner, 403–406
Allison AGT100, 412–413
developments in Japan, 413–414
EGT, 407–409
GE DLN combustor, 401
lean-lean, 403
premix, 403
primary, 402
secondary, 403
General Electric LM6000, 409–412
Rolls Royce RB211 industrial burner, 406–407
Siemens hybrid burner, 400–401
solar dry low-emissions concepts, 398–400
Dual-orifice atomizers, 19, 239–240
Dual-orifice fuel nozzles, 302
Dump diffusers, 14–15
computed and visualized flow fields, comparison of, 107
geometry, 97
divergence angles and area ratios, 98
dump gap ratio, 98
liner depth ratio, 98–99
splitter vanes, 99–100

E

Effervescent atomizers
advantages of, 249
drawback, 249
hollow-cone spray, 250
plain-orifice, 250
pressure and air/liquid ratio on mean drop size, 251
wide-angled spray, 250
Effusion cooling, 341–342
EGT, *see* European Gas Turbine company (EGT)
Emissions, 359
carbon monoxide (CO)
correlations, 432–434
catalytic combustion, 421
bed construction, 424–425
design and performance, 425–427
design approaches, 422
design constraints, 422–423
fuel preparation, 423–424
future, 428
postcatalyst combustion, 425
variable geometry, use, 427–428

concerns, 361
pollutants emitted by gas turbines, 360
conventional combustors, pollutants reduction, 382–383
CO and UHC, 383–384
nitrogen oxides (NO_x), 387–391
smoke, 384, 386–387
DLN combustors
ABB EV Burner, 403–406
Allison AGT100, 412–413
developments in Japan, 413–414
EGT, 407–409
GE DLN combustor, 401–403
General Electric LM6000, 409–412
Rolls Royce RB211 industrial burner, 406–407
Siemens hybrid burner, 400–401
solar dry low-emissions concepts, 398–400
flame temperature, control
staged combustion, 393–397
variable-geometry system, 391–393
lean premix prevaporize combustion, 415
fuel–air premixing, 416–418
mass quantities generated during flight of aircraft, 363
nitrogen oxides (NO_x)
correlations, 429–432
pollutant formation, mechanisms
carbon monoxide, 366–370
fuel atomization, influence, 381–382
nitric oxide (NO), 374–379
pressure effects on nitrogen oxides (NO_x), 379–381
smoke, 370–373
unburned hydrocarbons, 370
regulations
aircraft engines, 362–364
stationary gas turbines, 364–366
RQL combustor, 418–420
water injection, 170
Emissivity (gas), 318, 320
Emissivity (metal), 322
Energy conversion in diffuser, 83
Energy efficient engine, 99
Engine pressure ratio, historical trend in, 21
Environmental Protection Agency (EPA) emissions standards, 364–365
European Gas Turbine company (EGT), 407–409
EGT dry low-emissions combustor, 408
Evaporation-controlled systems
combustion efficiency, 161
combustor airflow, 161
effective evaporation constant, 162–163
rate-controlling step, 161
surface-tension values, 164
Evaporation rate, 46
convection, 49
convective effects, 52
effective evaporation constant, 52–54
evaporation constant, 51
heatup period, 50
influence of ambient gas temperature, 49
mean drop size, 48
spray evaporation, 54
EV-burner, 31
Exhaust smoke
influence of
fuel atomization, 372–373
fuel type, 371–372
pressure, 371
External convection, 323–324
External-mixing configuration, 242
External radiation, 321–322

F

Faired diffusers, 15, 93
aerodynamic flow pattern, 96
diffusion regions, 94–95
drawbacks, 96–97
pressure-balancing slots, 95
Film cooling; *see also* Wall-cooling
data, correlation of, 333
with augmented convection, 341–342
calculation of, 337–341
impingement cooling, 342–343
transpiration cooling, 343

turbulent boundary-layer model, theory, 334–335
wall-jet model, theory, 335–337
machined ring, 331
rolled ring, 331
splash-cooling ring, 331
stacked ring, 330
wigglestrips, 329–330
Z ring, 332
Fischer–Tropsch (FT) synthesis, 486–487
Flames
blowout mechanism, 40
classification, 36
drops and air, monodisperse spray of, 46
expression for speed, 46
laminar premixed
factors influencing, 41–42
mechanisms of stabilization, 179–180
in combustion chambers, 184–188
heterogeneous mixtures, 182–183
homogeneous mixtures, 181–182
propagation
heterogeneous mixtures in, 45
model for, 46
speed, influence of
fuel/air ratio and mean drop size, 47
mainstream velocity and mean drop size, 48
stability performance
bluff-body flameholders, 173–179
definition, 167–168
experimental findings, 179
measurement of, 168–170
water injection technique, 170–172
Flammability limits, 37–38
Flashback
CIVB flashback, 70
types, 69
Flash point, 38
Flat-vaned swirler, 142
Flow control
annulus, 89
diffuser, 82–83
liner holes, 120–122
Flow number, 266–268
Flow variables for ignition
air
pressure, 203–204
temperature, 204–205
velocity, 205–206
fuel
temperature, 209
type, 207
fuel/air ratio, 207–208
spray characteristics, 208–209
turbulence, 206
Flush fire igniter, 193
Free-stream mechanism, 69
FSII, *see* Fuel system icing inhibitors (FSII)
Fuel
evaporation rate, 52–53
temperature during drop lifetime variation, 49
continuous flow, 52
effective evaporation constant, 52–54
evaporation constant, 51
heat-up period, 50
spray evaporation, 54
Fuel–air mixture, 293–294
Fuel-injector instabilities, 302–303
Fuel nozzle coking, 254
design rules for reducing wetted-wall temperatures, 255–256
Fuel preparation
airblast atomizer, 19–20
gas injection, 20
pressure-swirl atomizers, 18–19
Fuel properties
distillation range, 458
ASTM D-86 distillation curves for, 459
flash point, 459
ignition temperatures for petroleum fuels, 461
significance and variation of, 459–460
freezing point, 462
IPK, 463
latent heat, 465
relative density, 456
API gravity formula, 457

molecular mass, 458
of typical gas turbine fuels, 457
specific heat, 463
freezing point on final boiling point, effect, 465
influence of temperature on, 466
paraffinic fuels, 464
surface tension, 462
temperature relations for various hydrocarbon fuels, 464
thermal conductivity of, 465
vapor pressure, 458–459
effect of temperature on, 460
viscosity, 460–461
of gas turbine fuels, 462
temperature characteristics of typical fuels, 462
and volatility, 460
relationship between, 461
Fuel spray, rate of evaporation, 54
Fuel system icing inhibitors (FSII), 454
Full-scale combustors, 170

G

Gaseous fuels; *see also* Alternative fuels
classification
coal gas, 478
coke-oven gas, 478, 480
common gaseous fuels, properties of, 479
impurities, 480–481
natural gas, 478
Gas injection, 256
Gas turbine combustor
geometry of, 8
requirements, 9–10
Gas turbines
ABB EV conical premix burner, operating principle, 404
ABB GT13E2 and ABB GT11N, 31
ABB GT11N gas turbine, 404
emissions characteristics of, 367
problems of oscillating combustion in, 302
GE CFM56-5B engine, 396
GE dual-annular combustor
fuel nozzle for, 395
GE90 dual-annular combustor, 396
General Electric GE DLN combustor, 395, 401
lean-lean, 403
premix, 403
primary, 402
secondary, 403
General Electric J33 tubular combustor, 7
General Electric LM6000, 409–412
dry low-emissions combustor, 99–100
Germany
combustor developments in
BMW 003, 6
Jumo 004, 5–6
turbojet engine, 6
Global reaction-rate theory
assumptions, 38
material balance equation, 38
for rich mixtures, 39–41
for weak mixtures, 39
Glow plug, 196
Godsave's evaporation constant, 52
Grid-or spoiler-generated turbulence, 93
Ground-level ozone, 361
Growl, 297–298
Gum, 449
antioxidants and peroxides, 450
fuel storage stability, 450
fuel thermal stability, 450
jet fuel use, 450

H

Hastelloy X nickel-based alloys, 23, 352
Heat transfer processes, 315–316
cold-side convection, 349
external convection, 323–324
external radiation, 321–322
film cooling
machined ring, 331
rolled ring, 331
splash-cooling ring, 331
stacked ring, 330
wigglestrips, 329–330
Z ring, 332
film cooling data, correlation, 333
with augmented convection, 341–342
calculation of, 337–341

impingement cooling, 342–343
transpiration cooling, 343
turbulent boundary-layer model, theory, 334–335
wall-jet model, theory, 335–337
internal convection, 322–323
internal radiation
from luminous gases, 320
from nonluminous gases, 318–320
liner failure modes, 354–355
materials, 351
ceramics, 353–354
mechanical integrity, 354
metal alloys, 352–353
requirements for, 352
model, 317
steady-state conditions in, 317
temperature gradient within, 317
thermal barrier coatings, 349–351
transpiration cooling, 343
effusion cooling, 345–346
Lamilloy, 345
Transply, 344–345
uncooled liner temperature, 324
method of calculation, 325–328
significance of, 328
wall-cooling methods
AFC, 346–347
measured and predicted values for GE F101 combustor, 341
tiles, 347–349
Heterogeneous combustion, 43
Heterogeneous spray combustion, 36
High-energy ignition unit, 190–191
High-pressure catalytic combustor, 426
Hot-surface ignition, 196–197
Howl, 298
Hybrid airblast atomizer, 246
Hybrid diffuser
arrangement of, 102
comparison with conventional diffuser performance, 103
feature of, 102
hybrid concept, 101
inlet Mach number, 102
overall area ratio, 102
Hydrocarbons
aromatics, 447–448
in fuel combustion, 38
naphthenes, 446
olefins, 445–446
paraffinic oils, 444–445

I

Ideal combustor, 14
Ideal diffuser, 80
Ideal pressure-recovery coefficient of diffuser, 84–85
IGCC, *see* Integrated gasification combined cycle (IGCC) plants
Ignition, 188
chemical, 198–199
delay time
drop-size distribution in spray, 67
effects of equivalence ratio on, 67–68
effects of fuel vaporization, 68
measurement technique, 65
stoichiometric value, 67
delay time, mixture strength effect on
methane–air mixtures, 66
delay time, pressure and temperature effect on
methane–air mixtures, 66
propane–air mixtures, 65
energy with mean drop size, 205
equipment, 27
experimental data correlation, 212
factors influencing
phase 1, 210
phase 2, 210–211
phase 3, 211
measured and predicted values
for F 100 combustor, 213
for J 79 combustor, 214
for J 85 combustor, 213
performance
assessment, 189
methods of improving, 211
types of failure, 212
performance, factors influencing
flow variables, 203–209
igniter location, 202–203
spark duration, 200–202
spark energy, 200

sparking rate, 202
test facility, 201
spark
high-energy ignition unit, 190–191
surface discharge igniter, 191–195
torch igniter, 195–196
Ignition theory
gaseous mixtures
highly turbulent mixtures, 55–56
low-turbulence, 55
minimum ignition energy, 56
quenching distance, 55
turbulence intensity, 56
heterogeneous mixtures, 56
propane–air mixtures
influence of turbulence intensity on, 58
propane, pressure and oxygen concentration, influence of, 58
quiescent and flowing mixtures, relationship between, 57
Impingement cooling, 331, 342–343
Indirect combustion noise, 293
Industrial chambers, 28
Industrial combustor
aeroderivative engines, 31–32
by General Electric, 29
liner construction, 29–30
single tubular combustor, 30
Initial jet angle, 123–124
Integrated gasification combined cycle (IGCC) plants, 443
Intermediate zone, 16–17
Internal convection, 322–323
Internal flow characteristics, 266
Internal-mixing configuration, 242
Internal radiation
from luminous gases, 320
from nonluminous gases, 318–320
International Civil Aviation Organization (ICAO)
regulations for civil subsonic turbojet/turbofan engines, 362
smoke emissions standards, 363
standards for gaseous emissions, 362
Isooctane fuel
water-fuel and equivalent pressure ratio, relationship between, 172
Iso-paraffinic kerosine (IPK), 463

J

Jet
initial jet angle, 123–124
mixing
annular ducts, 133
cylindrical ducts, 129–131
influence of density ratio, 132
rate of, 129
rectangular ducts, 131
plain-jet airblast, 243–244
and sheet breakup, 223
fuel jets, 223–226
fuel sheets, 226–227
trajectories
configurations, 126–128
in crossflow representation, 125
penetration data, 125–126
single jets, experiments on, 124–126
Jet Fuel Thermal Oxidation Tester (JFTOT), 450
Jumo 004 tubular combustor, 5–6

K

Kerosine (JP4/5)
adiabatic temperature rise curves for, 72
evaporation rate curves in, 50
Kidney-shaped profile, 124
Kinetic-energy coefficient of diffuser, 86

L

LAER, *see* Lowest available emission rate (LAER)
Lamilloy, 345
Laminar diffusion flames, 42–43
Laminar premixed flames
factors influencing
equivalence ratio, 41–42
initial temperature, 42
pressure, 42
Landing-takeoff cycle (LTO), 362
Laser ignition, 197–198
Laser-induced spark (LIS) ignition, 198
Lean premixed (LPM) combustor, 304

Lean premix prevaporize (LPP) combustor, 25, 68
Liner holes
  discharge coefficient, 120–121
  representation of flow through, 123
Liner-wall cooling, 21
Liquid fuels
  classification
    aircraft fuel specifications, 476
    aircraft gas turbine fuels, 475–476
    diesel fuels, 475
    gasolines, 475
    heavy residual fuels, 475
    industrial gas turbine fuels, 476, 478
    properties of, 474
Liquid fuels, production
  additives, 453–456
  contaminants, 449–450
    sodium, trace quantities of, 452
    vanadium, trace quantities of, 452–453
    water, 451–452
  conversion process, 448
  fuel properties, 456–457
    distillation range, 458
    flash point, 459–460
    freezing point, 462–463
    latent heat, 465
    relative density, 456–458
    specific heat, 463–465
    surface tension for, 462
    thermal conductivity, 465–466
    vapor pressure, 458–459
    viscosity, 460–462
    volatility point, 460
  separation process, 448
  sulfur compounds removal of
    clay treatment, 449
    hydrogen sulfide, 448–449
    mercaptans, 449
  upgrading process, 448
LIS ignition, *see* Laser-induced spark (LIS) ignition
Lockheed P-40, 4
Long-range cruise engine, 116
Loss coefficient of diffuser, 85
Louver cooling technique, 22
Lower flammable limit, 37
Lowest available emission rate (LAER), 365
Low-pressure-loss combustor, 116
LPM, *see* Lean premixed (LPM) combustor
LPP combustor, *see* Lean premix prevaporize (LPP) combustor
LTO, *see* Landing-takeoff cycle (LTO)
Lucas primary-zone airflow pattern, 16
Luminous radiation, 318
Lycoming T vaporizer, 252

M

Machined ring, 22, 331
Mach number for diffuser, 91–92
MBtu fuels, 404
Mean drop size
  equations, 256
    sheet thickness and spray cone angle, relationship between, 257
Metropolitan Vickers Beryl engine, 4
Metrovick annular combustor, 5
  feature of, 5
Minimum ignition energy, 59
  air pressure, effect on, 61
  chemical effects inclusion and, 62–63
  for quiescent heavy fuel oil and air mixtures, 60
Mixer designs, 411
Mixing-controlled systems
  evaporation and chemical reaction rates, 160
  mixing rate, 160
  subatmospheric pressures, 161
Mixing processes, 113
k-ε Model, 106–107
Monopole-type sound, 294–295
MS7001 combustion system, 29
Multi-can combustor arrangement, 11
Multiple fuel injectors, 30
Multiple-jet penetration correlation, 127
80 MW GE MS7001 gas turbine, 29

N

Naphthenes, 446
NASA design method, 137

Nimonic 263, 23
Nimonic 75 nickel-based alloys, 352
Nitride-bonded silicon carbide, 196
Nitrogen oxides (NO_x)
in conventional combustors
exhaust gas recirculation, 390–391
selective catalytic reduction, 390
water injection, 387–389
correlations
Lewis, 431
Odgers and Kretschmer, 429
Rizk and Mongia, 431–432
Rokke, 431
dependence on flame temperature, 375
exhaust gas recirculation, 390–391
formation
effect of pressure on, 380
as function of time and temperature, 375
influence of inlet air temperature on, 376
fuel, 378–379
nitrous oxide (N_2O), 378
in premixed fuel–air system, 377
prompt, 378
SCR, 390
thermal nitric oxide, 374
inlet air temperature, influence, 376
residence time, influence, 376–377
water injection, 387–389
Noise
active control noise
closed-loop system, 307
open-loop system, 307
combustion instabilities
LPM combustor noise, 304
direct combustion noise
core noise prediction methods, 296
radiated sound power, 294
theory, 294–295
thermoacoustic efficiency, 294
indirect combustion noise, 293
Nondimensional coefficient of ideal static pressure rise of diffuser, 84
Nonlinear model, 310

O

Ohnesorge number, 223, 258, 262
Olefins
formula, 445
molecules of, 446
Olympus engine
emissions standards, 364
Open-loop system, 307
Oscillating combustion, 303
Overall effectiveness of diffuser, 85
Over-penetrating jets, 130
Oxygen injection in surface discharge igniter, 199

P

Paraffinic oils, 444
current aviation fuels, 445
formula, 444
Passive control, 305–306
Pattern factor, 134
correlation
for annular combustors, 139
for tubular and tuboannular combustors, 139
Petroleum-based conventional jet fuels, 166
Piloted airblast atomizers, 245–246
Plain circular jet breakup, 224
Plain-jet airblast atomizers, 243–244
Plain-orifice atomizers, 238
discharge coefficient, 268–269
spray cone angle, 273–274
Plain-orifice effervescent atomizers, 250
Plasma jet, 197
Pollutants
control of flame temperature, reduction by
staged combustion, 393–397
variable-geometry system, 391–393
exposure to ozone concentrations, 361
formation mechanisms
carbon monoxide, 366–370
fuel atomization, influence, 381–382
nitric oxide (NO), 374–379

pressure effects on nitrogen oxides (NO_x), 379–381
smoke, 370–373
unburned hydrocarbons, 370
from gas turbines, 360
reduction, conventional combustors in, 382–383
CO and UHC, 383–384
nitrogen oxides (NO_x), 387–391
smoke, 384, 386–387
Power Jets W1 engine
first British turbojet-powered flight, 4
Pratt and Whitney axially staged combustor, 397
Pratt & Whitney J57 tuboannular combustor, 7
Pratt & Whitney V2500 tiled combustor, 348
Prediffuser, 103
Prefilming airblast atomizers, 244–245
Premixed flames, 36
Premixed fuel–air injection, 32
Premix-prevaporize atomizer, 19
Pressure atomizers, 237; *see also* Atomization
discharge coefficient of, 268
dual-orifice, 239–240
plain orifice, 238
simplex, 238–239
SMD equations for
plain orifice, 258
pressure swirl, 258–261
spill return, 240–241
Pressure effects
adiabatic flame temperature, 72
burning velocity, 42
combustion efficiency, 72
flame radiation, 72
flammability limits, 37–38
ignition delay time, 66–67
luminous emissivity, 318
on nitrogen oxides (NO_x), 379–381
non-luminous emissivity, 318
weak extinction limits, 174–177
Pressure loss factor, 114
Pressure-recovery coefficient of diffuser, 84
Pressure-swirl atomizers, 5, 164–165
circumferential fuel distribution, 280–282
discharge coefficient, 270
spray cone angle, 274–276
Primary atomization, 221
Primary zone, 15–16
configurations of, 184
Profile factor, 134
Prompt atomization
processes, 227
SMD equations for, 264–265
Propane-air flames, 45
Pyrophoric fuels, 198–199

Q

Quenching distance, 59
gaseous mixtures, 55–56
heterogeneous mixtures, 56–64
Quiescent and flowing mixtures, validity of general model, 62

R

Radial fuel distribution
ambient air pressure, influence of, 277
measurement of, 277
patternation and instruments, 276
Radial swirlers
airflow, 147
notation for, 146
Radiated sound power, 294
Radiation
beam length for, 319–320
blackbody, 318
external, 321–322
flame radiation, 72
internal
from luminous gases, 320
from nonluminous gases, 318–320
Rayleigh's theory, 225
RB211 annular combustor, 13
RB211 DLE combustor, 32
RB 199 vaporizer, 253
Reaction
global reaction-rate theory, 38
heat release rate, 159

order, 159
rate constant, 154
Reaction-controlled systems, 154
burning velocity model, 155–159
stirred reactor model, 159–160
Recirculation zone, 144–147
Rectangular ducts, 131
Refractory bricks, 23
Reverse-flow
annular combustor, 26
combustors, 406
Reynolds number for diffuser, 91
Reynolds-number index, 322
Rich-burn/quick-quench/lean-burn (RQL) combustor, 25, 130–131, 138, 381, 418–420
"Rich extinction," 168
Rich limit, *see* Upper flammable limit
Rolled ring, 331
Rolls Royce annular combustors, 15
Rolls Royce Industrial RB211 DLE combustor, 407
Rolls Royce Industrial Trent, 32
Rolls Royce RB211 industrial burner, 406–407
Rosin–Rammler relationship, 231–233; *see also* Atomization
expression, 231, 233, 235
modified, 233
RQL combustor, *see* Rich-Burn, Quick-Quench, Lean-Burn (RQL) combustor
RR Spey Mk 512 engine, Transply combustor, 345

S

"Sandwich" mixer, 4–5
SAR, *see* Slot aspect ratio (SAR) on jet mixing
Sauter mean diameter (SMD), 47, 54, 161, 165, 234
SMD equations for pressure atomizers
plain orifice, 258
pressure swirl, 258–261
SMD equations for prompt atomization, 264–265
SMD equations for twin-fluid atomizers, 261–264
Secondary atomization, 221
Sediment, 450
Selective catalytic reduction (SCR), 390
Selective fuel injection, 393
Settling length, 94, 104
Shell middle-distillate synthesis (SMDS), 487
Siemens hybrid burner, 30, 400–401
Siemens silo combustors, 30
Silicon carbide, 353
Silicon nitride, 353
Silo burner fitted with ABB EV burners, 405
Simplex atomizers, 18, 238–239
Simplex pressure-swirl atomizer, 230
Single-component fuel drop, 49
Slot aspect ratio (SAR) on jet mixing, 130
Slow-burning fuels, 42
Small combustors, 27
SMDS, *see* Shell middle-distillate synthesis (SMDS)
Smoke, 370
fuel
atomization, influence of, 372–373
type, 371–372
number, 508
point, 471–472
pressure, influence, 371
Sodium, trace quantities, 452
Solar dry low-emissions concepts, 398–400
Spark duration, 200–202
Spark ignition
high-energy ignition unit, 190–191
surface discharge igniter, 191–195
Sparking rate, 202
Spark kernel, 59
Spey aero engines, 31
Spill return atomizers, 240–241
Splash-cooling devices, 22
ring, 331
Splitter plate, 119–120
Splitter vanes, 99–100
Spontaneous ignition
model for, 68

propane–and methane–air mixtures, 65
Wolfer equation, 64
Spray cone angle
discharge orifice length/diameter ratio, influence of, 279
liquid viscosity and injection pressure, influence of, 278
plain-orifice atomizers, 273–274
pressure-swirl atomizers, 274–276
SST engines, *see* Supersonic transport (SST) engines
Stability limits, 211
Stability performance of aircraft combustor, 170
Stabilization mechanisms of flames, 179–180
in combustion chambers, 184–188
heterogeneous mixtures, 182–183
homogeneous mixtures, 181–182
Stacked ring, 330
Staged combustors
catalytic and conventional, combination of, 26
Stefan–Boltzmann constant, 318
Step-stabilized flames, 299
Stirred reactor model, 38
air temperature, 159
experimental data, 160
heat-release rates and, 159
Stoichiometry, 70–71
Straight-walled diffuser, 92
Sunken fire igniter, 193
Supersonic transport (SST) engines, 364
Surface discharge igniter, 191
development, 192
forms
chemical ignition, 198–199
gas addition, 199
glow plug, 196
hot-surface ignition, 196–197
laser ignition, 197–198
oxygen injection, 199
plasma jet, 197
torch igniter, 195–196
life of, 194–195
performance of, 193–194
types, 193
Swirler
aerodynamics, 140
flow recirculation, 141
flows and practical applications, 141
types, 142
vortex breakdown, 141
Swirl Flows, 141
Swirl number, 144
axial flow, 143
Swirl-stabilized combustors
CIVB in, 69
Q-Switched laser beam, 197–198
Synthetic and bio-based jet fuels, 166
Synthetic fuels
biofuels
attractions, 488–489
drawbacks, 489
gas chromatographs, 488
gas turbines, 488
heating value per unit, 488
combustion and emissions performance
biodiesel fuels, 498, 500–504
Fischer–Tropsch fuels, 490, 495–496
highly aromatic, 504–507
direct liquefaction of coal, 486
Fischer–Tropsch synthesis of coal/biomass
gas chromatographs, 487–488
modern coal gasification process, 487
SMDS, 487
fuel properties, 490
bio-based synthetic paraffinic kerosine, properties, 493–494
fuel density and fuel temperature for, 495
kinematic viscosity versus fuel temperature for, 496
nitrile tests, 489

T

T56 aero engine, 31
Test rig simulations, 304–305
Thermal barrier coatings (TBCs), 348–351

Thermoacoustic efficiency, 294
THM 1304 engine, 30–31
Three-vane diffusers, 99
Time lag models, 310
Torch igniter, 195–196
Transpiration cooling, 343
  effusion cooling, 345–346
  Lamilloy
    construction and airflow path, 345
  Transply
    constructional features of, 344
    requirement for, 345
Transply
  constructional features of, 344
  requirement for, 345
  transpiration, 345
  wigglestrip, 330
  Z-ring, 332
Trent DLE combustor, 32
Tuboannular combustors, 11, 104
  arrangement, 12
  configuration of, 105
  pattern factor correlation, 139
Tubular combustors, 11, 104–105
  correlation of data, 138
  pattern factor correlation, 139
Turbine entry temperature, historical trend in, 22
Turbine profile factor, 135
Turbojet engine by Whittle, 3
Turbulence
  defined, 36
  of diffuser, 92–93
Turbulent boundary-layer model theory, 334–335
Turbulent premixed flames
  Damkohler's theory, 43
  Kolmogoroff scale, 44
  large-scale turbulence, 43
  Schelkin's approach, 43
Twin-fluid atomizers
  SMD equations for, 261–264
Two-dimensional diffuser
  advantages, 88
  disadvantages, 89
  inlet boundary-layer thickness, 88
  regions, 92
Tyne aero engines, 31

U

Unburned hydrocarbons (UHC), 383–384
Uncooled liner temperature, 324
  method of calculation, 325–328
  significance of, 328
Under-penetrating jets, 130
United States
  combustor developments in
    aero-engine combustors, design features of, 7
    CFM56-B engine, 7
    dual-annular concept, 7
    J30 and J34, 7
    J35 and J47, 7
    J31 engine, 6–7
    Nene engine from Rolls Royce, 6
    W2B engine, 6
    W1 engine, 6
Upper flammable limit, 37

V

Vanadium
  trace quantities of, 452
    in petroleum fuels, contaminants concentrations, 453
Vaporizers, 251–254
  Lycoming T vaporizer, 252
  RB 199 vaporizer, 253
  vaporizing combustor, 253–254
  "walking stick" or "candy cane" vaporizers, 252
Variable-geometry combustors, 391–393
Vee-gutter flameholders, 175
Velocity profile at diffuser, 91
Vibrating-contact voltage generator, 191
Volume median diameter, 231
Vortex-controlled diffuser (VCD)
  air-bleed requirements, 101
  features of, 101
  flow mechanisms, 101
  with prediffuser, 103
  of tubular configuration, 100

W

"Walking stick" or "candy cane" vaporizers, 252

Wall-cooling, 20
angled effusion, 22–23
augmented convection, 341–342
continuous and cyclic high-temperature operation, 21
conventional effusion, 346–347
double-pass ring, 331–332
finned surfaces, 349
impingement, 343
influence of, engine pressure ratio, 354
Lamilloy machined ring, 345
requirements, 17
ribs, 349–350
rolled ring, 331
splash-cooling ring, 331
stacked ring, 330
technique
AEC, 22–23, 346–347
air metering, 22
louver cooling, 22
"machined-ring" or "rolled-ring" approach, 22
protective coating spray, 23
refractory bricks, 23
thermal barrier coatings, 323, 349–351
tiles, 347–349
turbine entry temperature, 17, 22
Wall-jet model, theory, 335–337
Water
dissolved water
water-saturated kerosine-type fuel, 451
emulsion, 452
free water
engine-filter icing, problem, 452
microorganisms, 452
subzero fuel temperatures, 451
water-contaminated fuels, problem with, 452
Water injection technique, 170
for NO_x reduction, 387–388
rig requirements for, 171
stability loop obtained using, 171
Weak extinction limits, influence
combustor size, 294
flameholder size, 174
fuel type, 299
geometric blockage, 182
inlet temperature, 176
mean drop size, 178
operating conditions, 185
pressure, 177
velocity, 178
"Weak" extinction points, 168–169
Weber number, 223
Welland engines, 16
Whittle atomizer combustor, 4
Whittle vaporizer combustor, 3
Wigglestrips, 22, 329–330

Z

Zeldovich mechanism, 374
Z ring, 332